McDONALD INSTITUTE MONOGRAPHS

Archaeogenetics: DNA and the population prehistory of Europe

Edited by Colin Renfrew & Katie Boyle

Published by:

McDonald Institute for Archaeological Research
University of Cambridge
Downing Street
Cambridge
CB2 3ER
(0)(1223) 339336

Distributed by Oxbow Books
  United Kingdom: Oxbow Books, Park End Place, Oxford, OX1 1HN.
   Tel: (0)(1865) 241249; Fax: (0)(1865) 794449
  USA: The David Brown Book Company, P.O. Box 511, Oakville, CT 06779, USA.
   Tel: 860-945-9329; FAX: 860-945-9468

ISBN: 1-902937-08-2
ISSN: 1363-1349

© 2000 McDonald Institute for Archaeological Research

All rights reserved. No parts of this publication may be reproduced, stored in a retrieval system, or transmitted, in any form or by any means, electronic, mechanical, photocopying, recording or otherwise, without the prior permission of the McDonald Institute for Archaeological Research.

Edited for the Institute by Chris Scarre (*Series Editor*) and Dora A. Kemp (*Production Editor*).

Film produced by Gary Reynolds Typesetting, 13 Sturton Street, Cambridge, CB1 2QG.
Printed and bound by Short Run Press, Bittern Rd, Sowton Industrial Estate, Exeter, EX2 7LW.

Contents

Contributors and participants		vi
Figures		xiv
Tables		xvi
Foreword		xix
Colin Renfrew & Katie Boyle		

Part I   Introductory

Chapter 1	Archaeogenetics: Towards a Population Prehistory of Europe Colin Renfrew	3
Chapter 2	Concepts in Molecular Genetics Bryan Sykes & Colin Renfrew	13
Chapter 3	Human Diversity in Europe and Beyond: From Blood Groups to Genes Bryan Sykes	23

Part II   The Archaeological and Environmental Background

Chapter 4	Where Received Wisdom Fails: the Mid-Palaeolithic and Early Neolithic Climates Tjeerd H. van Andel	31
Chapter 5	The History of European Populations as seen by Archaeology Marcel Otte	41
Chapter 6	Spatial and Temporal Patterns in the Mesolithic–Neolithic Archaeological Record of Europe Ron Pinhasi, Robert A. Foley & Marta Mirazón Lahr	45
Chapter 7	The Social Context of the Agricultural Transition in Europe Marek Zvelebil	57
Chapter 8	Expected Regional Patterns of Mesolithic–Neolithic Human Population Admixture in Europe based on Archaeological Evidence Marta Mirazón Lahr, Robert A. Foley & Ron Pinhasi	81
Chapter 9	Processes of Culture Change in Prehistory: a Case Study from the European Neolithic Mark Collard & Stephen Shennan	89

Part III   Genetic Markers: from the Global to the Continental

Chapter 10	The Demography of Human Populations inferred from Patterns of Mitochondrial DNA Diversity Laurent Excoffier & Stefan Schneider	101
Chapter 11	Nuclear Genetic Variation of European Populations in a Global Context Kenneth K. Kidd, Judith R. Kidd, Andrew J. Pakstis, Batsheva Bonné-Tamir & Elena Grigorenko	109
Chapter 12	Genetic Population Structure of Europeans inferred from Nuclear and Mitochondrial DNA Polymorphisms Guido Barbujani & Lounès Chikhi	119
Chapter 13	Spatial Variation of mtDNA Hypervariable Region I among European Populations Lucia Simoni, Francesc Calafell, Jaume Bertranpetit & Guido Barbujani	131
Chapter 14	Genetic Data and the Colonization of Europe: Genealogies and Founders Martin Richards & Vincent Macaulay	139

Chapter 15 The Phylogeography of European Y-chromosome Diversity — 153
ZOË H. ROSSER, MATTHEW E. HURLES, TATIANA ZERJAL, CHRIS TYLER-SMITH & MARK A. JOBLING

Part IV Regional Studies in the Molecular Genetic History of Western Europe

Chapter 16 Human Y-chromosomal Networks and Patterns of Gene Flow in Europe, West Asia and North Africa — 163
PATRIZIA MALASPINA, FULVIO CRUCIANI, ANTONIO TORRONI, LUCIANO TERRENATO, ANDREA NOVELLETTO & ROSARIA SCOZZARI

Chapter 17 Autosomal and Mitochondrial Genetic Diversity in Sicily — 167
VALENTINO ROMANO, FRANCESCO CALÍ, PETER FORSTER, ROSALBA P. D'ANNA, ANNA FLUGY, GIACOMO DE LEO, ALFREDO SALERNO, ORNELLA GIAMBALVO, GIUSEPPE MATULLO & ALBERTO PIAZZA

Chapter 18 MtDNA Variability in Extinct and Extant Populations of Sicily and Southern Italy — 175
OLGA RICKARDS, CRISTINA MARTÍNEZ-LABARGA, ROSA CASALOTTI, GIUSEPPE CASTELLANA, ANNA M. TUNZI SISTO & FRANCESCO MALLEGNI

Chapter 19 The Population History of Corsica and Sardinia: the Contribution of Archaeology and Genetics — 185
LAURA MORELLI & PAOLO FRANCALACCI

Chapter 20 Analysis of the Y-chromosome and Mitochondrial DNA Pools in Portugal — 191
LUÍSA PEREIRA, MARIA JOÃO PRATA, MARK A. JOBLING & ANTÓNIO AMORIM

Chapter 21 Morphological and Molecular Evolution in Prehistoric Skeletal Populations in the Basque Country — 197
NESKUTS IZAGIRRE, PATRICIA ARTIACH & CONCEPCION DE LA RUA

Chapter 22 Y-chromosome Variation and Irish Origins — 203
EMMELINE W. HILL, MARK A. JOBLING & DANIEL G. BRADLEY

Chapter 23 The 'Travellers': an Isolate within the Irish Population — 209
DAVID T. CROKE, ORNA TIGHE, CHARLES O'NEILL & PHILIP D. MAYNE

Part V Regional Studies in the Molecular Genetic Prehistory of Eastern Europe and Beyond

Chapter 24 Y-chromosome Diversity in the Eastern Mediterranean Area — 215
ANDREA NOVELLETTO, MARLA TSOPANOMICHALOU, APHRODITE LOUTRADIS, ELENI PLATA, MARILENA GIPARAKI, EMMANUEL N. MICHALODIMITRAKIS, NEJAT AKAR, NICHOLAS ANAGNOU, LUCIANO TERRENATO & PATRIZIA MALASPINA

Chapter 25 The Topology of the Maternal Lineages of the Anatolian and Trans-Caucasus Populations and the Peopling of Europe: Some Preliminary Considerations — 219
KRISTIINA TAMBETS, TOOMAS KIVISILD, ENE METSPALU, JÜRI PARIK, KATRIN KALDMA, SIRLE LAOS, HELLE-VIIVI TOLK, MUKADDES GÖLGE, HALIL DEMIRTAS, TAREKEGN GEBERHIWOT, SURINDER S. PAPIHA, GIAN FRANCO DE STEFANO & RICHARD VILLEMS

Chapter 26 Network Analysis of Y-chromosome Compound Haplotypes in Three Bulgarian Population Groups — 237
BORIANA ZAHAROVA, ANJA GILISSEN, SILVIA ANDONOVA-BAKLOVA, JEAN-JACQUES CASSIMAN, RONNY DECORTE & IVO KREMENSKY

Chapter 27 MtDNA Sequence Diversity in Three Neighbouring Ethnic Groups of Three Language Families from the European Part of Russia — 245
VLADIMIR OREKHOV, PAVEL IVANOV, LEV ZHIVOTOVKSY, ANDREY POLTORAUS, VIKTOR SPITSYN, EUGENY GINTER, ELSA KHUSNUTDINOVA & NIKOLAY YANKOVSKY

Chapter 28	Mitochondrial DNA Variability in Eastern Slavs BORIS A. MALYARCHUK & MIROSLAVA V. DERENKO	249
Chapter 29	Mitochondrial DNA Diversity among Nenets JULIETTE SAILLARD, IRINA EVSEVA, LISBETH TRANEBJÆRG & SØREN NØRBY	255
Chapter 30	Genetics and Population History of Central Asia FRANCESC CALAFELL, DAVID COMAS, ANNA PÉREZ-LEZAUN & JAUME BERTRANPETIT	259
Chapter 31	An Indian Ancestry: a Key for Understanding Human Diversity in Europe and Beyond TOOMAS KIVISILD, SURINDER S. PAPIHA, SIIRI ROOTSI, JÜRI PARIK, KATRIN KALDMA, MAERE REIDLA, SIRLE LAOS, MAIT METSPALU, GERLI PIELBERG, MAARJA ADOJAAN, ENE METSPALU, SARABJIT S. MASTANA, YIMING WANG, MUKADDES GÖLGE, HALIL DEMIRTAS, ECKART SCHNAKENBERG, GIAN FRANCO DE STEFANO, TAREKEGN GEBERHIWOT, MIREILLE CLAUSTRES & RICHARD VILLEMS	267

Part VI *Methodologies in the Application of Molecular Genetics to Archaeology*

Chapter 32	Maximum Likelihood Estimations of Genetic Diversity GUNTER WEISS	279
Chapter 33	Sampling Saturation and the European MtDNA Pool: Implications for Detecting Genetic Relationships among Populations AGNAR HELGASON, SIGRÚN SIGURÐARDÓTTIR, JEFFREY R. GULCHER, KÁRI STEFÁNSSON & RYK WARD	285
Chapter 34	Male and Female Differential Patterns of Genetic Variation in Human Populations MICHELE BELLEDI, LUCIA SIMONI, ROSA CASALOTTI & GIOVANNI DESTRO-BISOL	295
Chapter 35	Y Chromosomes Shared by Descent or by State PETER DE KNIJFF	301
Chapter 36	Lactase Haplotype Diversity in the Old World EDWARD J. HOLLOX, MARK POULTER & DALLAS M. SWALLOW	305
Chapter 37	History of Dairy Cattle-breeding and Distribution of LAC*R and LAC*P Alleles among European Populations ANDREW I. KOZLOV & DMITRY V. LISITSYN	309
Chapter 38	Mitochondrial DNA Diversity and Origins of Domestic Livestock DANIEL G. BRADLEY	315
Chapter 39	Wheat Domestication ROBIN ALLABY	321
Chapter 40	Inferring the Impact of Linguistic Boundaries on Population Differentiation and the Location of Genetic Barriers: a New Approach ISABELLE DUPANLOUP DE CEUNINCK, STEFAN SCHNEIDER & LAURENT EXCOFFIER	325
Chapter 41	Patterns of Genetic and Linguistic Variation in Italy: a Case Study FRANZ MANNI & ITALO BARRAI	333
Prospect	Concluding Remarks LUCA CAVALLI-SFORZA	339

Contributors and participants

* Conference participants
◊ Authors

AUGUSTO ABADE *
Departamento de Antropologia, Faculty of Sciences and Technology, Universidade de Coimbra, Coimbra, Portugal.

MAARJA ADOJAAN ◊
Estonian Biocentre and Department of Evolutionary Biology, Tartu University, 23 Riia Street, 51010 Tartu, Estonia.

NEJAT AKAR ◊
Pediatrics Department, Ankara University, Turkey.

SANTOS ALONSO *
Genetics Division, QMS, University of Nottingham, Nottingham, NG7 2UH, UK.

ROBIN ALLABY *◊
Biomolecular Sciences, UMIST, PO Box 88, Manchester, M60 1QD, UK.

MANUELA ALVAREZ *
Max Planck Institute for Evolutionary Anthropology, Inselstraße 22, D-04103 Leipzig, Germany.

ANTONIO AMORIM *◊
IPATIMUP, R Dr Roberto Faias, s/n, 4200 Porto, Portugal.

NICHOLAS ANAGNOU ◊
University of Crete and IMBB, Heraklion, Greece.

SILVIA ANDONOVA-BAKLOVA ◊
Laboratory of Molecular Pathology, University Hospital of Obstetrics and Gynaecology, Sofia, Bugaria.

PATRICIA ARTIACH *◊
Animali Biologia eta Genetica Saila, PO Box 644, 48080 Bilbao, Spain.

GIOVANNI AYALA *
Associazione Oasi Maria SS., Via C. Ruggero 73 - Troina (EN), Italy.

DAVID BALDING *
Department of Statistics, University of Reading, PO Box 240, Reading, RG6 6FN, UK.

HANS-JÜRGEN BANDELT *◊
Mathematisches Seminar, Universität Hamburg, Bundesstr. 55, 20146 Hamburg, Germany.

GUIDO BARBUJANI ◊
Departmento di Biologia, Università di Ferrara, via L. Borsari, 46, I-44100 Ferrara, Italy.

ITALO BARRAI ◊
Department of Biology, Section of Evolutionary Biology, University of Ferrara, via L. Borsari 46, 44100 Ferrara, Italy.

SANDRA BARRAL RODRIGUEZ *
Institute of Legal Medicine, University of Santiago de Compostela, E-15705 Spain.

MICHELE BELLEDI ◊
Dipartimento di Biologia Evolutiva e Funzionale, Università di Parma, Italy.

JAUME BERTRANPETIT *◊
Unitat de Biologia Evolutiva, Facultat de Ciécies de la Salut i de la Vida, Universitat Pompeu Fabra, Doctor Aiguader 80, Barcelona 08003, Spain.

WALTER BODMER *
Imperial Cancer Research Fund, Cancer and Immunogenetics Laboratory, Institute of Molecular Medicine, John Radcliffe Hospital, Headington, Oxford, OX3 9DS, UK.

BATSHEVA BONNÉ-TAMIR ◊
Department of Genetics, Sackler School of Medicine, Tel Aviv University, Tel Aviv, Israel 69978.

KATIE BOYLE *◊
McDonald Institute for Archaeological Research, Downing Street, Cambridge, CB2 3ER, UK.

DANIEL G. BRADLEY *◊
Department of Genetics, Smurfit Institute, Trinity College, Dublin 2, Ireland.

BERND BRINKMANN *
Institut für Rechtsmedizin, Universität Münster, Von-Esmarch-Str. 62, 48149 Münster, Germany.

JOACHIM BURGER *
Historische Anthropologie und Humanökologie, Universität Göttingen, Burgerstrasse 50, 37073 Göttingen, Germany.

FRANCESC CALAFELL *◊
Unitat de Biologia Evolutiva, Facultat de Ciències de la Salut i de la Vida, Universitat Pompeu Fabra, Doctor Aiguader 80, 08003 Barcelona, Spain.

FRANCESCO CALÍ *◊
Laboratorio di Genetica Molecolare, Instituto OASI Maria SS, Troina, Italy.

ANNE CAMBON-THOMSEN *
INSERM U518, Faculté de Médecine, Université de Toulouse III, 31073 Toulouse CEDEX, France.

DAVID CARAMELLI *
Instituto di Antropologia, Università degli Studi di Firenze, Florence, Italy.

DENISE REJANE CARVALHO-SILVA *
Departamento de Bioquímica e Imunologia,
Universidade Federal de Minas Gerais, Brazil.

ROSA CASALOTTI *◊
Centro di Antropologia Molecolare per lo Studio
del DNA Antico, Dipartimento di Biologia,
Università di Roma, Tor Vergata, Via della Ricerca
Scientifica n.1, 00133 Rome, Italy.

JEAN-JACQUES CASSIMAN ◊
Laboratory for Forensic Genetics and Molecular
Archaeology, Center for Human Genetics, Leuven,
Belgium.

GIUSEPPE CASTELLANA ◊
Museo Archeologico Regionale di Agrigento,
Contrada S. Nicola, 92100 Agrigento, Italy.

LUCA CAVALLI-SFORZA *◊
Department of Genetics, Stanford University, 300
Pasteur Drive, Stanford, CA 94305-5120, USA.

HELEN CHANDLER *
Institute of Molecular Medicine, University of
Oxford, Oxford, OX3 9DS, UK.

LOUNÈS CHIKHI ◊
School of Biological Sciences, Queen Mary and
Westfield College, University of London, Mile End,
London, E1 4NS, UK.

MIREILLE CLAUSTRES ◊
Laboratoire de Biochimie Genetique, Institut de
Biologie, Boulevard Henri IV, 34060 Montpellier
Cedex France.

MARK COLLARD ◊
Department of Anthropology, University College
London, Gower Street, London, WC1E 6BT, UK.

DAVID COMAS *◊
Unitat de Biologia Evolutiva, Facultat de Ciències
de la Salut i de la Vida, Universitat Pompeu Fabra,
Doctor Aiguader 80, 08003 Barcelona, Spain.

DAVID CROKE *◊
Department of Biochemistry, Royal College of
Surgeons in Ireland, 123 St Stephen's Green, Dublin
2, Ireland.

FULVIO CRUCIANI *◊
Dipartimento di Genetica e Biologia Molecolare,
Università degli Studi 'La Sapienza' di Roma, P. le
Aldo Moro 5, 00185 Rome, Italy.

ROSALBA P. D'ANNA ◊
Dipartimento di Biopatologia e Metodologie
Biomediche, Facoltà di Medicina e Chirurgia,
Università degli Studi di Palermo, Via Divisi 83,
Palermo.

ANDREW DAVIDSON *
Department of Sciences, Oaklands College, Welwyn
Garden City, Herts., AL8 6AH, UK.

PETER DE KNIJFF *◊
Forensic Laboratory for DNA Research,
Department of Human and Clinical Genetics,
Leiden University, PO Box 9503, 2300 RA Leiden,
The Netherlands.

CONCEPCION DE LA RUA *◊
Animali Biologia eta Genetica Saila, Euskal Herriko
Unibersitatea, PO Box 644, 48080 Bilbao, Spain.

GIACOMO DE LEO ◊
Dipartimento di Biopatologia e Metodologie
Biomediche, Facoltà di Medicina e Chirurgia,
Università degli Studi di Palermo, Via Divisi 83,
Palermo, Italy.

GIAN FRANCO DE STEFANO ◊
Department of Biology, University of Rome 'Tor
Vergata', Rome, Italy.

RONNY DECORTE ◊
Laboratory for Forensic Genetics and Molecular
Archaeology, Center for Human Genetics, Leuven,
Belgium.

HALIL DEMIRTAS ◊
Department of Medical Biology, Erciyes University,
Kayseri, Turkey.

MIROSLAVA V. DERENKO ◊
Institute of Biological Problems of the North,
Portovaya Str., 18, Magadan, 685000 Russia.

GIOVANNI DESTRO-BISOL *◊
Dipartimento di Biologia Animale e dell'Uomo,
Unitat di Antropologia, Università 'La Sapienza',
Rome, Italy.

PETER DONNELLY *
Department of Statistics, University of Oxford,
Oxford, OX1 3TG, UK.

ISABELLE DUPANLOUP *◊
Genetics and Biometry Laboratory, Department of
Anthropology, University of Geneva, Case Postale
24, 1211 Geneva 24, Switzerland.

A.W.F. EDWARDS *
Gonville and Caius College, University of
Cambridge, Cambridge, CB2 1TA, UK.

OLEG EVGRAFOV *
DNA Diagnostics Laboratory, Research Centre for
Medical Genetics, 1 Moskvorechie, Moscow 115478,
Russia.

IRINA EVSEEVA ◊
Arkhangelsk State Medical Academy, 51 Troitsky av., 163061, Arkhangelsk, Russia

LAURENT EXCOFFIER *◊
Genetics and Biometry Laboratory, Department of Anthropology, University of Geneva, 12, rue G. Revilliod, 1227 Geneva, Switzerland.

ANNA FLUGY ◊
Instituto di Biologia, Fac. di Medic. e Chir., Università degli Studi di Palermo, Via Divisi, 83, 90133 Palermo, Italy.

ALEXANDRA FOISSAC *
Département d'Epidemiologie, Economie de la Santé et Santé Communautaire, Faculté de Médecine, Université Paul Sabatier, 37 allées Jules Guesde, 31073 Toulouse CEDEX, France.

ROBERT FOLEY *◊
Department of Biological Anthropology, University of Cambridge, New Museums Site, Cambridge, CB2 3RF, UK.

LUCY FORSTER *
Institut für Rechtsmedizin, Universität Münster, Von-Esmarch-Str. 62, 48149 Münster, Germany.

PETER FORSTER *◊
McDonald Institute for Archaeological Research, Downing Street, Cambridge, CB2 3ER, UK.

PAOLO FRANCALACCI *◊
Dipartimento di Zoologia e Antropologia Biologica, Università degli Studi di Sassari, Corso Regina Margherita 15, 07100 Sassari, Italy.

SILVIA FUSELLI *
University of Bologna, Via Nadi 22, 40139 Bologna, Italy.

TAREKEGN GEBERHIWOT ◊
Karolinska Institutet, Stockholm, Sweden.

JULIA GERSTENBERGER *
Historische Anthropologie und Humanökologie, Universität Göttingen, Burgerstrasse 50, 37073 Göttingen, Germany.

ORNELLA GIAMBALVO ◊
Istituto di Statistica Sociale, Facoltà di Economia, Università degli Studi di Palermo, Viale delle Scienze, Palermo, Italy.

ANJA GILISSEN ◊
Laboratory for Forensic Genetics and Molecular Archaeology, Center for Human Genetics, Leuven, Belgium

EUGENY GINTER ◊
Medical-Genetic Scientific Centre, Moscow, 115478, Russia

MARILENA GIPARAKI ◊
Ministry of Health Center for Thalassemia, Athens, Greece.

DAVID GOLDSTEIN *◊
Department of Biology, Galton Laboratory, Univeristy College London, Wolfson House, 4 Stephenson Way, London, NW1 2HE, UK.

MUKADDES GÖLGE ◊
Department of Physiology, University of Kiel, Germany.

ANA GONZALEZ-NEIRA *
Institute of Legal Medicine, University of Santiago de Compostela, E-15705 Spain.

ELENA GRIGORENKO ◊
Psychology and Child Study Center, Yale University School of Medicine, PO Box 207900, New Haven, CT 06520-7900, USA.

JEFFREY R. GULCHER ◊
deCode Genetics Inc., Lynghalsi 1, IA-110 Reykjavik, Iceland.

KARIN HAACK *
Historische Anthropologie und Humanökologie, Universität Göttingen, Burgerstrasse 50, 37073 Göttingen, Germany.

ERIKA HAGELBERG *
Department of Biochemistry, University of Otago, PO Box 56, Dunedin, New Zealand.

CHARLOTTE HALLENBERG *
Department of Forensic Genetics, University of Copenhagen, 11 Fredrik Vs vej, DK 2100 Copenhagen, Denmark.

AGNAR HELGASON *◊
Institute of Biological Anthropology, Oxford University, 58 Banbury Road, Oxford, OX2 6QS, UK.

GODFREY M. HEWITT *
School of Biological Sciences, University of East Anglia, Norwich, NR4 7IJ, UK.

EILEEN HICKEY *
Institute of Molecular Medicine, Oxford University, Headly Way, Headington, Oxford, OX3 9AS, UK.

EMMELINE HILL *◊
Department of Genetics, Trinity College, Dublin 2, Ireland.

EDWARD HOLLOX *◊
MRC Human Biochemical Genetics Unit, Galton Laboratory, University College London, Wolfson House, 4 Stephenson Way, London, NW1 2HE, UK.

BASIL HORST *
Institut für Rechtsmedizin, Universität Münster, Von-Esmarch-Str. 62, 48149 Münster, Germany.

MATTHEW E. HURLES ◊
McDonald Institute for Archaeological Research, Downing Street, Cambridge, CB2 3ER.

CARLOS INFANTE *
Departamento de Genetica, Universidad de la Laguna, 38271 Tenerife, Spain.

MIKEL IRIONDO *
Animali Biologia eta Genetika Saila, Zientzi Fakultatea, Euskal Herriko Unibersitatea, 48940 Leioa, Spain.

PAVEL IVANOV ◊
Engelhardt Institute of Molecular Biology, Moscow 117894, Russia.

NESKUTS IZAGIRRE *◊
Animali Biologia eta Genetica Saila, PO Box 644, 48080 Bilbao, Spain.

MARK JOBLING *◊
Department of Genetics, University of Leicester, University Road, Leicester, LE1 7RH, UK.

MARTIN JONES *
Department of Archaeology, University of Cambridge, Downing Street, Cambridge, CB2 3DZ, UK.

KATRIN KALDMA ◊
Estonian Biocentre and Department of Evolutionary Biology, Tartu University, 23 Riia Street, 51010 Tartu, Estonia.

JURATE KASNAUSKIENE *
Human Genetics Centre, Santariskiu Clinics of the Vilnius University Hospital, Santariskiu Street 2, LT-2600 Vilnius, Lithuania.

LIISA KAUPPI *
Laboratory of Forensic Biology, Department of Forensic Medicine, University of Helsinki, PO Box 40, 00014 Helsinki, Finland.

ELIZA KHUSNUTDINOVA *◊
Department of Biochemistry & Cytochemistry, Ufa Research Centre, Pr.Oktyabria 69, Ufa, Russia.

JUDITH KIDD ◊
Department of Genetics, Yale University School of Medicine, PO Box 208005, 333 Cedar Street, New Haven CT 06520-8005, USA.

KENNETH KIDD ◊
Department of Genetics, Yale University School of Medicine, PO Box 208005, 333 Cedar Street, New Haven CT 06520-8005, USA.

TOOMAS KIVISILD *◊
Estonian Biocentre and Department of Evolutionary Biology, Tartu University, 23 Riia Street, 51010 Tartu, Estonia.

ANDREW KOZLOV *◊
ArctAn-C Innovative Research Laboratory, 7-27 J. Rainis blvd, Moscow 123363, Russia.

IVO KREMENSKY ◊
Laboratory of Molecular Pathology, University Hospital of Obstetrics and Gynaecology, Sofia, Bulgaria.

VAIDUTIS KUCINSKAS *
Human Genetics Centre, Santariskiu Clinics of the Vilnius University Hospital, Santariskiu Street 2, LT-2600 Vilnius, Lithuania.

SIVA KUMAR *
Max Planck Institut for Evolutionary Anthropology, Leipzig, Germany.

MARTA MIRAZON LAHR *◊
Department of Biological Anthropology, University of Cambridge, New Museums Site, Cambridge, CB2 3RF, UK.

SIRLE LAOS ◊
Estonian Biocentre and Department of Evolutionary Biology, Tartu University, 23 Riia Street, 51010 Tartu, Estonia.

SARAH LINDSAY *
Department of Molecular Sciences, UMIST, PO Box 88, Manchester, M60 1QD, UK.

DMITRY V. LISITSYN ◊
ArctAn-C Innovative Research Laboratory, 7-27 J. Rainis blvd, Moscow 123363, Russia.

APHRODITE LOUTRADIS ◊
Ministry of Health Center for Thalassemia, Athens, Greece.

JAN LUNDSTED *
Department of Forensic Genetics, University of Copenhagen, 11 Fredrik Vs vej, DK 2100 Copenhagen, Denmark.

VINCENT MACAULAY *◊
Department of Statistics, University of Oxford, Oxford, OX1 3TG, UK.

ELISABETTA MAIOLINI *
Department of Medicine and Public Health, via Irnerio 49, 40126 Bologna, Italy.

PATRIZIA MALASPINA ◊
Department of Biology, University of Rome 'Tor Vergata', Italy.

FRANCESCO MALLEGNI ◊
Sezione di Paleontologia Umana, Dipartimento di Scienze Archeologiche, Università degli Studi di Pisa, Via di S. Maria 53, 56100 Pisa, Italy.

BORIS MALYARCHUK *◊
Institute of Biological Problems of the North, Portovaya Str., 18, Magadan, 685000 Russia.

FRANZ MANNI *◊
Department of Biology, Section of Evolutionary Biology, University of Ferrara, via L. Borsari 46, 44100 Ferrara, Italy.

CONRADO MARTINEZ *
Institute of Biological Anthropology, University of Oxford, 58 Banbury Road, Oxford, OX2 6QS, UK.

CRISTINA MARTÍNEZ-LABARGA ◊
Centro di Antropologia Molecolare per lo Studio del DNA Antico, Dipartimento di Biologia, Università di Roma, Tor Vergata, Via della Ricerca Scientifica n.1, 00133 Rome, Italy.

SARABJIT MASTANA *
Department of Human Sciences (Human Genetics Lab), Loughborough University, Loughborough, LE11 3TU, UK.

EVA MATEU *
Unitat de Biologia Evolutiva, Universitat Pompeu Fabra, Doctor Aiguader 80, 08003 Barcelona, Spain.

GIUSEPPE MATULLO ◊
Dipartimento di Genetica, Biologia e Biochimica, Università di Torino, Via Santena 19, Torino, Italy.

PHILIP D. MAYNE ◊
Department of Pathology, The Children's Hospital, Temple Street, Dublin 1, Republic of Ireland.

PAUL MELLARS *
Department of Archaeology, University of Cambridge, Downing Street, Cambridge, CB2 3DZ, UK.

ENE METSPALU ◊
Estonian Biocentre and Department of Evolutionary Biology, Tartu University, Tartu, Estonia.

MAIT METSPALU ◊
Estonian Biocentre and Department of Evolutionary Biology, Tartu University, 23 Riia Street, 51010 Tartu, Estonia.

NICOLE MEYER *
Departamento de Genetica, Universidad de la Laguna, 38271 Tenerife, Spain.

SONJA MEYER *
Max Planck Institute for Evolutionary Anthropology, Inselstraße 22, D-04103 Leipzig, Germany.

EMMANUEL N. MICHALODIMITRAKIS ◊
University of Crete and IMBB, Heraklion, Greece.

HELEN MIDDLETON-PRICE *
Science Museum, Exhibition Road, London, SW7 2DD, UK.

LAURA MORELLI *◊
Dipartimento di Zoologia e Antropologia Biologica, Università degli Studi di Sassari, Corso Regina Margherita 15, 07100 Sassari, Italy.

JAYNE NICHOLSON *
CGC, Institute of Molecular Medicine, John Radcliffe Hospital, Oxford, OX3 9DS, UK.

SØREN NØRBY *◊
Institute of Forensic Medicine, University of Copenhagen, Frederik V's vej 11, 2100 Copenhagen, Denmark.

ANDREA NOVELLETTO *◊
Department of Cell Biology, University of Callabria, 87036 Rende, Italy.

CHARLES O'NEILL ◊
Department of Pathology, The Children's Hospital, Temple Street, Dublin 1, Republic of Ireland.

ONORA O'NEILL *
Newnham College, University of Cambridge, Cambridge, CB3 9DF, UK.

STEPHEN OPPENHEIMER *
Department of Pediatrics, University of Oxford.

VLADIMIR OREKHOV ◊
Vavilov Institute of General Genetics, Moscow 117809, Russia.

MARCEL OTTE *◊
Université de Liége - Préhistoire, Place du XX Aout 7 Bata 1, 1000 Liège, Belgium.

BRIGITTE PAKENDORF *
Institute of Human Biology, Allendeplatz 2, 20146 Hamburg, Germany.

ANDREW PAKSTIS ◊
Department of Genetics, Yale University School of Medicine, PO Box 208005, 333 Cedar Street, New Haven CT 06520-8005, USA.

SURINDER S. PAPIHA ◊
Department of Human Genetics, University of Newcastle-upon-Tyne, UK.

JÜRI PARIK ◊
Estonian Biocentre and Department of Evolutionary Biology, Tartu University, 23 Riia Street, 51010 Tartu, Estonia.

LUÍSA PEREIRA *◊
Instituto de Patologia e Imunologia Molecular da Universidade do Porto, Rua Dr. Roberto Frias, 4200 Porto, Portugal.

ANNA PÉREZ-LEZAUN ◊
Unitat de Biologia Evolutiva, Facultat de Ciències de la Salut i de la Vida, Universitat Pompeu Fabra, Doctor Aiguader 80, 08003 Barcelona, Spain.

HEIDI PFEIFFER *
Institut für Rechtsmedizin, Universität Münster, Von-Esmarch-Str. 62, 48149 Münster, Germany.

ALBERTO PIAZZA *◊
Dipartimento di Genetica, Biologia e Biochemica, Università di Torino, via Santena 19, 10126 Turin, Italy.

GERLI PIELBERG ◊
Estonian Biocentre and Department of Evolutionary Biology, Tartu University, 23 Riia Street, 51010 Tartu, Estonia.

RON PINHASI *◊
Department of Biological Anthropology, University of Cambridge, New Museums Site, Cambridge, CB2 3RF, UK.

LUIS MIGUEL PIRES *
Departamento de Antropologia, Faculty of Sciences and Technology, Universidade de Coimbra, 3000 - 276 Coimbra, Portugal.

ELENI PLATA ◊
Ministry of Health Center for Thalassemia, Athens, Greece.

ANDREY POLTORAUS ◊
Engelhardt Institute of Molecular Biology, Moscow 117894, Russia.

MARK POULTER ◊
MRC Human Biochemical Genetics Unit, Galton Laboratory, University College London, Wolfson House, 4 Stephenson Way, London, NW1 2HE, UK.

MARIA JOÃO PRATA ◊
Instituto de Patologia e Imunologia Molecular da Universidade do Porto, Rua Dr. Roberto Frias, 4200 Porto, Portugal.

LUIS QUINTANA-MURCI *
Unité d'Immunologie Humaine, Institut Pasteur, 25–28 rue du Dr. Roux, 75015 Paris, France.

MIRJA RAITIO *
Department of Human Molecular Genetics, National Public Health Institute, Mannerheiminte 166, 00300 Helsinki, Finland.

MAERE REIDLA ◊
Estonian Biocentre and Department of Evolutionary Biology, Tartu University, 23 Riia Street, 51010 Tartu, Estonia.

COLIN RENFREW *◊
McDonald Institute for Archaeological Research, Downing Street, Cambridge, CB2 3ER, UK.

MARTIN RICHARDS *◊
Department of Biology, Galton Laboratory, University College London, Wolfson House, 4 Stephenson Way, London, NW1 2HE, UK.

OLGA RICKARDS ◊
Dipartimento di Biologia, Università Tor Vergata, via della Ricerca Scientifica n.4, 00133 Rome, Italy.

VALENTINO ROMANO *◊
Instituto di Biologia, Fac. di Medic. e Chir., Università degli Studi di Palermo, Via Divisi, 83, 90133 Palermo, Italy.

CHIARA ROMUALDI *
Department of Statistics, Università di Padova, via S Francesco 33, Padua, Italy.

SIIRI ROOTSI ◊
Estonian Biocentre and Department of Evolutionary Biology, Tartu University, 23 Riia Street, 51010 Tartu, Estonia.

ZOË H. ROSSER *◊
Department of Genetics, University of Leicester, University Road, Leicester, LE1 7RH, UK.

JULIETTE SAILLARD *◊
Laboratory of Anthropological Biology, Blegdamsvej 3, 2100 Copenhagen, Denmark.

ALFREDO SALERNO ◊
Dipartimento di Biopatologia e Metodologie Biomediche, Facoltà di Medicina e Chirurgia, Università degli Studi di Palermo, Via Divisi 83, Palermo, Italy.

PAULA SANCHEZ DIZ *
Institute of Legal Medicine, University of Santiago de Compostela, E-15705 Santiago de Compostela, Spain.

WERA MARGARETE SCHMERER *
Historical Anthropology and Human Ecology, Göttingen.

ECKART SCHNAKENBERG ◊
Center for Human Genetics and Genetic Counselling, University of Bremen, Germany.

STEFAN SCHNEIDER ◊
Genetics and Biometry Laboratory, Department of Anthropology, University of Geneva, 12, rue G. Revilliod, 1227 Geneva, Switzerland.

ROSARIA SCOZZARI *◊
Dipartimento di Genetica e Biologia Molecolare, Università degli Studi 'La Sapienza' di Roma, P. le Aldo Moro 5, 00185 Rome, Italy.

STEPHEN SHENNAN ◊
Institute of Archaeology, University College London, Gordon Square, London, WC1E 6BT, UK.

SIGRÚN SIGURÐARDÓTTIR ◊
deCode Genetics Inc., Lynghalsi 1, IA-110 Reykjavik, Iceland.

JULIANA ALVES DA SILVA *
Max Planck Institute for Evolutionary Anthropology, Inselstraße 22, D-04103 Leipzig, Germany.

LUCIA SIMONI *◊
Dipartimento di Biologia Evoluzionistica Sperimentale, Unità di Antropologia, Università di Bologna, via Selmi 3, 40126 Bologna, Italy.

BO SIMONSEN *
Department of Forensic Genetics, University of Copenhagen, 11 Fredrik Vs vej, DK 2100 Copenhagen, Denmark.

VIKTOR SPITSYN ◊
Medical-Genetic Scientific Centre, Moscow, 115478, Russia.

KÁRI STEFÁNSSON *◊
deCode Genetics Inc., Lynghalsi 1, IA-110 Reykjavik, Iceland.

MICHELE STENICO *
Dipartimento di Biologia, Unitat de Biologia Evolutiva, Università di Ferrara, via L. Borsari 46, 44100 Ferrara, Italy.

MARK STONEKING *
Max Planck Institute for Evolutionary Anthropology, Inselstraße 22, D-04103 Leipzig, Germany.

PATRICK SULEM *
INSERM U518, Faculté de Médecine, Univeristé de Toulouse III, 31073 Toulouse CEDEX, France.

DALLAS SWALLOW *
Department of Biology, Galton Laboratory, University College London, Wolfson House, 4 Stephenson Way, London, NW1 2HE, UK.

BRYAN SYKES *◊
Institute of Molecular Medicine, University of Oxford, Oxford, OX3 9DS, UK.

KRISTIINA TAMBETS ◊
Estonian Biocentre and Department of Evolutionary Biology, Tartu University, Tartu, Estonia.

LARISSA TARSKAIA *
I Baltiski per., 3/25, k.197, Moscow, 125315 Russia.

LUCIANO TERRENATO ◊
Department of Biology, University of Rome 'Tor Vergata', Italy.

MARK THOMAS *
Centre for Genetic Anthropology, University College London, Darwin Building, Gower Street, London, WC1E 6BT, UK.

ORNA TIGHE ◊
Department of Biochemistry, Royal College of Surgeons in Ireland, 123 St Stephen's Green, Dublin 2, Ireland.

HELLE-VIIVI TOLK ◊
Estonian Biocentre and Department of Evolutionary Biology, Tartu University, Tartu, Estonia.

ANTONIO TORRONI ◊
Institute of Biochemistry, University of Urbino, Italy.

LISBETH TRANEBJÆRG ◊
Department of Medical Genetics, University Hospital of Tromsoe, 9038 Tromsoe, Norway.

MARIA TSOPANOMICHALOU *◊
Department of Forensic Sciences, Medical School, University of Crete, PO Box 71903, Stavrakia, Greece.

ANNA M. TUNZI SISTO ◊
Soprintendenza Archeologica della Puglia, Via Duomo 33, 74100 Taranto, Italy.

RICHARD TUTTON *
Department of Sociology, University of Lancaster, Lancaster, LA1 4YL, UK.

CHRIS TYLER-SMITH *◊
CRC Chromosome Molecular Biology Group, Department of Biochemistry, University of Oxford, South Parks Road, Oxford, OX1 3QU, UK.

PETER UNDERHILL *
Department of Genetics, Stanford University, 300 Pasteur Drive, Stanford, CA 94305-5120, USA.

TJEERD H. VAN ANDEL *◊
Department of Earth Sciences, University of Cambridge, Downing Street, Cambridge CB2 3EQ, UK.

RICHARD VILLEMS *◊
Estonian Biocentre and Department of Evolutionary Biology, Tartu University, 23 Riia Street, 51010 Tartu, Estonia.

RYK WARD ◊
Institute of Biological Anthropology, Oxford University, 58 Banbury Road, Oxford, OX2 6QS, UK.

MIKE WEALE *
Centre for Genetic Anthropology, University College London, Darwin Building, Gower Street, London, WC1E 6BT, UK.

GUNTER WEISS *◊
Max Planck Institute for Evolutionary Anthropology, Inselstraße 22, D-04103, Leipzig, Germany.

SPENCER WELLS *
Wellcome Trust Centre for Human Genetics, University of Oxford, Roosevelt Drive, Headington, OX3 7BN, UK.

EDWARD WHITEHEAD *
(sdme 3/47), 200 Rue de la Loi, B-1049 Brussels.

TOM WILKIE *
Wellcome Trust, London, NW1 2BE, UK.

JAMES WILSON *
Department of Biology, Galton Laboratory, University College London, Wolfson House, 4 Stephenson Way, London, NW1 2HE, UK.

NIKOLAY YANKOVSKY *◊
Genome Analysis Laboratory, Vavilov Institute of General Genetics RAS, Gubkina str., 3, 117333 Moscow, Russia.

BORIANA ZAHAROVA *◊
Laboratory of Molecular Pathology, University Hospital of Obstetrics, 2 Zdrave St, 1431 Sofia, Bulgaria.

TATIANA ZERJAL *◊
CRC Chromosome Molecular Biology Group, Department of Biochemistry, University of Oxford, South Parks Road, Oxford, OX1 3QU, UK.

LEV ZHIVOTOVSKY ◊
Vavilov Institute of General Genetics, Moscow 117809, Russia.

MAREK ZVELEBIL *◊
Department of Archaeology and Prehistory, Northgate House, West Street, Sheffield, S1 4ET, UK.

Figures

1.1.	Geographic map for the second Principal Component from the analysis of European gene frequencies.	5
1.2.	The proposed homeland of haplogroup V.	7
1.3.	The general position of some well-known hybrid zones for a range of plant and animals species in Europe.	8
2.1.	Gene frequency map for the Rhesus negative blood group factor.	18
2.2.	The taxonomic and phylogenetic perspectives compared.	19
2.3.	The network approach to the interpretation of data from human mitochondrial DNA.	20
3.1.	Frequencies of blood groups A and B from different ethnic groups.	24
3.2.	Selection against heterozygotes in the Rhesus blood group.	24
3.3.	The cline of the first principal component across Europe.	25
3.4.	Genetic tree of worldwide populations.	27
4.1.	SPECMAP, the widely used late Quaternary time-scale based on marine oxygen isotope records.	33
4.2.	Precipitation and temperature history for OIS 3 from a long pollen core at Les Echets, France.	34
4.3.	OIS 3 climate oscillations.	34
4.4.	Europe during the late mid-pleniglacial.	35
4.5.	The Greenland ice-core climatic record.	35
4.6.	Extent of the Indian Ocean summer monsoon.	36
4.7.	The solar energy curve for the interval from 20,000 years BP to the present.	37
4.8.	African and southwest Asian lake levels 9000 years ago.	37
5.1.	The arrival of anatomically modern men corresponds to the introduction of the bone component in technology.	43
6.1.	Plot of the latitude and longitude of archaeological sites (Late Palaeolithic and Mesolithic).	46
6.2.	Plot of the latitude and longitude of archaeological sites (Late Palaeolithic, Mesolithic and Neolithic).	47
6.3.	Plot of the latitude and longitude of archaeological sites (Late Palaeolithic, Mesolithic and Early Neolithic).	48
6.4.	Plot of the latitude and longitude of archaeological sites (Late Palaeolithic, Mesolithic and Early Neolithic).	49
6.5.	Plot of the latitude and longitude of archaeological sites (Mesolithic and Early Neolithic).	50
6.6.	Plot of the latitude and longitude of archaeological sites (Mesolithic and Neolithic).	51
6.7.	Plot of the latitude and longitude of archaeological sites (Mesolithic and Neolithic).	52
7.1.	'Colonist' and 'indigenous' regions of Europe at the agricultural transition.	61
7.2.	The origin and dispersal of the Linear Pottery Ware culture.	62
7.3.	Origins of the Neolithic communities in the Iberian peninsula: different views for different regions.	64
7.4.	Origins of the Neolithic communities on the North European Plain.	65
7.5.	Agricultural frontier zone and forager–farmer contacts in central and eastern Baltic 2500–1500 BC.	66
7.6.	Exchanges between foragers and farmers within an agricultural frontier zone: a general pattern.	67
8.1.	Pattern for a major Neolithic genetic component.	84
8.2.	Pattern for a major Neolithic genetic component with a minor Mesolithic contribution.	84
8.3.	Pattern for an equal Mesolithic and Neolithic genetic component.	84
8.4.	Pattern for a greater genetic contribution from the Mesolithic populations.	85
8.5.	Pattern for no genetic impact for the Neolithic.	85
8.6.	Summary map of Europe showing the expected genetic contributions of Neolithic and Mesolithic populations.	86
9.1.	Map of Merzbach sites.	90
9.2.	Decorative characters from the Merzbach pottery.	92
9.3.	Prediction of the phylogenesis hypothesis in the newly-founded settlements analyses.	94
9.4.	Prediction of the ethnogenesis hypothesis in the newly-founded settlements analyses.	94
10.1.	Expansion times of 62 populations analyzed for mtDNA HV1 diversity.	103
10.2.	Genetic affinities between 61 populations analyzed for mtDNA HV1 diversity.	105
10.3.	Potential scenario explaining the reduction in effective population size for hunter-gatherer populations.	106
11.1.	STRP heterozygosity and private alleles.	111
11.2.	An additive tree of 14 specific populations.	113
11.3.	A principal components analysis based on the same genetic distances as the tree in Figure 11.2.	113
13.1.	Geographic distribution of sampled populations from Europe, Near East and Caucasus.	133
13.2.	Spatial correlograms on mtDNA HVR-I in Europe.	135
14.1.	A schematic representation of the world mtDNA phylogeny.	142

Figures

14.2.	The full median network of a sample of haplogroup H mtDNAs.	144
14.3.	The principles of founder analysis.	145
14.4.	The phylogeography of haplogroup J1.	146
15.1.	Maximum parsimony tree of Y-chromosome haplogroups.	155
15.2.	Distribution of European population samples, and summary of haplogroup frequency data.	156
15.3.	Distribution of haplogroup 22 chromosomes within Iberia.	157
16.1.	A putative 7-locus microsatellite founder haplotype is supposed to undergo mutational events.	164
17.1.	Map of Sicily showing the four towns analyzed for autosomal microsatellites and mtDNA polymorphisms.	169
17.2.	Principal Components Analysis performed using the allele frequencies of four autosomal microsatellites.	172
18.1.	Size and geographical distribution of the study populations.	177
18.2.	Reduced median network showing the relationship among the different mtDNA haplotypes.	180
19.1.	Protocol used for mtDNA haplogroup assignment.	187
19.2.	Haplogroup frequency distribution in the samples studied and in control populations.	188
20.1.	Y-chromosome biallelic marker haplogroup distributions in Portugal.	193
20.2.	Nucleotide pairwise difference distributions observed in North, Central and South Portugal.	194
21.1.	Neighbour-joining trees.	199
22.1.	Approximate locative origins of surnames sampled in Ireland.	204
22.2.	Maximum Consensus parsimony networks of Irish Gaelic haplotypes.	206
23.1.	Neighbour-joining tree of Nei's genetic distance for the Traveller and other European populations.	211
24.1.	Occurrence of network 1.2 haplotypes in three sampling locations.	217
25.1.	Map of the area.	220
25.2.	22 haplogroup M mtDNAs found among 906 Turks, Ossetes, Georgians and Armenians.	224
25.3.	Topology and phylogeography of sub-haplogroup M1 in Africa and western Eurasia.	225
25.4.	Phylogenetic position of an internal node P*, occupied by one Indian.	226
25.5.	Phylogenetic position of an internal node O*.	226
25.6.	Topology and phylogeography of mtDNA sub-haplogroup U7.	227
25.7.	Sub-haplogroup T1 in Anatolians–Trans-Caucasians, Nile Valley populations, Ethiopians and in Europeans.	229
26.1.	Networks of 'adjacent' (one repeat difference) haplotypes of three Bulgarian population groups.	240
26.2.	Common networks of 'adjacent' haplotypes of Bulgarian population.	241
27.1.	Historical background of Eastern Europe (6th century) and sampling localities.	246
27.2.	Neighbour-joining tree built according to HVRI sequence data.	247
27.3.	The distribution of pairwise nucleotide differences of Russians, Maris and Tatars.	247
27.4.	Number of shared mitotypes between Russians, Mari and Tatars.	248
28.1.	Median network of 44 Eastern Slavonic mtDNA HVS-1 sequence types.	251
28.2.	Neighbour-joining tree for 18 European populations, based on mtDNA HVS-1 sequences.	252
29.1.	Location of blood sampling in the Nenets Autonomous Area.	256
29.2.	Reduced median network of 58 Nenets HVS I sequences.	257
30.1.	Location of the populations sampled.	261
30.2.	Proposed routes for the dispersion of anatomically modern humans out of Africa.	263
30.3.	Y-chromosome haplotype frequency distribution in four Central Asian populations.	265
31.1.	Two alternative routes for out-of-Africa migration.	269
31.2.	An unrooted tree relating mtDNA and Y-chromosomal haplogroups.	270
31.3.	Scheme combining the routes and expansions of modern human early dispersals.	273
32.1.	Coalescent genealogy of a sample of five sequences.	280
33.1.	The frequency spectrum of 418 German mtDNA HVS1 sequences.	288
33.2.	Sampling saturation of mtDNA HVS1 lineages from seven European populations.	289
33.3.	A median-joining phylogenetic network of haplogroup J lineages from different European regions.	291
34.1.	Geographic distribution of population samples analyzed in this study.	296
34.2.	Two different scenarios of female and male migration rates.	299
35.1.	The concept of reasoning and definitions explained.	302
35.2.	Reduced median networks connecting Y haplogroups (2a), Y haplotypes (2b) and Y lineages (2c).	303
36.1.	Allelic composition of the four common haplotypes.	306
36.2.	Frequencies of the four common haplotypes in different populations.	307

Figures

37.1.	LAC*R gene distribution in European populations.	310
37.2.	Northern borders of the dairy cattle-breeding in Europe.	311
38.1.	Neighbour-joining networks linking mtDNA sequences from four domestic ungulates.	317
38.2.	Approximate locations of suggested domestication centres for sheep, cattle, pig, and water buffalo.	318
40.1.	Regression of $F_{ST}/(1-F_{ST})$ on the logarithm of geographic distances between populations.	327
40.2.	Location of the linguistic boundary between populations of Afro-asiatic and Indo-european families.	328
40.3.	Location of the linguistic boundary between populations of Afro-asiatic and Indo-european families.	329
41.1.	Plan of the Province of Ferrara.	335
41.2.	Plan of the Province of Ferrara as in Figure 41.1.	335
41.3.	Bootstrap UPGMA tree based on seven enzymatic systems.	336

Tables

1.1.	A summary of the three main waves of European colonization, as derived from mitochondrial DNA analysis.	6
6.1.	Pearson correlations between radiocarbon dates, latitude and longitude by archaeological period.	53
6.2.	Pearson correlations between average radiocarbon dates and geographic location of Early Neolithic sites.	53
7.1.	Forms of contact between foragers and farmers.	58
7.2.	Outcomes of contact between foragers and farmers.	58
8.1.	Predictive model for Mesolithic/Neolithic population interaction.	82
8.2.	Demographic profiles by period and region based on archaeological data.	83
9.1.	Features at sites from which Merzbach pottery assemblages are derived, according to Frirdich's maps.	91
9.2.	Taxa used in phylogenetic analyses.	93
9.3.	Results of phase-by-phase analysis of assemblages from settlements occupied throughout the 10-phase period.	95
9.4.	Results of analyses focusing on the instances when a new assemblage is established during the 10-phase period.	95
12.1.	Expected consequences of the three main models of origin of the European gene pool.	122
12.2.	Summary of the data analyzed.	122
12.3.	Spatial autocorrelation analysis of molecular variation at seven loci.	123
12.4.	Estimated average divergence times between European populations.	125
13.1.	Populations considered in this study.	134
16.1.	Main features of 15 Y-chromosomal haplotype networks.	164
17.1.	MtDNA RFLP haplogroups in Sicily.	173
18.1.	Table of Sicilian and southern Italian HV1 haplotypes.	178
20.1.	Number of individuals screened in the three Portuguese samples.	192
20.2.	Haplogroups identified with the ten Y-chromosome biallelic markers studied.	192
20.3.	Population pairwise F_{ST} P values between samples from North, Central and South Portugal.	192
20.4.	MtDNA diversity parameters in North, Central and South Portugal, considering HVRI and/or HVRII.	193
20.5.	MtDNA haplogroup distributions (values in %) in North (NP), Central (CP) and South (SP) Portugal.	194
21.1.	Comparison via the χ^2 test between Basque populations and a set of Eurasian populations.	198
21.2.	MtDNA haplogroup frequencies at the three prehistoric sites analyzed in this study.	200
21.3.	MtDNA diversity based on mtDNA haplogroup frequencies in modern European populations.	200
22.1.	AMOVA test of population substructure.	205
22.2.	SNP haplogroup frequencies by population.	205
24.1.	Absolute and relative frequencies of three main Y-chromosomal lineages in six Mediterranean regions.	216
25.1.	MtDNA haplogroups in Anatolian-Trans-Caucasus populations (%).	222
25.2.	Distribution of haplogroup U varieties in Anatolians-Trans-Caucasians and in some other populations.	223
25.3.	Distribution of the 16,293–16,311 motif of haplogroup H.	230
29.1.	Frequency of the different haplogroups.	257
30.1.	Diversity parameters in Central Asian populations.	262
30.2.	Fraction of mtDNA sequences in four Central Asian populations.	264
31.1.	MtDNA haplogroup U and M frequencies worldwide.	271
33.1.	Summary statistics for mtDNA lineages from ten European populations.	286
33.2.	The proportion of rare HVS1 lineages in European populations.	290

Tables

34.1.	*Apportionment of genetic variation in different genetic systems in worldwide populations.*	297
34.2.	*Genetic variation (F_{ST}) and migration rate (Nv) in worldwide populations.*	297
34.3.	*Correlations between Y chromosome/mtDNA/geography.*	298
34.4.	*Migration rates in comparable data sets (f = female; m = male).*	298
35.1.	*Frequencies and sharing status of chromosome Y haplotypes.*	304
40.1.	*Results of the evaluation of the genetic impact of the linguistic boundary between populations.*	328
41.1.	*Scheme of the unit cost model through which linguistic differences were calculated.*	334
41.2.	*Key to the numbering of localities reported in Figures 41.1 and 41.2.*	337

Foreword

The application of molecular genetics to the study of human population history is a fast-growing field. The present volume is the product of a Conference held in Cambridge in September 1999 under the auspices of the McDonald Institute for Archaeological Research with the title: 'Human Diversity in Europe and Beyond: Retrospect and Prospect'. It constituted the Third Biennial Euroconference of the European Human Genome Diversity Project, the first having been held in Barcelona in 1995 and the second in Helsinki in 1997. Its aim was to undertake a wide review of the field, with the participation of archaeologists, anthropologists and climatologists as well as the molecular geneticists who have been the main participants at earlier such meetings.

The Euroconference was funded by the Training, Mobility and Research Programme of the European Union. These biennial conferences arose from a meeting on Human Genome Diversity held in Porto Conte, Sardinia, in 1993 with the encouragement of the Human Genome Organization, and it was on that occasion that the Standing Committee of the Human Genome Diversity Group for Europe was established under the chairmanship of Sir Walter Bodmer with the following members: Jaume Bertranpetit (Barcelona), Julia Bodmer (Oxford), Anne Cambon-Thomsen (Toulouse), Svante Pääbo (Leipzig), Alberto Piazza (Turin) and Colin Renfrew (Cambridge). With the encouragement of the Standing Committee, a United Kingdom Organizing Committee was established for the 1999 Conference with the following membership: Katie Boyle, Peter Forster, Mark Jobling, Colin Renfrew (chair) and Bryan Sykes advised also by Rob Foley and Leslie Knapp.

The wide range of papers submitted to and discussed by the Conference suggested the possibility of a publication which might offer a review of the current state of research in this area. Such is the objective of the present publication. It should be noted that, unlike previous symposium publications by the McDonald Institute, where all papers have been presented by invitation, the contributions here arose out of the conference itself, where many of the contributions came in response to the original call for papers made by the Organizing Committee. These offers of papers were then reviewed by the Committee on the basis of the abstracts submitted. Although the papers here have been subjected to critical scrutiny they have not been subjected to 'peer review' in the rigorous sense practised, for instance, by *Nature* or *Science* or by the *American Journal of Human Genetics*. The aim here is not so much to publish new data (although much in the way of new findings is in fact presented here) as to offer an overview of this new and important field of a kind which has not hitherto been available. For this reason the volume begins with an introductory section, which sets out to make some of the technical concepts involved more intelligible to non-specialist readers. The second part, 'The Archaeological and Environmental Background', reviews some of the basic environmental and archaeological information, as discussed at the meeting, which is indispensable in setting the succeeding molecular genetic information in a coherent historical context.

One important session of the Conference is not included here, since the papers were presented orally. This was the ethics session, ably chaired by Baroness Onora O'Neill, with papers by Kári Stefánsson of deCODE Genetics Inc., Reykjavik, Tom Wilkie of the Wellcome Trust, London, Anne Cambon-Thomsen (with A. de la Dorie, A. Thomas & A.M. Duguet) of the University of Toulouse III, and Richard Tutton, Department of Sociology, University of Lancaster. The ethical dimension is an important one, not only for questions of the informed consent of those giving samples, but also for the whole matter of the commercial exploitation of genetic information, and not least for issues surrounding patenting. It is to be hoped that future conferences will have the opportunity of exploring these issues further.

We are grateful to the following (in addition to Baroness O'Neill) for chairing sessions at the Conference: Hans-Jürgen Bandelt, Peter Donnelly, Laurent Excoffier, David Goldstein, Mark Jobling, Peter de Knijff, Mark Stoneking, and Bryan Sykes.

For financial assistance the Conference was primarily indebted to the European Commission, and also to the Alfred P. Sloan Foundation, Beckman Coulter UK Ltd. and Geneo BioProducts and to

Foreword

Edward Whitehead, formerly of the European Commission for helpful advice.

Thanks are due to the Judge Institute of Management Studies (whose Director is Professor Sandra Dawson) for allowing us to hold the Conference in their remarkable premises, and to the Sanger Centre at Hinxton for hosting a most agreeable and informative conference excursion. Peter Forster undertook a leading role in the organization of the Conference, in the preparation both of the programme and the volume of abstracts and in the accounting, and we wish to acknowledge his energetic and enthusiastic participation. Thanks are due also to Becky Coombs, Lila Janik, Oliver Jardine, Turi King, Jianjun Mei, Chau Pak-Lee, Neil Shailer, Caroline Adams, Patricia Salazar, Victor Paz and Anne Threkeld for their organizational assistance. In the production of this volume a major role has been played by Dora A. Kemp, the McDonald Institute's Production Editor to whom we are very grateful.

COLIN RENFREW & KATIE BOYLE

Part I

Introductory

Chapter 1

Archaeogenetics: Towards a Population Prehistory of Europe

Colin Renfrew

The term archaeogenetics *is proposed for the newly-emerged discipline which applies molecular genetics to the study of the human past. The first phase, using classical genetic markers has, during the 1990s largely been replaced by DNA-based analysis: the application of mitochondrial DNA and Y-chromosome analysis is now being conducted systematically. In Europe the fundamental contribution of the first phase was the application by Cavalli-Sforza and his colleagues of Principal Components analysis to classical genetic markers. The map of the first Principal Component, with clines from southeast to northwest, was originally seen as the result of the spread of farming to Europe from Anatolia; the map of the second Principal Component (from north to southwest) was difficult to explain. The major insight at the outset of phase 2 of research has been the suggestion that the most significant demographic processes and events took place in the Upper Palaeolithic period, to which much of the variation seen in the first two Principal Components may now be assigned. The enormous potential of archaeogenetics for the reconstruction of population prehistory, not least for that of Europe, is stressed.*

Archaeogenetics as an emerging discipline

Over the two past decades a new discipline has come into being that so far has lacked a distinctive name: it involves the application of molecular genetics to archaeology and towards the reconstruction of the population history of the world. And its ramifications go much wider. Here the term *archaeogenetics* is proposed for this new field of investigation: i.e. *the study of the human past using the techniques of molecular genetics*. In practice this is likely to involve the collaboration of molecular geneticists with archaeologists, anthropologists, historical linguists and climatologists. As the papers in this volume show, the study of DNA is central to this new field, although classical genetic markers have also been highly informative. So far most of the new information has come, perhaps surprisingly, not from ancient DNA, where the archaeological dimension is obvious, but rather from analyzing DNA obtained from living populations. Robust procedures of inference, as discussed in this volume, allow the genetic data obtained from living populations to yield conclusions about the relationships between these populations, relationships which have to be expressed in historical terms. Present relationships thus carry implications about common origins which are inevitably situated in the past. The recovery and analysis of ancient DNA is, however, technically more difficult and so far it has provided only a few significant new insights, although many more are to be expected as extractive and analytical techniques are further developed and methods for avoiding contamination are refined.

As the paper here by Sykes shows (Chapter 3), the study of blood groups and other genetically-determined markers, now considered 'classical', has a respectable history. In 1954 an authoritative survey entitled *The Distribution of the Human Blood Groups* (Mourant 1954) suggested to many researchers how spatial variation in such markers might correlate with modern ethnic and linguistic groupings. The high frequency of the Rhesus negative blood group in the Basque country of northern Spain promoted speculation on the possible correlations there between linguistic and genetic history (see Fig. 2.1). The first attempt at reconstructing human evolution on the basis of 'classical' genetic data based on samples taken from living populations was undertaken in 1963 by Cavalli-Sforza and Edwards in a pioneering paper entitled 'Analysis of human evolution' (Cavalli-Sforza & Edwards 1965). For Europe a key contribution, although one little noted at the time, in the field of what was to become archaeogenetics was the paper by Menozzi *et al.* (1978). In it a principal components analysis of data for classical genetic markers in Europe formed the basis for hypotheses about those historical processes which might have underlain each of the first four principal components observed. The same authors went on to compile their magisterial *The History and Geography of Human Genes* (Cavalli-Sforza *et al.* 1994), which relied primarily upon classical genetic markers, sampled on a worldwide basis. This notable volume may be taken to mark the end of the first phase in the development of archaeogenetics, at a time when the use of such classical genetic markers, in general, was coming to be replaced by DNA studies.

The second phase, currently in full spate, was initiated by the earliest papers utilizing DNA sequencing for the reconstruction of human population histories. One of the first of these, entitled 'Evolutionary relationships of human populations from an analysis of nuclear DNA polymorphisms' (Wainscoat *et al.* 1986) used nuclear DNA. The important paper by Cann *et al.* (1987) entitled 'Mitochondrial DNA and human evolution' was one of the first to utilize the potential for studying specific lineages offered by mitochondrial DNA for the female line. (Later work employed Y-chromosome analysis for the male line.) In promoting the 'out of Africa' hypothesis of human origins it was the starting point for such popular notions as that of the 'African Eve'. Trouble with the data base and with maximum parsimony trees gave rise to controversy, however, but with the concordance of further data (Vigilant *et al.* 1991), including Y-chromosome data, there is now what seems to be a general consensus among molecular geneticists that the basic premise of an African origin for *Homo sapiens* was correct. Already, nearly a decade ago, it was possible to predict the immense future impact which molecular genetics would soon have upon archaeology (Renfrew 1992).

As noted above, many of the interesting inferences have come about using genetic data from living populations, but an important and growing branch of archaeogenetics is the study of *ancient biomolecules* — that is to say of biochemical data obtained from ancient materials deriving mainly from archaeological excavations. The field of biomolecular archaeology (Hedges & Sykes 1992) is not quite coterminous with that of archaeogenetics, since ancient biomolecules can in addition provide data which are not exclusively genetic. For not all biomolecules contain DNA, and the study of lipids in food residues, for instance, can be of considerable interest in the reconstruction of diet, or in understanding the nature of the contents of pots found on archaeological sites. The study of ancient DNA, moreover, has had its disappointments as well as its successes (Renfrew 1998), but the remarkable achievement of extracting and analyzing mitochondrial DNA from Neanderthal remains (Krings *et al.* 1997; Ovchinnikov *et al.* 2000) is a promising indicator of what may be achieved in favourable circumstances.

It should be noted that the field of archaeogenetics is not restricted to the study of genetic data deriving from humans, whether living or ancient, nor to the reconstruction of human population history in the strictest sense. The papers here by Allaby (Chapter 39) and Bradley (Chapter 38) respectively indicate the relevance of the study of plant and animal DNA, particularly in relation to domestic plants and animals, in understanding the human past. The pioneering paper on the origins of einkorn by Heun and his colleagues (Heun *et al.* 1997) suggests, for instance, that einkorn wheat was first domesticated in a very restricted area of what is now southern Turkey and that its subsequent dispersal from that area may give valuable insights into the spread of farming from Anatolia to Europe.

The archaeogenetics of Europe

The current situation in the archaeogenetics of Europe is a complicated one, sometimes even a confusing one, as the attentive reader of the papers in this volume will readily conclude. The pessimistic reader may well take the view that the conflicting interpretations currently on offer simply reflect the uncertainties inherent in a discipline fraught with many pitfalls. Not least among these are the sometimes contradictory historical inferences which are made when studies based on classical genetic markers are compared with those based upon mitochondrial

DNA, or again when the latter are compared with the fruits of Y-chromosome analysis. The optimist, however, will note the rapid pace of research (see Renfrew in press) and the recent development of phylogenetic network methods which enable estimations of time depth which are not dependent upon external archaeological chronometrical factors but which instead employ molecular genetic inferences coupled with estimations of mutation rates. Both the optimist and the pessimist may well concede that the situation in Europe reflects some analogies with that in the Americas (see Renfrew 2000a). There too there are fundamental points of disagreement about the early peopling of the continent, and there too the pace of research is so rapid that within a decade the available picture is likely to have changed substantially.

It is to be observed also that some of the problems in the archaeogenetics of Europe are not restricted to the human species, but find parallels in 'the genetic legacy of the Quaternary ice ages' as applied to other species also (Hewitt 2000). The issue of ice age refugia, discussed below, is a general one for the fauna, and to some extent for the flora, of Europe.

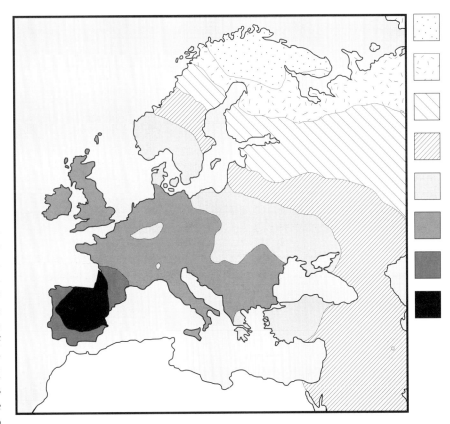

Figure 1.1. *Geographic map for the second Principal Component from the analysis by Cavalli-Sforza et al. (1994) of European gene frequencies. They suggested that since 'There is no known demic expansion from the Iberian peninsula . . . the simplest explanation may be . . . gene flow caused by one or more migrations of Mongoloid Uralic speakers from Northwestern Asia'.*

Phase one of research

The first phase of research in the archaeogenetics of Europe was dominated by the 1978 paper cited above, 'Synthetic maps of human gene frequencies in Europe' (Menozzi *et al.* 1978). That study was subsequently updated using a wider range of classical genetic markers, without any significant change in the patterning obtained (Cavalli-Sforza *et al.* 1994, 290–96). Seven principal components are listed, and maps are illustrated for the first six of these. The first three PCs are naturally the most significant, representing 28.1 per cent, 22.2 per cent and 10.6 per cent of the variance respectively. It should, of course, be clearly understood that a principal components analysis represents a palimpsest of all the processes which have taken place, from the earliest human settlement down to the present time. But the authors chose, rather arbitrarily it would seem, to identify each PC with a specific historical process or episode.

The explanation offered for the first PC, with its pronounced clines from southeast to northwest was suggested by the undoubted resemblance of this map to that reached in earlier studies by Ammerman & Cavalli-Sforza (1973) for the diffusion of early farming in Europe, using a model of 'demic diffusion', with a wave of advance from Anatolia into Europe. Such patterning had first been noticed by Grahame Clark, using radiocarbon determinations for early farming settlements in Europe (Clark 1965). As the authors note: 'The first synthetic map parallels very closely the map of the times of first arrivals of the Neolithic, or, archaeologically, of the cultivated forms of cereals domesticated in the Middle East and not present in Europe before the agricultural expansion' (Cavalli-Sforza *et al.* 1994, 291) (see Fig. 3.3).

The second PC (see Fig. 1.1) shows a strong

Table 1.1. *A summary of the three main waves of European colonization, as derived from mitochondrial DNA analysis of samples from living populations. The letters refer to the classification of haplotypes. It should be noted that the Neolithic contribution is estimated at c. 20 per cent of the total. (After Sykes 1999, 137.)*

Component	Dates (BP)	Main associated clusters	Contribution to modern gene pool
Neanderthal	300,000	unclassified	0%
Early Upper Palaeolithic	50,000	U5	10%
Late Upper Palaeolithic	11,000–14,000	H, V, I, W, T, K	70%
Neolithic	8500	J (+ more of H, T, K?)	20%

concentric gradient like that of the first but centred instead on the Iberian peninsula. As the authors remark (Cavalli-Sforza *et al.* 1994, 292): 'There is no known demic expansion from the Iberian peninsula, and an interpretation based on a migration from this area seems unlikely. The opposite pole of the second PC shows a strong peak among Lapps, which are certainly no candidate for an expansion.' They conclude: 'The simplest explanation may still be . . . gene flow caused by one or more migrations of Mongoloid Uralic speakers from Northwestern Asia.'

The third PC shows a strong peak in the European steppe north of the eastern part of the Black Sea with an approximately concentric gradient. The authors argue (Cavalli-Sforza *et al.* 1994, 293): 'The strongest candidate for an archaeological interpretation of the third map is an expansion of the Kurgan culture, which was associated by Gimbutas with the primary expansion of the Kurgan people.'

Subsequent studies, such as that of Sokal *et al.* (1991) served to confirm the spatial patternings observed, but in general were less clear-cut about the appropriate interpretations to be offered.

The beginning of phase two of research
A fundamental and very basic reappraisal of these interpretations, however, was suggested in a paper which we may regard as representing the inception of the second phase in the study of the archaeogenetics of Europe, inasmuch as it was based upon a study of mitochondrial DNA rather than upon classical genetic markers: 'Palaeolithic and Neolithic lineages in the European mitochondrial gene pool' (Richards *et al.* 1996).

That paper took a different analytical approach. In the first place it avoided classical genetic markers, but instead identified the frequency of the different mtDNA haplogroups in the various populations studied. It also broke new ground by using procedures (based upon the phylogenetic network method) for estimating, albeit approximately, the date of arrival in Europe of each of the mitochondrial types (and hence women) ancestral to today's European haplogroups. The great surprise was that, with the exception of one haplogroup (subsequently designated haplogroup J), the date of arrival of the incoming groups was set back in the Pleistocene period. These were, it was claimed, Palaeolithic or Mesolithic settlers, and only haplogroup J was assigned to the Neolithic.

This paper has proved highly controversial, coming in for severe criticism by Cavalli-Sforza and his colleagues (e.g. Cavalli-Sforza & Minch 1997). That critical approach is cogently presented in this volume by Barbujani & Chiki. They reassert (this volume, Chapter 12, 127) the original position:

> By and large . . . the overall genetic structure of the European population seems to reflect the consequences of a Neolithic process. This is also supported by a recent mitochondrial study, in which traces of Neolithic expansions were recognized in all European groups studied, except Saami. (Excoffier & Schneider 1999)

The contrary, *deep time* view is reiterated and developed here by Richards & Macaulay (Chapter 4).

Without presuming to offer a judgement in these matters it is relevant to point out how the 'deep time' position of Richards and colleagues can be reconciled with the principal components analysis of Cavalli-Sforza and his colleagues, although not with the interpretations which they originally placed upon the patternings observed. The chronological implications of the work by Richards and his colleagues, revised in the light of some later work, have been presented by Sykes (1999, 137, table 2) in a convenient table, seen here as Table 1.1.

The implication here is clear that the spread of farming, represented by haplogroup J, contributed only about 20 per cent to the modern European mitochondrial pool. It must therefore follow that the first PC, as derived from classical genetic markers, involves a substantial (and larger) Palaeolithic contribution. Naturally that first PC represents an aggregate — and this is a point which commentators seem slow to acknowledge. If there were indeed significant and persisting contributions of human gene flow from Anatolia to Europe in the Palaeolithic and Neolithic periods, the first PC would inevitably represent the sum over time of these contributions. Indeed that must be the case here, even if the relative scale of the input made in different periods represents a matter of controversy.

A further important insight is offered by another paper based upon mtDNA study: 'MtDNA analysis reveals a major Palaeolithic population expansion for southwestern to northeastern Europe' (Torroni *et al.* 1998). Here the authors recognize a significant contribution to the population of western and northwestern Europe made by what they term haplogroup V, an expansion from Iberia which they date to *c.* 15,000 years ago (Fig. 1.2).

The dating is important, since it allows one to draw upon the climatic history of Europe (see van Andel, this volume, Chapter 4) which is now increasingly well-dated by standard chronometric methods, principally radiocarbon dating (Houseley *et al.* 1997). It is now well established that the Late Glacial Maximum lasted from *c.* 20,000 to *c.* 15,000 BP, and that during that period the population of Europe retreated southwards, being concentrated in population refugia centred in Iberia (and southwest France) and in the Balkan–Ukraine area. The explanation for the patterning detected by Torroni and his colleagues may thus well be that the mutation which formed haplogroup V took place sometime around 20,000 BP, and that with the retreat of the ice and the repopulation of northern Europe, the relevant population gradually moved northwards, forming the basis for the patterning still seen today.

One interesting feature of this scenario is that it would account, much more clearly than any explanation hitherto offered, for the patterning seen in the second PC of the synthetic maps of Cavalli-Sforza

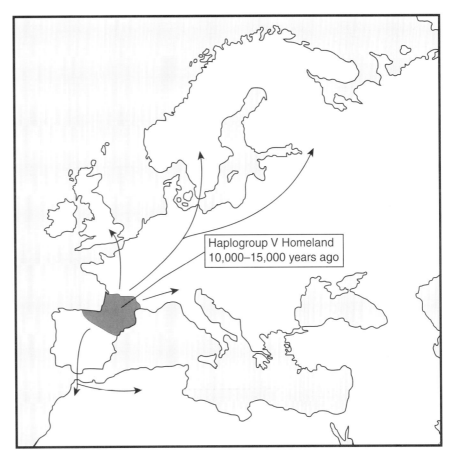

Figure 1.2. *The proposed homeland of haplogroup V, indicating the expansion of population after the end of the Late Glacial Maximum, c. 15,000 BP, as suggested by the study of the mitochondrial DNA of 419 contemporary individuals (after Torroni* et al. *1998). It is likely that this climate-induced demic diffusion underlies the patterning seen in Principal Component 2 of the analysis of Cavalli-Sforza and colleagues seen in Figure 1.1.*

and his colleagues. In a similar vein, the patterning of the third PC would no longer necessarily be assigned to a period as late as the Bronze Age and to the putative 'Kurgan' invasion of Gimbutas (1973). Once again PC 3 would be a palimpsest. One component would be the Palaeolithic population movements from east to west, along a trajectory passing north of the Black Sea, which Otte (this volume, Chapter 5) would associate with the initial Aurignacian population of Europe, and perhaps also with the later (middle phase of the Upper Palaeolithic) Gravettian cultural uniformity. But being a palimpsest the third PC will also represent the aggregate of any gene flows along this geographical axis, whether east-to-west or west-to-east, and of whatever date. It may be that the most significant of these are of Palaeolithic date (Underhill 1999). It is also to be noted, however, that the main Holocene coloniza-

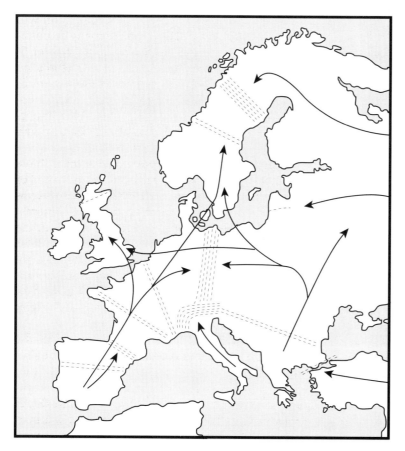

Figure 1.3. *The general position of some well-known hybrid zones for a range of plant and animals species in Europe (after Hewitt 2000), suggesting comparison with the spatial patterning of gene frequencies for humans in Europe obtained from haplogroup frequencies for mitochondrial DNA and Y-chromosome markers, and also for classical genetic markers. The zone boundaries seen here show major clustering in Scandinavia, central Europe and the Alps, and other clusters are apparent in the Pyrenees and the Balkans. As Hewitt remarks: 'These suture zones are caused by commonalities of ice-age refugia, rate of postglacial expansion and physical barriers. There is further subdivision in the southern regions'.*

ures 3.3, 1.1 and 1.2 with the map (Fig. 1.3) recently produced by Hewitt (2000) relating to the distributions of other species, in which pronounced 'suture zones' are seen, indicating the spatial extent of various migratory and other processes.

The broad analogies between this map and the human genetic patterns discussed here are striking. The caption to his figure 2 (not illustrated here; Hewitt 2000, 910) gives further food for thought:

> Three paradigm colonizations from southern Europe deduced from DNA differences for the grasshopper, *Corthippus parallelus*, the hedgehog, *Erinaceus europus/concolor*, and the bear, *Ursus arctos*. The main refugial areas, Iberia, Italy, the Balkans and Caucasus, contributed differently to the repopulation of the northern parts.

It is clear that the demographic processes underlying the modern genetic spatial patterning for humans have their counterparts among those generating the distributions of other species also. The comparison is likely to yield further fruitful insights in the future.

Current issues
The controversy outlined above will have to be settled before one can make significant progress with the archaeogenetics of Europe. Fortunately another body of data is now becoming available, namely the results of Y-chromosome analysis (e.g. Malaspina *et al.* 1998), represented in this volume by various papers including those by Rösser and colleagues (Chapter 15), by Malaspina and colleagues (Chapter 16) and by Novelletto and colleagues (Chapter 24). But there are several problems still to be resolved in the assimilation of their findings with those derived from mtDNA.

tion of the western Eurasian steppes seems to have taken place from west to east (Levine *et al.* 1999) rather than from east to west, as Gimbutas had argued. There is thus likely to be a Late Neolithic as well as a Palaeolithic contribution to the third PC.

It is thus possible to construct a perfectly plausible view of European population history which can reconcile the principal component analysis of the classical genetic data with the mitochondrial DNA data. But to assert this is one thing: to demonstrate the correctness of this convenient reconciliation is quite another.

It is interesting at this point, to compare Fig-

In the first place there is the rather puzzling observation (Seielstad *et al.* 1998) of an apparently higher migration rate for females than for males, using a comparison of data from mtDNA and from Y-chromosome analysis respectively. As far as actual migration events are concerned, one might expect more males than females to travel widely. But long-term processes are more likely to be significant than short-term events, and it has been argued that, if patrilocality were the norm, then it would be females who would move on the occasion of their marriage rather than males, thus generating the observed pattern. Another possible explanation is that there is more variability in the number of children among males in many societies, and especially in those in which polygyny is practised. That is to say that while most females would have a number of children close to the average, some males would father many more children than the average, and some distinctly fewer. This is an interesting area that needs to be researched further.

But there is another worrying feature. At present it seems that the age estimations for Y-chromosome mutation events, insofar as they can be compared with those for mtDNA mutation events, may not be in harmony. This is, of course, a difficult area, since in neither case are the age-determination methods very precise. It is a little disquieting, however, that in the study here by Rösser and colleagues (Chapter 15) an expansion is detected of a Y-chromosome haplogroup (hg 22) of Iberian origin, and that this does not correlate at all with the out-of-Iberia mitochondrial DNA haplogroup V expansion discussed above. The latter is dated to some 15,000 years ago, while the Y-chromosome haplogroup 22 expansion is given an age range of between 1600 and 3500 years ago. The explanation (Chapter 15, 157): '[haplogroup] 22 chromosomes in Europe and in South America . . . may reflect the activities of Basque cohorts of the Roman army, and the conquistadores respectively' does not entirely carry conviction!

The future

Over the next decade or so we may expect many of these issues to be resolved. And in this volume we may see already the range of case studies which, when the big picture discussed above has taken definitive shape, will give valuable information at a more local level. It is becoming clear that the archaeogenetic approach does have the capacity to give information about demographic process in the Late Palaeolithic and Mesolithic periods, information of a kind which few archaeologists have previously dreamt of (but see Binford 2000). It may well become possible to find genetic markers for the Aurignacian occupation of Europe, and to decide more clearly whether the main gene flow at that time was from the Levant via Anatolia or from Iran via the route north of the Black Sea. It may well be possible to find genetic markers for other such episodes of gene flow, just as it now seems likely that the demographic processes following the Late Glacial Maximum will be clearly represented. It should certainly be possible to evaluate the suggestion by Otte (Chapter 5) that the Solutrean culture of Iberia represents an incursion from North Africa across the Straits of Gibraltar, a route not recently favoured by palaeoanthropologists (see Rando *et al.* 1998). The importance of palaeoclimatology, as exemplified by van Andel (Chapter 4) will be enhanced.

Certainly, when current controversies have abated, it should be possible to evaluate the extent of gene flow to be associated with the coming of farming to Europe. And while the extreme isolationist position of Whittle (1996) may not be sustained, the significance of the Mesolithic contribution to later European prehistory, as outlined for instance by Zvelebil (1986), will probably find considerable support. More sophisticated models for the spread of farming such as that of Zilhão (1993) and of Zvelebil (Chapter 7) will be developed.

This is not the place to develop in detail the various questions of historical linguistics as they relate to Europe. It should be noted, however, that among linguists there is now a more general acceptance that, for major language families, a time depth of the order of 10,000 years may be realistic (Kaufman & Golla 2000; Renfrew *et al.* 2000). Indeed it may prove possible to relate the language-farming hypothesis in its European manifestation (Renfrew 1973; 1987; 1999) with the more general, almost worldwide distribution of 'spread zone' languages (Renfrew 2000b).

When the basic outline of European population history has been established along the lines of the studies seen in Part III of this volume, it will be possible to evaluate and refine a wide range of more local population histories in a manner already initiated in Parts IV and V. This is a task already being undertaken in the Americas (e.g. Merriwether *et al.* 2000). We see hints in the chapters relating to Ireland (Chapters 22 & 23), as in the papers at our Conference discussing the role of the Vikings in Orkney (Wilson *et al.* 1999), that more recent demographic episodes will also be susceptible to study.

These are early days in the archaeogenetics of Europe. Phase two in its development as defined here is less than five years old. In another decade, when various current controversies have been resolved we may begin to find that a 'new synthesis' (Renfrew 1992) will indeed become possible between the data of archaeology, of molecular genetics and of historical linguistics. Clearly the study of ancient DNA will have a greater role to play than hitherto, when technical problems have been overcome. It will of course be necessary to look outside of Europe at neighbouring areas (e.g. Kivisild 2000), if an integrated perspective is to be achieved. Perhaps most exciting of all, the insights obtained are likely to contribute as much to our understanding of the Upper Palaeolithic period as they do to more recent times.

References

Ammerman, A.J. & L.L. Cavalli-Sforza, 1973. A population model for the diffusion of early farming in Europe, in *The Explanation of Culture Change*, ed. C. Renfrew. London: Duckworth, 343–57.

Binford, L.R., 2000. Time as a clue to cause. *Proceedings of the British Academy* 101, 1–36.

Boyle, K. & P. Forster (eds.), 1999. *Human Diversity in Europe and Beyond — Retrospect and Prospect: Final Program and Abstracts* (The Third Biennial Euroconference of the European Human Genome Diversity Project, 9–13 September 1999). Cambridge: McDonald Institute for Archaeological Research.

Cann, R.L., M. Stoneking & A.C. Wilson, 1987. Mitochondrial DNA and human evolution. *Nature* 325, 31–6.

Cavalli-Sforza, L.L. & A.W.F. Edwards, 1965. Analysis of human evolution, in *Genetics Today: Proceedings of the Eleventh International Conference of Genetics, The Hague, September 1963*, vol. 3, ed. J. Geerts. New York (NY): Pergamum Press, 923–33.

Cavalli-Sforza, L.L. & E. Minch, 1997. Paleolithic and Neolithic lineages in the European mitochondrial gene pool. *American Journal of Human Genetics* 61, 247–51.

Cavalli-Sforza, L.L, P. Menozzi & A. Piazza, 1994. *The History and Geography of Human Genes*. Princeton (NJ): Princeton University Press.

Clark, J.G.D., 1965. Radiocarbon dating and the spread of the farming economy. *Antiquity* 39, 45–8.

Excoffier, L. & S. Schneider, 1999. Why hunter-gatherer populations do not show signs of Pleistocene demographic expansions. *Proceedings of the National Academy of Sciences of the USA* 96, 10,597–602.

Gimbutas, M., 1973. The beginning of the Bronze Age in Europe and the Indo-Europeans 3500–2500 BC. *Journal of Indo-European Studies* 1, 163–214.

Hedges, R.E.M. & B.C. Sykes, 1992. Biomolecular archaeology: past, present and future. *Proceedings of the British Academy* 77, 267–84.

Heun, M., R. Schäfer-Pregel, D. Klawan, R. Castagna, M. Accerbi, B. Borghi & F. Salamini, 1997. Site of einkorn wheat domestication identified by DNA fingerprinting. *Science* 278, 1312–14.

Hewitt, G., 2000. The genetic legacy of the Quaternary ice ages. *Nature* 405, 907–14.

Houseley, R.A., C.S. Gamble, M. Street & P. Pettitt, 1997. Radiocarbon evidence for the late glacial human recolonisation of northern Europe. *Proceedings of the Prehistoric Society* 63, 25–54.

Kaufman, T. & V. Golla, 2000. Language groupings in the New World: their reliability and usability in cross-disciplinary studies, in Renfrew (ed.) 2000a, 47–57.

Kivisild, T., 2000. *The Origins of Southern and Western Eurasian Populations: an mtDNA Study*. (Dissertationes Biologicae Universitatis Tartuensis 59.) Tartu: Tartu University Press.

Krings, M., A. Stone, R.-W. Schmitz, H. Kranitzki, M. Stoneking & S. Pääbo, 1997. Neanderthal DNA sequences and the origin of modern humans. *Cell* 90, 19–30.

Levine, M., Y. Rassamakin, A. Kislenko & N. Tatarintseva, 1999. *Late Prehistoric Exploitation of the Eurasian Steppes*. (McDonald Institute Monographs.) Cambridge: McDonald Institute for Archaeological Research.

Malaspina, P., F. Cruciani & A. Novelletto, 1998. Network analyses of Y-chromosomal types in Europe, northern Africa and western Asia reveal specific patterns of geographic distribution. *American Journal of Human Genetics* 63, 847.

Menozzi, P., A. Piazza & L.L. Cavalli-Sforza, 1978. Synthetic maps of human gene frequencies in Europe. *Science* 210, 786–92.

Merriwether, D.A., B.M. Kemp, D. Crews & J.V. Neel, 2000. Gene flow and genetic variation in the Yanomama as revealed by mitochondrial DNA, in Renfrew (ed.) 2000a, 89–124.

Mourant, A.E., 1954. *The Distribution of the Human Blood Groups*. Oxford: Blackwell Scientific.

Ovchinnikov, I.V., A. Götherström, G.P. Romanova, V.M. Kharitonov, K. Liden & W. Goodwin, 2000. Molecular analysis of Neanderthal DNA from the northern Caucasus. *Nature* 404, 490–93.

Rando, J.C., F. Pinto, A.M. Gonzalez, M. Hernandez, J.M. Larruga, V.M. Cabrera & H.-J. Bandelt, 1998. Mitochondrial DNA analysis of Northwest African populations reveals genetic exchanges with European, Near-Eastern, and sub-Saharan populations. *Annals of Human Genetics* 62, 530–50.

Renfrew, C., 1973. Problems in the general correlation of archaeological and linguistic strata in prehistoric Greece: the model of autochthonous origin, in *Bronze Age Migrations in the Aegean*, eds. R.A. Crossland & A. Birchall. London: Duckworth, 263–76.

Renfrew, C., 1987. *Archaeology and Language: the Puzzle of Indo-European Origins*. London: Jonathan Cape.

Renfrew, C., 1992. Archaeology, genetics and linguistic diversity. *Man* 27, 445–78.

Renfrew, C., 1998. Applications of DNA in archaeology: a review of the DNA studies of the Ancient Biomolecules Initiative. *Ancient Biomolecules* 2, 107–16.

Renfrew, C., 1999. Time depth, convergence theory and innovation in Proto-Indo-European: 'Old Europe' as a PIE linguistic area. *Journal of Indo-European Studies* 27, 257–93.

Renfrew, C. (ed.), 2000a. *America Past, America Present: Genes and Languages in the Americas and Beyond*. (Papers in the Prehistory of Languages.) Cambridge: McDonald Institute for Archaeological Research.

Renfrew, C., 2000b. At the edge of knowability: towards a prehistory of languages. *Cambridge Archaeological Journal* 10(1), 7–34.

Renfrew, C., in press. Genetics and language in contemporary archaeology. *Proceedings of the British Academy* forthcoming.

Renfrew, C., A. McMahon & L. Trask (eds.), 2000. *Time Depth in Historical Linguistics.* (Papers in the Prehistory of Languages.) Cambridge: McDonald Institute for Archaeological Research.

Richards, M.R., H. Côrte-Real, P. Forster, V. Macaulay, H. Wilkinson-Herbots, A. Demaine, S. Papiha, R. Hedges, H-J. Bandelt & B. Sykes, 1996. Palaeolithic and Neolithic lineages in the European mitochondrial gene pool. *American Journal of Human Genetics* 59, 185–203.

Seielstad, M.T., E. Minch & L.L. Cavalli-Sforza, 1998. Genetic evidence for a higher female migration rate in humans. *Nature Genetics* 20, 278–80.

Sokal, R.R., N.L. Oden & C. Wilson, 1991. Genetic evidence for the spread of agriculture in Europe by demic diffusion. *Nature* 351, 143–5.

Sykes, B., 1999. The molecular genetics of European ancestry. *Philosophical Transactions of the Royal Society, Biological Sciences* 354, 131–40.

Torroni, A., H.-J. Bandelt, L. D'Urbano, P. Lahermo, P. Moral, D. Sellito, C. Rengo, P. Forster, M.L. Savontaus, B. Bonné-Tamir & R. Scozzari, 1998. MtDNA analysis reveals a major Palaeolithic population expansion from southwestern to northeastern Europe. *American Journal of Human Genetics* 62, 1137–52.

Underhill, P.A., 1999. Y-chromosome biallelic haplotype diversity: global and European perspectives, in Boyle & Forster (eds.), 35–6.

Vigilant, L., M. Stoneking, H. Harpending, N. Hawkes & A.C. Wilson, 1991. African populations and the evolution of human mitochondrial DNA. *Science* 253, 1503–7.

Wainscoat, J.S., A.V.S. Hill, L. Boyce, J. Flint, M. Hernandez, S.L. Thein, J.M. Old, J.R. Lynch, A.G. Falusi, D.J. Weatherall & J.B. Clegg, 1986. Evolutionary relationship of human populations from an analysis of nuclear DNA polymorphisms. *Nature* 319, 491–3.

Whittle, A., 1996. *Europe in the Neolithic: the Creation of New Worlds.* Cambridge: Cambridge University Press.

Wilson, J.F., D.A. Weiss, M.G. Thomas & D.B. Goldstein, 1999. Y-chromosomal variation in Orkney: identification of Viking signature haplotypes in the British Isles, in Boyle & Forster (eds.), 35.

Zilhão, J., 1993. The spread of agro-pastoral economies across Mediterranean Europe: a view from the far west. *Journal of Mediterranean Archaeology* 6, 5–63.

Zvelebil, M. (ed.), 1986. *Hunters in Transition.* Cambridge: Cambridge University Press.

Chapter 2

Concepts in Molecular Genetics

Bryan Sykes & Colin Renfrew

Some of the basic concepts of molecular genetics are concisely reviewed for the benefit of the non-specialist: from the chromosomes to the genes to DNA, and to the analytical methods which are useful in the study of human genetic variability. DNA sequencing is, however, only one part of the task. The genetic data need to be interpreted, whether by spatial analysis, by the techniques of numerical taxonomy or by phylogenetic network methods. These are summarized. The role of ancient DNA is also briefly considered.

In reconstructing the past it is natural to look for clues in stone tools, pottery and grave goods. These are the tangible artefacts of the past — but they are not the only survivors. In every cell of our bodies we all carry DNA which has been passed down almost unchanged from our earliest ancestors. This volume describes the many ways in which geneticists are attempting to reconstruct past events using DNA. The purpose of this chapter is to provide an introduction to genes and genetics for those readers who might welcome it before embarking on the more detailed material.

We will accept as our starting point that DNA is the messenger of heredity. There are many helpful metaphors but one of the most useful is to imagine that the instructions on how to build and run a human, or any other living thing come to that, can be reduced to a set of written instructions. These instructions might be, for example, how to make a blood cell or a nerve or a bone. Not surprisingly these instructions are immensely complicated and nowhere near fully understood yet. Nonetheless, the *language* of the instructions is very straightforward. Like many languages, the meaning is contained within a sequence of symbols or letters — English has twenty-six, computer binary only two. The genetic language is only slightly more complicated with four symbols. The physical embodiment of these symbols are not marks on a piece of paper but four fairly simple organic chemicals adenine, cytidine, guanine and thymidine, always referred to by their initial letters A, C, G and T. These four chemicals, the nucleotide bases, are joined together one after another in a long molecular necklace with the uncomfortable name of *deoxyribonucleic acid*, better known by its acronym DNA. In fact the DNA molecule consists of two strands, the famous double-helix, each one containing the same information in its sequence of bases but in a complementary way. When A appears in one strand, it is always opposite a T in the other. G and C are similarly matched.

There is a good reason for this. When cells divide, DNA must be copied so that each daughter cell receives a full set of instructions. This is accomplished by unwinding the helix and, using each single strand as a template, making two new identical double helices. Because of the complementarity of the bases, the sequence remains intact. The copying mechanism is remarkably exact, but there are occasional mistakes, called mutations. It is these mutations, introduced quite randomly, that, on the one hand, cause inherited disease if they distort a vital message and, on the other, are the engine of evolution if, as occasionally happens, they introduce beneficial changes which become the target of natural selection. Most mutations, however, have little or no effect and it is these neutral changes that are neither eliminated nor promoted by natural selection, that are the most useful for our purpose.

Before we go on, a word about the human genome — the collective name for all the DNA in each cell. Reviving the metaphor for a moment, rather like the instructions for any piece of complex machinery, the human genome is organized into separate volumes called *chromosomes*, tightly packed away into the cell nucleus. Between them these volumes contain the three billion symbols of the human genome. There are twenty-four different chromosomes in the human genome but — and this is where the metaphor begins to show signs of strain — we have two sets of most of them, one from each parent. Twenty-two of the chromosomes fall into this category and are collectively known as the *autosomes* to distinguish them from the X and Y sex chromosomes. Females have a pair of X chromosomes but males have both an X and a Y chromosome. There are scarcely any genes on the male-specific Y chromosome, and only one which is responsible for an embryo turning into a male. If this gene is missing or isn't working properly, the embryo pursues its natural course and becomes a female. In other words, the Y chromosome prevents femaleness.

The human genome contains one other piece of DNA contained not in the nucleus but in small particles in the cell cytoplasm called mitochondrial DNA. It is very much smaller than the nuclear genome with only just over 16,500 bases compared with 3000 million. Its peculiar genetic characteristics, however, have made it extremely important to molecular archaeology. Mitochondria are probably remnants of bacteria which have co-existed within cells in a symbiotic relationship for many millions of years. They contain the essential enzymes necessary for aerobic metabolism — the ability to produce energy using oxygen. Their probable origin is recalled by the organization of the mitochondrial DNA (mtDNA for short) which is circular as in bacteria.

There are several elements that make mtDNA special. The first is that unlike nuclear DNA it is inherited from only one parent — the mother. This is because human eggs have a large cytoplasm full of mitochondria while sperm contains only a few — and those either do not get into the fertilized egg or are eliminated shortly afterwards. This material inheritance has two entirely logical yet awe-inspiring consequences. The first is that you, the reader, will have inherited all your mtDNA from your mother. She inherited it from her mother, who got it from hers and so on back in time. This means that at any time in the past one cares to choose — 100, 1000, 10,000 or even 100,000 years ago — only one woman alive at the time was your maternal and hence mtDNA ancestor. Compare this to the situation with nuclear DNA where the number of potential nuclear ancestors doubles at every generation. Only 20 generations ago — about 500 years — you could have had 2^{20} or a million ancestors, although in practice many will have been the same person.

The other simplifying feature of mtDNA is that it does not undergo genetic recombination. Recombination is the device used by chromosomes to shuffle their genes at every generation, which has the evolutionary advantage that new, favourable gene combinations do occasionally emerge. However, it makes tracing ancestry much more complex. However, some aspects of genetics, like the admixture of two different populations, might well benefit from studying recombination in their chromosomes.

Fortunately, the Y chromosome also shares two of the useful features of mtDNA, uniparental inheritance and a lack of recombination, and so provides an equivalent handle on the past for males. All males have inherited their Y chromosomes from their father, who got it from their father etc. etc. Combining information from both mtDNA and the Y chromosome is proving very productive as a means of distinguishing the male and female counterparts of population movement and holds great promise for work in tracing events of interest to archaeologists.

Tracing mtDNA and Y chromosomes or any other genes in populations is only interesting if there is variation. If all human mtDNA sequences were exactly the same, there would be no point at all in studying it genetically. Variation arises through mutation and recombination but since we have already realized that the latter process is more of a nuisance than a blessing and since it does not occur in mtDNA or the male-specific part of the Y chromosome we will concentrate on mutation. Most mutations arise during the DNA-copying process prior to cell division. The simplest type of mutation is the replacement of one base by another. This always happens in one individual cell in one individual. To be passed on to the next generation the mutations must occur in the so-called germ line cells that are the precursors of either eggs or sperm. All sorts of mutations could happen in other body cells — the somatic cells — but they are irrelevant to us because they won't be passed on to the next generation. Furthermore the mutations will have to increase to be noticed at all. If the new mutation does not alter the biological fitness of the individuals carrying it — in other words it is a neutral change — then the process by which it spreads, or is eliminated, is governed purely by chance and referred to as *genetic drift*. Drift is all about the random chances of the new version making it to the next generation. So, taking the Y chromosome as an example, suppose a mutation happened in the germ line of a male. If he did not have any sons then that is the end of the line for the new mutation — it is extinct. MtDNA is more complicated since a new mutation arising in a female germ line will only be in one molecule to begin with. There are thousands of mtDNA circles in each cell but along the line of cell divisions to the mature egg cell the number of mitochondria are successfully reduced so there is a chance for this new mutation to slip through the cellular 'bottleneck' and get into the next generation. By a succession of similar lucky escapes the new allele might manage to reach a reasonable proportion in the egg cells sufficient to be noticed in, say, a blood sample of the individual concerned. This transition state is known as *heteroplasmy* and persists for half a dozen generations before the new mutation either takes over the entire germ line (fixation) or recedes into oblivion. Even when fixed, though, the new version is far from secure because if the women who carry it do not have any daughters then that is the end. We need not concern ourselves here with the mathematics of genetic drift but, as a general rule, it has a strong effect when the population size is low and its influence diminishes as the population size increases.

Returning to mutation, single base replacements are the commonest source of variation in mtDNA. The circular mitochondrial genome contains the genes for making the components of aerobic metabolism leaving about 1000 bases which do not code for anything. Mitochondria are not very good at correcting copying errors, so the rate of mutation in mtDNA is perhaps 20 times faster than in nuclear DNA. Furthermore, the 1000 bases of non-coding DNA — the so-called control region — accumulate mutations even faster than the rest of the mtDNA, making it by far the most variable stretch of DNA in the whole human genome. Sequencing this 1000 base segment — or even just two segments within it called hypervariable segments I and II (HVS I & II) — is a very productive way of getting at mtDNA variation and many papers in this volume refer to mtDNA control region sequences.

There are other variable bases outside the control region and a selection of these are often recruited to clarify the control region variation. Because they are rather thinly spread they tend not to be sequenced directly but detected by their ability to create or destroy so-called restriction sites. Restriction sites are short DNA sequences, typically 4–6 bases in length, which are the targets for a host of bacterial enzymes available from suppliers in purified form. When the restriction enzyme encounters its precise recognition sequence it cuts the double helix at that point. Simple techniques can reveal the fragment sizes of the DNA after this has happened. A single base change at the restriction site will destroy the enzyme's ability to cut and the fragmentation patterns will be visibly altered. Though it sounds complicated, in practice it is fast and straightforward to detect these so-called RFLPs — short for Restriction Fragment Length Polymorphisms.

The Y chromosome also has, spread along its length, a useful selection of RFLPs, now often referred to as the *bi-allelic* markers, so named because they distinguish two alleles — one where the restriction site is intact and the enzyme cuts and the other where the site is disrupted and the enzyme cannot cut. But there is a second useful source of variation in the Y chromosome, the *microsatellites*. For some reason certain very short DNA sequences of 2, 3 or 4 nucleotides in length have a tendency to grow. The short sequence is repeated several, sometimes hundreds of times. In some, but not all, the number of repeats is unstable and several versions of different lengths are to be found. This is an excellent source of variation, especially when combined with the bi-allelic data from the same individual. Variable *minisatellites*, tandemly repeated units often much longer than microsatellites and frequently with a complex internal structure, are found more rarely, but when they are, such as *MSY1* located on the Y chromosome, they can be extremely useful.

This chapter has concentrated on mtDNA and the Y chromosome because of their simple inheritance and immediate application which finds its expression in many contributions to this book. It is a frequent criticism that the statistical fluctuations associated with studying single genetic loci are so great as to make such studies an unreliable witness of past events. Though this is certainly possible and should always be borne in mind, in practice there are very few, if any, studies where mtDNA or the Y chromosomes — or preferably both together — have led to patently misleading conclusions.

* * *

So far in this introductory chapter we have been speaking of DNA as it survives in the genes of living people. For it turns out that the best record of the population history of Europe, as of that of the world, is retained in the genes of living people. As we shall see, and as the papers in this volume document, a vast amount of information about population history, dealing also with events and processes in the very remote past, hundreds of thousands of years ago, is there, in the DNA of living populations. But there are limitations to this approach. The DNA in the living populations of the world relates to the survivors — to the existing descent lines, whether male as represented in the Y chromosomes, or female as recorded in the mtDNA. Those numerous lineages which became extinct are by definition not recorded in what we may term the living record.

For this reason and for others it would be very desirable to gain information from ancient DNA, from DNA extracted from the remains — bone, hair, any surviving tissue — of past populations. But the study of ancient DNA faces a number of difficulties (Pääbo 1999; Renfrew 1998). Not surprisingly the DNA in living tissues degrades after death with the passage of time. The long chains of bases which constitute the DNA break down into shorter lengths, and information is lost. Ancient DNA is difficult to extract in the laboratory. Moreover the quantities surviving are small, and require amplification by the PCR (polymerase chain reaction) process. This in turn makes it very difficult to exclude the possibility of contamination from modern DNA, including the DNA of those working in the laboratory

It has in fact proved possible to extract and identify ancient mtDNA from various ancient humans, including some Egyptian mummies and also from 'Ötzi' the ancient 'Ice Man' discovered in the Italian/Austrian Alps (Handt et al. 1994). But these individual analyses do not usually tell us very much: they are not sufficiently numerous to create a meaningful pattern. Very often they are indeed sufficient to show that lineages present at the time in question survive to the present day — but that is not really a very surprising answer. We can be confident, however, that techniques of extraction will improve, and that the study of ancient human DNA will become increasingly important. Already there has been success in North America with the study of a cemetery of Oneota American Indians some centuries old (Stone & Stoneking 1999). But in every successful case so far it has been mtDNA which has been used (just occasionally with some nuclear DNA), and Y-chromosome DNA, being relatively less abundant in the body is a more difficult target. There has been just one dazzling success with ancient human (or hominid) DNA: the successful recovery of Neanderthal mtDNA, from remains some 40,000 years old. That is further discussed below. The other field where ancient DNA is beginning to give us information not otherwise available is the case of plant and animal DNA. Interesting work is being undertaken on the origins of the domesticated plants and animals — notably for cattle (Bradley et al. 1996) and for wheat (Heun et al. 1997). The study of ancient DNA is playing its part in this process (Brown 1998).

The interpretation of molecular genetic data is far from straightforward (Renfrew 1992). Indeed for the non-specialist it is not so much the basic facts about genes and DNA, as outlined above, which present difficulties, but the various quantitative techniques — which naturally lie within the field of mathematics and statistics — which offer initial obstacles to understanding. The production of the relevant genetic data is to a large extent a laboratory matter. It involves choosing a number of human populations for sampling, the procurement of the samples (whether in the form of blood, or increasingly just of hair or of mouth swabs) and their analysis in the laboratory. The laboratory analysis of these samples is increasingly becoming standardized and automated for a wide range of markers.

The first key choice, however, and perhaps the most important element of the project design, is the selection of the populations. The aim is generally to compare different human groups, and by establishing the mtDNA and Y-chromosome DNA specificities (and sometimes the autosomal features as well) of each individual within the sample, to begin to make statements about population history. The analysis, of course, has the purpose of going on to make inferences about the past. The groups from which the sample populations are drawn may be selected simply on the basis of their geographical location. But in such cases it is wise to enquire also about the places of birth of the parents and grandparents of the individuals sampled. For if one is seeking to establish the characteristics of a population from a specific region, for instance of south Wales, it is prudent to ensure not only that the individual offering the sample was born in south Wales but also that the same goes for parents and grandparents. Sometimes samples are chosen on the basis of tribal affiliation, sometimes on the basis of the language spoken by the individual in question (which can amount to much the same thing). The first step in any project design is to be clear about the purposes of the study, and to ensure that the sampling of populations is appropriate to that purpose. Consideration will be given also to the number of individuals to be studied from each sampling location — twenty or thirty are often considered sufficient for many purposes. In practice, however, it may not be necessary for the researcher to go out and collect fresh samples. These may already be available in blood banks or other collections previously compiled for medical purposes (although it is important to remember that the 'informed consent' of those sampled, in relation to the type of research being undertaken, must have been obtained).

Already decades ago, before the structure of

DNA was discovered, it was understood that a whole series of biochemical systems, for instance blood groups, blood sera, proteins, antigens etc. are under close genetic determination. Within a given system, for instance a specific blood group system, phenotypic differences (e.g. that between Rhesus positive and Rhesus negative), are determined by genetic differences of a specific nature, as described above. *Polymorphism*, the condition that within a population there exist such difference, implies the presence of two or more *alleles*: alternative variants, similarly but not identical, located at a particular position or locus on a chromosome. The classical biochemical approach (Cavalli-Sforza *et al.* 1994) consists in taking samples, usually blood samples, from a well-defined human population and testing these to determine for each individual the presence or absence of the alleles of the given polymorphisms under investigation. For instance specific blood groups (such as Rhesus negative) may be studied. The number of individuals within the given sample of the population who possess a particular blood group, or immune reaction or whatever, is then expressed as a frequency. The matter becomes interesting when these frequencies, often termed *gene frequencies*, are compared for the different sample populations. A comparable approach applies when one is dealing with DNA data. Initially the data may be displayed in tabular form, with a matrix of frequencies for the different alleles of the different blood or immune systems considered (or of the difference DNA haplotypes), as seen in the various human populations sampled.

In general there are three major approaches towards the interpretive analysis of such raw data. The first approach is *spatial*, with the production of gene frequency maps (Fig. 2.1). Statistical methods, including principal components analysis are available to allow the pooling of all the data and to permit the production of a single map, or of a series of maps, which gives an accurate picture of the spatial variation in the data (e.g. Menozzi *et al.* 1978). In this pioneering study of the genetic variation in Europe the map for the First Principal Component (Fig. 3.3) showed a clear trend from southeast to northwest, recognizable by the *clines* (lines marking the direction of change in frequency). When this paper was published it was proposed that this southeast to northwest picture was the result of the spread of a farming population across Europe, taking place by a process of demic diffusion, at the inception of the Neolithic period. Later it was realized that the picture is rather more complicated, and that a large part of the first principal component may have been due to still earlier gene flow processes, and specifically to the first population event undertaken by our species *Homo sapiens sapiens* in Europe at the beginning of the Upper Palaeolithic period: the initial colonization and subsequent colonization episodes.

A second analytical approach is to create a *phenetic dendrogram*, a tree based on the computation of similarities and differences among the populations sampled. A good example is offered by the work of Excoffier and his colleagues on a range of African populations (Excoffier *et al.* 1987, 170). The frequencies of the various *haplotypes* (i.e. varieties) of Gamma-globulin in the blood of various African populations (sampled on the basis of language families) were determined and a phenetic dendrogram constructed. A notable correspondence was observed between the grouping of the populations as seen on the tree derived entirely from the genetic data, and the language families to which these populations belonged.

The method of comparing similarities and differences between living populations in order to produce a phenetic dendrogram is a very straightforward procedure. Any selection of populations can be compared in this way, in terms of just about any convenient set of variables seen within them. But there now comes a very important next step. By making a number of important (but sometimes unrecognized) assumptions, it is possible to regard the phenetic dendrogram as a sort of family tree or *phylogenetic tree*.

The first diagram in Figure 2.2, a phenetic dendrogram is just a classification, possibly a very arbitrary classification, and of itself it says nothing of history or descent. But if we are able to make those key assumptions, and hence regard the diagram as constituting a phylognetic tree, then this can be claimed as giving information about the past.

The first of these necessary assumptions is that all the populations in question are in fact descended from a single common ancestor — that is indeed an assumption, not something which can readily be inferred from the data. And second it has to be assumed that there is some sort of regular process at work which may be seen to introduce variation, so that after a period of time and the succession of a number of generations, the descendent phenotype differs somewhat from the ancestor. That is of course a process of innovation (or of loss of initial characteristics). In genetics we speak here of the process of mutation. In historical linguistics the equivalent processes are word-loss and linguistic innovation.

Thirdly it is necessary, in order to make reliable inferences about phylogeny, to assume that there is some regularity and constancy in this innovation

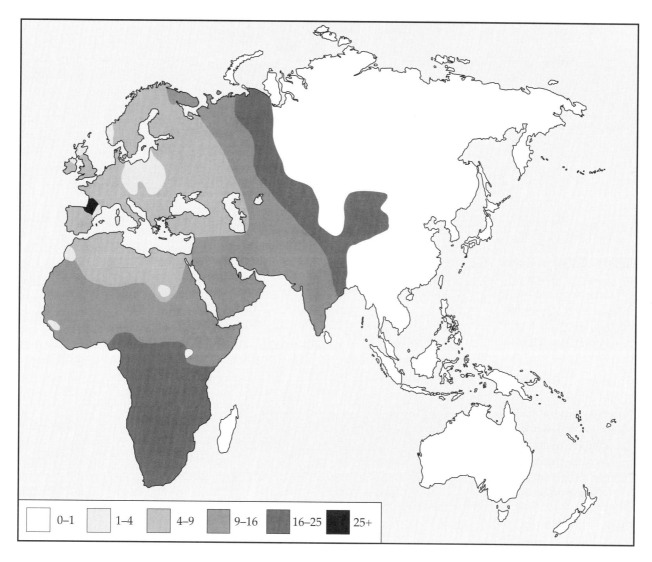

Figure 2.1. *Gene frequency map for the Rhesus negative blood group factor (based on Mourant et al. 1976; see also Cavalli-Sforza et al. 1994, 290 & 503). This shows a maximum in northern Spain in the region where the Basque language is spoken. This molecular-genetic/linguistic correlation, one of the first to be noted, was interpreted as suggesting that the Basque population and language represent the descendants of a Late Palaeolithic population which was not assimilated by the arrival of the first farmers.*

process — in genetics one speaks of constant mutation rates. And fourthly it is important that the innovations between the different branches of the emerging tree arise independently one from another. The system of inference does not work if the same innovations occur with more than random frequency in the different branches of the tree. That would imply the operation of the phenomenon of convergence. Convergence of that kind is not a usual occurrence in genetics. But it is, of course, a common enough happening in the field of human culture — where we often speak of the diffusion of innovations, or of acculturation. And it is common also in languages, where one speaks of convergence, of 'borrowing' and of 'area effects'.

It is on the basis of such reasoning that the 'Out of Africa' model for human origins has been formulated using molecular genetic data, in the first place using mtDNA (Cann *et al.* 1987). In their pioneering study they compared the mtDNA of 147 human individuals. Where lineage methods are used, in the case of mtDNA or Y-chromosome research, it is feasible to compare individuals as well as populations. The basic matrix of data from which the phenetic dendrogram (which can then be claimed as a phylogenetic tree) is constructed is in each case a compari-

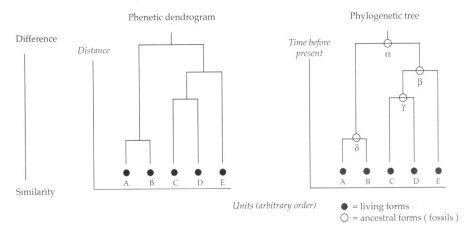

Figure 2.2. *The taxonomic (left) and phylogenetic (right) perspectives compared. Individuals or groups can always be classified in terms of the degree of similarity or difference between them, and the classification shown in tree form (left). Under certain circumstances and making several assumptions taxonomic similarity can imply genealogical proximity, and the phenetic dendrogram can become a phylogenetic tree (right).*

son of the degree of similarity or difference between all the populations (or between all the individuals) under consideration, tabulated on a pairwise basis. Although the work undertaken in the pioneering paper by Cann and colleagues was later subjected to criticism, on the basis that the dendrogram they constructed was not necessarily the only or even the best (most parsimonious) such tree diagram which could be constructed from the data, the basic arguments still seem to hold. It was seen that the first branching separated many of the sub-Saharan African individuals from the rest, and the greatest range of genetic variability was indeed found in sub-Saharan Africa. From this work and on the basis of other arguments it was possible to infer that our species emerged first in Africa and then dispersed to the other continents. Similar arguments have been developed on the basis of autosomal and especially of Y-chromosome data.

In the papers in this volume we see repeated reference to these two analytical approaches — to the geographical study of gene (or haplotype) frequencies, and to inference made from phenetic dendrograms which those producing them claim as phylogenetic trees. It should be noted, moreover, that the phylogenetic tree approach does offer the possibility of offering some dating for the successive branchings in the tree so long as it is possible to assume some approximate standard rate for the mutation process.

The third interpretive approach, which has offered considerable advances in the analysis and interpretation of DNA lineage data (both mtDNA and Y chromosome), uses network-joining methods. Here the data are no longer presented in the form of dendrograms compiled on the basis of the overall similarities and differences between individuals which are being taken in aggregate. Instead the evolutionary pathways of mutations between and within the different haplogroups are considered. The first approach of this technique to the population history of Europe was offered by Martin Richards, Bryan Sykes and their colleagues (Richards *et al.* 1996; Sykes 1999) studying the sequence variation in mtDNA from some 757 individuals in Europe. Through the consideration of some 350 base pairs from the control region of the mtDNA of each individual they compiled the diagram seen in Figure 2.3.

The data are shown here in network form, representing the phylogenetic pathways for the development of the of the haplotypes. In the diagram each haplotype which occurred more than once in the sample is shown, and the sizes of the circles are proportional to the number of occurrences of each specific haplotype in the sample. One advantage of this method is that it offers the possibility of dating the haplogroup cluster. Using certain assumptions it is possible to suggest which specific haplotype may be regarded as the cluster founder. The average number of mutations which have accumulated over the millennia from the original cluster ancestor is measured, and using the mutation rate for the relevant mtDNA control region (estimated at 1 mutation per 20,000 years) it is possible to calculate the approximate age of the cluster. This dating procedure represents a considerable conceptual breakthrough, since it offers the remarkable possibility of a DNA chronology which is effectively independent of direct archaeological considerations. Of course the method is not precise, and it makes a number of assumptions, not least that there is a constant mutation rate which is known. That is matter for argument, but at least it does not depend on archaeological considerations. Comparable arguments have now

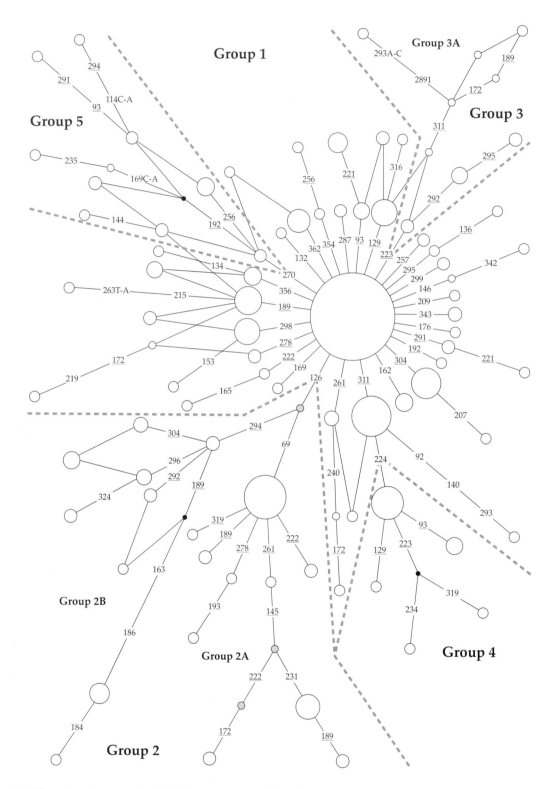

Figure 2.3. *The network approach to the interpretation of data from human mitochondrial DNA from 821 European and Near Eastern individuals: showing the mtDNA phylogeny obtained using the control region sequence haplotypes occurring more than once in the sample. The size of circle shows the number of individuals assigned to the specific haplotype. Dotted lines separate the different groups identified, of which Group 2a (later renamed haplogroup J) reflects the Neolithic population event. (From Sykes 1999, 134.)*

been developed for the analysis of Y-chromosome data. The two approaches taken together offer the most powerful tool now available for the investigation of human population prehistory.

The papers which follow in this volume make repeated reference to the use of all these methods. One very important outcome, so far as Europe is concerned, is that they offer the possibility of investigating population processes which took place as early as the Upper Palaeolithic period. Indeed one of the liveliest controversies of recent years, reflected in some of the papers which follow, is the extent to which the gene frequencies seen in contemporary Europe are largely the result of demographic processes taking place at the onset of the Neolithic period (with the coming of a farming population by a process of demic diffusion) and to what extent the greater part of the variation goes right back to the Palaeolithic period. That is a controversy which will ultimately be resolved by the detailed work now in progress.

The importance of the investigation of Neanderthal DNA was mentioned above as an outstanding utilization of ancient DNA, and one which takes the study back well before the outset of the Upper Palaeolithic period in Europe and the appearance there of *Homo sapiens*. This pioneering work (Krings *et al.* 1997; also Ovchinnikov *et al.* 2000) has shown that the mtDNA extracted from the bone of a Neanderthal dated some 40,000 years ago, differs to much greater extent than had been expected from that of *Homo sapiens* humans. The Neanderthalers are thus much more distant cousins of ourselves than had been thought. Indeed it now seems likely that they made no genetic contribution at all to the living human population of the world.

It is certainly to be hoped that more information about human population history will emerge from the study of ancient DNA. But the broad patterning of human genetic variability is currently emerging from the study of samples from living population. Our past is within us. For the archaeologist it seems indeed paradoxical that we are likely to learn as much about human population history from our living selves as from the archaeological record of the past.

References

Bradley, D.G., D.E. Machugh, P. Cunningham & R.T. Loftus, 1996. Mitochondrial diversity and the origins of African and European cattle. *Proceedings of the National Academy of Sciences of the USA* 93, 5131–5.

Brown, T.A., 1998. How ancient DNA may help in understanding the origins and spread of agriculture. *Philosophical Transactions of the Royal Society*, Series B 354, 88–98.

Cann, R.L., M. Stoneking & A.C. Wilson, 1987. Mitochondrial DNA and human evolution. *Nature* 325, 31–6.

Cavalli-Sforza, L.L., P. Menozzi & A. Piazza, 1994. *The History and Geography of Human Genes*. Princeton (NJ): Princeton University Press.

Excoffier, L.B., B. Pellegrini, A. Sanchez-Masas, C. Simon & L. Langaney, 1987. Genetics and the history of Sub-Saharan Africa. *Yearbook of Physical Anthropology* 30, 151–94.

Handt, O., M. Richards, M. Trommsdorff, C. Kilger, J. Simanainen, O. Georgiev, K. Bauer, A. Stone, R. Hedges, W. Schaffner, G. Utermann, B. Sykes & S. Pääbo, 1994. Molecular genetic analyses of the Tyrolean Ice Man. *Science* 264, 1775–8.

Heun, M., R. Schafer-Pregl, D. Klawan, R. Castagna, M. Accerbi, B. Borghi & F. Salamini, 1997. Site of einkorn wheat domestication identified by DNA fingerprinting. *Science* 278, 1312–14.

Krings, M., A. Stone, R-W. Schmitz, H. Krainitzki, M. Stoneking & S. Pääbo, 1997. Neanderthal DNA sequences and the origin of modern humans. *Cell* 90, 19–30.

Menozzi, P., A. Piazza & L.L. Cavalli-Sforza, 1978. Synthetic map of human gene frequencies in Europe. *Science* 210, 786–92.

Mourant, A.E., A.C. Kopec & K. Domaniewska-Sobczak, 1976. *The Distribution of the Human Blood Groups and Other Polymorphisms*. Oxford: Oxford University Press.

Ovchinnikov, I.V., A. Götherström, G.P. Romanova, V.M. Kharitonov, K. Liden & W. Goodwin, 2000. Molecular analysis of Neanderthal DNA from the northern Caucasus. *Nature* 404, 490–93.

Pääbo, S., 1999. Ancient DNA, in *The Human Inheritance*, ed. B. Sykes. Oxford: Oxford University Press, 119–34.

Renfrew, C., 1992. Archaeology, genetics and linguistic diversity. *Man* 27, 445–78.

Renfrew, C., 1998. Applications of DNA in archaeology: a review of the DNA studies of the Ancient Biomolecules Initiative. *Ancient Biomolecules* 2, 107–16.

Richards, M.R., H. Côrte-Real, P. Forster, V. Macaulay, H. Wilkinson-Herbots, A. Demaine, S. Papiha, R. Hedges, H-J. Bandelt & B. Sykes, 1996. Palaeolithic and Neolithic lineages in the European mitochondrial gene pool. *American Journal of Human Genetics* 59, 185–203.

Stone, A.C. & M. Stoneking, 1999. Analysis of ancient DNA from a prehistoric Amerindian cemetery. *Philosophical Transactions of the Royal Society, Biological Sciences* 354, 153–9.

Sykes, B., 1999. The molecular genetics of European ancestry. *Philosophical Transactions of the Royal Society, Biological Sciences* 354, 131–40.

Chapter 3

Human Diversity in Europe and Beyond: From Blood Groups to Genes

Bryan Sykes

This paper places the current effort of genetics to disentangle European biological prehistory in its historical context and previews future prospects. The first impact of genetics on human evolution sprang from the demonstration that ABO blood groups were Mendelian traits and that their frequencies differed appreciably between 'races', simultaneously embodying the controversial concept of the 'population' as a biological unit. Blood groups and other classical markers revealed frequency clines across Europe and these, taken together with dates for agricultural dispersal, formed the basis of the demic diffusion model. This stressed the importance of the Neolithic in the demography of Europe which has been interpreted, among other things, as a vehicle for linguistic change. Cladistic loci, such as mtDNA, suggested a different history with a greater emphasis on the Palaeolithic. I shall discuss how these differences arose and might be resolved. A frequent criticism of single locus studies is that they are unrepresentative of the genome in general. Is this a valid comment in Europe and how can we tell? Finally, I shall anticipate the contributions that autosomal loci might be expected to make, and at what cost, to the debate on the biological history of the Europeans.

Both to the east and to the west, towards Europe and Africa, a broad stream of Indians poured out, ever lessening in its flow, which finally, although continuously diminishing, penetrated to Western Europe.

With these words, genetics made its entrance onto the stage of human evolution. They come from a paper first delivered to the Salonika Medical Society on June 5th 1918, later translated and published in 'The Lancet' (Herschfeld & Herschfeld 1919). The authors were a husband and wife team working in the Balkans. Ludwik Herschfeld had some years before shown that the human blood groups A, B and O conformed to Mendel's laws of inheritance. It doesn't matter to the story that the theory needed a later correction. What was important was that the blood groups were shown to be *bona fide* genetic characters beyond the influence of age or disease.

The First World War found Hanka Herschfeld in the Balkans at the Central Bacteriological Laboratory of the Royal Serbian Army — part of the Allied force fighting the Germans. The Allies drew soldiers from many parts of the world and the Herschfelds set out to type their blood. From each 'race' they took a blood sample from 500–1000 individuals. A lot of work, but easier in wartime than later when the research would, as they put it, 'have necessitated long years of travel'. The Herschfelds found substantial differences in blood group frequencies between the 'races' (Fig. 3.1) and their explanation was that *Homo sapiens* was composed of two different 'biochemical' races with different points of origin — one in India as the opening quotation suggests, the other somewhere in North or Central Europe. Seen through modern eyes eighty years later this bold speculation on human origins, which completely overlooks Africa, looks wildly wrong. Even so, their

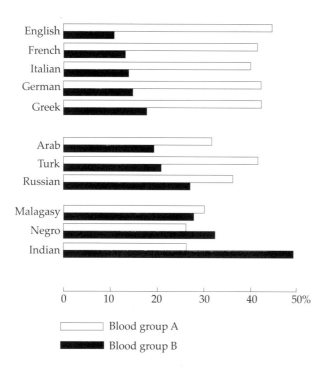

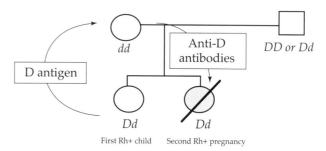

Figure 3.2. *Selection against heterozygotes in the Rhesus blood group.*

Figure 3.1. *Frequencies of blood groups A and B from different ethnic groups. (Data from Herschfeld & Herschfeld 1919.)*

anticipation of the wider use of genetics in anthropology and the study of human origins was absolutely correct.

After the war it fell to the American physician William Boyd to compile the abundant data coming from blood transfusion centres throughout the world. By then the inconsistencies of the kind revealed by the Herschfeld data, for example between the Siberian Russians and the Malagasy which confounded any historical interpretation, had been seen again and again. 'I tried to see what blood groups would tell me about ancient man and found the results very disappointing' he wrote in 1950 (Boyd 1950). Even so, the unsuccessful attempts to explain ethnogenesis in terms of blood groups had its compensations for the liberal-minded Boyd. He writes:

> In certain parts of the world an individual will be considered inferior if he has, for instance, a dark skin but in no part of the world does possession of a blood group A gene exclude him from the best society.

After the Second World War, William Boyd's baton was passed to the Englishman Arthur Mourant. Mourant, a native of Jersey in the Channel Islands, had experienced a strict Methodist upbringing which led him in later life to a desire to become a psychoanalyst. This required reading medicine and he enrolled at St Bartholemews just before the outbreak of the Second World War. This medical school was evacuated from London to Cambridge and it was here that he met the R.A. Fisher, the most influential geneticist of his day. Fisher and his colleagues had been engaged in unravelling the genetics of the minor blood groups which were now being discovered. In particular, Fisher had proposed a complex scheme for the Rhesus blood group. Mourant discovered a large family which provided the practical proof of the theory. A delighted Fisher found him a job at the Galton Laboratory Serum Unit at Cambridge, and the meticulous Mourant spent the rest of his working life compiling and interpreting the most detailed blood group frequency distribution maps.

As well as being instrumental in settling Mourant into his outstanding career, the Rhesus blood group system also played a key role in formulating ideas about the origins of modern Europeans and identifying its most influential genetic population — the Basques. The Rhesus system is a very unusual example of selection operating against the heterozygotes. When a Rhesus negative mother, who must have the genotype (*dd*) bears the children of a Rhesus positive father then the children have, depending on the father's genotype, either a 50 per cent or 100 per cent chance of being Rhesus positive (*Dd*) heterozygotes (Fig. 3.2). As the mother gives birth to her first Rhesus positive baby, a few red blood cells can get into her circulation and initiate an immune response against the D antigen molecules on their surface which eventually leads to the production of circulating antibodies. These antibodies cross the placenta in subsequent pregnancies and, if the foetus is again Rhesus positive, will bind to the foetal red cells causing a haemolytic anaemia which is frequently fatal. Though no longer a clinical problem thanks to anti-D injections during the first Rh⁺ pregnancy, the heterozygotes had been at a clear selective disadvantage.

The significance of all this to the thinking about European prehistory stems from the prediction from basic population genetics that, since one of each allele is lost from the gene pool for each case of fatal haemolytic disease one allele will eventually be eliminated and the other one become fixed in the population. But the overall frequencies in Europe are 60 per cent (*D*) and 40 per cent (*d*), a theoretically unstable mixture presumably on the way to fixation in favour of (*D*). Since (*D*) is at higher frequencies than (*d*) in other parts of the world it occurred to Mourant that modern Europe might be a recent hybrid population of Rh+ people from the Near East, probably the Neolithic farmers, and the descendants of earlier Rh- Palaeolithic and Mesolithic people (Mourant 1995). But who were the Rh-negatives? Mourant came across an observation by the French anthropologist H.V. Vallois who described the skeletons of contemporary Basques as resembling Late Palaeolithic rather than modern Europeans. It was already known that Basques had a very low frequency of blood group B. Mourant arranged through two Basque representatives, in London attempting to form a provisional government, to type a panel of French and Spanish Basques who turned out, as hoped, to have a very high frequency of Rh-negative, in fact the highest in the world. From that moment, the Basques assumed the mantle of the population against which all ideas about European genetic prehistory were, and still are, to be judged. The fact that they alone of all the Western Europeans spoke a non-Indo-European language enhanced their special position and led directly to the investigation of other gene/language correlations which continue to this day.

Although the ABO and Rhesus blood groups had easily identified the Basques and also the Saami (*aka* Lapps) as genetic outliers in Europe there was clearly a need to devise a statistical treatment which

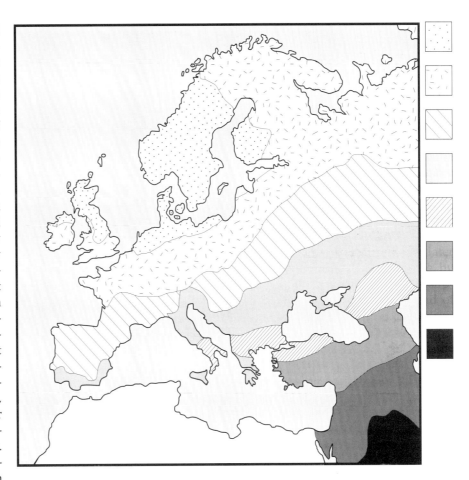

Figure 3.3. *The cline of the first principal component across Europe. The range of PC values has been divided into the eight equal classes indicated by different shading. (Redrawn from Cavalli-Sforza* et al. *1994.)*

would amalgamate the data from the several different loci then being investigated — Kell, Duffy, MNS and other blood groups, protein polymorphisms revealed by electrophoresis and the HLA immune-response genes. This was accomplished by the man who has dominated the field for the past thirty years, Luca Cavalli-Sforza. Cavalli-Sforza, working with the statistician Anthony Edwards, achieved this amalgamation with the help of the earliest punched-card computing machines (Cavalli-Sforza & Edwards 1967). For the first time the condensed data from several loci could be taken through a series of different treatments of which one of the most enduring has been the distribution of the principal components. In these famous diagrams the values for the first and subsequent principal components (PCs) are computed for the available populations and the maps generated by linking together geographical locations with similar values. In Europe the first PC accounts for 26–28 per cent of the total variance and shows a cline from the Near East to Northwest Europe or *vice versa* (Fig. 3.3). As is well known, this has been interpreted as the result of the spread into Europe of Neolithic agriculturalists from Anatolia and the Near East, the same event which Mourant held responsible for the unstable Rhesus frequencies (Ammerman & Cavalli-Sforza 1984).

As well as constructing PC maps, the amalgamated frequency data were also used to draw population trees. This began with a remarkable article by Anthony Edwards in which he imagined the genetical consequences of population fission (Edwards 1966). A tribe carried with it a pole along which were arrayed 100 discs either black or white. Every year, one disc, chosen at random, was changed to the other colour. When the tribe split, each group carried with it a copy of the pole and discs in their current order and, after separation, they continued the custom of annual random change. From this simple principle diagrams describing the order of fissions were drawn linking, eventually, all the populations of the world. As Edwards says, 'The resultant evolutionary trees will certainly not provide the last word on human evolution', but he offered the diagrams as a way of providing the genetic information in an understandable form. Unfortunately, the population trees first drawn with this admirable and modest intention have become widely misunderstood and a source of contention for several reasons. The first of these touches on the thorny issue of ethnicity. For a population tree to be drawn at all it follows that the population has to be defined, a process which in itself segregates groups of people in ways that can tend to perpetuate racial classifications. Zegura argues that 'Objectively defined races simply do not exist — the quintessentially arbitrary and subjective race concept is moribund as a unit of human biological analysis' (Zegura *et al.* 1990).

Almost forty years earlier, Mourant had come to much the same conclusion but expressed it in a very different style:

> Rather does a study of blood groups show a heterogeneity in the proudest nation and support the view that the races of the present day are but temporary integrations in the constant process of mutation, selection and mixing that marks the history of every living species (Mourant 1954).

The second misapprehension about the trees comes from their appearance. 'It is probably unreasonable' writes Quicke in his textbook on taxonomy 'to expect the majority of people to interpret these trees as anything other than suggestions of evolutionary histories' (Quicke 1993). They would only work as such if human evolution really were a succession of fissions. Then and only then would the nodes really represent a real entity, a true proto-population. But the construction of trees in Europe, for instance, reveals the shortcomings of this literal interpretation. No-one seriously imagines, for example, a proto-Anglo-Danish population which divided, never to meet again, and became the modern inhabitants of England and Denmark (Fig. 3.4).

The trend to literal interpretation of population trees, never intended by their invention, has also had an influence on sampling strategy which can be summarized as follows. Is it better to have a small sample and test it for a large number of markers? Or is it better to use a large sample at the expense of analyzing only a few markers? If the desired outcome is to produce a population tree with the highest statistical support then the answer is simple — small sample, largest possible number of loci tested. I am sure it is this sentiment which initially prejudiced many in the field against the reconstructions of prehistory using only a single locus — mitochondria.

Mitochondrial DNA (mtDNA for short) differs radically from its predecessors as a genetic marker in several respects of which the most important is that the phylogeny can, usually, be reconstructed with confidence. Now, for the first time, the evolutionary relationship between alleles can be known and, with an accurate mutation rate estimate, the new dimension of time can be added. These time depths, despite their wide confidence intervals and their dependence on the as yet unclear influence of demography, has revised interpretations of prehistory. The best known example is the the overall genetic impact of the Neolithic in Europe with a greater emphasis now being placed on Palaeolithic and Mesolithic events in shaping the current gene pool (Sykes 1999).

Largely because of this reinterpretation, mtDNA has been subjected to intense scrutiny as a marker. The control region sequence, used by many as an unbiased report of variation, has been criticized as being unreliable and subject to unacceptable levels of recurrent mutation. However, recent work showing the excellent concordance with markers outside the control region has done a lot to stabilize confidence in the locus (Macaulay *et al.* 1999b). Fears that the mutation rate used in estimating time depth could have been wrong by almost an order of magnitude rose then subsided (Howell *et al.* 1996; Macaulay *et al.* 1997). Heteroplasmy, the inevitable transition state during the fixation of a new variant when an indi-

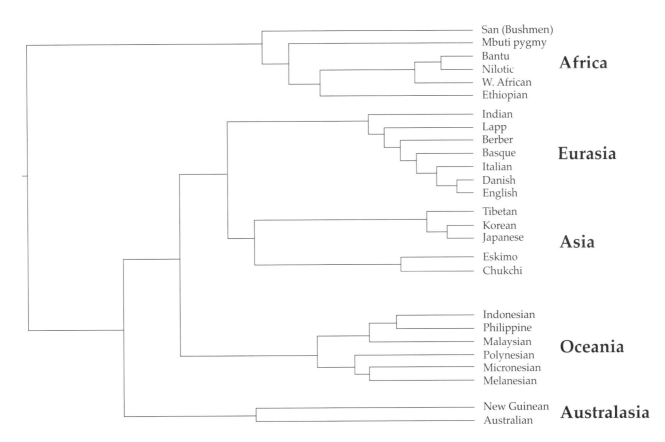

Figure 3.4. *Genetic tree of worldwide populations. (Redrawn from Cavalli-Sforza* et al. *1994.)*

vidual will contain a mixture of old and new alleles, was seen as a widespread source of inaccuracy for a while (Cavalli-Sforza & Minch 1997). Mitochondrial recombination has become the latest panic fuelled by two papers seductively coupled in the same journal, the first seemingly providing a theoretical argument in favour of recombination (Eyre-Walker *et al.* 1999), the second apparently providing evidence of the real thing in the small South Pacific island of Nguna (Hagelberg *et al.* 1999).

Very quickly, the theoretical argument became less persuasive when it was shown that many of the sequences used in the paper were inaccurate (Macaulay *et al.* 1999a). There is scarcely a mtDNA sequence data set in existence which is completely accurate, despite the best endeavours of their compilers. A systematic sequencing error has been suggested as the basis for the Nguna claim (Merriwether & Kaestle 1999) — a situation that will eventually be resolved if and when the sequences are independently replicated. By and large mtDNA has come through these quite proper challenges with flying colours and will, I hope, have an easier ride from now on.

Y-chromosome variation is rapidly comple-

menting what we know about the maternal side of human evolution with its paternal counterpart. I sense that the chorus 'but it's only a single locus' is slightly less shrill as the comparisons of the two uniparental loci show what to do appear to be substantial, yet comprehensible, differences in male and female patterns of behaviour and migration (Seilestad *et al.* 1998). Some of these differences must be due to the random forces of drift acting on markers with low effective population size but the empirical evidence is persuasive that we really are being told the truth, or something close to it, by these two markers.

As to the future, it will not be long before the two uniparental loci are fully understood, with complete phylogenies. The eccentricities of the control region will be tolerated and a common standard adopted for the Y-chromosome markers. But what of the autosomes? The technology to score large numbers of single nucleotide polymorphisms (SNPs) throughout the genome in an individual is already available and will before long be cheap enough for mass application. What benefits will come from analysis of the complex genotypes — or rather *genomotypes*?

Each nucleotide has its own history and new

phylogenetic treatments could be able to be developed which will identify the changing character of each base pair during human evolution. Phylogenies of haplotypes which have helped the uniparental loci so much will obviously be confounded by recombination though the rate at which this process homogenizes genomotypes might itself be a source of useful information about population structure. However, it would be a great waste if this vast bank of knowledge were used merely to refine population trees. They cannot, usefully, become any more exact. There is a danger, in the words of Lewis Wolpert referring to another branch of biology, that geneticists will 'know everything and explain nothing'. It is an exciting time and the rewards will go to the teams who manage to integrate this flood of genetic information with the achievements of the other disciplines — archaeology, anthropology, linguistics — with an interest in the past.

References

Ammerman, A.J. & L.L. Cavalli-Sforza, 1984. *The Neolithic Transition and the Genetics of Populations in Europe.* Princeton (NJ): Princeton University Press.

Boyd, W.C., 1950. Use of blood groups in human classification. *Science* 112, 187–97.

Cavalli-Sforza, L.L. & A.W.F. Edwards, 1967. Phylogenetic analysis: models and estimation procedure. *American Journal of Human Genetics* 19, 233–57.

Cavalli-Sforza, L.L. & E. Minch, 1997. Paleolithic and Neolithic lineages in the European mitochondrial gene pool. *American Journal of Human Genetics* 61, 247–51.

Cavalli-Sforza, L.L., P. Menozzi & A. Piazza, 1994. *The History and Geography of Human Genes.* Princeton (NJ): Princeton University Press.

Edwards, A.W.F., 1966. Studying human evolution by computer. *New Scientist* 30, 438–40.

Eyre-Walker, A., N.H. Smith & J. Maynard Smith, 1999. How clonal are mitochondria? *Proceedings of the Royal Society of London* B 266, 477–83.

Hagelberg, E., N. Goldman, P. Lio, S. Whelan, W. Schiefenhovel, J.B. Clegg & D.K. Bowden, 1999. Evidence for mitochondrial DNA recombination in a human population of island Melanesia. *Proceedings of the Royal Society of London* B. 266, 485–92.

Herschfeld, L. & H. Herschfeld, 1919. Serological differences between the blood of different races. *Lancet* 2, 675–8.

Howell, N., I. Kubacka & D.A. Mackey, 1996. How rapidly does the human mitochondrial genome evolve? *American Journal of Human Genetics* 59, 501–9.

Macaulay, V.A., M.B. Richards, P. Forster, K.E. Bendall, E. Watson, B. Sykes & H-J. Bandelt, 1997. MtDNA mutation rate: no need to panic. *American Journal of Human Genetics* 60, 983–6.

Macaulay, V.A., M.B. Richards & B.C. Sykes, 1999a. Mitochondrial recombination: no need to panic. *Proceedings of the Royal Society of London* B 266, 2037–40.

Macaulay, V.A., M. Richards, E. Hickey, E. Vega, F. Cruciani, V. Guida, R. Scozzari, B. Bonné-Tamir, B.C. Sykes & A. Torroni, 1999b. The emerging tree of West Eurasian mtDNAs: a synthesis of control region sequences and RFLPs. *American Journal of Human Genetics* 64, 232–49.

Merriwether, D.A. & F.A. Kaestle, 1999. Mitochondrial recombination? *Science* 285, 837.

Mourant, A., 1954. *The Distribution of the Human Blood Groups.* Oxford: Blackwell Scientific Publications.

Mourant, A.F., 1995. *Blood and Stones: an Autobiography.* Jersey: La Haule Books.

Quicke, D.L.J., 1993. *Principles and Techniques of Contemporary Taxonomy.* London: Blackie Academic Publishers.

Seielstad, M., E. Minch & L.L. Cavalli-Sforza, 1998. Genetic evidence for a higher female migration rate in humans. *Nature Genetics* 20, 278–80.

Sykes, B.C., 1999. The molecular genetics of European ancestry. *Philisophical Transactions of the Royal Society of London* B 354, 131–9.

Zegura, S.L., W.H. Walker, K.K. Stout & J.D. Diamond, 1990. More on genes, language and human phylogeny. *Current Anthropology* 32, 420–26.

Part II

The Archaeological and Environmental Background

was used to compare the frequency distribution of archaeological dates with the calendrical climate history displayed by the GRIP ice core record (Fig. 4.5). Notwithstanding wide (and poorly known) error margins and other problems associated with the use of sets of dates as proxies for population density (Rick 1987), the graph clearly encourages further pursuit of the potential relations between the human history of the mid-glacial interval and its palaeoenvironments.

The causes of the millennial-scale climate variations have not yet been firmly established, but are probably related to drastic changes in pole-ward heat transport by the Gulfstream and the removal of colder water at depth by the so-called 'Atlantic Conveyor' which are part of a global surface and deep ocean circulation system that is the main carrier of equatorial solar heat across the world (Broecker 1994).

Whatever their causes, the existence of a highly unstable mid-glacial climate is firmly established and we have a reasonable and rapidly improving idea as to what the environmental conditions were to which Neanderthals and modern humans were exposed. An interdisciplinary study called the Stage Three Project, sponsored among others by the McDonald Institute for Archaeological Research and the Leverhulme Trust, and co-ordinated by this author, is in the final stages of modelling the climate, vegetation and faunal resources of a warm and a cold phase of OIS 3 in the context of human dispersal, success and failure, using existing palaeoenvironmental data as input to and for validation of the simulations.

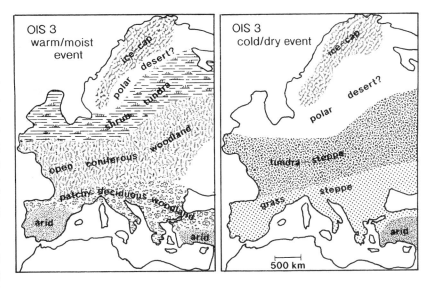

Figure 4.4. *Europe during the late mid-pleniglacial. Sketch maps of a typical warm event at c. 40 ka and a major cold one at c. 30 ka BP. Coastline based on the –70 m isobath for the warm and –85 m for the cold event. Baltic sea shore omitted as too speculative. (After van Andel & Tzedakis 1998, fig. 6.6, modified with more recent data.)*

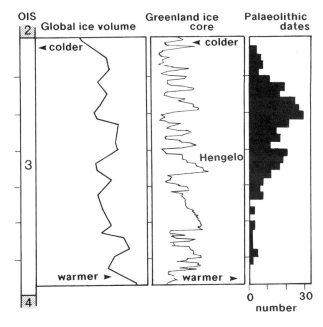

Figure 4.5. *The Greenland ice-core climatic record (GRIP Members 1993, fig. 1) compared with the frequency distribution of dates of late Middle and early Upper Palaeolithic sites. The ^{14}C dates in the set have been converted to calendrical dates following van Andel (1998). (See van Andel & Tzedakis 1998 for sources.)*

Early–Middle Holocene pastoralist migrations from the Middle East into and across Africa north of the Equator

My second example also deals with rapidly changing climatic conditions, in this case of the Early and Middle Holocene in the Near and Middle East and in Africa north of the equator recently summarized by Hassan (1996; 1997; pers. comm. 1999) in the archaeological context of their relation to the migrations of early pastoralists. The questions that concern us here are: i) what precisely were those climate changes?; ii) by which climatic systems were they driven?; and, crucially, iii) whether and to what extent we can correlate the climatic record which is based on geological proxies with the archaeological data as a preliminary to considering a causal relationship.

The food producers revolution
The earliest food producers came from Anatolia, the Levant and elsewhere in the Middle East, regions of marginal environments and unstable climates, and moved to regions of similarly unstable conditions. Once there, those predominantly cultivating plants responded to local conditions in ways quite different from those depending mostly on pastoralism. The former required suitable arable soil and a dependable water supply, conditions only rarely satisfied in Anatolia, Greece and the southernmost Balkans, and once established on good-quality land, they lacked the mobility to respond to climate changes such as drought. Pastoralists, on the other hand, had little investment in land, depended mainly on adequate rainfall and readily responded to regionally shifting patterns of precipitation by moving with their herds.

Holocene climate changes in southwest Asia and Africa north of the equator
Southwestern Asia and Africa north of the equator lie in a climatologically complex region where several major climatic systems meet and interact, the Indian Ocean Monsoon, itself intermittently strongly influenced by the El Niño-Southern Oscillation (ENSO) in the Pacific, the Atlantic Monsoon on the west side of Africa and, as a usually minor participant, the Mediterranean climate system.

The North Atlantic–European climate system has also been invoked as a major influence especially as regards the Near and Middle Eastern palaeoclimates, perhaps because its history has been familiar for so long to so many as part of the folklore of the Holocene climate of western and northwestern Europe. Recently, the millennial oscillations that strongly mark the deglaciation after the last glacial maximum (e.g. Alley *et al.* 1997; Bond *et al.* 1997), such as the famous Younger Dryas cold event, have been tempting those seeking to elucidate the Late Pleistocene climate of southeastern Europe and beyond. Yet this extension of the maritime regime of western and northwestern Europe into the continental regions of southeastern Europe and western Asia is neither self-evident nor solidly documented and highly controversial (cf. Bottema 1995; Lamb *et al.* 1995; Rossignol-Strick 1995; 1999; Willis 1994), and should be viewed with a great deal of scepticism.

The main player in the late Pleistocene climate of southwestern Asia and adjacent Africa is the Indian Ocean Monsoon (e.g. Gasse & van Campo 1994; Kutzbach & Street-Perrott 1985; Sirocko *et al.* 1996) which between roughly 15,000 and 5000 years BP extended much farther to the northwest and west than it does today, thus providing increased rainfall to large parts of the Middle East and far westward into north-equatorial Africa (Fig. 4.6).

The monsoon is driven by the temperature contrast in low latitudes between any large continental area and the adjacent ocean. In summer the continent heats up more than the ocean and the rising continental air draws in a cooler, moist oceanic air flow that turns into the famous monsoon rains as it passes onto land. In winter, the opposite happens: strong cooling of the land produces the sinking and

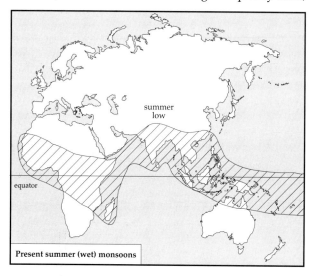

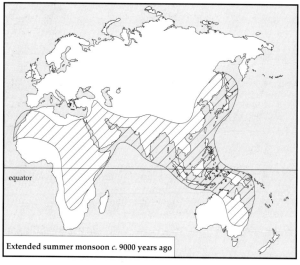

Figure 4.6. *Extent of the Indian Ocean summer monsoon. Top: at present; bottom: 9000 years ago. (Adapted from Roberts 1998, fig. 4.11 and miscellaneous sources.)*

therefore dry airflow towards the ocean that produces the cool, dry monsoon.

The strength of the monsoon depends on the level of solar heat input at low latitudes. This strength varies over time because of cyclic changes in the attitude and path of the earth as it orbits around the sun. The orbital ellipse is eccentric and solar warming is maximal when the earth is closest to the sun in the summer. However, under the simultaneous pull of sun and moon the earth wobbles like a top. In consequence, the equinoxes shift (precess) slowly clockwise around the orbit with a period of 22,000 years; the time when the earth finds itself closest to the sun in summer thus returns every 22,000 years. Today, in low latitudes on the northern hemisphere the level of insolation at low is close to its minimum, whereas 11,000 years ago it reached its maximum (Fig. 4.7). Because the earth's axis is tilted, the opposite is true for the southern hemisphere.

What is the evidence for a significantly wetter climate during the very latest Pleistocene and Early–Middle Holocene from about 15,000 to 5000 years ago? The increased rainfall in western Asia and in northern Africa, which benefits in addition from the Atlantic monsoon, should have raised lake levels across the entire region (Fig. 4.8: top) to a degree that can be predicted by models providing the level of increased precipitation corrected for evaporation due to increased warmth (Fig. 4.8: bottom).

Is there evidence for full lakes in the latest Pleistocene and Early Holocene in this vast region? Indeed, measurements of the elevations of former lake beaches and determinations of their ages clearly show the extent to which, in the Early Holocene, southwestern Asia and much of Africa north of the equatorial zone were much wetter than they are today (Kutzbach & Street-Perrott 1985; Roberts & Wright Jr 1993; Street-Perrott & Perrott 1993) or were during the glacial maximum (van Andel & Tzedakis 1996). It is worth noting, however, that the compilations of lake levels encompass an interval of some 5000 years and are therefore generalizations; the lake levels themselves are fairly easy to measure, but dating them is a different matter and the top graph in Figure 4.8 may well conceal something much more complex than the simple insolation curve of Figure 4.6.

And indeed, a provisional compilation of many Holocene records of events of intensified rainfall in Africa (Fekri A. Hassan pers. comm. 1998) which clearly shows the expected, precession-induced gradual increase followed by a decrease of the intensity of the monsoon, is embellished by many sharp individual peaks of wetness, that are uncomfortably

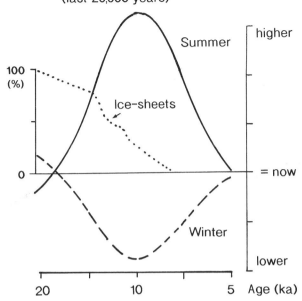

Figure 4.7. *The solar energy curve for the interval from 20,000 years BP to the present, calculated from known temporal changes of the properties of the earth's orbit, the so-called Milankovitch curve. (Modified from Kutzbach & Webb 1993, fig. 2.2; Kutzbach & Street-Perrott 1985.)*

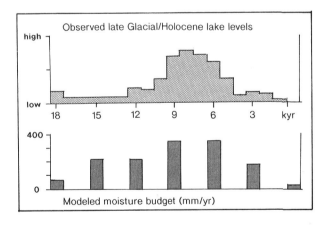

Figure 4.8. *African and southwest Asian lake levels 9000 years ago. Top: observed (from Street-Perrott & Perrott 1993, fig. 13.19; Roberts & Wright 1993, and miscellaneous sources); bottom: modelled.*

reminiscent of the warm/wet, cold/dry oscillations of the glacial North Atlantic. Whether this is a valid comparison is questionable, not only because of doubt that the North Atlantic system could exert its influence at such distance and into a totally different climate zone, but also because the Holocene Greenland ice-core record shows none of those oscillations, — although interestingly, it does show them during the previous interglacial.

Of course, there may not actually have been quite so many oscillations because, given the vast region from where the data have been obtained, the very brevity of the events and the often inadequate resolving power of the dates, it is difficult to establish with confidence which excursions mark separate events and which are merely evidence for the same evident in widely spaced locations. Nonetheless, a substantial temporal and spatial variability of the climatic that is probably not due to the simple insolation curve of Figure 4.6 appears to be likely. If it can not be attributed without serious doubts to oscillations of the North Atlantic system, to what may these instabilities be due?

The answer is that we do not yet know. Our understanding of the interaction of the major global climate systems today has only recently begun to improve significantly. Therefore we are some distance away from being able to grapple seriously with their Holocene histories on various time-scales, and even farther away from spelling out their interactions over that time. That they interact is clear; the inverse of the El Niño condition, for example, cleverly named La Niña has long ago been show to be clearly related to a long record of Nile floods (Quinn 1992). How and how much they interact remains a question for the future.

Here, then, is the reason why I chose my second example when demonstrating the current state of the art in palaeoclimatological studies. In the case of OIS 3, the fact that we do not fully understand the causes of the high-frequency climate changes does not much matter to the student of early human history because their reality and impact are clear. In the case of the Early Holocene African pastoralists who found themselves trying to make a go of it at the meeting point of several climatic systems, our still scant knowledge of Holocene climatic processes prevents us from making useful predictions. At the same time, the vastness of the territory and the great difficulty of obtaining the very precise dates needed to define operationally the individual brief events and determine their durations will for some time yet stand in the way of making reliable statements regarding cause and consequence.

References

Alley, R.B., P.A. Mayewski, T. Sowers, M. Stuiver, K.C. Taylor & P.U. Clark, 1997. Holocene climatic instability: a prominent widespread event 8200 yr ago. *Geology* 25, 483–6.

Bigg, G.R., 1996. *The Oceans and Climate*. Cambridge: Cambridge University Press.

Bond, G.C., W.D. Broecker, S. Johnsen, J. McManus, L. Labeyrie, J. Jouzel & G. Bonani, 1993. Correlations between climate records from North Atlantic sediments and Greenland ice. *Nature* 365, 143–7.

Bond, G., W. Showers, M. Cheseby, R. Lotti, P. Almasi, P. Demenocal, P. Priore, H. Cullen, I. Hajdas & G. Bonani, G., 1997. A pervasive millennial-scale cycle in North Atlantic Holocene and glacial climates. *Science* 278, 1257–66.

Bottema, S., 1995. The Younger Dryas in the Eastern Mediterranean. *Quaternary Science Reviews* 14, 883–92.

Broecker, W.S., 1994. Ocean circulation: an unstable superconveyor. *Nature* 367, 414–15.

Coope, G.R., P.L. Gibbard, A.R. Hall, R.C. Preece, J.E. Robinson & A.J. Sutcliffe, 1997. Climatic and environmental reconstructions based on fossil assemblages from Middle Devensian (Weichselian) deposits of the River Thames at South Kensington, central London. *Quaternary Science Reviews* 16, 1163–96.

Gamble, C.S., 1986. *The Palaeolithic Settlement of Europe*. Cambridge: Cambridge University Press.

Gamble, C.S., 1987. Man the shoveler: alternative models for Middle Pleistocene colonization and occupation in northern latitudes, in *The Pleistocene Old World: Regional Perspectives*, ed. O. Soffer. New York (NY): Plenum Press, 81–98.

Gamble, C.S., 1993. People on the move: interpretations of regional variation in Palaeolithic Europe, in *Cultural Transformations and Interactions in Eastern Europe*, eds. J. Chapman & P. Dolukhanov. Avebury: Aldershot, 36–55.

Gasse, F. & E. van Campo, 1994. Abrupt post-glacial climatic events in West Asia and North Africa monsoon domains. *Earth and Planetary Science Letters* 126, 435–56.

GRIP (Greenland Ice-core Project) Members, 1993. Climate instability during the last interglacial period recorded in the GRIP ice core. *Nature* 364, 203–7.

Guiot, J, A. Pons, J.-L. de Beaulieu & M. Reille, 1989. A 140,000-year continental climate reconstruction from two European pollen records. *Nature* 338, 309–13.

Hassan, F.A., 1996. Abrupt Holocene climatic events in Africa, in *Aspects of African Archaeology*, eds. G. Pwiti & R. Soper. Harare: University of Zimbabwe Publications, 83–6.

Hassan, F.A., 1997. Holocene palaeoclimates of Africa. *African Archaeological Review* 14, 213–30.

Hays, J.D., J. Imbrie & N.J. Shackleton, 1976. Variations in the earth's orbit: pacemaker of the Ice Ages. *Science* 194, 1121–32.

Imbrie, J. & K.P. Imbrie, 1979. *Ice Ages: Solving the Mystery*. Short Hills (NJ): Enslow.

Kutzbach, J.E. & J.A. Street-Perrott, 1985. Milankovitch forcing of fluctuations in the level of tropical lakes from 18 to 0 kyr BP. *Nature* 317, 130–34.

Kutzbach, J.E. & T. Webb III, 1993. Conceptual basis for understanding late-Quaternary climates, in Wright *et al.* (eds.), 5–11.

Lamb, H.F., F. Gasse, A. Benkaddour, N. El Hamouti, S. van der Kaars, W.T. Perkins, N.J. Pearce & C.N. Roberts, 1995. Relation between century-scale Holocene arid intervals in tropical and temperate zones. *Nature* 373, 134–7.

Lowe, J.J. & M.J.C. Walker, 1997. *Reconstructing Quaternary Environments*. 2nd edition. London: Longman.

Martinson, D., N.G. Pisias, J.D. Hays, J. Imbrie, T.C. Moore Jr. & N.J. Shackleton, 1987. Age dating and the orbital theory of the Ice Ages: development of a high-resolution 0–300,000-year chronostratigraphy. *Quaternary Research* 27, 1–29.

Quinn, W.H., 1992. A study of Southern Oscillation-related climatic activity for AD 622–1990 incorporating Nile River flood data, in *El Niño: Historical and Paleoclimatic Aspects of the Southern Oscillation*, eds. H.F. Diaz & V. Markgraf. Cambridge: Cambridge University Press, 119–50.

Rick, J.W., 1987. Dates as data: an examination of the Peruvian pre-ceramic radiocarbon record. *American Antiquity* 52, 55–73.

Roberts, N., 1998. *The Holocene: an Environmental History*. Oxford: Blackwell.

Roberts, N. & H.E. Wright Jr, 1993. Vegetational, lake-level and climatic history of the Near East and southwestern Asia, in Wright *et al.* (eds.), 194–220.

Roebroeks, W., N.J. Copnard & T. van Kolfschoten, 1992. Dense forests, cold steppes and the Palaeolithic settlement of northern Europe. *Current Anthropology* 33, 551–86.

Rossignol-Strick, M., 1995. Sea–land correlation of pollen records in the eastern Mediterranean for the glacial-interglacial transition: biostratigraphy versus radiometric time-scale. *Quaternary Science Reviews* 14, 893–916.

Rossignol-Strick, M., 1999. The Holocene climatic optimum and pollen records of Sapropel 1 in the eastern Mediterranean 9000–6000 BP. *Quaternary Science Reviews* 18, 515–30.

Sirocko, F., M. Sarnthein, H. Erlenkeuser, H. Lange, M. Arnold & J.C. Duplessy, 1993. Century-scale events in monsoonal climate over the past 24,000 years. *Nature* 364, 322–4.

Sirocko, F., M. Garbe-Schönberg, A. McIntyre & B. Molfino, 1996. Teleconnections between the subtropical monsoons and high latitude climates during the last deglaciation. *Science* 277, 526–9.

Street-Perrott, F.A. & R.A. Perrott, 1993. Holocene lake levels and climate of Africa, in Wright *et al.* (eds.), 318–56.

van Andel, T.H., 1998. Middle and Upper Palaeolithic environments and the calibration of 14C dates beyond 10,000 BP. *Antiquity* 72, 26–33.

van Andel, T.H. & P.C. Tzedakis, 1996. Palaeolithic landscapes of Europe and Environs, 150,000–25,000 years ago. *Quaternary Science Reviews* 15, 481–500.

van Andel, T.H. & P.C. Tzedakis, 1998. Priority and opportunity: reconstructing the European Middle Palaeolithic climate and landscape, in *Science in Archaeology: an Agenda for the Future*, ed. J. Bayley. London: English Heritage, 37–46.

Willis, K.J., 1994. The vegetational history of the Balkans. *Quaternary Science Reviews* 13, 769–88.

Wright, H.E., Jr, J.E. Kutzbach, T. Webb III, W.F. Ruddiman, F.A. Street-Perrott & J. Bartlein (eds.), 1993. *Global Climates Since the Last Glacial Maximum*. Minneapolis (MN:) University of Minnesota Press.

Chapter 5

The History of European Populations as seen by Archaeology

Marcel Otte

1. The original foundation of the European Palaeolithic (from a million years ago onward) is essentially Asiatic, both culturally and anatomically. Around 500,000 years ago, there was an influx — limited to the west of the continent — coming from Africa ('Acheulian'). The two populations were superimposed and became totally uniform over the whole of Europe to form, about 300,000 years ago, the 'Middle Palaeolithic' and the 'Neanderthals', who were typical natives.

2. A migration from the east broke up this unity between 40,000 and 35,000 years ago. Culturally, these traditions are linked to Central Asia and the Zagros. Anatomically, this was a population that had evolved towards modern morphology. This wave of migration imposed new customs and genes.

3. Very complex processes of acculturation then took place following this opposition of radically different ideas and populations. These hotbeds of acculturation developed in the marginal geographical areas: the far west, the northern plains and the eastern steppes of Europe.

4. The middle phase of the Upper Palaeolithic ('Gravettian') corresponds to a profound standardization in Europe, from the Atlantic to the Urals, both culturally and anatomically. These were specifically European peoples and cultures, a little like the Neanderthals but this time of 'modern' type.

5. A 'crisis' phase seems to correspond to the Ice Age's phase of maximum severity (from about 20,000 to 15,000 years ago). This is the only time when climatic modifications seem to have had an influence on human behaviour. One can see an ethnic dispersal, towards the southwest (France and Spain) and the southeast (Balkans), where henceforth the traditions persisted independently. At the same 'moment', an African influence appears via Gibraltar and produces the 'Middle Solutrean'. As a result one can sense a frenetic and cultural influence from outside Europe in the southern fringe of the continent (from Portugal to Provence).

6. Next, one sees a spatial 'shattering' of European traditions, because around 18,000 years ago ('Magdalenian') a migratory movement begins, towards the plains of the northwest (from the Paris Basin to Scandinavia). This northward 'stretching' of the areas of occupation (reoccupation of the British Isles) is the equivalent of a migration, on the Russian plains, towards the Baltic Sea, on the eastern flank ('Epi-Gravettian').

7. All the 'provinces' of the final European Palaeolithic (from about 12,000 years ago) correspond to the future areas of the local Mesolithic where sedentism begins well before food production. In our view, these are the true sources of the European populations, within which the various forms of Neolithic were to develop.

Archaeological information has the advantage of being datable and, hence, of basing historical reconstructions on firm foundations. However, the relations between their variable aspects and the traditions or peoples from earliest times are extremely questionable.

Since the mid-nineteenth century, the general information available has remained sufficiently stable and coherent for us to be able to draw up a highly probable account of the different populations that inhabited Europe.

Two processes can be seen to alternate in these movements of peoples: waves of migration and local evolution. The former can be detected in 'breaks', interpreted as rapid influxes from outside. In these cases, the task is to perceive the original movement compared with the European assemblages. In contrast, phases of internal evolution show up in slow transitions, which are also detectable through archaeology. Finally, we also observe phases of contact or acculturation during which different populations have a reciprocal influence. The history of Europe is marked by all these phenomena, the rhythm of which can be specified thanks to dating methods.

The major event in this general development is the appearance of modern humans, around 45,000 years ago. Everything before that is the result of acculturation between 'primitive' populations (*Homo erectus*) originating in Asia and Africa. Everything after this event corresponds, in our view, to the slow evolution of these modern populations within the European continent.

In order to define the origins of modern humans in Europe, archaeological analogies (techniques and ways of life) point to Central Asia rather than Africa, because it is in the Altai and the Zagros that the oldest non-European assemblages are to be found. Africa may have been a biological 'reservoir' for the whole of humanity, but its cultural contribution cannot be seen at all at this time of transition.

Origins

In comparison to Africa or Asia, the first European populations are 'latecomers'. There is nothing definite there before a million years ago. The first diffusion out of Africa must have occurred as a lateral expansion, towards the south and east of Asia, from two million years ago. Out of this tropical belt, the expansion continued through adaptation to the northern steppic areas. The European appendage was then populated, for the first time, from Asia, along the latitudinal axis, around a million years ago. The first European '*erectus*' specimens evoke Asia, both in their morphology (Petralona, Atapuerca) and in their industries, which are always on pebbles and flakes in this phase (the African continent was 'already' Acheulian at this stage).

Europe was settled through several successive waves of occupation, the last of which — from Africa via Gibraltar — was limited to the south and west. The Acheulian biface industries are not found in the centre and east of the continent. On the other hand, they are abundant, albeit late, in the west from about 500,000 years ago.

The end of the Middle Palaeolithic

After that, a complete fusion began: during the Middle Pleistocene, industries and anatomical forms display purely native aspects, in the form of Neanderthals and the 'facies' of the European Mousterian. There is now immense potential, in terms both of techniques (wooden tools) and of culture (burials), but it took the arrival of a 'historical' phenomenon to crystallize the different traditions into new, recognizable expressions. This event was a migration of new populations and values. The closest analogies, at the sources of this movement, are to be found in Central Asia: the Altai and the Zagros contain the oldest assemblages of the Aurignacian culture which finally diffused to Europe around 40,000 years ago. Hence, modern humans in Europe seem to originate in the east and not in Africa, because the dates obtained in the 'Levantine corridor' are younger than those in Europe, while the African continent has never yielded industries equivalent to the Aurignacian.

The evolved Upper Palaeolithic

Following this primary migration, a period of acculturation began, and everywhere one can see the Mousterian substrate being modified under external influence: the Châtelperronian to the west, the leaf-point industries to the north, the Szeletian in Central Europe. All display these mixtures of traditions and new impulses.

The middle phase that follows ('Middle Upper Palaeolithic' or Gravettian) is the result of such an acculturation, which can be seen particularly in religious images (Aurignacian) and in sharpening techniques (Mousterian); after this stage, Palaeolithic cultures display coherent evolution and continue in such a way that the cultures of the local Mesolithic seem to be connected to them in an unbroken fash-

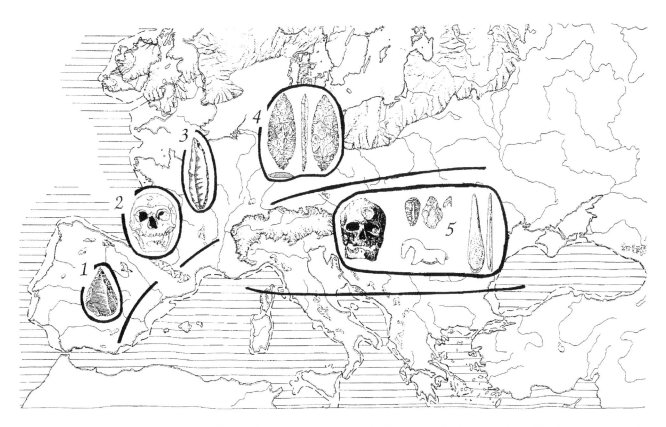

Figure 5.1. *The arrival of anatomically modern men corresponds in Europe to the introduction of the bone component in technology (5). This adaptation shows the steppic origin — i.e. eastern and not African — of this migration. At contact with this movement, various reactions or inertias can be seen in the northern and western fringes of the continent: continuation of Mousterian traditions (1) and of the local fossil race (2), Châtelperronian acculturation (3) and development of the leaf-point blade industries in northern Europe (4).*

ion. Only southwest Europe — once again — was colonized by another influence of African origin: the Solutrean, grafted onto the regional Gravettian substrate.

Neolithic

The economic modifications peculiar to the Neolithic gradually diffused outwards from the Balkans, across Central Europe, through successive acculturations. Hence the local ethnic substrata of the Mesolithic of central (LBK) and then northern Europe (TRB) progressively found themselves acquiring the new economy, while conserving their 'style' in the technical domain. With the diffusion of domestic cereals, a series of values was gradually imposed on Europe in the course of the adaptation of these ways of life to the landscapes successively encountered. In particular, the use of wood, rather than clay or mud when building structures corresponds to these forms of adaptation. The changeover from Mediterranean ovicaprines to bovids which were better adapted to forests also illustrates these pauses by the Neolithic across Europe, as it was acquired by the native populations.

DNA connections

Although the family tree produced by studies of human DNA seems to show an African convergence, this may be more a matter of an origin of the species rather than of its evolved, so-called 'modern' form (Cavalli-Sforza *et al.* 1994). One can indeed see convergences, with a trend towards contemporary morphology, in all places and all periods — so much so that this morphological disposition does not define a biological criterion (as produced by DNA) but an evolutionary stage that is more or less developed within this species. Therefore, an ultimately African origin does not rule out a far more recent migration by a population that was both evolved and adapted to the Eurasian steppes.

Moreover, Tyler-Smith has observed the microbiological analogies that exist between European populations and the present-day population of Iran, where a separation may have occurred around 40,000 years ago (Zerjal *et al.* 1999). If we consider the archaeological extent of cultural traces at the origins of modern humans, it is indeed the mountainous borders formed by the chains of the Zagros that constitute the clearest candidate for this diffusion.

The connection with the Indian sub-continent was also observed by research in molecular biology (Kivisild 1999). The far eastern limit of the Aurignacian, according to our interpretation, is to be found in the Pakistan area, close to the territory defined by DNA (Kozlowski & Otte in press).

Finally, the analogies displayed by the populations of the Caucasus, depending on whether they are in the north or south, clearly fit our explanatory scheme (Stoneking *et al.* 1999). The movement that we believe we can see between Central Asia and Europe must have left 'biological traces' in the northern areas of this mountain chain (as the traces of the Aurignacian also indicate). The southern margins seem more closely linked to the semitic populations, both today and archaeologically.

Conclusion

Although the archaeological data, considered globally in this way, provide a considerably more complex picture than that generally accepted by biologists, at the detailed level there nevertheless remain numerous points of contact whose significance must be taken seriously and with the mutual respect of the two disciplines.

The theory of African Eve does not correspond to anything in the data that are available today from archaeology: there is no connection of any kind between Africa and the industries that were to accompany modern humans in Europe. A prolonged 'detour' through Central Asia is at least indispensable to uphold this theory. Another point of view that is equally fruitful consists of differentiating the biological components (as revealed by DNA) from the formal aspects (revealed by palaeontology). In our opinion, the confusion that has arisen between the different usages of the expression 'modern men' is the source of innumerable misunderstandings between archaeologists and biologists who use the same words: we are not speaking about the same thing, and nothing is more misleading than an anatomical form, that is evolving like a kind of plastic and supple envelope and responding to cultural and environmental demands (Otte 1998).

Nevertheless, there are so many precise data today in both disciplines that both sides need to listen attentively in order to make any progress in the possible significance of contemporary observations. This illumination is provided by the Cambridge meeting, and we need to seize this chance.

Translated by Paul Bahn

References

Boyle, K. & P. Forster (eds.), 1999. *Human Diversity in Europe and Beyond: Retrospect and Prospect, Final Program and Abstracts of the Third Biennal Euroconference of the European Human Genome Diversity Project*. Cambridge: McDonald Institute for Archaeological Research.

Cavalli-Sforza, L.L., P. Menozzi & A. Piazza, 1994. *The History and Geography of Human Genes*. Princeton (NJ): Princeton University Press.

Kivisild, T., 1999. An Indian ancestry: a key for human diversity in Europe and beyond, in Boyle & Forster (eds.), 18.

Kozlowski, J.K. & M. Otte, in press. La formation de l'Aurignacian en Europe, *L'Anthropologie*.

Otte, M., 1994. Origine de l'homme moderne: approche comportementale. *Comptes-rendus de l'Académie des Sciences de Paris* t. 318, série II, 267–73.

Otte, M., 1996a. Le Mésolithique du Bassin Panonnien et la formation du Rubané. *Actes du Colloque de Szolnok 1996*.

Otte, M., 1996b. *Le Paléolithique inférieur et moyen en Europe*. Paris: Armand Collin.

Otte, M., 1998. La mise en place des populations européennes. *Rivista di Scienze Preistoriche* XLIX, 625–34.

Stoneking, M., I.S. Nasidze & M. Batzer, 1999. Genetic diversity in Caucasus populations, in Boyle & Forster (eds.), 31.

Zerjal T., R. Qamar, Q. Ayub, A. Mohyuddin, S.Q. Mehdi, W. Bao, S. Zhu, J. Xu, Q. Shu, R. Du, H. Yang, N. Pearson, N. Yuldasheva, R. Ruzibakiev, S. Wells, M.A. Jobling & C. Tyler-Smith, 1999, Y-chromosomal DNA diversity in Asia and Europe, in Boyle & Forster (eds.), 37.

Chapter 6

Spatial and Temporal Patterns in the Mesolithic–Neolithic Archaeological Record of Europe

Ron Pinhasi, Robert A. Foley & Marta Mirazón Lahr

The use of molecular genetics to explore problems in prehistory has brought into sharp relief the role that archaeological data can play in testing hypotheses about population dispersals and expansions. This has been the case particularly for the spread of farming into Europe. This paper uses a quantified data base of the distribution of archaeological sites from the Late Palaeolithic, the Mesolithic and the Neolithic in relation to radiocarbon dates to gain independent insights into the pattern of human occupation of Europe during the Early Holocene. Although the archaeological data cannot directly address the question of genetic continuity or replacement, it does provide a detailed context for testing hypotheses. The results of this study show that there is clear evidence for a spread, by whatever means, of Neolithic farming practices, but that this occurs in the context of an equally dynamic and geographically variable pattern of Mesolithic occupation. The patterns of these population distributions are discussed, and it is suggested that the complexities of the archaeological record can be used to generate testable hypotheses of relevance to molecular genetics.

The spread of agriculture into Europe has been a subject of debate for a considerable period of time, and this has been fuelled recently by contributions from both classical and molecular genetics. On the one hand, Cavalli-Sforza and co-workers (1993; 1994) have argued that there is evidence in gene frequencies for a strong demic diffusion across Europe, from the southeast to the northwest, that is indicative of a considerable level of genetic replacement. On the other hand, Richards *et al.* (1996) have suggested, on the basis of mtDNA data, that there is a much stronger case for population continuity between the Palaeolithic and Neolithic populations of Europe.

The role of archaeology in this debate has been largely focused on the relationship between agricultural dispersals and language (Renfrew 1987; 1998), and the ecological and social processes of interaction between hunter-gatherers and farmers. Archaeology has tended to emphasize that such interactions are likely to be complex and highly variable, and to suggest that the events associated with the spread of agriculture are unlikely to be either a straightforward population replacement or a simple cultural diffusion, but a mixture of the two (Zvelebil 1986b).

However, another aspect of archaeology that can be used to contribute to this problem lies in the overall temporal and spatial distribution of archaeological sites. The archaeological record of Holocene Europe shows that the change from the Mesolithic to the Neolithic was not a simple shift, but rather a series of changes in population distributions. These are associated with the post-glacial recolonization of Europe, expansions of Mesolithic populations, the spread of farming and formation of hunter-gatherer refugia. The archaeological record, which can pinpoint such distributions in time and space, provides a powerful framework for setting up hypotheses about variation in patterns of demic and cultural diffusion.

In this paper, we present an analysis of the

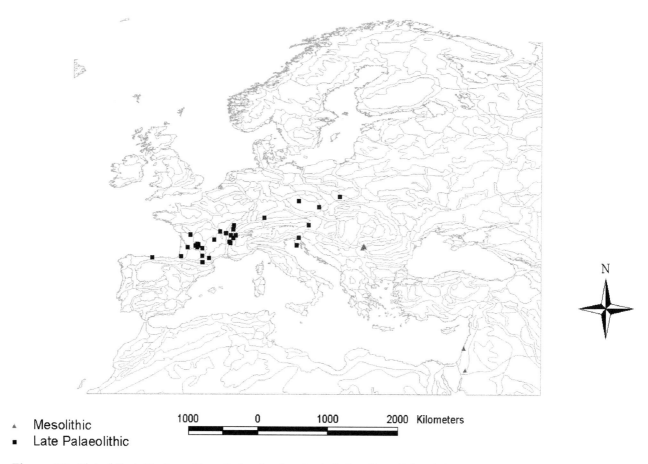

Figure 6.1. *Plot of the latitude and longitude of archaeological sites (Late Palaeolithic and Mesolithic) for which radiocarbon dates are available for the period 14,000–12,000 BP.*

archaeological record of human occupation of Europe in the Early Holocene. This analysis is based on a quantified data base of archaeological sites, their location, date and diagnostic properties regarding subsistence. These are then used to show the patterns of human population expansions and contractions through the earlier parts of the Holocene. Millennium-based maps of human distributions are then used to provide a basis for estimating the scale of demographic interactions between Mesolithic and Neolithic populations in different parts of Europe.

Archaeology of Late Pleistocene and Early Holocene Europe: the data base

In order to investigate the pattern of geographical distribution of farmers during the Early Holocene in Europe, data on Late Palaeolithic, Mesolithic and Neolithic sites across the Near East, Anatolia and Europe were collected and transformed into a systematic data base. Data were collected for a total of 2600 sites, and the information entered using the MS Excel spreadsheet application. Information on each site was collected from the literature, and entered into the data base following a categorization scheme:

1. *Site name*
2. *Date:* Radiocarbon dates were entered by site and stratigraphic context. All dates were entered in uncalibrated years before present (BP uncal.). Average date and standard deviations were calculated where appropriate.
3. *Geographic location:* The location of sites was extrapolated using physical maps of scale 1:400,000 against the description and maps available from site reports, and then entered as latitude and longitude coordinates.
4. *Archaeological category:* The archaeological category as it appears in the prevalent literature on a given site, e.g. Neolithic, Mesolithic, etc. These were then codified for use in statistical and GIS analysis.
5. *Subsistence data:* Information on subsistence was derived from quantified archaeozoological and palaeobotanical data for the site. The archaeozoological data were entered in percentages per

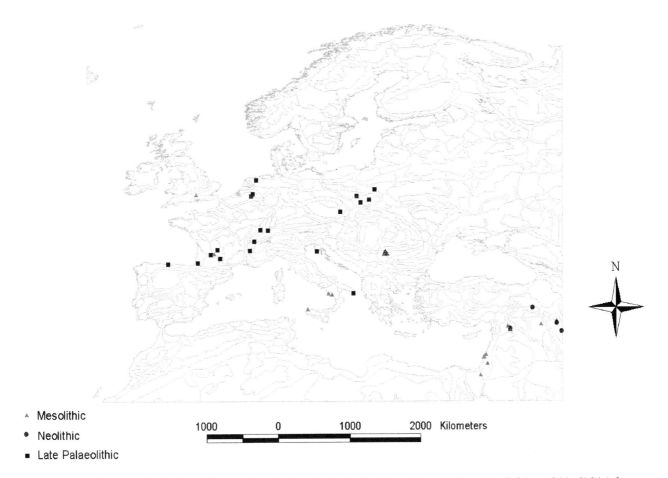

Figure 6.2. *Plot of the latitude and longitude of archaeological sites (Late Palaeolithic, Mesolithic and Neolithic) for which radiocarbon dates are available for the period 12,000–10,000 BP.*

species. The palaeobotanical data were entered using a presence/absence dichotomous system relating to the presence or not of specific domesticates.

6. *Material culture categories:* Data on artefact and structure types were entered using a presence/absence dichotomous system.

Two types of analysis were used to explore the information in the data base. The first of these was a spatial analysis, in which the data base was sub-sampled in order to select all entries with both geographic location and radiocarbon dates. The resulting subset contained 1217 entries, of which 80 were Late Palaeolithic sites, 502 Mesolithic sites, and 635 were Neolithic sites. This data set was then imported into ArcView for Windows and selected by millennium (with the exception of the first two queries, which were of two millennia periods). Each subset was then plotted by its archaeological code on a map of Europe. The results are chronologically divided maps of archaeological data that provide evidence for population distribution over Europe during the period of post-glacial recolonization and the spread of farming.

The second analysis involved statistically testing for correlations between the temporal and spatial distribution of sites in each subset, using the SPSS for Windows (V 7.5) package. These correlation analyses were carried out for Late Palaeolithic, Mesolithic and Neolithic sites. The first analysis performed Pearson correlations between average first appearance dates and geographical location (latitude and longitude) for each subset of sites. The second analysis aimed at investigating how the spatial and temporal patterns of Neolithic sites relate to the pattern of spread of domesticates and the appearance of permanent structures. In order to do this, a subset of sites which encompassed all those considered Neolithic by archaeological category was selected. This sub-set was then sub-sampled three times according to subsistence and material culture criteria — in the first instance sampling all Neolithic sites with evidence for domestic wheat; then all sites with evidence for domestic caprovines; and finally, all sites with evidence of permanent living structures. In each sampling procedure, only the earliest ap-

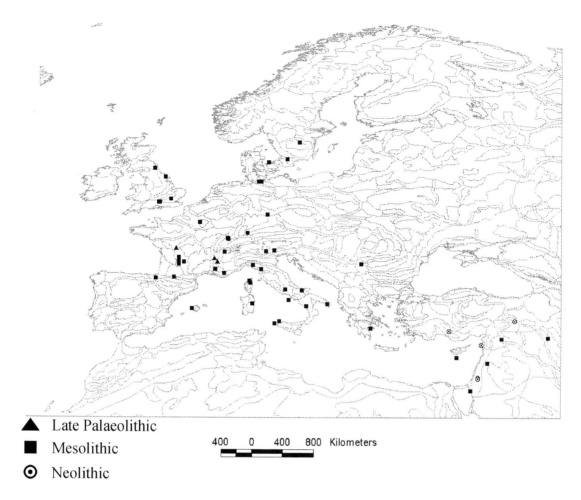

Figure 6.3. *Plot of the latitude and longitude of archaeological sites (Late Palaeolithic, Mesolithic and Early Neolithic) for which radiocarbon dates are available for the period 10,000–9000 BP.*

pearance of one of the above criteria was selected, thus yielding only one entry per site (in the case of sites with various levels). These sub-samples were neither all inclusive, nor all exclusive, as some sites may appear in all three sets, while others in which there were, for example, permanent structures but no domestic wheat or caprovines, would be selected only in the last sampling. Pearson correlations between archaeological date and latitude and longitude were calculated for each subset in the same manner described above.

The temporo-spatial distribution of archaeological traditions in Late Pleistocene and Early Holocene Europe

The results of the spatial analysis are presented as a series of maps showing the distribution of archaeological sites according to category by millennium (Figs. 6.1–6.7). The observable patterns for each period are briefly described below. Although it is assumed that there is a relationship between the distribution of archaeological sites and the distribution of human populations, factors such as taphonomy, excavation effort and relative material richness may contribute to the picture obtained. However, it is likely that the broad patterns will be relatively robust and can thus be extracted from the available data.

14,000–12,000 BP (Fig. 6.1)
During this period, Mesolithic and Late Palaeolithic populations concentrate in the Eastern Adriatic coast, the Danube Gorge, southern Poland and Slovakia, the Elbe region, the outskirts of the German Alps next to Munich, southern France and in the northern coast of Spain.

12,000–10,000 BP (Fig. 6.2)
This period witnesses the appearance of the first Early Neolithic sites in the Near East and Mesopota-

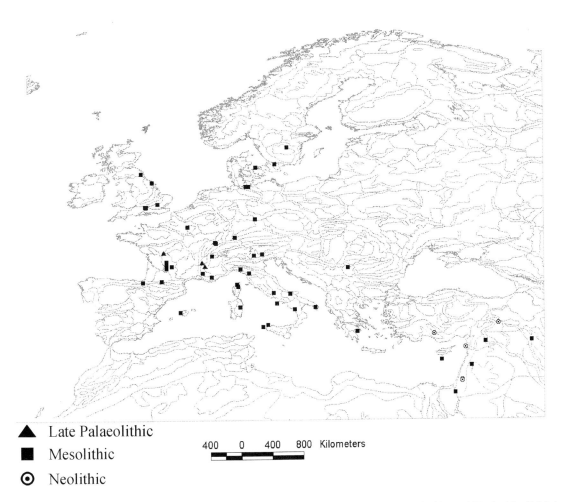

▲ Late Palaeolithic
■ Mesolithic
⊙ Neolithic

Figure 6.4. *Plot of the latitude and longitude of archaeological sites (Late Palaeolithic, Mesolithic and Early Neolithic) for which radiocarbon dates are available for the period 9000–8000 BP.*

mia. Mesolithic occupation in Europe now extends to southern Italy and Sicily. Some Mesolithic and Late Palaeolithic sites appear in southern Poland, Belgium and northern Germany. There is also a growth in the number of Mesolithic and Late Palaeolithic sites in southern France and northern Spain close to the Pyrenees. In southeast Europe there is continuity in the Mesolithic occupation along the Danube Gorge.

10,000–9000 BP (Fig. 6.3)
This is a period of great expansion of Mesolithic populations into new areas. These include the northward colonization of the British Isles (including Ireland), southern Scandinavia and the North Sea coast of Germany. There is also an expansion of Mesolithic populations into the mountainous regions of the Alps and the Pyrenees and the appearance of new Mesolithic sites in the Massif Central and the Paris Basin. This pattern contrasts with the scarcity of occupation in the Iberian Peninsula, Balkans and western Anatolia.

9000–8000 BP (Fig. 6.4)
During this period farming reaches Central Anatolia and mainland Greece. The settlement pattern during this millennium suggests little geographical overlap between the European Mesolithic groups and the expanding agricultural populations. The Mesolithic habitation becomes denser in northern Europe and expands eastwards into southern France. New sites also appear in the mountainous regions of the Alps and Pyrenees. This is a time of local Mesolithic expansions rather than colonizations of new areas as in the case of the previous millennium.

8000–7000 BP (Fig. 6.5)
The expansion of farming from Anatolia into Europe is now dramatically apparent. By 7000 BP Neolithic sites are found in Greece, Macedonia, the Danube

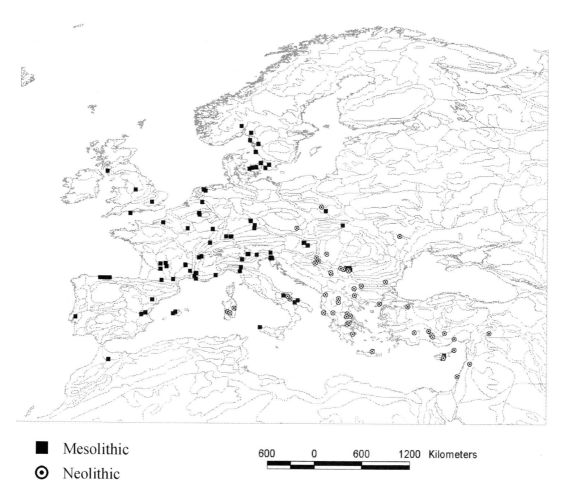

Figure 6.5. *Plot of the latitude and longitude of archaeological sites (Mesolithic and Early Neolithic) for which radiocarbon dates are available for the period 8000–7000 BP.*

Valley, the Hungarian Plains, and the Adriatic coast of Italy. The rapid expansion of farming throughout these areas may be related not only to the increased reproduction rates of farmers generally, but also to the fact that the Mesolithic occupation of this region was particularly sparse. There are few areas within this region in which interaction between Neolithic and Mesolithic ways of life would have been a major component of competition and trade. Among these, we may highlight the Tisza river in Hungary and the banks of the Danube Gorge, two areas that were occupied by Mesolithic hunter-gatherers when farming became established.

7000–6000 BP (Fig. 6.6)
Between 7000 and 6000 BP farming undergoes a second wave of expansion, but in two separate directions from the area occupied in the previous millennium. The first of these takes agriculture along the western Mediterranean coast to southern France and the Iberian Peninsula. Neolithic sites are now found in the Italian Alps, and spread westward along river valleys reaching the Massif Central, the Dordogne and the Mediterranean Spanish coast. The second direction of expansion of farming originated from the Hungarian Plains, advancing northwards towards the Czech Republic and Slovakia, and then along the tributaries of the Elbe, Warta, and Odra rivers. This wave eventually reached northern Germany and the Netherlands. A concomitant change in the Mesolithic settlement of Europe is apparent. All over southern Europe Mesolithic sites become sparser, while in the north, both in the British Isles and southern Scandinavia, the density of archaeological sites, all lacking any evidence of domesticates, becomes significantly greater.

6000–5000 BP (Fig. 6.7)
This period witnesses the consolidation of agriculture as the main form of subsistence throughout most

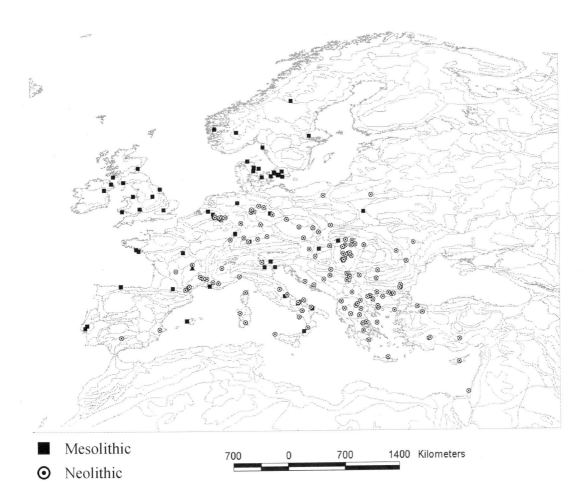

Figure 6.6. *Plot of the latitude and longitude of archaeological sites (Mesolithic and Neolithic) for which radiocarbon dates are available for the period 7000–6000 BP.*

of Europe. Further expansion of farming leads to the disappearance of Mesolithic sites in southeastern Europe, the northern European plains and central France, bringing together spatially the two waves of agricultural expansion that had taken place in the preceding millennia along the Mediterranean coast and north of the Alps. Farming also is found, for the first time, in the British Isles. During the same period, the distribution of Mesolithic sites extends northwards and eastwards, reaching the northern areas of Scandinavia — Finland and the Baltic Sea coast.

Statistical testing of the temporo-spatial distribution of Late Palaeolithic, Mesolithic and Neolithic sites in Europe

These analyses tested for correlations between time and geographical distribution of archaeological sites in Europe divided into three categories — Late Palaeolithic, Mesolithic and Neolithic. The results, in terms of Pearsons' correlation coefficients, are shown in Table 6.1. In the case of Late Palaeolithic sites, no statistically significant correlation between time (from radiocarbon dates) and spatial distribution (latitude and longitude) was found. For Mesolithic sites, however, there is a significant negative correlation between both radiocarbon dates and latitude ($r = -0.643$, at $p > 0.01$), and a weaker positive correlation between radiocarbon date and longitude ($r = 0.266$ at $p > 0.01$). These correlations, particularly that between earliest radiocarbon date and latitude, imply that subsequent to the establishment of full interglacial conditions, the human population of Europe, mainly Mesolithic hunter-gatherers, underwent an important expansion, expressed geographically by the colonization of the northern, temperate regions.

The distribution of Neolithic sites in time and space reveals two strong significant correlations — a negative one between earliest radiocarbon date and

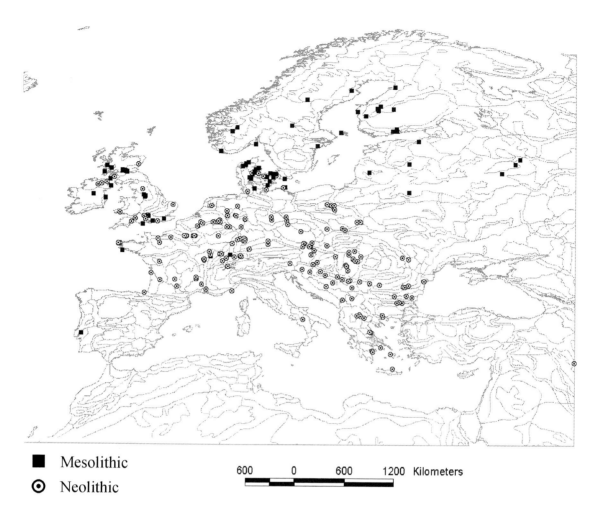

Figure 6.7. *Plot of the latitude and longitude of archaeological sites (Mesolithic and Neolithic) for which radiocarbon dates are available for the period 6000–5000 BP.*

latitude ($r = -0.644$ at $p > 0.01$), and a positive one between date and longitude ($r = 0.574$ at $p > 0.01$). These results show that the northward and westward expansion of farming from their southeastern Anatolian source is very strongly represented in the distribution of archaeological sites in time and space. In other words, the nature of these expansive movements was not random in its sequence or direction.

Statistical testing of the temporal and spatial distribution of domesticates and permanent structures

In the previous analyses the distinction between Neolithic and earlier archaeological sites was based on a general typological categorization, and this may not reflect the actual spread of farming practices and subsistence changes. It is possible that what might be considered the cultural presence of the Neolithic does not reflect the economic basis of it. To test for this we looked at the correlations between first appearance of wheat, caprovines and permanent structures in European archaeological sites and their location (latitude and longitude) (Table 6.2). All three items, used here as indicators of a farming and a more sedentary way of life, show similar Pearson correlation coefficients, suggesting that their geographical establishment was similar in both time and direction. As was found for the distribution of Neolithic sites in general, independent of whether they had evidence for domesticated wheat, or caprovines or permanent structures, the first appearance of these three indicators shows a strong negative correlation with latitude and a strong positive one with longitude, describing the clear uni-directional expansion of farming products or co-products in time.

Discussion of the pattern of early post-glacial and Mid-Holocene population expansions

When farming first appeared in Europe, the continent was occupied by hunter-gatherers. The history of this occupation is important, because not only the population density in different regions, but also where people were present, is intimately related to the climatic changes at the end of the Pleistocene. It is generally accepted that, besides those areas actually covered by ice-sheets, the northern plains of Europe became uninhabited during the Last Glacial Maximum (LGM) (Gamble & Sofer 1990). Prior to 13,000 BP, there is meagre evidence for human occupation north of a line running north of the Paris Basin and eastwards along the southern part of the North European Plain (Dennell 1985). These observations are supported by results of the spatial analysis carried out here, which show that before 13,000 BP there are no sites north of 48°N in western Europe. While there are a few sites at the eastern edge of the North European plain (north Poland), settlements at this time concentrate in the southwestern part of Europe, the Adriatic coast and the Danube Gorge. Therefore, understanding the re-colonization of Europe by hunter-gatherers is necessary to understand the nature of the potential interaction between them and the arrival of farmers and their subsistence practices, and thus whether the process was characterized by population replacement or cultural diffusion.

Dennell (1985) argues that two population expansions occurred during the Mesolithic — a Late Glacial expansion after 13,000 BP, and an Early Holocene expansion starting around 10,000 BP. According to this author, the first of these led to the colonization of much of northern France, the northern European plain, Denmark, northern England, the higher parts of the Alps, the Pyrenees and Moravia. The results of an analysis of Accelerator Mass Spectrometry (AMS) dating of the Late Glacial cultures of northern Europe by Housley and co-workers (1997) supports this view. This analysis investigated the distribution of AMS dates for eight regions of northern Europe — the Upper Rhine, the Thuringian Basin in Eastern Germany, the Meuse basin in Belgium, the Paris Basin in northern France, northern Germany and the British Isles. Not only is the earliest date of occupation of these regions consistent with a Late Glacial expansion of Mesolithic hunter-gatherers after 13,000 BP, but their analysis of the moving sum averages for the dates from these regions suggests that, with the exception of the Upper Rhine region, the largest number of dates falls 400–600 years after the earliest date in which the region was colonized. This is a strong indication that the first occupation of these areas was followed by a flourishing of Mesolithic groups locally.

Housley and colleagues (1997) believe that this pattern implies that the re-colonization of northern Europe was a dynamic two-stage process. They posit an initial pioneer phase, when only a few hunter-gatherer bands explored these newly available areas. This colonizing phase began around 14,200 BP in the Upper Rhine, and spread into most of Germany, the Meuse Basin in Belgium and the Paris Basin in northern France by 13,200 BP. The British Isles were first re-colonized 200 years later, around 13,000 BP. This phase would have been followed by a residential phase in which occupation of these regions by hunter-gatherers became more permanent.

Dennell (1985) argued that the northward expansion of Mesolithic bands was closely tied to changes in the distribution of both terrestrial and

Table 6.1. *Pearson correlations between radiocarbon dates, latitude and longitude by archaeological period.*

	Late Palaeolithic (n = 80)	Early Mesolithic (n = 402)	Early Neolithic (n = 470)
Latitude	−.094	−.643**	−.644**
Longitude	−.215	.266**	.574**

** Correlation is significant at the 0.01 level (2-tailed).

Table 6.2. *Pearson correlations between average radiocarbon dates and geographic location (latitude and longitude) of Early Neolithic sites with evidence for (a) wheat, (b) domestic caprovines, and (c) permanent structures.*

	N	Latitude	Longitude
Wheat	83	−.764**	.682**
Domesticated caprovines	48	−.724**	.748**
Permanent structures	55	−.637**	.700**

** Correlation is significant at the 0.01 level (2-tailed)

marine resources following the deglaciation process. These include the northward dispersal of herds of elk, horse and reindeer, and the southward dispersal of whales and seals. Similarly, Housley and co-workers (1997) argue that the pattern of re-colonization of Europe by Mesolithic hunter-gatherers is likely to be a reflection of changes in availability of hunting areas, which in turn reflect changes in the migrations of herds and seasonality of reindeer and horse. The changes in seasonality would have enabled human populations to reside more permanently in these areas of Europe without having to resort to long-distance seasonal travelling. However, the relationship between this early Late Glacial expansion and the migratory behaviour of game animals may be undermined in two ways. First, the scarcity of Mesolithic settlements suggests comparatively low population densities at the time, and thus a wealth of resources. Second, the nature of the deglaciation process was such that changes in fauna and flora were not only gradual, but also lagged behind changes in sea level, precipitation and seasonal temperatures (Bogucki 1988), implying that corresponding fluctuations in the availability of game and plants were slow. Nevertheless, independent of whether the first expansion wave of Mesolithic hunter-gatherers was strongly influenced spatially by game availability or not, it was followed by a more significant and extensive one at the beginning of the Holocene.

A second wave of Mesolithic expansion is apparent between 10,000 to 9000 BP, shown by the first appearance of sites in the northward part of the British Isles (including Ireland) and Southern Scandinavia, as well as the mountainous regions of the Alps and the Pyrenees. This period corresponds to the onset of fully interglacial conditions, which would have exposed these areas to human occupation (Dennell 1985). In spite of this clear expansion of hunter-gatherers towards the north, continental Europe at this time is not homogeneously occupied. On the one hand, the Massif Central and the Paris Basin seem to have been relatively densely populated. On the other, Mesolithic populations in the Iberian Peninsula, as well as the Balkans and the Greek peninsula seem to have been very scarce.

During the following millennia, there are no significant changes in the geographical distribution of Mesolithic populations throughout Europe. The Mesolithic occupation becomes denser in northern Europe, expands further into Italy and France, and particularly, it becomes prominent in the mountainous regions of the Alps and Pyrenees. In addition, there is a gradual expansion eastwards into the Czech Republic.

At the time when farming first appears in Europe, the Mesolithic settlement pattern is that of pockets of recently expanded populations. While certain areas, such as parts of France, the northern European plains and southern Scandinavia are relatively densely populated, other areas in central and eastern Europe were scarcely occupied. This implies that in each of the various regions under scrutiny, the expansion of a farming way of life faced different human circumstances, from areas where it must have displaced local hunter-gathering practices, to those that were virtually empty for the establishment of newly arrived farmers. It is possible that this pattern is partly an artefact of a biased density of archaeological investigation. However, the number of sites known from the different areas of most of Europe suggest that the observed distribution of Mesolithic sites is, in most cases, a reliable indicator of pre-Neolithic settlement patterns.

When considering the population substrate for the expansion of farming in Europe, another element besides the actual spatial distribution of hunter-gatherers needs to be examined. The ecological parameters for a successful foraging subsistence are considerably different from those of settled agriculture (Vencl 1986; Bogucki 1988). Mesolithic foragers searched for open, less-forested locations, with easy reach to watercourses. On the other hand, the earliest agriculturalist settlements are found next to fertile soils, in areas of secondary importance to those groups that lived off hunting and gathering. The different ecological distribution of optimal occupation by both subsistence economies, would have minimized their initial competition. This may have created the contingent circumstances that made demographic growth of the first farmers only intrinsically constrained by its higher potential reproductive survivorship. The earliest farming communities in Anatolia, as well as the subsequent ones in Greece, would have faced almost no competition from Mesolithic foragers, who were virtually absent from these areas at the time. This would have allowed these early agriculturalists to increase their economic potential and build up their populations locally (Vencl 1986), accounting for the period between 9000–7500 BP during which farming was restricted to Anatolia and eventually the Greek peninsula. This regional population growth would have given impetus for the expansion of farmers and their subsistence economy, but one that now involved the colonization of lands of optimal foraging. It is possible that the initial displacement of foraging as the established economy resulted directly from the demographic pressure posed by the accumulated growth of the southeastern European farming population, and this may explain the rapid spread of Neolithic sites towards the north and west between 7000 and 6000 BP (Zvelebil & Rowley-Conwy 1984; Rowley-Conwy 1986; Dolukhanov 1982). However, this rapid wave, which may have acted as a release from demographic pressure, was followed by a significantly different process. Along the western and northern margins of continental Europe — the Atlantic coast, the British Isles, and Scandinavia, Mesolithic foragers were relatively dense and more recently established. In these areas, the establishment of farming was slower and characterized by a slower disappearance of Mesolithic sites.

Therefore, it appears that, as with the earlier Mesolithic expansions, the expansion of farming in Europe was also characterized by several stages. These may be described as: 1) the initial phase, in which farming became established in southeastern Europe and agricultural populations grew in numbers locally; 2) a second phase, in which significant expansion of farming takes place from the focal area into the immediately surrounding region to the north and west, and eventually towards the extremes of the western Mediterranean, the Alps and the northern European plains; 3) and a final phase, in which farming becomes progressively established in northwestern and northern Europe. In many ways, the expansion of farming mirrors that of Mesolithic hunter-gatherers before, but while the latter was characterized by a northward movement of people into empty lands from several post-glacial refugia, the Neolithic expansion had a clear east to west and northwest direction.

Conclusions

This paper has used the spatio-temporal distribution of Mesolithic and Neolithic populations in Europe to explore the historical record for the colonization of Europe after the last glaciation. A number of conclusions can be drawn from the analyses of this distribution that have important implications for our understanding of the genetic structure of European populations. First, the Mesolithic occupation of certain areas at the time of the appearance of farming was negligible from a demographic point of view, while in others it was not only dense, but increased during the succeeding millennia. In these areas, the genetic effect of immigrant Neolithic farmers must have been either major or negligible, and this should be modelled at a regional level (see Lahr *et al.* Chapter 8). Second, that the dynamic pattern of Mesolithic colonization will tend to obscure any clinal trends that are presently attributed to the Neolithic 'wave of advance'. The combination of the differing regional Mesolithic palaeodemographies and the actual contemporaneous expansion of foragers across the continent may invalidate the interpolation of average gene frequencies across large areas as a means of obtaining the main historic signature in Europe.

Perhaps the main observation from the archaeological analysis of the present study is that the population history of both Mesolithic and Neolithic populations in Europe was dynamic and variable. In discussions of the relationships between genetics and the archaeological or linguistic record, the focus has been on the problem of the spread of farming populations. However, it should be stressed that during the period when farming was developing in the Near East, there was a very dynamic process of expansion of Late Palaeolithic and Mesolithic populations in Europe, and that these expansions overlapped with those of farming populations, at least during the period between 8000 to 5000 BP. Thus, while genetic studies may identify a relatively constant rate of expansion, this could be an artefact of the actual history of the region, whose genetic map was being formed by the growth of several populations with differing geographical origins and economic strategies at the time. In locating genetic estimates of the timing of events in the past it is necessary to take into account this temporally and spatially dynamic context. We would argue that it would be worthwhile to turn the scientific procedure around. Instead of generating historical hypotheses from the present genetic patterns, and trying to fit the archaeological record onto these, it would be useful to build hypotheses from the actual record of the past in time and space, which could then be tested with genetic data.

References

Bogucki, P., 1988. *Forest Farmers and Stock Herders: Early Agriculture and its Consequences in North-Central Europe*. Cambridge: Cambridge University Press.

Cavalli-Sforza, L.L., P. Menozzi & A. Piazza, 1993. Demic expansions and human evolution. *Science* 259, 639–46.

Cavalli-Sforza, L.L, P.Menozzi & P.A. Piazza, 1994. *The History and Geography of Human Genes*. Princeton (NJ): Princeton University Press.

Dennell, R.W., 1985. *European Economic Prehistory: a New Approach*. London: Academic Press.

Dolukhanov, P.M., 1982. Upper Pleistocene and Holocene cultures of the Russian Plain and Caucasus: ecology, economy, and settlement pattern, in *Advances in World Archaeology*, vol. 1, eds. F. Wendorf & A.E. Close. New York (NY): Academic Press, 323–57.

Gamble, C. & O. Soffer, 1990. Pleistocene polyphony: the diversity of human adaptations at the Last Glacial Maximum, in *The World at 18,000 BP*, vol. 1: *High Altitudes*, eds. O. Soffer & C. Gamble. London: Unwin Hyman, 1–23.

Housley, R.A., C.S. Gamble, M. Street & P. Pettitt, 1997. Radiocarbon evidence for the Late glacial human recolonisation of northern Europe. *Proceedings of the Prehistoric Society* 63, 25–54.

Renfrew, C., 1987. *Archaeology and Language: the Puzzle of the Indo-European Origins*. London: Penguin Books.

Renfrew, C., 1998. World linguistic diversity and farming dispersals, in *Archaeology and Language*, vol. I: *Theoretical and Methodological Orientations*, eds. R. Blench & M. Spriggs. London: Routledge, 82–90.

Richards, M.R., H. Côrte-Real, P. Forster, V. Macaulay, H. Wilkinson-Herbots, A. Demaine, S. Papiha, R. Hedges, H-J. Bandelt & B. Sykes, 1996. Palaeolithic and Neolithic lineages in the European mitochondrial gene pool. *American Journal of Human Genetics* 59, 185–203.

Rowley-Conwy, P., 1986. Between cave painters and crop planters: aspects of the temperate European Mesolithic, in Zvelebil (ed.) 1986a, 17–32.

Vencl, S., 1986. The role of hunting-gathering populations in the transition to farming: a central-European perspective, in Zvelebil (ed.) 1986a, 43–51.

Zvelebil, M. (ed.), 1986a. *Hunters in Transition: Mesolithic Societies of Temperate Eurasia and their Transition to Farming*. Cambridge: Cambridge University Press.

Zvelebil, M., 1986b. Mesolithic prelude and Neolithic revolution, in Zvelebil (ed.) 1986a, 5–16.

Zvelebil, M. & P. Rowley-Conwy, 1984. Transition to farming in northern Europe: a hunter-gatherer perspective. *Norwegian Archaeological Review* 17, 104–28.

Chapter 7

The Social Context of the Agricultural Transition in Europe

Marek Zvelebil

The agricultural transition in Europe and the origin of Neolithic societies have been a subject of concentrated attention among archaeologists, geneticists and linguists. In this paper I first review the archaeological evidence for the transition to farming and evaluate different archaeological interpretations. I then go on to discuss the archaeogenetic evidence and its integration with archaeological data. Although the genetic evidence appears at first to be at variance with archaeology, this is probably due to methodological problems affecting both disciplines. In the final section I pose some questions for the geneticists and suggest some solutions which should help us to advance our understanding of the agricultural transition and its consequences from both the perspectives of population history and cultural change.

In my contribution, I address the origin and development of Neolithic societies in Europe, with a special attention paid to the meaning and the role of the genetic evidence in this process. My point of departure is that neither the introduction of farming through contact, nor by migration can alone explain the establishment of Neolithic societies. More sophisticated processes, which include both movement and contact, must have been responsible for the regional variation characteristic of the Neolithic (Tables 7.1 & 7.2).

The basic premise of my argument is that the dispersal of farming and the process of Neolithization were embedded in the existing, pre-Neolithic social and historical conditions of each region, in the history of contacts with communities which had already adopted farming (beginning in the Levant or Anatolia), and in the inter-generational transmission of knowledge. In this sense, the social context of the agricultural transition in Europe had its structure and agency. The structure was set by the network of social relationships and contacts, and by tradition: the socially and culturally defined normative rules for the transmission of knowledge and practical skill from one generation to another. People, through contact and colonization, provided the agency for such transmissions, for the incorporation of innovations such as cultigens and domesticates, and for changing the structural framework of society.

Agro-pastoral dispersals

There can be little doubt that agro-pastoral (Neolithic) farming originated in the Levant and Anatolia some 10,000 years ago. But how was it introduced to Europe?

This question is most commonly debated in terms of deceptively simple dichotomy: introduction through contact or population movement. However, the situation is not so simple. Considered more thoughtfully, the following mechanisms of diffusion can be suggested (Table 7.1):
a) *Folk migration* is a directional and major population movement to a previously identified region (causing sudden gene replacement).
b) *Demic diffusion* is a sequential colonization of a region by small groups or households. It occurs over many generations and involves slowly expanding farming populations, colonizing new areas by the 'budding off' of daughter hamlets from the old agricultural settlements in a non-directional pattern (causing gradual gene replacement)
c) *Élite dominance* involves the penetration of an area by social élite and subsequent imposition of con-

Table 7.1. *Forms of contact between foragers and farmers.*

a) **Folk migration** - directional movement of a population to a previously defined region.

b) **Demic diffusion** - sequential colonization of a region by small groups or households, non-directional.

c) **Élite dominance** - penetration of an area by social élite and subsequent imposition of control over the native population.

d) **Infiltration** - gradual penetration by small, usually specialist groups of a region, who fill a specific economic or social niche (i.e. itinerant smiths, tinkers, leather workers, livestock herders).

e) **Leapfrog colonization** - selective colonization of an area by small groups, who target optimal areas for settlement, thus forming an enclave, or colony, among native inhabitants.

f) **Frontier mobility** - small-scale movement of population within contact zones between foragers and farmers, occurring along the established social networks, such as trading partnerships, kinship lines, marriage alliances.

g) **Contact** - through trade, exchange, within the framework of regional or extra-regional trading networks which served as channels of communication through which innovation spread.

Table 7.2. *Outcomes of contact between foragers and farmers.*

Replacement

Annihilation. Hunter-gatherer communitites are annihilated by farming communities in a violent conflict or through disease.

Assimilation. Hunter-gatherer communities disintegrate and their members join farming communities, introducing some aspects of (material) culture into farming communities.

Adoption. Hunter-gatherer communities adopt a farming way of life in the main, and accept farming ideology and a farming sense of identity, whilst retaining some aspects of traditional hunter-gatherer existence.

Acquisition. Hunter-gatherer communities adopt farming practices selectively, whilst retaining significant elements of traditional hunter-gatherer existence thereby producing new, or mixed (hybrid) cultural traditions.

Integration

Infiltration. Gradual penetration by small, usually specialist groups of a region, who fill a specific economic or social niche (i.e. hunters, fishermen, honey collectors, leather workers, livestock herders).

Absorption of farmers by foragers. Hunter-gatherer communities absorb farming households by force or peacefully within their communities and way of life, whilst at the same time adopting some aspects of farming existence.

Survival

Isolation. Hunter-gatherer communities remove themselves from contact with farming communities, usually by moving away and imposing 'no man's land' between foragers and farmers; this results in a 'closed static frontier'.

Encapsulation. Foragers are forced by farmers to move into suboptimal areas where they survive in relative isolation and impoverishment.

Commercialization. Hunter-gatherers reorganize their economy in response to demands by farming communities and commerical interests further afield (e.g. fur trade).

Reversion

Reversion. Farmers return to hunting, fishing and gathering as the principal means of subsistence.

trol over the native population (causing gene mixing, genetic continuity with genetic *ad stratum*, and the retention of genetic markers of intrusive population).

d) *Infiltration* involves a gradual penetration by small, usually specialist groups of a region, who fill a specific economic or social niche (i.e. itinerant smiths, tinkers, leather workers, livestock herders). (This may be undetectable genetically if there is no inter-group gene flow, if gene flow occurs, then small-scale genetic signature as in (c) can be expected.)

e) *Leapfrog colonization* denotes selective colonization of an area by small groups, who target optimal areas for exploitation, thus forming an enclave settlement among native inhabitants (causing gene replacement which is regionally variable, genetic 'islands' which may be diffused in time through gene mixing with local population).

f) *Frontier mobility* denotes small-scale movement of population within contact zones between foragers and farmers, occurring along the established social networks, such as trading partnerships, kinship lines, marriage alliances and so on (causing gene mixing marked by graded or discontinuous patterning in gene frequencies between genetically distinct populations, but if population were genetically similar, this would be undetectable).

g) *Contact* through trade, exchange, within the framework of regional, or extra-regional trading networks which served as channels of communication through which innovations, including domesticated plants and animals, spread (there is no gene replacement due to migration, genetic continuity prevails).

Agricultural transition: interpretations of the archaeological evidence

The migrationist position
Ever since Childe (1925; 1957), it has become an established view to regard the adoption of farming in Europe as a case of replacement of indigenous hunter-gatherers by farmers immigrating from the Near East and, over the generations, colonizing hitherto unfarmed areas of Europe: a process driven by a rapid population growth experienced by the Neolithic farming populations (Piggott 1965; Case 1969; Lichardus & Lichardus-Itten 1985; Vencl 1986; Aurenche & Cauvin 1989; Cauvin 1994; van Andel & Runnells 1995; Cavalli-Sforza & Cavalli-Sforza 1995, etc.). This colonization process is thought to have shaped the genetic map of Europe (Ammerman & Cavalli-Sforza 1984; Cavalli-Sforza *et al.* 1994 with refs.; Cavalli-Sforza & Cavalli-Sforza 1995; Cavalli-Sforza 1997), and to have been responsible for the introduction of Indo-European languages to the continent (Renfrew 1987; but see Renfrew 1996 and 2000 for recent modifications).

This school of thought holds processes 1–5 in Table 7.1 exclusively or primarily responsible for the introduction of farming into Europe, although the relative contribution of each is a matter of debate. Earlier scholars (i.e. Childe 1957; Piggott 1965) tended to favour migration, but more recent workers favour demic diffusion (i.e. Ammerman & Cavalli-Sforza 1984; Renfrew 1987). Élite dominance is discounted by some (i.e. Renfrew 1987), while others accept infiltration as a part of the Neolithization process (Neustupný 1982). Leapfrog colonization has recently been introduced as a more realistic alternative to other forms of movement (Arnaud 1982; Zilhão 1993; Renfrew 1996). The migrationist view is most readily accepted among the public, among non-archaeologist scholars, and commands a favoured position among archaeologists on the continent.

The rationale most often cited for the immigration of Neolithic farmers from the Near East to Europe for the demic diffusion of farming populations is the rapid population growth brought about by the emergence and development of farming (i.e. Renfrew 1987; 1996), regarded by some as 'demographic explosion' (Cavalli-Sforza & Cavalli-Sforza 1995, 133–4). The shift to agriculture brought about increasingly sedentary existence, improved diet, and rise in the economic value of child labour. This in turn reduced the need for population controls and made having more children both possible and desirable. In consequence, farming populations grew rapidly, colonized adjacent regions, and replaced hunter-gatherer communities, whose population growth was negligible or nil.

The indigenist position
This school of thought believes that the adoption of farming into Europe, and the origins of the Neolithic, came about exclusively through frontier contact and cultural diffusion (processes 6 and 7: Table 7.1). Migration from the Near East had little or no role to play. Genetically, then, populations of Near Eastern origin had little or no contribution to make. This view is based on strict interpretation of archaeological evidence, where the burden of proof is placed on the presence of clear archaeological markers of migration.

'Indigenists' fall into two groups, depending on their perceived importance of innovations which were spreading with cultural diffusion. Dennell (1983; 1992) and Barker (1985) regard the spread of agro-pastoral farming and Neolithic technology as the defining features of the Neolithic. Tilley (1994) and Thomas (1988; 1996) perceive the eventual shift from hunting-gathering to farming communities as internal social and ideological restructuring of Mesolithic communities that also — almost incidentally — involved farming. Whittle (1996) and Pluciennik (1998) adopt an intermediate position. The indigenist position has almost no support outside Britain and Scandinavia.

The integrationist position
This group regards processes of leapfrog colonization, frontier mobility and contact responsible for the agricultural transition (Zvelebil 1986a,b; 1989; 1995; 1996; Chapman 1994; Thorpe 1996; Price 1987; 1996; Price & Gebauer 1992; Zilhão 1993; 1997; Auban 1997; Renfrew 1996), although the relative contribution of each differs from author to author. A good number of archaeologists in Britain as well as in North America and continental Europe adhere to this view, although it is less popular outside the

profession (but see Willis & Bennett 1994; Richards *et al.* 1996). Although the differences of interpretation between the three groups are of a degree rather than categorical, the implications for the population history and genetic patterning at the agricultural transition are quite major. In adopting an intermediate and regionally specific position, integrationists, in my opinion, offer the best explanation for the transition to farming in Europe based on the current archaeological evidence.

The indigenist scenario places emphasis on archaeological evidence which shows lack of support for any kind of population movement. The problem here is the resolution of archaeological data: we cannot expect clear and unequivocal signatures for human behaviour, including migration. Past human behaviour is merely one among many factors which structure the archaeological record. Without going into the details of the debate about the variables which constitute the archaeological record (i.e. Binford 1983; Clarke 1968; 1972; Barrett 1994; Hodder 1978; 1999), archaeological cultures seem best regarded as cultural traditions of multivariate origin, including most recent variables of taphonomy and modern hermeneutics.

The specific relationship between archaeological cultures and human migration has also been much discussed recently, without resolution (Renfrew 1987; Mallory 1989; Anthony 1990; Chapman & Dolukhanov 1992; Bellwood 1996). The problem lies in specifying the relationship between population movement, normative (ethnically-identified, see below) concept of culture and archaeological signatures of these phenomena. Despite the fuzziness between past human identities, behaviour and their archaeological signatures, there are four developments, which, if coeval, are likely to indicate population movement:
a) the introduction of new cultural traits into a region in more than one cultural 'subsystem' (or, aspects of culture);
b) their discrete and coeval distribution;
c) the lack of earlier traditions for such traits within the region;
d) and the existence of an adjacent donor culture where such traits occur.

Gordon Childe has already drawn attention to such signifiers of population movement in the material culture (1957). In here, they are accepted as indicators of population movements (processes 1–5 in Table 7.1) without the corresponding ethnic connotations of a 'folk' or 'people'. The more precise form of population movements then has to be identified on the basis of other historical observations.

Bearing in mind this argument, and taking into account archaeological evidence for continuity and discontinuity at the time of the agricultural transition, the indigenist explanation throughout Europe seems untenable. Too many new traits are introduced coevally in parts of the east and west Mediterranean, southeast Europe and central Europe (Fig. 7.1).

Equally, the migrationist hypothesis does not find unequivocal support in either the archaeological, ecological, or demographic evidence. Archaeologically, there is no evidence for sustained and wide-ranging immigration that would support either the demic diffusion hypothesis or a major continent-wide migration (Dolukhanov 1979; Dennell 1983; 1992; Barker 1985; Zvelebil 1986; 1989; 1995; Thomas 1996; Midgley 1992; Larsson 1990). Demographically, there is no evidence for population pressure which would encourage first farmers to migrate, nor is there evidence for rapid population growth (i.e. van Andel & Runnels 1995[1]). Ecologically, there is no evidence for sustained woodland clearances after the initial phase and for environmental degradation that would indicate extensive agriculture on one hand, and provide a rationale for relocation on the other before the Late Neolithic (Willis & Bennett 1994; Willis *et al.* 1998; Smith 1981; Edwards & Whittington 1997; Bergelund 1990). At the same time, the ecology of Europe was favourable to supporting greater-than-average densities of hunter-gatherer populations, especially in coastal and lacustrine regions and along major rivers (Clarke 1976; Price 1987; Zvelebil 1986a; 1996).

Ethnographically, the choice by the migrationist school of examples as analogues for the historical situation at the Mesolithic–Neolithic transition is inappropriate (i.e. Piggott 1965; Ammerman & Cavalli-Sforza 1984; Cavalli-Sforza & Cavalli-Sforza 1995). In fact, pertinent ethno-historical evidence shows that there is a wide overlap in population densities between hunter-gatherers and subsistence farmers, further eroding the demographic basis of the farming colonization hypothesis. The ethnographic sample shows that hunter-gatherer population densities range from 0.02 to about 100 per square kilometre (Hassan 1975) with coastal, more sedentary foragers having the greater population densities. Given their economic and mobility patterns, Mesolithic communities were likely to approximate the higher population densities found among the Californian and Northwest Coast Indians. By comparison, the population densities of subsistence farmers engaged in swidden agriculture ranged from 3 per km^2 in Laos and Zimbabwe, to 30 in the Philippines and to 300 in

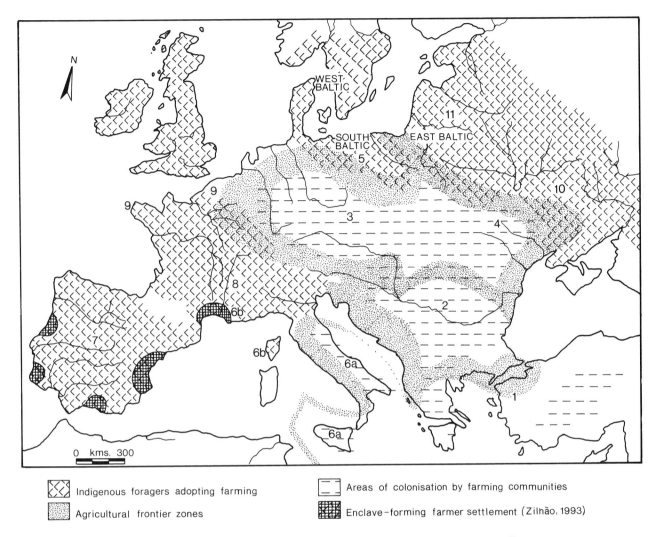

Figure 7.1. *'Colonist' and 'indigenous' regions of Europe at the agricultural transition according to one (integrationist) interpretation of the archaeological evidence. (Base map after Renfrew (1987) with additional information from Zvelebil & Zvelebil (1988) and Zilhão (1993).)*

New Guinea, while the rural population of Lorraine and of Belgium in the mid-fifteenth century was 10–25 and 30–70 people per km² respectively, and the population of England in 1086 was calculated as 78 per km² (Hassan 1978). Hammel (1996, 228) notes that the current evidence suggests no major change in mortality rates between the Palaeolithic and the eighteenth century AD, and that rapid population growth took off only 300 years ago, 'when doubling times generally dropped below a millennium' (Hammel 1996, 221). Finally, recent genetic studies in Africa also show the lack of any great differences in population dynamics between hunter-gatherers and subsistence farmers (Bandelt & Forster 1997). Even though one cannot make much of these figures, they suggest in aggregate a more even demographic playing field between foragers and farmers in prehistoric Europe. These considerations remove a central plank from arguments in favour of the migrationist hypothesis. Although population growth rates for farmers were likely to be greater than for hunter-gatherers, the difference must have been considerably smaller than originally postulated. The population densities of prehistoric foragers and farmers in Europe may have partly overlapped as they do in the ethnographic sample.

In summary, then, the assumption of marked population differences between prehistoric hunter-gatherers and Neolithic farmers is based on misunderstanding of hunter-gatherers as always mobile and organizationally simple: yet in Mesolithic Europe they tended toward socio-economic complexity

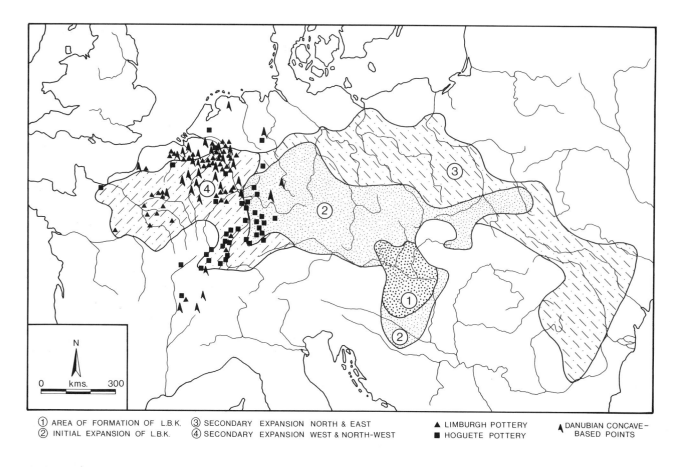

① AREA OF FORMATION OF L.B.K. ③ SECONDARY EXPANSION NORTH & EAST   ▲ LIMBURGH POTTERY   ▲ DANUBIAN CONCAVE-
② INITIAL EXPANSION OF L.B.K.   ④ SECONDARY EXPANSION WEST & NORTH-WEST ■ HOGUETE POTTERY   BASED POINTS

Figure 7.2. *The origin and dispersal of the Linear Pottery Ware culture. (Base map after Lunning* et al. *1989 and Bogucki 1995, with additional information from Guillaine & Manen 1995; Verhart & Vansleeben 1997; Gronenborn 1998).)*

and sedentism; and of Neolithic farmers as always sedentary and super-productive: yet in Neolithic Europe they were often transhumant or mobile, with mixed hunting-farming economy.[2]

Bearing this in mind, I would argue that the agricultural transition in Europe was, in the main, accomplished by the local hunter-gatherer communities, with varying degrees of gene flow between the hunter-gatherer communities and the settlements of Neolithic farmers. Enduring contact and exchange between the foraging and farming communities led to the development of agricultural frontier zones, manifested in the archaeological record by enduring cultural boundaries, for example between the Balkan Neolithic cultures and the Mesolithic/LBK of central Europe, or the LBK and derived communities in central Europe and the Mesolithic/TRB cultures of North Temperate Europe (Figs. 7.1, 7.2 & 7.4). From an integrationist perspective, two patterns can be discerned.

Within southeast and central Europe, colonization by farmers occurred through 'leapfrog colonization', which I find more convincing process of population movement than the demic diffusion model. Even though the idea of leapfrog colonization was originally applied by Arnaud (1982) and Zilhão (1993) to explain sea-borne colonization of the West Mediterranean from the East, a similar process could be used to explain the rapid spread of farming communities through the fertile lowland basin and river valleys in the Balkans and Central Europe.

Within such a scenario, the farming groups would target patches of fertile soil — for example loess in Central Europe — for 'enclave-forming' settlement. At the same time, local adoption of farming occurred through contact in the frontier zones around the initial farming settlement. Such a combination of colonization and contact can perhaps explain the origins of the Neolithic in the Balkans and in Central Europe. Here, the genesis of the LBK culture can be explained as the adoption and the adaption of the First Balkan Neolithic farming by the local hunter-gatherers at the periphery of the Körös culture (Fig. 7.2). With the adaption of farming practices to local conditions, hunter-gatherers turned farmers were in a position to expand quite rapidly within their own ecological region or culture area, in a 'star-burst' pattern of local adoption of farming, integration with local hunter-gatherer communities, and regional demographic expansion. This did not require any major population explosion, only a shift in settlement pattern and moderate population growth associated with the initial opening of a new economic niche. Genetically, then, the people who were colonizing these habitats mainly originated from the area of present-day Hungary, rather than from southeast Europe.

Similar processes of contact and colonization may have been responsible for the origins of the Neolithic in southeast Europe and parts of the Mediterranean: Greece, Istria and Dalmatia, Danube Gorges, southern Italy and the Iberian peninsula, for example (Radovanovic 1996; Bujda 1993; Chapman & Muller 1990; Auban 1997; Zilhão 1993; 1997). However, in some regions of the west Mediterranean, as in modern Languedoc or Tuscany, local cultural continuity and staggered introduction of farming practices and technology would argue in favour of a local adoption through contact and frontier mobility, rather than any form of colonization (Guillaine 1976; Lewthwaite 1986; Vaquer 1989; Barnett 1995; see Fig. 7.3).

In other parts of Europe, I see the transition to farming occurring through contact and frontier mobility. In either case, such exchanges were socially contextualized: they happened within an established framework of social networks, such as kinship ties, marriage alliances, trading/exchange partnerships and other social ties of reciprocity and obligation between the hunter-gatherers and the first farming settlements in a region. Within this scenario, the direction and the pace of the adoption of farming reflected as much the existing Mesolithic social context and routes of communication, as it did the conditions of the Neolithic communities and the regional ecological circumstances. The outcomes of such contacts between the foragers and farmers, documented ethnographically, are listed in the Table 7.2. Although such information can only serve as a rough guide to prehistoric situations, it is this form of contact, of socially embedded mobility unfolding between the two kinds of communities — foragers and farmers — which in my view was mostly responsible for the formation of the Neolithic in most regions of Europe.

Forager–farmer contacts and the social context of the agricultural transition

From my review so far, it is clear that contacts between foraging and farming communities, and the social context of such contacts are fundamental to our understanding of the cultural, genetic and linguistic history of communities undergoing the transition to farming. How can we recognize the operation of social networks, with all its genetic and linguistic implications, in the archaeological record?

At the Mesolithic–Neolithic transition, the social context for such networks would have been provided by the agricultural frontier zones. Such frontier zones can be either static or mobile, and open to contact or closed (Alexander 1978; Dennell 1983; 1992). The role of contact between foragers and farmers across this frontier could have been both supportive (Gregg 1988; Bogucki 1988), and disruptive for the foragers (Moore 1985; Keeley 1992). I suggest that in the early phase of forager–farmer contact, co-operation would prevail. At this stage, the effect of the frontier would have been largely supportive: the exchange of raw materials, foodstuffs, tools and prestige items across the frontier would reduce unpredictable variation in food supply and the risk of failure for both the hunting and farming communities (Fig. 7.6).

Contacts between foragers and farmers may have also occurred in terms of client–patron relationships, in which foragers acted as providers of specialist services or as rented herders of livestock for farming communities (Fewster 1996). Typically, foragers derive economic benefit from livestock or its products, while farmers are able to extend the grazing area and increase the size of their herds through renting out to client foragers. Such a system has been in operation as a part of forager–farmer relationships in southern Africa (Fewster 1996).

The movement of livestock may also have been of major importance in regional exchange systems. Such exchange in cattle would pass, as Sherratt (1982, 23) suggested, 'as transactions between acephalous groups linked by alliances and as symbols of competitive prestige'.

There is a growing body of evidence for such exchanges between foragers and farmers: this evidence comes from all parts of Europe (Figs. 7.2, 7.4 & 7.5). Let us take the frontier zone between foragers and farmers across the North European Plain as an example (Fig. 7.4). The date is fifth and fourth millennium BC. The imports from farming societies include the technology of pottery-making and the pots themselves, such as the Baalberg and Michelsberg

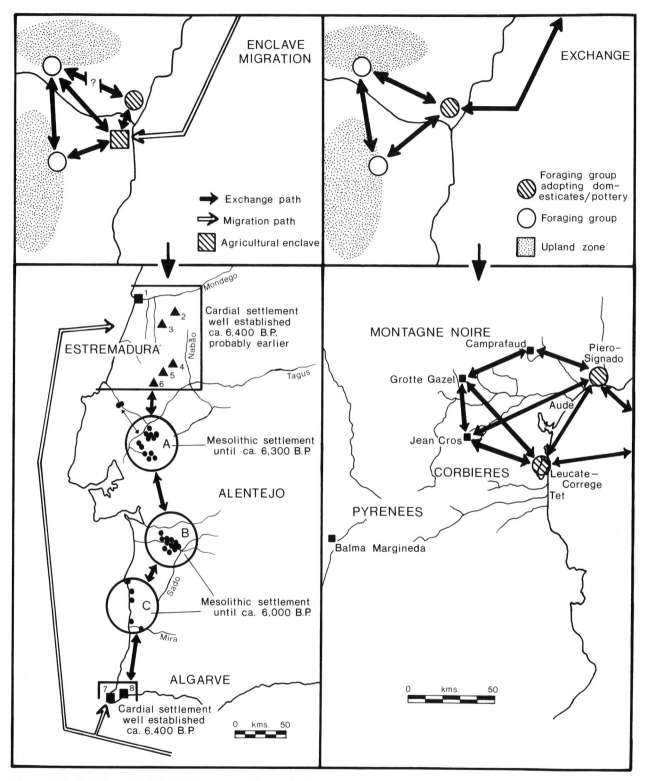

Figure 7.3. *Origins of the Neolithic communities in the Iberian peninsula: different views for different regions. Left: Leapfrog colonization (or enclave migration) along the Atlantic coast of Iberia. Right: Contact, exchange and frontier mobility in the Languedoc region of the northwest Mediterranean. (Redrawn after Barnett 1995; Zilhão 1997; Auban 1997.)*

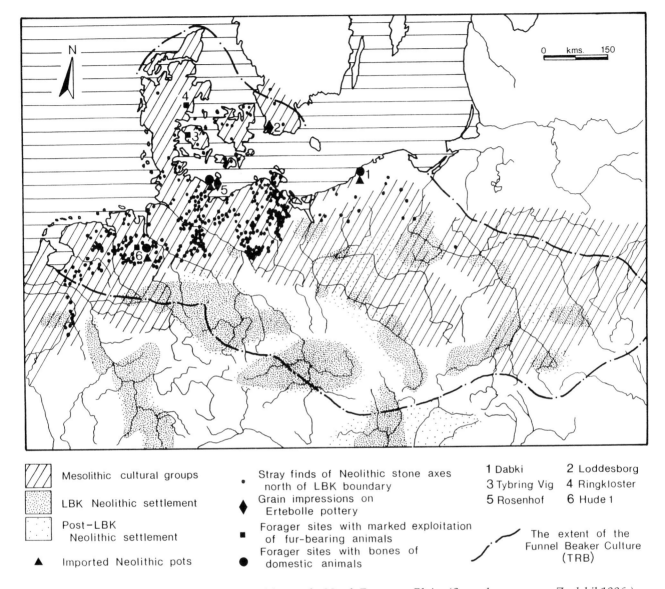

Figure 7.4. *Origins of the Neolithic communities on the North European Plain. (Several sources, see Zvelebil 1996.)*

pottery at Rosenhof (Schwabedissen 1981). They also include shoe-last adzes and other stone axe imports, while t-shaped antler axes, bone combs, and rings appear to be Ertebølle imitations of Neolithic artefacts (Solberg 1989; Price & Gebauer 1992). Bones of cattle which are found in small quantities on Late Mesolithic sites in Denmark, Scania and northern Poland are also probably the results of trade, traded perhaps as prestige items as well as food.

These products may have been exchanged for furs, seal fat, and forest products such as honey. The evidence for the specialized exploitation of fur animals, and their use for fur rather than meat, at such sites as Tybrind Vig and Ringkloster (Andersen 1975; 1987; Rowley-Conwy 1999) offers at least some support for this suggestion.

A similar exchange system existed within the frontier zone in the Central and East Baltic, where we have clear evidence for trade in amber (Vankina 1970) and other prestige items (axes, pots), and possibly also agricultural imports (Dolukhanov 1979; 1993) and trade in seal fat (Fig. 7.5; Zvelebil 1981; Rowley-Conwy & Zvelebil 1989). Local pottery shows the influence of ornamental motifs from Early Neolithic sites in the Dnieper basin (Zvelebil & Dolukhanov 1991) and from the western Baltic (Dolukhanov 1979; Timofeev 1987; 1990), giving rise to hybrid ceramic traditions in northeast Poland and Lithuania (Timofeev 1987). Such a network of contact and exchange reached out over a wide area of the

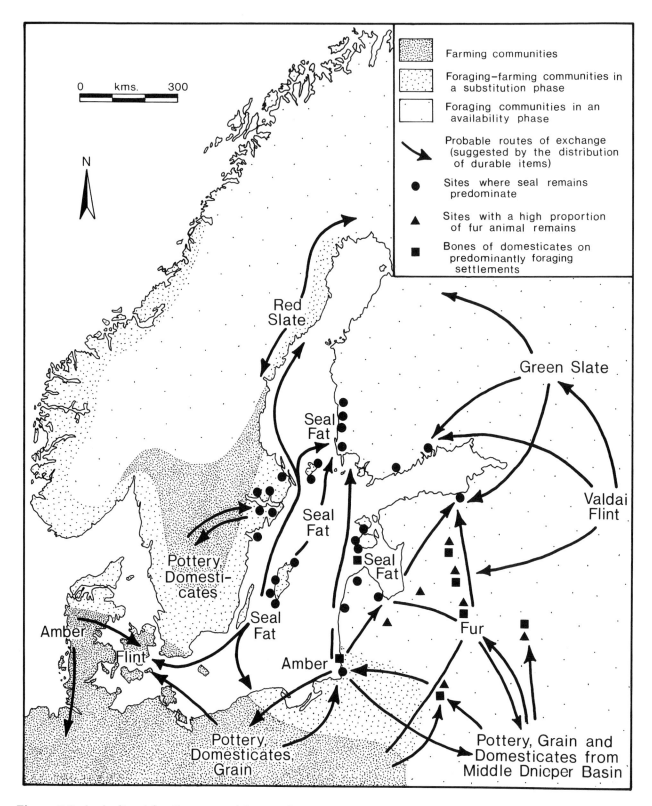

Figure 7.5. *Agricultural frontier zone and forager–farmer contacts in central and eastern Baltic 2500–1500 BC. (After Zvelebil 1996.)*

Baltic and eastern Europe, creating a pathway for new ideas and cultural innovations, which, in the later stages, may have been manifested archaeologically in the Corded Ware/Boat Axe horizon (Zvelebil 1993).

With the increasing stability of the agricultural frontier, disruptive effects gained the upper hand (Fig. 7.6). This would have been marked by the following developments:

1. Internal disruption of the social fabric among hunter-gatherers arising from increased circulation of prestige items and increased social competition.
2. Opportunistic use of hunter-gatherer lands by farmers, which, as Moore has shown, can cause serious interference in hunter-gatherer foraging strategies and information exchange (Moore 1985) and initiate an ecological change disruptive for foraging strategies.
3. Direct procurement of raw materials and wild foods by farmers establishing their own 'hunting lands' in hunter-gatherer territories as part of a secondary agricultural expansion.
4. Ecological change and over-exploitation consequent upon the development of commercially-oriented hunting and gathering.
5. Hypergyny: loss of women through marriage, voluntary departure or appropriation from hunting-gathering to farming communities, thereby generating an excess of women among farmers (hypergyny), and a shortage among hunter-gatherers (hypogyny). This is an ideologically conditioned practice, occurring in situations where among women farming is perceived as being of greater advantage than a hunting-gathering existence.
6. Transmission of disease between the two communities

There are several indicators of conflict and competition within the agricultural frontier zone in northern Europe. These include marks of increased social competition, territoriality, and violence among the Late Mesolithic hunter-gatherers around the perimeter of the agricultural frontier on the North European Plain (Whittle 1996; Keeley 1992) and southern Scandinavia (Persson & Persson 1984; Bennike 1985; Meiklejohn & Zvelebil 1991; Price & Gebauer 1992), the presence of fortified farming villages on the farming side of the frontier and, in some areas such as in Limburgh and Brabant, the existence of a 'no man's land' (Keeley 1992). Similar areas of apparently unoccupied land around 20–40 km in width can be detected between the agricultural Bronze Age and forager inland Neolithic sites during the first millennium BC in Finland, again suggesting antagonistic relations prior to the transformation of the hunter-gatherer

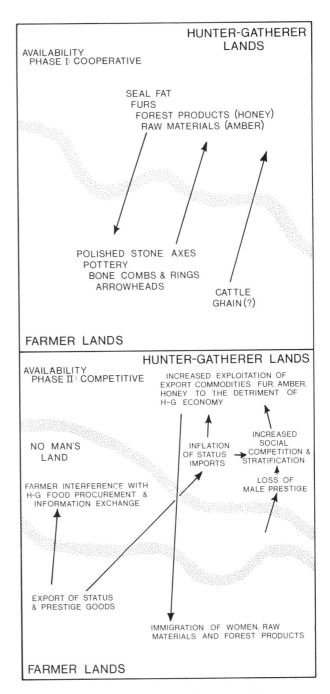

Figure 7.6. *Exchanges between foragers and farmers within an agricultural frontier zone: a general pattern. (After Zvelebil 1996.)*

communities there (Zvelebil 1981). Similarly, the presence of Mesolithic armatures for arrows in Neolithic assemblages in Poland, Germany and the Low Countries could be explained as a manifestation of conflict between foragers and farmers, while Neolithic artefacts could be seen as loot rather than imports (Tomaszewski 1988; Keeley 1992; Gronenborn 1990).

Some regional examples

It is my belief that contacts and exchanges such as those outlined here were principally responsible for the emergence of the Neolithic communities in Europe through cultural transformations of the kind illustrated in Table 7.2. We are now beginning to reconstruct regional histories of the emergence of the Neolithic communities in various parts of Europe. This includes social and ideological, not just economic contexts. For example, Radovanovic (1996) argues convincingly that ideological integration and a shift from individual to collective identity in the Iron Gates region extended the existence of hunter-gatherer communities there and enabled their eventual assimilation to the surrounding world of farmers. Similar arguments were used to explain the constitution of Neolithic societies in northwest Europe (Armit & Finlayson 1992; Tilley 1994; Thomas 1996).

Similarly, if we turn to the Baltic Sea basin as an example, it is clear that hunter-gatherers, as individuals and as communities, played an active part in the introduction of agro-pastoral farming and the appearance of the first Neolithic communities on the North European Plain. In so doing they have contributed to the generation of the Neolithic in two ways: by the transformation of their own communities and by their influence on the established farming settlements (Zvelebil 1986b; 1993; 1998; Bogucki 1988; Midgley 1992; Janik 1998; Whittle 1996; see also Thomas 1996 and contrast with Thomas 1988). The remarkable cultural diversity which characterizes the first Neolithic of the TRB tradition there (*Trichterbecherkultur* or Funnel Beaker) and of the subsequent cultural groups is a reflection of the divergent ways in which Neolithic communities developed through contact and native transformation.

Western Baltic region

The historical situation of the west Baltic region is marked by the extended delay and then a rapid adoption of farming: long availability, short substitution. As hunter-gatherers of relative social and economic complexity (Rowley-Conwy 1983; 1999; Price 1985; 1987; Larsson 1990; Tilley 1994) the inhabitants of the coastal zone were better equipped demographically and technologically to interact with the farming communities on a more equal basis than the foragers of the interior. Here, the erosive effects of the competition may have never gained the upper hand. The early and extended phase of contact between forager and farmer communities in the fourth millennium BC may have established enduring kinship ties, and resulted in associated transferral of exchange from the inter-tribal to tribal context, i.e. from negative to generalized/balanced reciprocity. Such relations were also likely to result in inter-marriage rather than loss of women to farming communities (hypergyny for farmers/hypogyny for hunter-gatherers), and consequently in the blending of cultural traits and the genesis of a new archaeological culture. In terms of cultural developments, listed in Table 7.2, these considerations suggest processes of acquisition, absorption, and then adoption of the farming way of life in this region (Table 7.2, Fig. 7.5).

Southern Baltic region

The genesis of the TRB culture east of the Odra (Oder) river on the North European (Polish) Plain shows similar patterns of change and continuity. One of the most striking features of the conditions prevailing on the Polish Plain is the long co-existence of farming and hunting-gathering communities, co-existence that lasted for more than 2500 years between 4400 and 1700 BC. In some areas, such as Kuyavia or Pomerania, hunter-gatherers and farmers — both of the TRB and the Danubian tradition — lived side by side only a few kilometres apart (Zvelebil *et al.* 1998). Despite the coarse spatial and temporal resolution of the evidence available today, such patterning suggests a very gradual incorporation of foraging communities with those of farmers after an extended history of contact, occurring within some established and effective framework. Such a framework may have been created by hunter-gatherers responding to the needs of the farming settlements and to their own social needs by commercializing their operations. Within such a framework, hunter-gatherers would play the role of suppliers of specialized goods and services, such as products of hunting, fishing, and sealing, and act perhaps as herders in client–patron relationships. The inter-marriage between the two communities would result in the breakdown of the early farming (LBK and Lengyel) social and ideological structure, witnessed, for example, in the final stage of the Brzesc Kujawski settlement in Kuyavia (Bogucki 1995; 1998), and a subsequent development of a new foraging-farming community, identified archaeologically as TRB (Midgley 1992). This process would have been accomplished inter-generationally, as one generation replicated and combined the cultural traditions of earlier foraging and farming generations, in an act of cultural creolization. These considerations suggest the processes of commercialization followed by integration of farmers as the basis of the cultural transformation responsible for the emergence of the TRB Neolithic (Table 7.2, Fig. 7.4).

East Baltic region
In the eastern Baltic, the picture was different again. Instead of generations of separate co-existence and creolization, we can identify the slow and staggered adoption of cultural traits and innovations, traditionally associated with the Neolithic, by communities of indigenous hunter-gatherers. The use of ceramics was adopted first, between 4500 and 4000 bc (see Timofeev 1987; 1998; 1999; Dolukhanov 1979; 1986; 1996; Zvelebil & Dolukhanov 1991). Elements of agro-pastoral farming were adopted at a very slow rate over the following three thousand years: the decisive shift to an agro-pastoral economy occurred between 1300 and 600 bc. In between, there was a society based principally on hunting and gathering for subsistence, yet making some occasional use of domesticates and possibly cultigens from about 2500 bc (Rimantiené 1992; Vuorela & Lempiäinen 1988). The presence of domesticates in such low numbers can be explained as a result of wide-ranging trading networks, operating within the context of the Corded Ware/Boat Axe culture (Dolukhanov 1979; Zvelebil 1993); while their limited use, which continued until the end of the second millennium BC, fits with the notion of their ritual and symbolic, rather than economic significance (Hayden 1990). The picture emerging here, then, is one of acquisition of Neolithic technology by hunter-gatherers and commercialization of hunter-gatherer communities during some 3000 years before the final adoption of farming (Table 7.2, Fig. 7.5).

Agricultural transition: interpretations of genetic evidence

A wide range of genetic studies, relating to the agricultural transition in Europe and the origins of the Neolithic, has been carried out to date (i.e. Ammerman & Cavalli-Sforza 1984; Cavalli-Sforza 1991; 1997; Cavalli-Sforza & Piazza 1993; Cavalli-Sforza *et al.* 1994; Cavalli-Sforza & Cavalli-Sforza 1995; Richards *et al.* 1996; 1998; Calafell & Betranpetit 1993; Barbujani & Sokal 1990; Sokal *et al.* 1989; 1991; 1992; 1998; Torroni *et al.* 1998, etc.) These studies include human DNA, as well as the DNA of domestic plants and animals (i.e. Bailey *et al.* 1996; Bradley 1997; Chapter 38). They involve mostly modern but also ancient samples. Indeed many papers at this conference addressed the issue (for example those by van Andel, Lahr, Bradley, Allaby, Bandelt, Bertranpetit, Underhill, Piazza, Pinhasi and Foley, Richards, Barbujani, Cavalli-Sforza and others). Most of this work is at the cutting edge of research and of enormous importance to our understanding of the cultural, genetic and linguistic history of populations in Europe and elsewhere.

At the same time, genetically-driven explanations are usually used to argue the case for the introduction of the Neolithic into Europe through migration or demic diffusion — both forms of population movement. Consequently, such explanations are often at variance with the archaeological interpretation of the evidence. In particular, the question of social context and of socially embedded, small scale genetic exchanges at the agricultural transition represent a problematic issue. In the critical appraisal below I address questions of methodology to my colleagues in palaeogenetics and argue that the conditional pattern and structure identified in the genetic patterning of European populations through principal component analyses and other methods can, to my mind, be explained in ways other than migration or demic diffusion.[3]

March of the genes: the case of Europe
Based on published genetic evidence and the papers given at the 1999 HUGO conference at Cambridge (see this volume), five major migratory events contributed to shaping the demographic history of modern populations in Europe:

1. Initial colonization by *anatomically* modern humans from North Africa/Near East by all or any of three routes: from North Africa, from Anatolia into the Balkans, and by a Circum-Pontic route north of the Black Sea. Date, based on mutation rates (dating by 'molecular clock'), falls between 50,000 and 30,000 BP. This migration horizon is indicated by mito-chondrial and Y- chromosomal evidence (Otte Chapter 5; Richards *et al.* 1996; 1998).

2. Later intrusion into Europe during the Upper Palaeolithic, perhaps associated with the Gravettian culture, dated between 25,000 and 20,000 BP from Eastern Europe/Near East. This is based principally on mitochondrial evidence (Richards *et al.* 1998; Torroni *et al.* 1998; Evison 1999; Evison *et al.* 1999).

3. Late Glacial population expansion and colonization of areas freed by deglaciation in northern Europe. Thought to originate from southwest France/northern Spain, Late Palaeolithic hunter-gatherers of the Magdalenian tradition moved north between 15,000 and 10,000 bp (13,000–8000

bc), colonizing areas hitherto covered by ice, water or polar desert. This is based on mitochondrial, Y-chromosomal and classical marker evidence (Torroni *et al.* 1998, 1149). The modern composition of European genepool reflects this movement more strongly than any other demographic event (according to Richards *et al.* (1996; 1998): around 85 per cent of European mitochondrial sequences are thought to originate in the Upper Palaeolithic). It provides the best correlation with archaeological data (Richards *et al.* 1996; 1998; Torroni *et al.* 1998; Evison 1999; Evison *et al.* 1999).

4. Early post-glacial 'demic diffusion' into Europe by the first farmers from the Near East, ushering the Neolithic into Europe. Identified initially through 'classical markers', this notion is now supported to the extent of 'pioneer' or 'leapfrog' colonization by mitochondrial DNA and Y-chromosomal DNA: dated to 8500–5500 bp (6500–5500 bc) (Richards *et al.* 1996).[4]

5. Late prehistoric intrusion from eastern Europe, thought to represent nomadic and pastoral Indo-European speakers, moving into Central Europe and adjacent regions in the north, west and southeast. This horizon is dated to 6000–4500 bp (4000–2500 bc) and supported principally by the principal component analysis of classical markers (Ammerman & Cavalli-Sforza 1984; Cavalli-Sforza *et al.* 1994; Cavalli-Sforza & Cavalli-Sforza 1995; contributors to this volume).

6. Later movements of the Classical and early medieval 'migration' period, which are more geographically restricted in character, and much better documented historically. They are held to explain only a small amount of modern genetic variation in Europe, yet the genetic evidence for gene flow in the first millennium AD is more compelling than any other (see papers in this volume; Cavalli-Sforza *et al.* 1994; Cavalli-Sforza & Cavalli-Sforza 1995; Laan & Pääbo 1997; Torroni *et al.* 1998; Richards *et al.* 1998, 253, 258).

It is clear there is disagreement among geneticists themselves on the relative contribution of each of these demographic events to the genetic history of European populations (compare and contrast, for example: Cavalli-Sforza *et al.* 1994 and Richards *et al.* 1996; 1998; Evison 1999 about the Neolithic dispersals; Richards *et al.* 1996; 1998; Torroni *et al.* 1998 and Cavalli-Sforza & Minch 1997; Izagirre & de la Rua 1999 about the late glacial migrations, or see Calafell & Bertranpetit 1993; Lalueza Fox 1996; Jackes *et al.* 1997 about the genetic history of the Iberian peninsula). There are also different degrees of correspondence with archaeological and historical data, the late glacial and the early historic (first millennium AD) perhaps commanding the best support. Against this background, I would like to focus now on the genetic support for the demic diffusion at the beginning of the Neolithic period (the fourth major demographic event).

Population movements at the agricultural transition: a closer look

The genetic evidence for the post-glacial 'demic diffusion' of Neolithic farmers is based on three sets of data:

a) Principal component analysis of the 'classical markers'. The first principal component explains, according to Cavalli-Sforza (Ammerman & Cavalli-Sforza 1984; Cavalli-Sforza & Cavalli-Sforza 1995), about 26–28 per cent of the modern genetic variation of Europe, mapped as a gradual distribution in values between the Near East and northwest Europe the directionality of spread indicated could be from either margin).

b) Mitochondrial DNA analysis, which seems to be more reliable than the component analysis of 'classical markers' because of fewer assumptions involved, shows a similar trend, but this accounts only for 9–14 per cent of mitochondrial sequences (Richards *et al.* 1996; 1998).

c) Y-chromosomal DNA analysis confirm the mitochondrial evidence: the frequency of Y-chromosome haplotypes originating in the Near East average about 15 per cent, with *c.* 25 per cent in the Balkans, and less than 10 per cent in western Europe (Underhill, HUGO 1999 conference; Semino *et al.* 1996).

In my understanding of these patterns, two other explanations are more plausible than the demic diffusion model:

a) 'Star-Burst' pattern of regional demic expansion, which I outlined above (in-filling or locally available niches by a genetically mixed population comprising local hunter-gatherers and some immigrant farmers). Arguably, this might produce the graded variation pattern observed in modern genome more faithfully than would the demic diffusion.

b) 'Incremental palimpsest' whereby the pattern we see today is a palimpsest of small-scale population movements progressing from southeast Europe to the northwest over millennia. This would not be surprising given that Europe is a northwestern peninsular extension of Asia.[5]

No reasoned refutal of these more recent interpretations, incorporating the conclusions from the archaeological evidence, has been advanced, although a critique of the integrationist position, often misconstrued as 'indigenist', has been made (Ammerman 1989; Cavalli-Sforza *et al.* 1994). Some felt that archaeologists have failed to offer 'a convincing alternative explanation for the southeast to northwest genetic gradient in Europe' (Evison *et al.* 1999, 5), without further elaboration. One problem is that archaeogenetics is no more accurate in the reconstruction of past events than archaeology (i.e. see Straus 1998, 400). Below, I list what I perceive as the weaknesses of the genetic evidence and its interpretation as it relates to the agricultural transition, although the points I wish to raise do have a broader implication for linking genetic evidence with population movements and demographic history in general.

Critique of genetic interpretations: category 1 errors and category 2 errors
In my opinion, one can group the uncertainties regarding the understanding and the historical interpretation of genetic evidence into two types of potential errors.

Category 1 error is a group of potential errors internal to archaeogenetic analysis of the human genome as a methodological procedure. Reconstructing genetic history from modern population genetics (i.e. tracing ancestry of modern populations back into the remote past, reconstructing their lines of descent) appear to have the following potential sources of error (Table 7.3):

1. *The size of the sample.* This is often too small for the size of the sampled population unit, itself often defined in a questionable way (see below) (Moore 1994; MacEachern 2000).
2. *Dating of genetic changes* within samples by mutation rates, or molecular clock. As some have noted, 'molecular clock models are full of questionable assumptions' (Clark 1997; Lewin 1988a,b). The mutation rates, held to account for gene or gene-derived polymorphisms, are assumed to be constant but apparently are not always so. The constant rate of accumulation of genetic changes is based on the assumptions of demographically stable populations and on adaptively neutral role of genetic traits. These assumptions are rarely if ever met in reality for reasons outlined below. The result is that the dating of genetic changes, and, by implication, demographic events, such as gene flow (migration) have very broad confidence limits and may be in error altogether.
3. *Genetic drift.* It is assumed that genetic drift in small isolated populations will result in marked genetic heterogeneity relative to other populations and in the expression of signature mutations through founder effect. Hunter-gatherer populations in general are often quoted as examples of such populations, for example by Cavalli-Sforza *et al.* (1994, 15) in the case of the European Mesolithic. Yet as many have recognized, exogamy is a common feature of such populations to keep them as viable interbreeding networks (i.e. Wobst 1974; Cavalli-Sforza & Cavalli-Sforza 1995, 19–20). Moore (1994, 934) has shown that intermarriage between separate ethnic groups of North American hunter-gatherers was likely to equalize any distinct genetic signatures and homogenize genetic patterning across large areas such the Plains of North America 'within a few hundred years'. In reality, many if not all small-scale populations share in large inter-breeding networks for reasons of survival. This appears to violate the assumption of stable population units (see also MacEachern 2000). Would this not homogenize the genetic landscape of small, low-density populations and obscure genetic signatures of population units defined by language or ethnicity (i.e. Amorin 1999; Moore 1994)?
4. *Natural selection* and environmental factors are not given the full role in the explanation of genetic variability. Although genes are assumed to be adaptively neutral, or at least non-directional (in that stochastic variation neutralizes any patterning), it is clear that the presence or absence of specific haplotypes may be related to disease resistance, or otherwise, conferring selective advantages or disadvantages on an individual in specific ecological and/or cultural circumstances. The HLA complex (Cavalli-Sforza *et al.* 1994; Evison *et al.* 1999), or genetic mutations controlling for thalassemia (Cavalli-Sforza *et al.* 1994) or for lactase tolerance (Harrison 1975; McCracken 1971; Simoons 1979; Hollox Chapter 36) can all be used as examples. Given the well-known selective role of some genetic variants, one is tempted to ask why is the role of selection apparently minimized in archaeogenetics?
5. *Age–sex structure of the reproducing population.* Mutation rates can be expected to increase as the child-bearing population gets older. This would indicate that mutation rates should have speeded up in the last few generations (*c.* 300 years: see Hammel 1996), rendering the 'molecular clock' faster. This is at variance with the assumption of the constant rate of mutation changes which forms the basis for the dating of demographic events by

molecular clock.[5]

6. There is a wide range of *statistical problems* such as spatial auto-correlation, associated with the principal component analysis and other forms of correlation between genes and geography, weakening the statistical treatment of genetic evidence and reducing the probability of the conclusions being correct (Clark 1997; Bandelt *et al.* 1995; Amorin 1999). Failure to address weaknesses inherent in some of the assumptions operationally necessary for the performance of statistical tests is leading to the loss of confidence:

> Cavalli-Sforza uses principal component analysis (PCA) to ransack correlation coefficient matrices for pattern in genetic polymorphisms and isolates a number of principal components, expressed geographically, which are interpreted as time-successive, quasi-historical, migration events... This form of argument from induction is called *post hoc* accommodation... a weak form of inference. (Clark 1997, 407, for similar critique, see also Moore 1994; MacEachern 2000)

Are the critics wrong or should the geneticists adhere to a more sober form of statistically-supported interpretation?

7. *The overall representativeness of the sample.* All the assumptions discussed above bear on the representativeness of the investigated sample. In addition, there is the problem of relationships between the different units of analysis within the population as an interbreeding unit. This is true somatically of different genetic units within an individual, as well as extra-somatically, when it comes to specifying the relationship between the individual and the population. As Moore put it:

> It is misleading for synthesists to treat the nodes of genetic cladograms as if they were tribes or demes, not to mention regional or continental populations. Even if we had a complete mitochondrial cladogram for all human beings, it would say nothing about where the individual carriers of the genotypes lived or what the genetic variability in local populations might have been. Individual pedigrees and histories of populations are two entirely different matters. Nevertheless, certain syntheses continue to treat ancestral sequences as if they were characteristic of populations all carrying the same genotype as the reconstructed individual. (Moore 1994, 934)

8. *Inter-demic genome similarities, the dating of demographic events by molecular clock and the palimpsest effect.* All the 'category 1 errors' noted above combine to reduce the reliability of reconstructing population histories from genetic evidence. This is particularly true if the representativenes of the sample is statistically compromised and if the dating of demographic events depends on mutation rates within a single class of genetic data. Genetic variation described by the principal component analysis and other diversity measures reflects not only demographic events such as migrations, but also the genetic distance between incomers and the native population, as well as the genetic distances between incomers at any one time and subsequent population movements (Cavalli-Sforza & Cavalli-Sforza 1995; Zvelebil 1995; Zvelebil 1998; Richards *et al.* 1996; 1998). Most human genetic diversity is intra-populational, with only a very small proportion of genome accounting for differences between populations (Amorin 1999, 18). The consequence of this realization appears to be at least twofold. On the one hand, principal components such those used to argue for the Neolithic colonization of Europe from the Near East may in fact reflect a diachronic incremental palimpsest of small-scale intrusions into Europe, the patterning of which is set by the geography of Europe as a peninsular extension of Asia.[6] On the other hand, a migration of Neolithic farmers into Europe may not be detected genetically if the donor and target gene pools were sufficiently similar.

Category 2 errors are relational, arising from presumed relationships between the genetic population (gene pool) and its related components, such as language, material culture, and ethnicity. We are back to the notion of human societies whose organization is predicated on the ideology of ethnic nationalism, and on the normative definition of ethnicity, based on descent (and therefore genetic uniformity). But these relational components are not corresponding units, in either the analytical sense, or in a conceptual sense (Moore 1994; Pluciennik 1996; MacEachern 2000). As MacEachern notes:

> Probably the most obvious of these problems is one that bedevils all interdisciplinary investigations of the human past: to what extent are the very different analytical units in these various disciplines comparable? Under what circumstances may we expect that ethnicity, language, material culture and gene pool will co-vary in the past, and when can we expect that they will differ in extent and characteristics? (MacEachern 2000, 359)

As a number of studies in archaeology, ethnography and linguistic anthropology have shown, the answer is that co-variance very rarely, if ever, can be demonstrated (for example, see Clarke 1968; Hodder 1978;

Shennan 1989; Graves-Brown *et al.* 1996; Jones 1997 for archaeology; Ehret 1988; Bateman *et al.* 1990; Thomason & Kaufman 1988 for linguistics; and Moore 1994; MacEachern 2000; or Terrell & Stewart 1996 for ethnography; see also Renfrew 2000 for further discussion).

Analytically, it is a matter of size and definition. Different population sizes pose different sampling and methodological problems. Related to this is the definition of demes, as groups whose members share greater genetic similarity because of greater frequency of interbreeding relative to non-members. How to define these units operationally? As many studies have shown, ethnic identity or shared language is a poor indicator of demes genetically defined (e.g. Bateman *et al.* 1990; Moore 1994; MacEachern 2000). If this is the case, where do we go from here? Is there a case for random sampling of the gene pool, irrespective of cultural attributes?

Conceptually, it is a matter of meaning and temporality. It is often assumed that human society is organized in culturally meaningful corresponding units ('analogous taxonomies': MacEachern 2000), giving us a normative definition of a genetic population as linguistically and culturally uniform ethnic unit so: population = language unit = cultural unit = ethnic unit (tribe). Yet the analytical units used are not comparable. It cannot be assumed that language, ethnicity, material culture and gene pool will co-vary in the past, and we do not know how such co-variation might work. At best, we can assume a broader relationship approximately as follows: deme = speech community = social network = shared material culture, but not exclusively so.

When we add the historical dimension, it becomes clear that both the analytical and the underlying cultural units of genetic analysis are often treated as static in time and space, as unchanging, bounded entities. Yet it is clear that such categories have been changing through time. The notion of ethnicity, as a historically contingent and relational phenomenon is a good example: there is nothing permanent about ethnic identification. Perceptions of ethnicity varied through time. Moreover, at any one time, in addition to the commonly acknowledged collective self-perception of identity (Jones 1997; Renfrew Chapter 1) at least two other perspectives play a role: others' perception, individual self-perception. The modern definition of ethnicity is one of ethnic groups as culture-bearing units, a notion central to the idea of ethnically defined nationalism. Within this framework, ethnic groups are perceived as fixed, homogeneous, bounded entities, extending deep into the past. This is a re-definition of earlier notions, a modern invention, linked to the rise of nationalism (Eriksen 1993; Jones 1997; Jenkins 1997).

The historical reality shows that ethnic groups are subjective, constructed and situational, deeply embedded in economic and political relations. As Barth (1969; 1994) pointed out, ethnicity is a changing phenomenon, which tends to attain greatest expression in situations of conflict, competition and cultural change. As such, ethnic groups can be characterized as interest groups competing for economic and political resources and territory. This adduces a degree of opportunism to ethnicity as a temporal and strategic resource.

According to Barth, the primary emphasis is placed on the categories of ascription and identification by the actors themselves. This is confirmed by my own personal experience. In addition to collective self-definition, we need to add two other perspectives: self-definition within a group and others' definition of a group or an individual. For example, I can subjectively regard myself as an Englishman on the grounds of having lived in England for almost thirty years, but will not be regarded as one by (other) English people: my self-definition within a group may be at variance with others' definition. Conversely, I may no longer regard myself as a Czech, but other Czechs may perceive me as one, no matter how I feel: others' definition may be at variance with self-definition. Needless to say, the same applies to groups as to individuals. The unfortunate adherence to ethnicity as a normative construct defined by a single dimension (collective self-definition) is a constant source of confusion and conflict.

Conclusions

To summarize my argument, the agricultural transition in Europe, and the origin of the Neolithic communities can only be understood in its social and historical context, which involved both the resident hunter-gatherer Mesolithic populations, as well as immigrating communities of early farmers. The degree of mobility and the mechanism of dispersal were regionally variable across Europe, as was the genetic contribution of each of the foragers and farmers to the subsequent Neolithic populations of Europe. To date, genetic evidence can be interpreted to accommodate several mechanisms of dispersal, while archaeological evidence shows hunter-gatherer continuity and contact across the agricultural transition in western, northern and eastern Europe.

From what I argued here we should expect that

the gene pool of the Mesolithic and the Neolithic populations was largely the same in western, northern and eastern Europe, while in the European continental interior we can expect a mixed gene pool comprising both the indigenous and immigrant elements. In central and southeastern Europe, this would involve a limited gene flow between the initial farming settlements and the indigenous hunter-gatherers, and in some regions such as Danubian basin, the farmers themselves could be expected to originate mostly in the same region as the foragers, if one accepts that it was the local foragers who adopted farming and then undertook regionally specific dispersals through Central Europe, archaeologically recognizable as the Linear Pottery Ware Culture.

I would like to conclude with three equally untruthful but relevant statements. It is significant that none have been made by specialists in research: 'History is bunk' (attributed to Henry Ford); 'There is no such thing as society' (attributed to Margaret Thatcher); 'Our genes make us what we are' (John Hands, reporting on recent archaeogenetic research, *The Independent*, 20.10.96).

Far from it. Our behaviour, even our physical characteristics are determined in a large measure by our history and our society. The resources placed at our disposal by culture enable us to change and transform the conditions set by our genes and make us into something else than our genes would. That is the essential point if we want to understand the way we were in the prehistoric past, as well as who we are today.

Acknowledgements

I would like to thank Professor Renfrew for an invitation to address the HUGO conference in Cambridge upon which this contribution is based, and to Dr Andrew Chamberlain for reading and commenting on a final draft of this paper. All errors and omissions are my own.

Notes

1. Archaeological evidence does not record any evidence for rapid saturation of areas colonized by Neolithic farmers, or for demographic expansion, with the single possible exception of the Linear Pottery Culture in central Europe. Even in the presumed core area for such expansion, southeast Europe, the saturation process was slow and incomplete. This is shown, for example, through the work of van Andel and Runnels in Thessaly. Even though they argue in favour of the demic diffusion for the spread of farming (1995, 494–8), their own calculations fail to substantiate the population growth rates necessary for such models to operate. They conclude that the Early and Middle Neolithic periods 'seemed to have been a time of steady but not very rapid population growth' so that 'even the Larisa basin, region of major growth, required some 1500 years, from about 9000 to 7500 BP to reach saturation' (1995, 497). This is a far cry from the 'demographic explosion' of Cavalli-Sforza, but in complete agreement with the recent palynological work carried out by Willis & Bennett (1994) showing that even in southeast Europe (including Greece) the impact of agriculture is not evident until *c.* 6000 BP, suggesting that the introduction of farming 'was not of sufficient intensity to be detected upon a landscape scale' (1994, 327).

2. Archaeological evidence for the Mesolithic in much of Europe (except central and southeast Europe) records stable, relatively affluent, often semi-sedentary communities which would have maintained relatively high population densities see Rowley-Conwy 1983; 1999; Price 1987; Price & Brown 1985a; Zvelebil 1986; 1996; Tilley 1996; Finlayson & Edwards 1997; Voytek & Tringham 1989, etc.). Archaeological evidence for the Early Neolithic in much of Europe records partly mobile communities which relied on a mixture of farming, hunting, gathering and animal husbandry (except for southeast and central Europe: Barker 1985; Bogucki 1988; Tilley 1994; 1996; Thomas 1991; Whittle 1996; Thorpe 1996; Barclay 1997, etc.). Consequently, the differences in economy and sedentism between hunters and farmers, which are held responsible for differences in population growth of the two types of communities, were much reduced during the time in question, removing the rationale for 'demographic explosion' and 'the growth-migration cycle' (Cavalli-Sforza 1997, 386).

3. As a non-geneticist, I am all too aware of my own incomplete understanding of methodological issues involved as well as of the implications for the interpretation of broader patterns of human behaviour. My critique should therefore be viewed as a series of questions posed on behalf of archaeological community (or at least those as confused as I), rather than criticisms based on clear understanding of the issues. I apologize in advance for any errors and naïve assumptions committed in the course of this exercise.

4. For example, Richards and his co-workers:
   Of the three models for the spread of agriculture outlined earlier, our interpretation favours the pioneer colonization model whereby there was selective penetration by fairly small groups of Middle Eastern agriculturists of a Europe numerically dominated by the descendants of the original Palaeolithic settlements. The ensuing conversion of this population from a hunter-gatherer-fishing economy to one based on agriculture would then have been achieved by technology transfer rather than large-scale population

(Richards et al. 1996, 197).

Or, 'It is not true to say that the farmers overwhelmed the indigenous hunter-gatherers . . .' Most of us who call ourselves Europeans will have had ancestors who were here long before the advent of farming' (Hickey, in Shreeve 1999, 75).

5. The effect of aging of the reproducing population will differ between men and women. Although mutation rates do increase very rapidly as men get older, this effect is not as marked in females, because their reproductive span is limited by the menopause (Chamberlain pers. comm.).

6. Incremental palimpsest of directional small-scale migrations over millennia may provide one explanation for the first principal component of genetic variability noted by Cavalli-Sforza (Ammerman & Cavalli-Sforza 1984; Cavalli-Sforza et al. 1994), provided that source populations in the Near East overlapped sufficiently in genetic composition to contribute to the aggregate of genetic traits expressed in the first principal component. Europe, after all, is a peninsular extension of Asia. Freedom of movement is restricted by seas in the south, west and north, and once the shores of the Atlantic were reached, there was nowhere else to go — until very recently. The first principal component may therefore reflect graduated immigration of small groups of people over millennia as they entered the continent from southeast, the one of two major routes into Europe. The Neolithic farmers may have formed one contributory element in this process; other later prehistoric and historical migrations created the palimpsest. Mixing with the local population and the repetition of the pattern over the millennia would reduce the local variation in the pattern and graduate the percentage representation of the first component from southeast to the northwest, producing 'a very clear genetic gradient of nearly circular shape' observed by Cavalli-Sforza (1997, 387).

Similarly, recent DNA studies on domestic cattle clearly identify the origin of cattle domestication in the Near East and the subsequent dispersal of domestic cattle into Europe from there (with possible secondary imput from North Africa). But such dispersal could be achieved through exchange and contact, particularly as cattle were invested with emblematic symbolism of high social value. The variation in genetic patterning of cattle across Europe allows for several variables being active in its dispersal, including genetic drift, climatic change, and possible genetic contribution from the local wild aurochsen stock to some coastal and insular populations (Bailey et al. 1996; Bradley 1997; Bradley Chapter 38).

References

Alexander, J., 1978. Frontier studies and the earliest farmers in Europe, in *Social Organisation and Settlement*, eds. D. Green, C. Haselgrove & M. Spriggs. (British Archaeological Reports International Series 47.) Oxford: BAR, 13–29.

Ammerman, A.J., 1989. On The Neolithisation in Europe: a comment on Zvelebil and Zvelebil (1988). *Antiquity* 63, 162–5.

Ammerman, A.J. & L.L. Cavalli-Sforza, 1984. *The Neolithic Transition and the Genetics of Population in Europe*. (Addison-Wesley Module in Anthropology 10.) Princeton (NJ): Princeton University Press.

Amorin, A., 1999. Archaeogenetics. *Journal of Iberian Archaeology* 1, 15–25.

Andersen, S.H., 1975. Ringkloster: En jysk indlands-boplats med Ertebollekultur. *Kuml* 1973–74, 11–108.

Andersen, S.H., 1987. Tybring Vig: a submerged Ertebølle settlement in Denmark, in *European Wetlands in Prehistory*, eds. J.M. Coles & A.J. Lawson. Oxford: Clarendon Press, 253–81.

Anthony, D.W., 1990. Migration in archaeology: the baby and the bathwater. *American Anthropologist* 92, 895–914.

Armit, I. & W. Finlayson, 1992. Hunters-gatherers transformed: the transition to agriculture in northern and western Europe. *Antiquity* 66, 664–76.

Arnaud, J.M., 1982. Neolithique ancien et processus de néolithisation dans le sud du Portugal. *Archéologie en Languedoc*, no. special, Actes du Colloque International de Prehistoire, 29–48.

Auban, J.B., 1997. Indigenism and migrationism: the neolithization of the Iberian Peninsula. *Porocilo* XXIV, 1–18.

Aurenche, O. & J. Cauvin (eds.), 1989. *Neolithisations*. (British Archaeological Reports International Series 516.) Oxford: BAR.

Bailey, J.F., M.B. Richards, V.A. Macaulay, I.B. Colson, I.T. James, D.G. Bradley & R.E.M. Hedges, 1996. Ancient DNA suggests a recent expansion of European cattle from a diverse wild progenitor species. *Proceedings of the Royal Society of London* Series B 263, 1467–73.

Bandelt H-J. & P. Forster, 1997. The myth of hunter-gatherer mismatch distributions. *American Journal of Human Genetics* 61, 980–83.

Bandelt, H-J., P. Forster, B.C. Sykes & M.B. Richards, 1995. Mitochondrial portraits of human populations using median netwroks. *Genetics* 141, 743–53.

Barbujani, G. & R. Sokal, 1990. Zones of sharp genetic change in Europe are also linguistic boundaries. *Proceedings of the National Academy of Sciences of the USA* 87, 1816–19.

Barclay, G.J., 1997. The Neolithic, in Edwards & Ralston (eds.), 127–50.

Barker, G., 1985. *Prehistoric Farming in Europe*. Cambridge: Cambridge University Press.

Barnett, W.K., 1995. Putting the pot before the horse: early pottery and the Neolithic transition in the western Mediterranean, in Barnett & Hoopes (eds.), 79–88.

Barnett, W.K. & J. Hoopes (eds.), 1995. *The Emergence of Pottery*. Washington (DC): Smithsonian Institution Press.

Barrett, J.C., 1994. *Fragments from Antiquity: an Archaeology*

of Social Life in Britain, 2900–1200 BC. Oxford: Blackwell.
Barth, F. (ed.), 1969. *Ethnic Groups and Boundaries.* Boston (MA): Little, Brown.
Barth, F., 1994. Enduring and emerging issues in the analysis of ethnicity, in *The Anthropology of Ethnicity: Beyond 'Ethnic Groups and Boundaries'*, eds. H. Vermeulen & C. Govers. Amsterdam: Het Spinhuis, 11–32.
Bateman, R., I. Goddard, R. O'Grady, V.A. Funk, R. Mooi, W.J. Kress & P. Cannell, 1990. Speaking of forked tongues: the feasibility of reconciling human phylogeny and the history of language. *Current Anthropology* 31(1), 1–13.
Bellwood, P., 1996. Phylogeny vs reticulation in prehistory. *Antiquity* 70, 881–90.
Bennike, P., 1985. *Palaeopathology of Danish skeletons: a Comparative Study of Demography, Disease and Injury.* Copenhagen: Akademisk Forlag.
Bergelund, B.E. (ed.), 1990. *The Cultural Landscape During 6000 years in Southern Sweden: the Ystad Project.* (Ecological Bulletin 41.) Copenhagen: Munksgaard.
Binford, L.R., 1983. *In Pursuit of the Past.* London: Thames & Hudson.
Bogucki, P.I., 1988. *Forest Farmers and Stock Herders: Early Agriculture and its Consequences in North-Central Europe.* (New Studies in Archaeology.) Cambridge: Cambridge University Press.
Bogucki, P.I., 1995. The Linear Pottery culture of central Europe: conservative colonists?, in Barnett & Hoopes (eds.), 89–98.
Bogucki, P.I., 1998. Holocene climatic variability and early agriculture in temperate Europe, in Zvelebil *et al.* (eds.), 77–86.
Bradley, D.G., R.T. Loftus, P. Cunninham & D.E. MavHugh, 1997. Genetics and domestic cattle origins. *Evolutionary Anthropology* 6, 79–86.
Bradley, R., 1997. Domestication as a state of mind. *Analecta Praehistorica Leidensia* 29, 13–18.
Bradley, R., 1998. Interpreting enclosures, in *Understanding the Neolithic of Northwestern Europe*, eds. M. Edmonds & C. Richards. Glasgow: Cruithne Press, 188–203.
Bujda, M., 1993. Neolitizacija Evrope: Slovenska Perspectiva. [The Neolithization of Europe: Slovenian Aspect]. *Prorocilo* 21, 163–93.
Calafell, F. & J. Betranpetit, 1993. The genetic history of the Iberian peninsula: a simulation. *Current Anthropology* 34, 735–45.
Case, H., 1969. Neolithic explanations. *Antiquity* 43, 176–86.
Cauvin, J., 1994. *Naissance des divinités. Naissance de l'agriculture.* Paris: CNRS.
Cavalli-Sforza, L.L., 1991. Genes, peoples and languages. *Scientific American* 263, 72–8.
Cavalli-Sforza, L.L., 1997. Genetic and cultural diversity in Europe. *Journal of Anthropological Research* 53, 383–404.
Cavalli-Sforza, L.L. & F. Cavalli-Sforza, 1995. *The Great Human Diasporas: the History of Diversity and Evolution.* New York (NY): Addison-Wesley.
Cavalli-Sforza, L.L. & E. Minch, 1997. Paleolithic and Neolithic lineages in the European mitochondrial gene pool. *American Journal of Human Genetics* 61, 247–51.
Cavalli-Sforza, L.L. & A. Piazza, 1993. Human genomic diversity in Europe: a summary of recent research and prospects for the future. *European Journal of Human Genetics* 1, 3–18.
Cavalli-Sforza, L.L., P. Menozzi & A. Piazza, 1994. *The History and Geography of Human Genes.* Princeton (NJ): Princeton University Press.
Chapman, J., 1994. The origins of farming in south east Europe. *Préhistoire Européenne* 6, 133–56.
Chapman, J. & P. Dolukhanov, 1992. The baby and the bathwater: pulling the plug on migrations. *American Anthropologist* 94, 169–75.
Chapman, J. & P. Dolukhanov (eds.), 1993. *Cultural Transformations and Interactions in Eastern Europe.* Aldershot: Avebury Press.
Chapman, J. & J. Muller, 1990. Early farmers in the Mediterranean Basin: the Dalmatian evidence. *Antiquity* 64, 127–34.
Childe, V.G., 1925. *The Dawn of European Civilisation.* 4th edition. London: Kegan Paul.
Childe, V.G., 1957. *The Dawn of European Civilisation.* 6th edition. London: Routledge and Kegan Paul.
Clark, G.A., 1997. Multivariate pattern searches, the logic of inference, and European prehistory: a comment on Cavalli-Sforza. *Journal of Anthropological Research* 54(3), 406–11.
Clarke, D.L., 1968. *Analytical Archaeology.* London: Methuen.
Clarke, D.L. (ed.), 1972. *Models in Archaeology.* London: Methuen.
Clarke, D.L., 1976. Mesolithic Europe: the economic basis, in *Problems in Economic and Social Archaeology*, eds. I. Sieveking, I.J. Longworth & K.E. Wilson. London: Duckworth, 449–81.
Dennell, R., 1983. *European Economic Prehistory.* London: Academic Press.
Dennell, R., 1985. The hunter-gatherer/agricultural frontier in prehistoric temperate Europe, in Green & Perlman (eds.), 113–40.
Dennell, R., 1992. The origin of crop agriculture in Europe, in *The Origins of Agriculture: an International Perspective*, eds. C.W. Cowan & P.J. Watson. Washington (DC): Smithsonian Institution Press, 71–100.
Dolukhanov, P.M., 1979. *Ecology and Economy in Neolithic Eastern Europe.* London: Duckworth.
Dolukhanov, P.M., 1986. The late Mesolithic and the transition to food production in Eastern Europe, in Zvelebil (ed.) 1986b, 109–20.
Dolukhanov, P.M., 1993. Foraging and farming groups in north-eastern and north-western Europe: identity and interaction, in Chapman & Dolukhanov (eds.), 122–45.
Dolukhanov, P.M., 1996. The Mesolithic/Neolithic transi-

tion in Europe: the view from the East. *Prorocilo* XXIII, 49–60.

Edwards, K.J. & I.B.M. Ralston (eds.), 1997. *Scotland: Environment and Archaeology, 8000 BC–AD 1000*. Chichester: Wiley.

Ehret, C., 1988. Language change and the material correlates of language and ethnic shift. *Antiquity* 62, 564–74.

Eriksen, T.H., 1993. *Ethnicity and Nationalism: Anthropological Perspectives*. London: Pluto Press.

Evison, M., 1999. Perspectives on the Holocene in Britain: human DNA. *Quaternary Proceedings* 7, 615–23.

Evison, M., N.R.J. Fieller & D.M. Smillie, 1999. Ancient HLA: a preliminary survey. *Ancient Biomolecules* 3, 1–28.

Fewster, K., 1996. Interaction Between Basarwa and Bamangwato in Botswana: an Ethnoarchaeological Approach to the Study of the Transition to Agriculture in the West Mediterranean. Unpublished Ph.D. dissertation, University of Sheffield.

Finlayson, B. & K.J. Edwards, 1997. The Mesolithic, in Edwards & Ralston (eds.), 109–26.

Gebauer, A.B. & T.D. Price (eds.), 1992. *Transitions to Agriculture in Prehistory*. (Monographs in World Archaeology 4.) Madison (WI): Prehistory Press.

Graves-Brown, P., S. Jones & C. Gamble (eds.), 1996. *Identity and Archaeology: the Construction of European Communities*. London: Routledge.

Green, S. & S.M. Perlman (eds.), 1985. *The Archaeology of Frontiers and Boundaries*. New York (NY): Academic Press.

Gregg, S., 1988. *Foragers and Farmers*. Chicago (IL): University of Chicago Press.

Gronenborn, D., 1990. Mesolithic/Neolithic interactions: the lithic industry of the earliest Bandkeramik cultures — site at Friedburg-Bruchenbrüchen, West Germany, in *Contributions to the Mesolithic in Europe*, eds. P.M. Vermeersch & P. van Peer. Leuven: Leuven University Press, 173–82.

Gronenborn, D., 1998. Altestbandkeramische Kultur, La Hoguette, Limburg, and . . . What else?: contemplating the Mesolithic–Neolithic transition in southern central Europe. *Documenta Praehistorica* XXV, 189–202.

Guillaine, J., 1976. The earliest Neolithic in the West Mediterranean: a new appraisal. *Antiquity* 53, 22–30.

Guillaine, J. & C. Manen, 1995. Contacts sud-nord au Neolithique ancien: temoignages de la grotte de Gazel en Languedoc, in *Le Néolithique Danubien et ses Marges entre Rhin et Seine*, ed. C. Jeunesse. Strasbourg: Cahiers de l'Association pour la Promotion de la Recherche Archeologique en Alsase, 301–11.

Hammel, G., 1996. Demographic constraints in population growth of early humans. *Human Nature* 7(3), 217–55.

Harris, D. (ed.), 1996. *The Origin and Spread of Agriculture and Pastoralism in Eurasia*. London: UCL Press, 346–63.

Harrison, G.G., 1975. Primary adult lactase deficiency: a problem in anthropological genetics. *American Anthropologist* 77, 812–35.

Hassan, F., 1975. Determination of the size, density, and growth rate of hunting-gathering populations, in *Population Ecology and Social Evolution*, ed. S. Polgar. Chicago (IL): Mouton Publishers.

Hassan, F., 1978. Demographic archaeology, in *Advances in Archaeological Method and Theory*, vol. l, ed. M. Schiffer. New York (NY): Academic Press, 49–103.

Hayden, B., 1990. Nimrods, piscators, pluckers and planters: the emergence of food production. *Journal of Anthropological Archaeology* 9, 31–69.

Hodder, I. (ed.), 1978. *The Spatial Organisation of Culture*. London: Duckworth.

Hodder, I., 1999. *The Archaeological Process: an Introduction*. Oxford: Blackwell.

Izagirre, N. & C. de la Rua, 1999. An mtDNA analysis in ancient Basque populations: implications for haplogroup V as a marker for a major Palaeolithic expansion from southwest Europe. *American Journal of Human Genetics* 65, 199–207.

Jackes, M., D. Lubell & C. Maiklejohn, 1997. On physical anthropological aspects of the Mesolithic–Neolithic transition in the Iberian Peninsula. *Current Anthropology* 38, 839–46.

Janik, L., 1998. The appearance of food producing societies in the southeastern Baltic region, in *Agricultural Frontier and the Transition to Farming in the Baltic*, eds. L. Domanska, R. Dennell & M. Zvelebil. Sheffield: Sheffield Academic Press, 237–44.

Jenkins, R., 1997. *Re-Thinking Ethnicity*. London: Routledge.

Jones, S., 1997. *The Archaeology of Ethnicity*. London: Routledge.

Keeley, L.H., 1992. The introduction of agriculture to the western North European Plain, in Gebauer & Price (eds.), 81–97.

Kretchmer, N., 1972. Lactose and lactase. *Scientific American* 227(4), 71–8.

Laan, M. & S. Pääbo, 1997. Demographic history and linkage disequilibrium in human populations. *Nature Genetics* 17, 435–8.

Lalueza Fox, C., 1996. Physical anthropological aspects of the Mesolithic–Neolithic transition in the Iberian Peninsula. *Current Anthropology* 37, 689–95.

Larsson, L., 1990. The Mesolithic of southern Scandinavia. *Journal of World Prehistory* 4(3), 257–91.

Lewin, R., 1988a. Conflict over DNA clock results. *Science* 241, 1598–600.

Lewin, 1988b. DNA clock conflict continues. *Science* 241, 1756–9.

Lewthwaite, J., 1986. The transition to food production: a Mediterranean perspective, in Zvelebil (ed.) 1986b, 53–66.

Lichardus J. & M. Lichardus-Itten (eds.), 1985. *La Protohistoire de l'Europe: Le Néolithique et le Chalcolithique*. Paris: P.U.F.

Lunning, J., U. Kloos & S. Albert, 1989. Westliche Nachbarn der bandkeramischer Kultur: La Hoguette und Limburg. *Germania* 67, 355–93.

McCracken, R., 1971. Lactase deficiency: an example of dietary evolution. *Current Anthropology* 12, 479–577.

MacEachern, S., 2000. Genes, tribes and African history. *Current Anthropology* 41(4), 357–84.

Mallory, J.P. 1989. *In Search of the Indo-Europeans: Language, Archaeology and Myth*. London & New York (NY): Thames and Hudson.

Meiklejohn, C. & M. Zvelebil, 1991. Health status of European populations at the agricultural transition and the implications for the adoption of farming, in *Health in Past Societies: Biocultural Interpretations of Human Skeletal Remains in Archaeological Contexts*, eds. H. Bush & M. Zvelebil. (British Archaeological Reports International Series 567.) Oxford: BAR, 129–45.

Midgley, M., 1992. *TRB Culture*. Edinburgh: Edinburgh University Press.

Moore, J.A., 1985. Forager/farmer interactions: information, social organization, and the frontier, in Green & Perlman (eds.), 93–112.

Moore, J.H, 1994. Putting anthropology back together again: the ethnogenetic critique of cladistic theory. *American Anthropologist* 96, 925–48.

Neustupný, E., 1982. Prehistoric migrations by infiltration. *Archeologické rozhledy* 34, 278–93.

Persson, O. & E. Persson, 1984. *Anthropological Report on the Mesolithic Graves from Skateholm, Southern Sweden: Excavation Seasons 1980–1982*. (Report Series 21.) Lund: University of Lund, Institute of Archaeology.

Piggot, S., 1965. *Ancient Europe*. Edinburgh: Edinburgh University Press.

Pluciennik, M., 1996. Genetics, archaeology and the wider world. *Antiquity* 70, 13–14.

Pluciennik, M., 1998. Deconstructing 'The Neolithic' in the Mesolithic–Neolithic transition, in *Social Life and Social Change in North-western Europe*, eds. M.R. Edmonds & C. Richards. Edinburgh: Cruithne Press.

Price, T.D., 1985. Affluent foragers of Mesolithic southern Scandinavia, in Price & Brown (eds.) 1985b, 341–60.

Price, T.D., 1987. The Mesolithic of western Europe. *Journal of World Prehistory* 1, 225–332.

Price, T.D., 1996. The first farmers of southern Scandinavia, in Harris (ed.), 346–63.

Price, T.D. & J.A. Brown, 1985a. Aspects of hunter-gatherer complexity, in Price & Brown (eds.) 1985b, 3–20.

Price, T.D. & J.A. Brown (eds.), 1985b. *Prehistoric Hunter-gatherers: the Emergence of Cultural Complexity*. Orlando (FL): Academic Press.

Price, T.D. & A.B. Gebauer, 1992. The final frontier: foragers to farmers in Southern Scandinavia, in Gebauer & Price (eds.), 97–115.

Radovanovic, I., 1996. Mesolithic/Neolithic contacts: a case of the Iron Gates region. *Porocilo* XXIII, 39–48.

Renfrew, C., 1987. *Archaeology and Language: the Puzzle of Indo-European Origins*. London: Jonathan Cape.

Renfrew, C., 1996. Prehistory and the identity of Europe, or don't let's be beastly to the Hungarians, in Graves-Brown *et al.* (eds.), 125–37.

Renfrew, C., 2000. At the edge of knowability: towards a prehistory of languages. *Cambridge Archaeological Journal* 10(1), 7–34.

Richards, M., H. Côrte-Real, P. Forster, V. Macaulay, A. Demaine, S. Papiha, R. Hedges, H-J. Bandelt & B. Sykes, 1996. Palaeolithic and Neolithic lineages in the European mitochondrial gene pool. *American Journal of Human Genetics* 59, 185–203.

Richards, M., V.A. Macaulay, H-J. Bandelt & B.C. Sykes, 1998. Phylogeography of mitochondrial DNA in western Europe. *Annals of Human Genetics* 62, 241–60.

Rimantiené, R., 1992. Neolithic hunter-gatherers at Šventoji in Lithuania. *Antiquity* 66, 367–76.

Rowley-Conwy, P., 1980. Continuity and Change in the Prehistoric Economies of Denmark, 3700–2300 BC. Unpublished Ph.D. thesis, University of Cambridge.

Rowley-Conwy, P., 1983. Sedentary hunters, the Ertebølle example, in *Hunter-Gatherer Economy in Prehistory*, ed. G.N. Bailey. Cambridge: Cambridge University Press.

Rowley-Conwy, P., 1999. Economic prehistory in southern Scandinavia. *Proceedings of the British Academy* 99, 125–59.

Rowley-Conwy, P. & M. Zvelebil, 1989. Saving it for later: storage by prehistoric hunter-gatherers in Europe, in *Bad Year Economics*, eds. P. Halstead & J. O'Shea. Cambridge: Cambridge University Press, 40–56.

Schwabedissen, H., 1981. Ertebølle/Ellerbek: Mesolithilum oder Neolithikum?, in *Mesolithikum in Europa*, ed. B. Gramsch. Berlin: VEB Deutcher Verlag der Wissenschaften, 129–43.

Semino, O., G. Passarino, A. Brega, M. Fellos & A.S. Santachiara-Benerecetti, 1996. A view of the Neolithic demic diffusion in Europe through two Y chromosome-specific markers. *American Journal of Human Genetics* 59, 964–8.

Shennan, S.J., 1989. *Archaeological Approaches to Cultural Identity*. London: Routledge.

Sherratt, A.G., 1982. Mobile resources: settlement and exchange in early agricultural Europe, in *Ranking, Resources and Exchange: Aspects of the Archaeology of Early European Society*, eds. C. Renfrew & S. Shennan. Cambridge: Cambridge University Press, 13–26.

Sherratt, A.G., 1990. The genesis of megaliths: monumentality, ethnicity and social complexity in Neolithic north-west Europe. *World Archaeology* 22(2), 148–67.

Shreeve, J., 1999. Secrets of the gene. *National Geographic* 196(4), 42–75.

Simoons, F.J., 1979. Dairying, milk use and lactose malabsorbtion in Eurasia: a problem in culture history. *Anthropos* 74, 61–80.

Smith, A.G., 1981. The Neolithic, in *The Environment in British Prehistory*, eds. I.G. Simmons & M. Tooley. London: Duckworth, 125–8, 133–83, 199–209.

Sokal, R., N. Oden, P. Legendre, M-J. Fortin, J. Kim & A. Vador, 1989. Genetic differences among language families in Europe. *American Journal of Physical Anthropology* 90, 393–407.

Sokal, R., N. Oden & C. Wilson, 1991. Genetic evidence for

the spread of agriculture in Europe by demic diffusion. *Nature* 351, 143–5.

Sokal, R., N. Oden & B. Thomson, 1992. Origins of the Indo-Europeans: genetic evidence. *Proceedings of the National Academy of Sciences of the USA* 89, 7669–73.

Sokal, P., N. Oden & B. Thompson, 1998. A Problem with Synthetic Maps. A paper presented in the symposium 'Genetics and Origins of Human Diversity' at the annual Meeting of the American Association of Physical Anthropologists, Salt Lake City, March 1998.

Solberg, B., 1989. The Neolithic transition in southern Scandinavia: internal development or migration? *Oxford Journal of Archaeology* 8(3), 261–96.

Straus, L.G., 1998. The peopling of Europe: A J.A.R. debate. *Journal of Anthropological Research* 54, 399–419.

Terrell, J. & P. Stewart, 1996. The paradox of human population genetics at the end of the twentieth century. *Reviews in Anthropology* 26, 13–33.

Thomas, J., 1988. Neolithic explanations revisited: the Mesolithic–Neolithic transition in Britain and south Scandinavia. *Proceedings of the Prehistoric Society* 54, 59–66.

Thomas, J., 1991. *Rethinking the Neolithic*. Cambridge: Cambridge University Press.

Thomas, J., 1996. The cultural context of the first use of domesticates in continental Central and Northwest Europe, in Harris (ed.), 310–22.

Thomason, S.G. & T. Kaufman, 1988. *Language Contact, Creolization, and Genetic Linguistics*. Berkeley (CA): University of California Press.

Thorpe, I.J., 1996. *The Origins of Agriculture in Europe*. London: Routledge.

Tilley, C., 1994. *A Phenomenology of Landscape*. Oxford: Berg.

Tilley, C., 1996. *An Ethnography of the Neolithic*. Cambridge: Cambridge University Press.

Timofeev, V.I., 1987. On the problem of the Early Neolithic of the East Baltic area. *Acta Archaeologica* 58, 207–12.

Timofeev, V.I., 1990. On the links of the East Baltic Neolithic and the Funnel Beaker culture, in *Die Trichterbeckerkultur*, ed. D. Jankowska. Poznan: Poznan University Press, 135–49.

Timofeev, V.I., 1998. The beginning of the Neolithic in the eastern Baltic', in Zvelebil *et al.* (eds.), 225–36.

Timofeev, V.I., 1999. The east–west relations in the Late Mesolithic and Neolithic Baltic region, in *Beyond Balkanization Poland*, ed. A. Kosko. Poznan: Baltic Pontic Studies, 44–58.

Tomaszewski, A.J., 1988. Foragers, farmers and archaeologists: a comment on B. Olsen's paper 'Interaction between hunter-gatherers and farmers': ethnographical and archaeological perspectives from the viewpoint of Polish archaeology. *Archeologia Polski* 33, 434–40.

Torroni, A., H-J. Bandelt, L. D'Urbano, P. Laherno, P. Moral, D. Sellito, C. Rengo, P. Forster, M-L. Savontaus, B. Bonné-Tamir & R. Scozzari, 1998. MtDNA analysis reveals a major Late Palaeolithic population expansion from southwestern to northeastern Europe. *American Journal of Human Genetics* 62, 1137–52.

van Andel, T.H. & C.N. Runnells, 1995. The earliest farmers in Europe. *Antiquity* 69, 481–500.

Vankina, L.V., 1970. *Torfyanikovaya Stoyanka Sarnate*. Riga: Zinatne.

Vaquer, J., 1989. Innovation et inertie dans le processus de néolithisations en Languedoc occidental, in Aurenche & Calvin (eds.), 187–97.

Vencl, S., 1986. The role of hunting-gathering populations in the transition to farming: a central European perspective, in Zvelebil (ed.) 1986b, 43–52.

Verhart, L. & M. Vansleebenk, 1997. Waste and prestige: the Mesolithic and Neolithic transition in the Netherlands from a social perspective. *Analecta Praehistorica Leidensia* 29, 65–73.

Voytek, B. & R. Tringham, 1989. Rethinking the Mesolithic: the case of south-east Europe, in *The Mesolithic in Europe*, ed. C. Bonsall. Edinburgh: John Donald, 492–500.

Vuorela, I. & T. Lempiäinen, 1988. Archaeobotany of the oldest cereal grain find in Finland. *Annales Botanici Fennici* 25, 33–45.

Whittington, G. & K.J. Edwards, 1997. Climate change, in Edwards & Ralston (eds.), 11–22.

Whittle, A., 1996. *Europe in the Neolithic*. Cambridge: Cambridge University Press.

Willis, K. & K.D. Bennett, 1994. The Neolithic transition — fact or fiction? Palaeoecological evidence from the Balkans. *The Holocene* 4, 326–30.

Willis, K., P. Sümeg, M. Braun, K.D. Bennett & A. Toth, 1998. Prehistoric land degradation in Hungary: who, how and why? *Antiquity* 72, 101–13.

Wobst, M., 1974. Boundary conditions for Palaeolithic social systems: a simulation approach. *American Antiquity* 39, 147–78.

Woodburn, J., 1982. Egalitarian societies. *Man* (N.S.) 17, 431–51.

Zilhão, J., 1993. The spread of agro-pastoral economies across Mediterranean Europe: view from the far west. *Journal of Mediterranean Archaeology* 6, 5–63.

Zilhão, J., 1997. Maritime pioneer colonisation in the Early Neolithic of the west Mediterranean: testing the model against the evidence. *Porocilo* 24, 19–42.

Zvelebil, M., 1981. *From Forager to Farmer in the Boreal Zone*. (British Archaeological Reports, International Series 115.) Oxford: BAR.

Zvelebil, M., 1986a. Postglacial foraging in the forests of Europe. *Scientific American* 254(5), 104–15.

Zvelebil, M. (ed.), 1986b. *Hunters in Transition: Mesolithic Societies of Temperate Eurasia and their Transition to Farming*. Cambridge: Cambridge University Press.

Zvelebil, M., 1989. On the transition to farming in Europe, or what was spreading with the Neolithic: a reply to Ammerman (1989). *Antiquity* 63, 379–83.

Zvelebil, M., 1993. Hunters or farmers? The Neolithic and Bronze Age societies of north-east Europe, in field Academic Press.

Chapter 8

Expected Regional Patterns of Mesolithic–Neolithic Human Population Admixture in Europe based on Archaeological Evidence

Marta Mirazón Lahr, Robert A. Foley & Ron Pinhasi

While it is generally accepted that farming spread into Europe from the Near East, the extent to which this occurred through population movements or cultural diffusion or a combination of the two remains uncertain. One of the primary conditions for determining the role of dispersals versus cultural spread is the scale of the European Mesolithic populations. Here we use conclusions drawn from an archaeological data base analysis (Pinhasi et al. Chapter 6) to generate a population density based model for determining the expected degree of admixture or replacement that occurred across the Mesolithic–Neolithic transition in Europe. Five regional patterns are described: 1) areas with very sparse Mesolithic populations and dense Neolithic, where the Neolithic would have had a major genetic impact on the succeeding gene pool (Anatolia, Balkans); 2) areas with sparse or moderate Mesolithic populations and moderately dense Neolithic settlement, where there would be minor genetic contribution from the Mesolithic populations (Central Europe, North European Plains, northern Italy); 3) areas where both Mesolithic and Neolithic populations were at least moderately dense, and so where there would be expected an equal contribution from the two populations (southern Italy, western Continental Europe); 4) areas where the Neolithic was relatively sparse and the Mesolithic populations significant, and so a greater contribution from the latter would be expected (Britain, Scandinavia); and areas where the Neolithic would have had little or no impact (Alps). This model of regional variation emphasizes the dynamic and geographically specific nature of prehistoric human population distributions.

Recent work in molecular genetics has rekindled interest in the question of the extent to which the spread of farming into Europe involved demic replacement, cultural diffusion, or a mixture of the two processes. The genetic data have provided conflicting evidence, with some analyses suggesting major genetic replacement (Ammerman & Cavalli-Sforza 1971; 1973; 1984; Barbujani *et al.* 1995; Sokal *et al.* 1991), and others showing greater levels of continuity (Richards *et al.* 1996). Other lines of evidence, principally the relationship between the spread of Indo-European languages and agriculture (Renfrew 1987; 1998), while still controversial, certainly support the view that the spread of farming was a major demographic event, but have not contributed significantly to the question of the extent of genetic change. Morphological evidence has also not proven to be conclusive, partly because the crucial remains in terms of spatial and temporal provenance have not been studied in an appropriate comparative framework, and partly because the recent common ancestry of all European and adjacent populations, together with significant Palaeolithic gene flow, have resulted in comparatively small levels of morphological diversity within the continent (Lahr 1996).

Pinhasi *et al.* (Chapter 6) have shown that the quantitative use of archaeological data can show regional patterns of population history during the Holocene. This confirms the work of Housley *et al.* (1997) and Dennell (1985) that the Early Holocene human population history of Europe is dynamic, and involves expansions and contractions of Mesolithic populations as much as of those of Neolithic farmers. In particular, it is clear that the demographic landscape on to which farming expanded was by no means uniform. In some areas, hunter-gatherers were dense and had been present for considerable time; in others populations were sparse; in yet others their expansions occurred at the same time as the Neolithic in other regions of Europe, while in some regions hunter-gatherers seem to have disappeared prior to the appearance of farming.

This complex pattern of Mesolithic occupation has major implications for understanding the genetic contribution of any Neolithic populations that may have spread into Europe. Where there were no hunter-gatherers, there could be no admixture. Where the Mesolithic populations were dense, there was a greater potential. In this chapter we develop a model of potential interactions between hunter-gatherers and farmers based on relative population density, and then explore regional variation across Europe. The aim of the chapter is to provide molecular geneticists with a more realistic set of demographic expectations for estimating the variable patterns of admixture and replacement that occurred with the spread of farming into Europe.

A demographic model for population replacement and genetic admixture in European Early Holocene history

The key insight derived from the archaeological data base is that the distribution of populations across Europe during the Early Holocene is highly dynamic, and that this dynamism is not simply related to the appearance of the Neolithic. In population terms, the major inference is that when farming appeared in Europe, it met a number of very different demographic conditions, which may underlie the differences in the scale, as well as timing, of Neolithic settlements throughout the continent.

In order to try and produce testable models concerning the levels of admixture among successive populations, a general demographic model is necessary. This general model uses the spatio-temporal patterns emerging from the archaeological data base described in Pinhasi *et al.* (Chapter 6) to predict genetic admixture between local hunter-gatherers and incoming farmers. As a simple set of initial assumptions, it can be argued that the probability of admixture will be influenced by the relative sizes of the populations involved. More precisely, our ability to detect admixture will be related to the probability of each of the two populations inter-breeding, which in turn will be at least partly dependent upon relative population size. From this simple assumption we can posit a number of theoretical possibilities (Table 8.1). First, that in regions where there was a very sparse Mesolithic occupation and a very dense Early Neolithic one, the Neolithic population would largely carry the expanding Middle Eastern farmers' genes. Second, that in those regions where the Mesolithic occupation was dense and the Neolithic one sparse, the resulting agricultural population would have been largely composed genetically of local foragers. And third, that in those areas where there was a dense Mesolithic occupation, followed by a dense Neolithic one, the agricultural population would have had more equal levels of genetic contribution from the indigenous groups. The possible interactions are shown in Table 8.1.

Regional variation in Mesolithic and Neolithic populations in Europe
The basic regional demography according to subsistence economy and period is shown in Table 8.2. The information is derived from the available archaeological data for the major regions of Europe, categorizing the level of human population density

Table 8.1. *Predictive model for Mesolithic/Neolithic population interaction.*

Neolithic demography	Mesolithic demography		
	Absent	*Sparse*	*Dense*
Sparse	Replacement/No admixture	Minimal admixture (outcome dependent upon relative population growth rates)	High levels of admixture, resulting in genetic continuity
Dense	Replacement/No admixture	Low levels of admixture	Either replacement or admixture

Table 8.2. *Demographic profiles by period and region based on archaeological data. Pattern derived from Pinhasi et al. (Chapter 6). Thick vertical lines indicate the timing of introduction of farming in the different areas of Europe between 10–5 Kya.*

	Late Palaeolithic	Early, Middle & Late Mesolithic 10–5 Kya	Early, Middle & Late Neolithic 9–5 Kya
Anatolia	No data	Area uninhabited	Moderately dense / No data
Balkans	Very sparse	Very sparse	Moderately dense / Dense
Southern Italy	Very sparse	Moderately dense / Very sparse	Moderately dense / Dense
Northern Italy	Very sparse	Very sparse	Very sparse / Sparse with pockets
Central Europe	Sparse with pockets	Sparse with pockets / Very sparse	Very sparse / Dense
Northeast Plains	Very sparse	Very sparse	Very sparse / Dense
Scandinavia	Very sparse	Sparse with pockets	Very sparse
Alps	Very sparse	Very sparse	Very sparse
Western Europe	Very sparse	Moderately dense / Very sparse	Moderately dense
Southern France	Very sparse	Sparse with pockets	Sparse with pockets
Iberian Peninsula	Very sparse	Sparse with pockets	No data
British Isles	Very sparse	Sparse with pockets / Very sparse	Sparse with pockets

Legend:
- — No data
- Area uninhabited
- Very sparse occupation
- Sparse occupation with occasional pockets of denser occupations
- Moderately dense occupation
- Dense occupation

of foragers and agriculturalists during different periods. This table clearly identifies a number of scenarios of population density across the transition, which underlie the genetic outcome.

From these data the following regional patterns can be discerned.

1. Those parts of Europe where there was only a very sparse Mesolithic and Late Palaeolithic occupation, and where the first Neolithic populations were moderately dense across the entire region, followed by a very dense occupation by farmers and no surviving hunter-gatherers by 5000 BP. This is the case for Anatolia and the Balkans, where the impact, both demographic and genetic, of the Neolithic farmers is thus likely to have been very significant.
2. The second scenario describes three cases, in which the genetic impact of the Neolithic would have been major, but in which small amounts of admixture with local foragers would have been expected. These cases describe those parts of Europe where the Mesolithic and Late Palaeolithic populations were very sparse or concentrated in dense pockets, and where the first Neolithic groups appear as a sparse population that grows significantly in density or as a dense population already. The main aspect of these regions is the fact that foragers do not disappear, and may have increased slightly in number after the establishment of farming in these areas. This is the case of Central Europe, the North European plains and northern Italy.
3. The third scenario describes those areas of Europe where the genetic contribution of immigrant farmers and the local Mesolithic population would have been similar. This would have been the case where a moderately dense foraging population is followed by a moderately dense agricultural one,

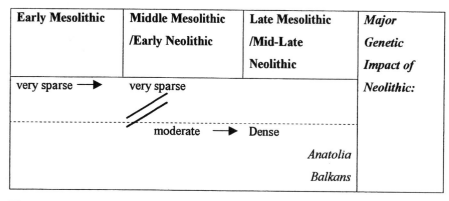

Figure 8.1. *Pattern for a major Neolithic genetic component.*

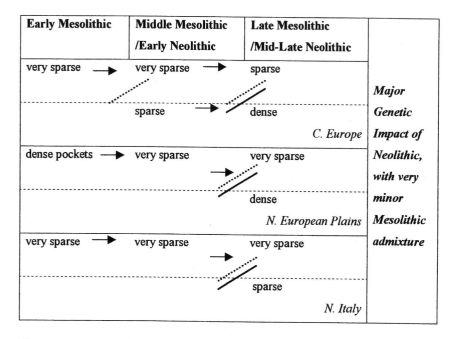

Figure 8.2. *Pattern for a major Neolithic genetic component with a minor Mesolithic contribution.*

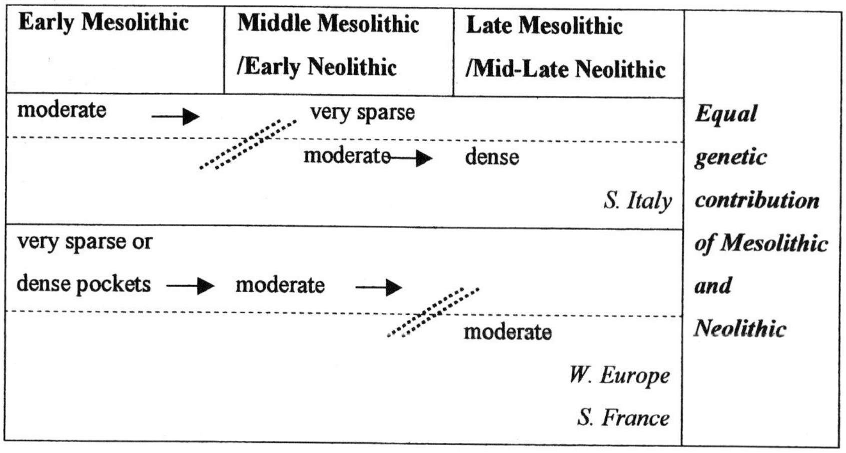

Figure 8.3. *Pattern for an equal Mesolithic and Neolithic genetic component.*

without the survivorship of a hunter-gatherer way of life. This is the case of southern Italy, and western Europe.

4. The fourth case describes those areas where the contribution of the local Mesolithic groups to the resulting population would have been greater than those of the immigrant Neolithic farmers. In these areas, the foraging populations increase after the appearance of farming, while the latter establish a moderately dense occupation in some areas. This is the case of the British Isles and Scandinavia.

5. Finally, there are those areas of Europe where the impact of the Neolithic was virtually none. This is the case of the Alps, where in the period between 7000 and 5000 BP the number of foraging sites increased, while there is no evidence for farming communities at all.

Implications for genetic studies

Figure 8.6 shows a map of Europe which summarizes the regional patterns and expectations discussed above. The darker the shading, the greater the expected Neolithic genetic contribution if the archaeologically derived demographic model is correct.

Over two decades ago, Menozzi and colleagues (1978) introduced a series of maps of gene frequencies of classical genetic systems across Europe. The map based on the loadings on the first principal component in their statistical analysis displayed a southeast to northwest cline which is compatible with the model of the spread of farming into Europe from the Near East. Since then, similar results have been obtained by Sokal and colleagues (1991) in their study of 26 genetic systems from 3373 localities in Europe, using spatial autocorrelation statistics, by Semino and colleagues (1996) in the frequencies of two distinct Y-chromosome markers, and by Chikhi and colleagues (1998) based on nuclear DNA variations.

The existence of a cline from southeast to northwest Europe was originally interpreted as being the outcome of the 'demic diffusion' of farmers from the Near East, who would have absorbed along their migration route, many of the local Mesolithic hunter-gatherer bands (Cavalli-Sforza *et al.* 1993; Cavalli-Sforza 1996). The nature of these studies, drawing their inferences from the pattern found in the living population of Europe, precluded the establishment of the routes and timing of the proposed Neolithic diffusion across the continent. These genetic studies also did not include in their assessment of the impact of such a demic diffusion the varying population densities of the local Mesolithic population, either prior, during or after the 'wave of advance'.

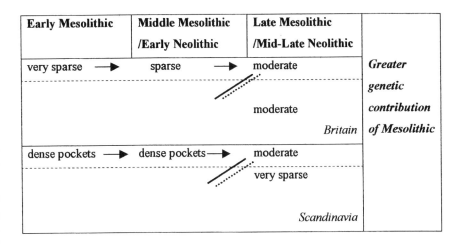

Figure 8.4. *Pattern for a greater genetic contribution from the Mesolithic populations.*

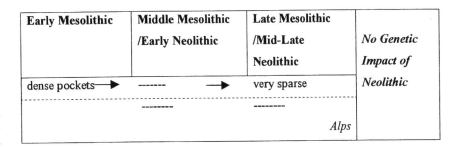

Figure 8.5. *Pattern for no genetic impact for the Neolithic.*

Barbujani and colleagues (1995) took a more theoretical approach to the study of the spread of farming into Europe. They conducted a simulation study that compared five models of microevolution in European populations. The simulations were based on a combination of possible factors explaining the observed clinal distribution of genetic systems across the continent. Their first model was based on the assumption that European populations evolved under conditions of isolation by distance (IBD). Under such a model, current differentiation in allele frequencies among European populations would be simply the outcome of random factors such as genetic drift, dispersal and isolation. Their second model (OAG — isolation by distance plus effects of the origins of agriculture) assumed that local hunter-gatherer populations were replaced without admixture by the incoming Neolithic farmers. The third model (OAC — isolation by distance plus the effects of the origins of agriculture and cultural transmission) adds the hypothetical effects of cultural transmission between the local hunter-gatherer bands and the Neolithic farmers. The underlying assumption in this case is that in certain regions, local hunter-gatherers adopted farming, and thus some of their genes would have contributed to the gene pool of the Neolithic population. This is equivalent to the 'demic diffusion' model described above. The fourth model (ATC — as above plus archaeological time constraints) added the effects of archaeological time constraints by incorporating a temporal sequence of when farmers arrived in the various localities in Europe. The fifth and last model (GIM — as above plus effects of later Neolithic migrations) adds to the former ones the effect of Late Neolithic migrations of the proto-Indo-European people. Barbujani and colleagues (1995) found that the fit of the classical genetic data to simpler models, such as OAG and OAC, is better than to the more complex ones. Thus, in their view, the addition of archaeological data (radiocarbon dates) and Late Neolithic population movements added nothing to the understanding of

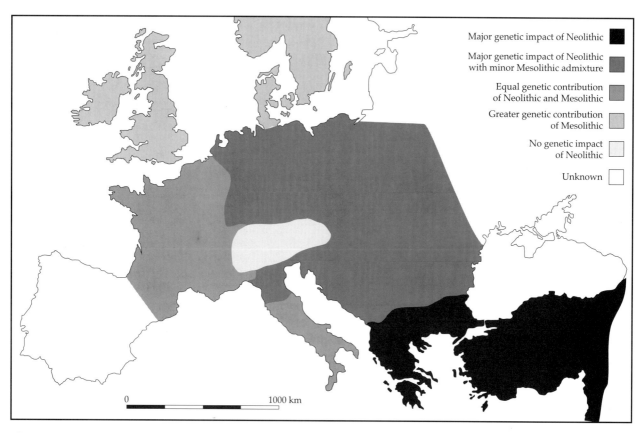

Figure 8.6. *Summary map of Europe showing the expected genetic contributions of Neolithic and Mesolithic populations based on the demographic model and archaeological evidence.*

the pattern of genetic differences in Europe today. The OAC model, which is consistent with the 'demic diffusion' model of Ammerman & Cavalli-Sforza (1971; 1984), and the OAG model, that argues for no admixture with local populations, show similar congruence with actual genetic data. Barbujani and colleagues (1995) conclude that this may be explained by the fact that a series of successive founder effects and admixture between geographically separated farming groups would have affected the clinal distribution of genes in a similar manner had admixture between expanding farmers and local hunter-gatherers taken place.

In contrast, Richards *et al.* (1996) have used mtDNA data to construct a somewhat different model. The mtDNA network analyses suggest a much less spatially structured pattern of European genetic variation, with less evidence for a geographical gradient from the southwest to the northeast. Furthermore, their analyses suggest that there is little evidence for a coalescence of mitochondrial lineages at the time of Neolithic expansions across Europe, and that the existing Mesolithic and Palaeolithic populations made significant contributions to the extant gene pool.

The analyses of archaeologically inferred population densities presented here suggest that neither extreme is likely to be correct, but that regional variation should be expected. The impact of the Neolithic is dependent upon three factors. The first is the density of the Mesolithic populations. Significant contributions from Mesolithic populations can only occur where there were sufficient numbers of such peoples, regardless of the size of any incoming farmers. It is this regional demographic variation in Mesolithic populations that has not been taken into account in most previous work. Pinhasi *et al.* (Chapter 6), and the models developed here indicate that it was only the north and northwestern parts of the continent and in parts of southern/western France and southern Italy that such dense populations existed. Elsewhere, the dynamics of the Mesolithic populations during the earlier parts of the Holocene make a significant contribution of Mesolithic genes to the present gene pool less likely.

The second factor is the scale of the Neolithic expansion. Large and rapid expansions should lead to a genetic signature that is distinctive. This would be expected primarily in the Balkans and Anatolia, and to a lesser extent across Central Europe, particularly the Danube basin. If the genetic data do show continuity, then this would imply that any incoming population was already closely related to the populations of Europe, and thus perhaps show evidence for a common Later Pleistocene population in the Eastern parts of the Mediterranean, with continuous gene flow leading to older estimated coalescence ages.

The final factor is the nature of any interaction between Mesolithic and Neolithic populations. These would ultimately be dependent upon social and ecological factors, which have not been considered here. Two considerations are essential. One of these is whether any interactions are competitive or co-operative. In either case this may lead to admixture, either through exchange of women in particular between groups, or through the absorption of individuals from displaced groups into the farming communities. The other is that ethnographic experience suggests that female mobility between such groups is likely to be higher than that of males (Foley & Lahr in press). This differential sex bias might be the primary cause of the differences between the evidence derived from mtDNA and that from other gene systems. In the end it may be that Mesolithic women are better represented in the modern populations of Europe than Mesolithic men.

Finally, it is worth stressing that this model takes no account of subsequent movements of populations in Europe. We know from both historical and archaeological evidence that there have been substantial population movements over the last five millennia, and thus there is a question mark to be placed over any assumptions about the geographical stability of populations derived from Neolithic expansions.

Conclusions

In this chapter we have argued that the nature of genetic continuity between Mesolithic and Neolithic populations of Europe depends upon demographic conditions prevailing at the time. The archaeological record, as discussed by Pinhasi *et al.* (Chapter 6) can provide information about these conditions. The archaeological evidence shows that the Mesolithic populations were very sparse in some parts of Europe, and considerably more densely packed in others. Any arriving farmers would thus have met very different competitive and interactive conditions, with divergent potential for gene flow. Furthermore it is clear from the archaeological record that the Neolithic expansion also occurred at very different population densities. Farmers, whatever their genetic affinities, were much denser in the Balkans and Anatolia, and parts of Central Europe, but were never, during the Neolithic, very abundant to the north and west. These archaeologically derived models can be used to provide better expectations for the nature of any replacement or admixture that may have occurred at the time of the first farming economies in Europe.

References

Ammerman, A.J. & L.L. Cavalli-Sforza, 1971. Measuring the rate of spread of early farming in Europe. *Man* (N.S.) 6, 674–88.

Ammerman, A.J. & L.L. Cavalli-Sforza, 1973. A population model for the diffusion of early farming in Europe, in *The Explanation of Culture Change*, ed. C. Renfrew. London: Duckworth, 343–57.

Ammerman, A.J. & L.L. Cavalli-Sforza, 1984. *The Neolithic Transition and the Genetics of Populations in Europe*. Princeton (NJ): Princeton University Press.

Barbujani, G., R.R. Sokal & N.L. Oden, 1995. Indo-European origins: a computer-simulation test of five hypotheses. *American Journal of Physical Anthropology* 96, 109–32.

Cavalli-Sforza, L.L., 1996. The spread of agriculture and nomadic pastoralism: insights from genetics, linguistics, and archaeology, in *The Origins and Spread of Agriculture and Pastoralism in Eurasia*, ed. D. Harris. London: University College London, 51–69.

Cavalli-Sforza, L.L., P. Menozzi & A. Piazza, 1993. Demic expansions and human evolution. *Science* 259, 639–46.

Chikhi, L., G. Destro-Bisol, V. Pascall, V. Baravelli, M. Dobosz & G. Barbujani, 1998. Clinal variation in the nuclear DNA of Europeans. *Human Biology* 70, 643–57.

Dennell, R.W., 1985. *European Economic Prehistory: a New Approach*. London: Academic Press.

Foley, R.A. & M.M. Lahr, in press. Anthropological and ecological parameters for modelling human genetics, in *Genes, Fossils and Behaviour*, ed. P. Donnelly & R.A. Foley. Brussels: NATO Advanced Seminar Publications.

Housley, R.A., C.S. Gamble, M. Street & P. Pettitt, 1997. Radiocarbon evidence for the Late Glacial human recolonisation of northern Europe. *Proceedings of the Prehistoric Society* 63, 25–54.

Lahr, M.M., 1996. *The Evolution of Modern Human Diversity*. Cambridge: Cambridge University Press.

Menozzi, P., A. Piazza & L.L. Cavalli-Sforza, 1978. Synthetic maps of human gene frequencies in Europeans. *Science* 201, 786–92.

Renfrew, C., 1987. *Archaeology and Language: the Puzzle of the Indo-European Origins*. London: Penguin Books.

Renfrew, C., 1998. World linguistic diversity and farming dispersals, in *Archaeology and Language*, vol. I: *Theo-*

retical and Methodological Orientations, eds. R. Blench & M. Spriggs. London: Routledge, 82–90.

Richards, M., H. Côrte-Real, P. Forster, V. Macaulay, H. Wilkinson-Herbots, A. Demaine, S. Papiha, R. Hedges, H-J. Bandelt & B. Sykes, 1996. Paleolithic and Neolithic lineages in the European mitochondrial gene pool. *American Journal of Human Genetics* 59(1), 185–203.

Semino, O., G. Passarino, A. Brega, M. Fellous & A.S. Santachiara-Benerecetti, 1996. A view of the Neolithic demic diffusion in Europe through two Y chromosome-specific markers. *American Journal of Human Genetics* 59(4), 964–8.

Sokal, R.R., N.L. Oden & C. Wilson, 1991. Genetic evidence for the spread of agriculture in Europe by demic diffusion. *Nature* 351, 143–4.

Chapter 9

Processes of Culture Change in Prehistory: a Case Study from the European Neolithic

Mark Collard & Stephen Shennan

It has been claimed that ethnogenesis is far more important than phylogenesis in generating the material culture patterns recorded by archaeologists. We have tested this assumption by applying phylogenetic techniques from biology to assemblages of pottery from Neolithic sites in the Merzbach valley, Germany. Our results indicate that both ethnogenesis and phylogenesis were involved in the generation of the pottery assemblages. This suggests that archaeologists should not simply assume that ethnogenesis is the process responsible for the assemblages they study. Rather, they need to resolve the issue empirically on a case by case basis.

It is obvious — indeed it is so obvious that it bears repeating — that identifying the processes of culture change, and determining their relative contribution to the patterns in the archaeological record, are crucial for an understanding of prehistory, and also for linking archaeological data to genetic and linguistic patterns. Recent discussions regarding culture change have focused on two processes that J.H. Moore (1994a,b) has termed 'phylogenesis' and 'ethnogenesis'. In phylogenesis a new cultural assemblage is the result of descent with modification from an ancestral assemblage, whereas in ethnogenesis a new cultural assemblage arises through the blending of elements of two or more contemporaneous assemblages. Currently, most authors consider ethnogenesis to be far more important than phylogenesis in the generation of cultural assemblages. However, most assessments of the relative importance of phylogenesis and ethnogenesis in human cultural affairs have so far been theoretical and/or qualitative (e.g. Kirch & Green 1987; Terrell 1988; Moore 1994a,b; Rowlands 1994; Dewar 1995; Bellwood 1996; Boyd *et al.* 1997; Terrell *et al.* 1997); only a few attempts have been made to address the problem in a quantitative fashion (e.g. Welsch *et al.* 1992; Mace & Pagel 1994; Moore & Romney 1994; Guglielmino *et al.* 1995; Holden & Mace 1999). Moreover, most of the work carried out to date has focused on ethnographic data rather than archaeological evidence (e.g. Kirch & Green 1987; Durham 1992; Welsch *et al.* 1992; Mace & Pagel 1994; Moore 1994a,b; Moore & Romney 1994; Guglielmino *et al.* 1995; Holden & Mace 1999).

How can we assess the relative contribution of phylogenesis and ethnogenesis to the patterns in the archaeological record? In this chapter, we argue that this archaeological problem is related to problems that have been successfully confronted by biologists, linguists and stemmatists. We then present a case study, in which we use a technique that was developed to tackle the aforementioned biological problem to assess the roles of phylogenesis and ethnogenesis in producing the patterns of variation in a group of pottery assemblages from the Central European Neolithic. Lastly, we consider the implications of our findings for current archaeological approaches to evidence for culture change.

Related problems in other disciplines

The problem of determining the relative contribution of phylogenesis and ethnogenesis to the patterns in the archaeological record is, we suggest, related to problems that have been successfully tackled by biologists, linguists and stemmatists. These problems are, respectively, estimating the phylo-

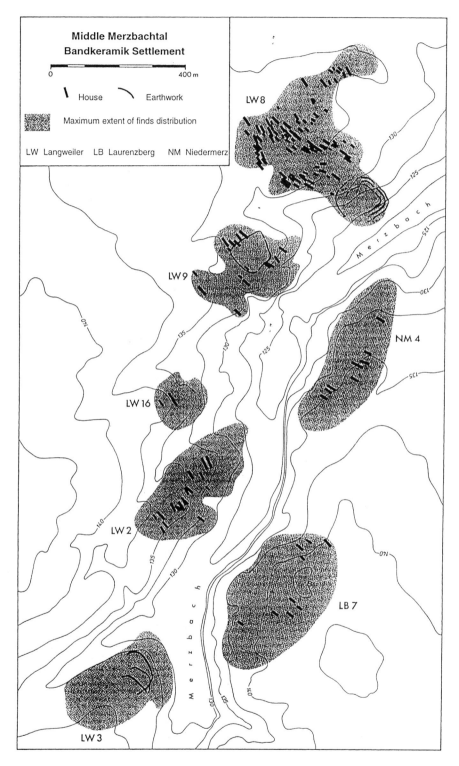

Figure 9.1. *Map of Merzbach sites.*

have in common is that they require the similarities exhibited by a group of taxa to be divided into those that are the result of shared ancestry (homologies) and those that are the result of mechanisms other than shared ancestry (homoplasies).

In biology, linguistics and stemmatics, this task is accomplished by generating a tree-structure which links the taxa in such a way that the number of hypotheses of change required to account for the observed distribution of similarities is minimized. Using this tree-structure it is then possible to classify the similarities as either homologous or homoplasious. Homologous similarities suggest relationships that are compatible with the tree-structure, whereas homoplasious similarities support relationships that conflict with the tree-structure. This procedure, which appears to assume what needs to be demonstrated, can be defended in relation to the principal of parsimony, the methodological injunction which states that explanations should never be made more complicated than is necessary (Sober 1988).

We suggest that the parsimony approach should be adopted in relation to the problem of determining the relative contribution of phylogenesis and ethnogenesis to the patterns in the archaeological record. If a statistically-robust tree-structure can be derived from a group of archaeological assemblages, then phylogenesis can reasonably be inferred to have played a more important role than ethnogenesis in the generation of the assemblages. Conversely, if such a tree-structure cannot be identified, then ethnogenesis can be inferred to be the most important process.

genetic relationships between species (Minelli 1993), delineating the genealogical relationships among languages (Ross 1997), and reconstructing ancient texts (Gjessing & Pierce 1994). What all four problems

Materials and methods

The data set for the case-study was taken from Frirdich (1994). It comprised the frequencies of decorative bands on ceramic vessels from the Linearbandkeramik (LBK) settlements of the Merzbach valley, western Germany (Frirdich 1994).

The settlements, which represent the first farming communities in this part of Europe, consist of groups of longhouses and pits that are scattered along the banks of the Merzbach stream, covering an area of about 3 km^2 (Fig. 9.1; Table 9.1). The number of houses in occupation varies through time, but altogether the settlement sequence covers nearly 500 years, from *c.* 5300 to 4850 BC. A chronological sequence has been defined for the Merzbach settlements on the basis of two different sets of criteria: a detailed stratigraphic and spatial analysis of the sites (Stehli 1994) and a seriation of the pottery (Frirdich 1994). These two sequences have been correlated with one another (Frirdich 1994), a process which involved grouping the seriation intervals into the independently defined phases. Because the whole area was excavated prior to its destruction by lignite mining, and there is no evidence for large-scale taphonomic bias, the settlement picture can be considered to be undistorted.

It appears that there was an initial founding settlement in the area (Langweiler 8) which was occupied more or less continuously from beginning to end, and that subsequently new settlements were founded, which often had gaps in occupation. It is usually assumed that the new settlements were established by members of existing settlements in the micro-region, although it has been suggested that one (Langweiler 2) represents a movement of people into the area from somewhere outside (Mattheusser 1994). The Merzbach micro-region is so small that distance cannot have been an obstacle to extensive interaction between the communities, and it seems reasonable to assume that there were extensive kinship links between them.

The vessels are broadly ovoid in shape and take the form of deep bowls. The body of each vessel is decorated with a series of bands made up of incised lines, strokes or indentations (Fig. 9.2). The decoration is highly distinctive and stylized, comprising a variety of distinct but clearly related motifs

Table 9.1. *Features at sites from which Merzbach pottery assemblages are derived, according to Frirdich's maps.*

Phase	Site	Feature	Phase	Site	Feature
6	Laurenzberg 7	Houses and pits	11	Langweiler 8	Houses and pits
6	Langweiler 2	No feature shown	11	Langweiler 9	Houses and pits
6	Langweiler 8	Houses and pits	11	Niedermerz 4	Houses and pits
6	Langweiler 9	Houses and pit	12	Laurenzberg 7	House and pits
6	Langweiler 16	House and pits	12	Langweiler 2	Houses and pits
7	Laurenzberg 7	House and pits	12	Langweiler 8	Houses and pits
7	Langweiler 2	Houses and pits	12	Langweiler 9	Houses and pits
7	Langweiler 8	Houses and pits	12	Langweiler 16	Pits
7	Langweiler 9	No feature shown	12	Niedermerz 4	Houses
7	Langweiler 16	House	13	Laurenzberg 7	House and pits
8	Laurenzberg 7	House and pits	13	Langweiler 2	Houses and pits
8	Langweiler 2	No feature shown	13	Langweiler 8	Houses and pits
8	Langweiler 8	Pits	13	Langweiler 9	Enclosure and pits
8	Langweiler 9	Houses and pits	13	Langweiler 16	No structure shown
8	Langweiler 16	Pits	13	Niedermerz 4	Houses and pits
9	Laurenzberg 7	House and pits	14	Laurenzberg 7	No structure shown
9	Langweiler 2	Houses and pit	14	Langweiler 2	Houses and pit
9	Langweiler 8	Houses and pits	14	Langweiler 8	Enclosure, houses and pits
9	Langweiler 9	House and pits	14	Langweiler 9	Pits
10	Laurenzberg 7	House and pits	14	Niedermerz 4	Houses
10	Langweiler 2	Houses	15	Laurenzberg 7	No structure shown
10	Langweiler 8	Houses and pits	15	Langweiler 2	No structure shown
10	Langweiler 9	Houses and pits	15	Langweiler 8	No structure shown
10	Langweiler 16	Pits	15	Langweiler 9	Pits
11	Laurenzberg 7	House and pits	15	Niedermerz 4	No structure shown
11	Langweiler 2	Houses and pits			

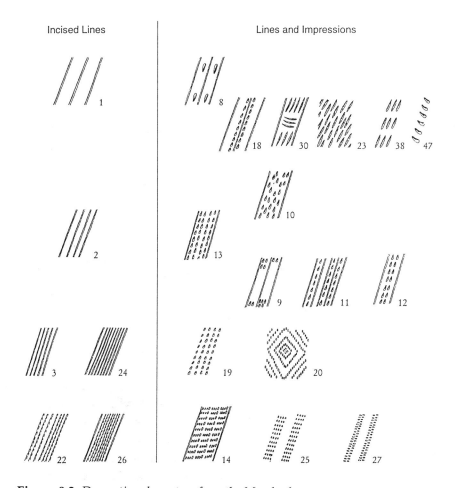

Figure 9.2. *Decorative characters from the Merzbach pottery.*

which have been defined by the excavation team. A total of 35 different band types were recorded for the vessels, the most frequent of which are shown in Fig. 9.2. The vertical dimension of the typology in Fig. 9.2 embodies multiplicity of lines, which increases towards the bottom. The horizontal dimension signifies a tendency towards fragmentation of the continuous linear patterns into rows of spatulate, punctuate marks; in extreme forms these entirely supplant the linear incisions.

Frirdich's (1994) catalogue tabulates the data for six settlements within the Merzbach micro-region, in terms of the number of vessels possessing a particular band type, by pit, site and seriation interval. The counts from the pits within the sites were amalgamated to produce a count for each seriation interval for each site. The seriation intervals were then amalgamated to produce a phase-by-phase count for each settlement. Thereafter, a table was compiled which indicated the number of times a given band type occurs at a given site in a given phase. For the purposes of phylogenetic analysis, each band type was considered to be a character, and each pottery assemblage from a single phase at one site was considered to be a taxon (Table 9.2). Thus, the phylogenetic analyses used the frequencies of decorative characters to reconstruct the relationships between pottery assemblages.

The analyses were carried out using the biological phylogenetic technique of maximum parsimony bootstrapping, which is a procedure for estimating the statistical likelihood of a given phylogenetic relationship being real (Felsenstein 1985; Swofford 1991; Sanderson 1995). In this form of analysis, a matrix is constructed in which the row headings comprise the taxon names, the column headings consist of the character names, and the cells indicate the states of the characters exhibited by the taxa. Next, a large number of new matrices (normally 50 to 1000) are created by randomly sampling with replacement from the original matrix. The new matrices are then subject to branch-and-bound parsimony analysis, which employs an exact algorithm to identify all optimal cladograms for a given character state data matrix. The optimality of a cladogram is assessed in relation to the sum of the lengths of its branches. The length of a branch connecting a pair of taxa on a cladogram is computed as the sum of the character state differences between the taxa under a given model of character state evolution (e.g. ordered, unordered, irreversible). The shortest cladogram is considered to be optimal, because it minimizes the number of hypotheses of change that are required to explain the distribution of character states among the taxa. In other words, the shortest cladogram is considered to be optimal because it is the most parsimonious cladogram. Lastly, a list of the clades that comprise the optimal cladograms is compiled, and the percentage of the bootstrap cladograms in which each clade appears is calculated. Currently there is no consensus as to the percentage of bootstrap cladograms in which a clade should occur for it to be

Table 9.2. *Taxa used in phylogenetic analyses.*

Taxon	Composition	Taxon	Composition
LB7_6	Assemblage from phase 6 at Laurenzberg 7	LW8_11	Assemblage from phase 11 at Langweiler 8
LW2_6	Assemblage from phase 6 at Langweiler 2	LW9_11	Assemblage from phase 11 at Langweiler 9
LW8_6	Assemblage from phase 6 at Langweiler 8	NM4_11	Assemblage from phase 11 at Niedermerz 4
LW9_6	Assemblage from phase 6 at Langweiler 9	LB7_12	Assemblage from phase 12 at Laurenzberg 7
LW16_6	Assemblage from phase 6 at Langweiler 16	LW12_12	Assemblage from phase 12 at Langweiler 2
LB7_7	Assemblage from phase 7 at Laurenzberg 7	LW8_12	Assemblage from phase 12 at Langweiler 8
LW2_7	Assemblage from phase 7 at Langweiler 2	LW9_12	Assemblage from phase 12 at Langweiler 9
LW8_7	Assemblage from phase 7 at Langweiler 8	LW16_12	Assemblage from phase 12 at Langweiler 16
LW9_7	Assemblage from phase 7 at Langweiler 9	NM4_12	Assemblage from phase 12 at Niedermerz 4
LW16_7	Assemblage from phase 7 at Langweiler 16	LB7_13	Assemblage from phase 13 at Laurenzberg 7
LB7_8	Assemblage from phase 8 at Laurenzberg 7	LW2_13	Assemblage from phase 13 at Langweiler 2
LW2_8	Assemblage from phase 8 at Langweiler 2	LW8_13	Assemblage from phase 13 at Langweiler 8
LW8_8	Assemblage from phase 8 at Langweiler 8	LW9_13	Assemblage from phase 13 at Langweiler 9
LW9_8	Assemblage from phase 8 at Langweiler 9	LW16_13	Assemblage from phase 13 at Langweiler 16
LW16_8	Assemblage from phase 8 at Langweiler 16	NM4_13	Assemblage from phase 13 at Niedermerz 4
LB7_9	Assemblage from phase 9 at Laurenzberg 7	LB7_14	Assemblage from phase 14 at Laurenzberg 7
LW2_9	Assemblage from phase 9 at Langweiler 2	LW2_14	Assemblage from phase 14 at Langweiler 2
LW8_9	Assemblage from phase 9 at Langweiler 8	LW8_14	Assemblage from phase 14 at Langweiler 8
LW9_9	Assemblage from phase 9 at Langweiler 9	LW9_14	Assemblage from phase 14 at Langweiler 9
LB7_10	Assemblage from phase 10 at Laurenzberg 7	NM4_14	Assemblage from phase 14 at Niedermerz 4
LW2_10	Assemblage from phase 10 at Langweiler 2	LB7_15	Assemblage from phase 15 at Laurenzberg 7
LW8_10	Assemblage from phase 10 at Langweiler 8	LW2_15	Assemblage from phase 15 at Langweiler 2
LW9_10	Assemblage from phase 10 at Langweiler 9	LW8_15	Assemblage from phase 15 at Langweiler 8
LW16_10	Assemblage from phase 10 at Langweiler 16	LW9_15	Assemblage from phase 15 at Langweiler 9
LB7_11	Assemblage from phase 11 at Laurenzberg 7	NM4_15	Assemblage from phase 15 at Niedermerz 4
LW2_11	Assemblage from phase 11 at Langweiler 2		

considered statistically significant. Some workers favour Felsenstein's (1985) original ≥95 per cent criterion, while others have suggested that clades can occur in 70 per cent of bootstrap cladograms and still be real (e.g. Hillis & Bull 1993).

Two sets of analyses were carried out. The first set focused on the four settlements that have evidence for occupation throughout the 10-phase period (Laurenzberg 7, Langweiler 2, Langweiler 8 and Langweiler 9). We conjectured that, if the phylogenesis hypothesis is correct, phase-by-phase bootstrap analyses of the frequency data for the four settlements should separate the settlements into the same groups in consecutive phases. On the other hand, if the ethnogenesis hypothesis is correct, such analyses should separate the settlements into different groups in consecutive phases. Ten taxon-by-character matrices were generated, each of which comprised just the data for Laurenzberg 7, Langweiler 2, Langweiler 8 and Langweiler 9 from one of the phases. Next, each matrix was coded using the procedure described by Baum (1988). The values for each character were ranked in ascending order and a new character-by-taxon matrix produced in which each cell displayed the rank of a given taxon for a given character. Lastly, each matrix was bootstrapped using the cladistics program Phylogenetic Analysis Using Parsimony (PAUP) 3.0s (Swofford 1991). Because the characters were metrical and their states could therefore be assumed to have evolved serially, the characters were treated as freely-reversing, linearly-ordered variables (Slowinski 1993). The matrices were resampled 10,000 times.

The second set of analyses focused on the three instances in the 10-phase period in which a new pottery assemblage appears. We reasoned that, if the

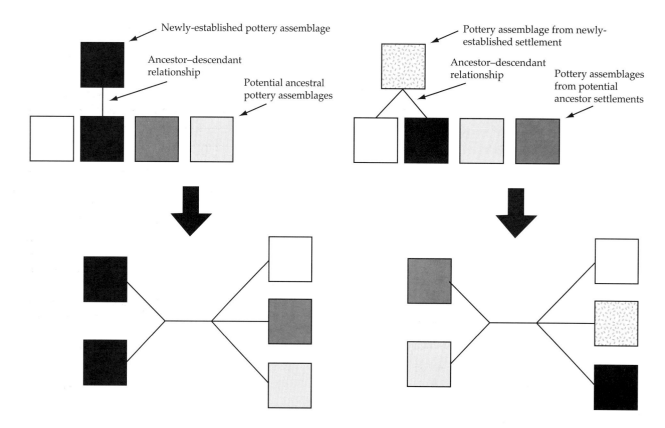

Figure 9.3. *Prediction of the phylogenesis hypothesis in the newly-founded settlements analyses. If the hypothesis is correct, the bootstrap analysis should group the newly-established pottery assemblage with just one of the potential ancestral assemblages.*

Figure 9.4. *Prediction of the ethnogenesis hypothesis in the newly-founded settlements analyses. If the hypothesis is correct, the bootstrap analysis should group the newly-established pottery assemblage with never less than two of the potential ancestral assemblages.*

phylogenesis hypothesis is correct, a bootstrap analysis of the band frequencies for one of the newly-founded pottery assemblages and its potential ancestors should group the newly-established assemblage with just one of the potential ancestral assemblages (Fig. 9.3). Conversely, if the ethnogenesis hypothesis is correct, such an analysis should group the newly-established assemblage with never less than two of the potential ancestral assemblages (Fig. 9.4). Three character-by-taxon matrices were generated, each of which comprised the band frequencies recorded for one of the newly-founded assemblages, plus the band frequencies for the preceding phase of the potential ancestral settlements. The taxa in the first matrix were LW16_10, LB7_9, LW2_9, LW8_9 and LW9_9. The taxa in the second matrix were LW16_12, LW2_11, LW8_11, LW9_11 and LW16_11 and NM4_11. The taxa in the third matrix were NM4_11, LB7_10, LW2_10, LW8_10 and LW9_10. As in the first analysis, each matrix was coded using Baum's (1988) procedure and then bootstrapped using PAUP 3.0s (Swofford 1991). Again, the characters were treated as freely-reversing, linearly-ordered variables, and the matrices were resampled 10,000 times.

Results and discussion

The first set of analyses focused on the assemblages from the four settlements that have evidence for occupation throughout the 10-phase period. It was conjectured that, if the phylogenesis hypothesis is correct, phase-by-phase bootstrap analyses of the assemblages should divide them into the same groups in consecutive phases. On the other hand, if the ethnogenesis hypothesis is accurate, the analyses should separate the settlements into different groups in consecutive phases.

The results of the first set of analyses are summarized in Table 9.3. The four settlements are divided into the same groups in six of the instances in which consecutive phases can be compared (phases

6 & 7, 7 & 8, 8 & 9, 9 & 10, 12 & 13, 14 & 15). In the remaining three instances, the settlements are divided into different groups in consecutive phases (phases 10 & 11, 11 & 12, 13 & 14). These results are not wholly compatible with either hypothesis. Rather, they indicate that phylogenesis and ethnogenesis were both involved in the generation of the Langweiler pottery assemblages.

The second set of analyses focused on the three instances in the 10-phase period in which a new pottery assemblage appears. We reasoned that, if the phylogenesis hypothesis is correct, a bootstrap analysis of the band frequencies for one of the newly-founded pottery assemblages and its potential ancestors should group the newly-established assemblage with just one of the potential ancestral assemblages. Conversely, if the ethnogenesis hypothesis is correct, such an analysis should group the newly-established assemblage with never less than two of the potential ancestral assemblages.

Well-supported divisions were returned in all the analyses (Table 9.4). Two such divisions were identified in the analysis that focused on the origins of the assemblage that appears at Langweiler 16 in phase 10. The first grouped the newly-established assemblage with LW2_9 and LW9_9 to the exclusion of LB7_9 and LW8_9. The second division grouped the newly-established assemblage with LW2_9 to the exclusion of LB7_9, LW8_9 and LW9_9. This result is in line with the prediction of the phylogenesis hypothesis that the newly-founded assemblage should be grouped with just one of the potential ancestral assemblages, since the second division groups the newly-founded assemblage with LW2_9 to the exclusion of the other potential ancestral assemblages. Significantly, this result also supports the excavators' contention that the pits at Langweiler 16 in phase 10 are outliers of the Langweiler 2, which is just across a small valley (Stehli 1994).

Two well-supported divisions were also returned in the analysis that focused on the origin of the assemblage from phase 12 at Langweiler 16. One of these grouped the newly-established assemblage with LB7_11 and NM4_11 to the exclusion of the LW2_11, LW8_11 and LW9_11. The other grouped the newly-established assemblage with NM4_11 to the exclusion of LB7_11, LW2_11, LW8_11 and LW9_11. This result is also in line with the phylogenesis hypothesis, since there is strong support for a division among the assemblages that groups the newly founded assemblage (LW16_12) with just one of the potential ancestral assemblages (NM4_10).

The third analysis, which focused on the origin of NM4_11, identified two well-supported divisions

Table 9.3. *Results of phase-by-phase analysis of assemblages from settlements that are occupied throughout the 10-phase period.*

Phase	Group 1	Group 2	Support for division
6	LB7_6 & LW8_6	LW2_6 & LW9_6	65%
7	LB7_7 & LW8_7	LW2_7 & LW9_7	98%
8	LB7_8 & LW8_8	LW2_8 & LW9_8	100%
9	LB7_9 & LW8_9	LW2_9 & LW9_9	100%
10	LB7_10 & LW8_10	LW2_10 & LW9_10	67%
11	LW8_11 & LW9_11	LB7_11 & LW2_11	70%
12	LB7_12 & LW8_12	LW2_12 & LW9_12	100%
13	LB7_13 & LW8_13	LW2_13 & LW9_13	94%
14	LW8_14 & LW9_14	LB7_14 & LW2_14	66%
15	LW8_15 & LW9_15	LB7_15 & LW2_15	96%

Table 9.4. *Results of analyses that focused on the instances in which a new assemblage is established during the 10-phase period. Analysis 1 concentrated on the origin of the assemblage from phase 10 at Langweiler 10. Analysis 2 examined the origin of the assemblage from phase 12 at Langweiler 16. Analysis 3 focused on the origin of the phase 11 assemblage at Niedermerz 4.*

Analysis	Group 1	Group 2	Support for division
1	LB7_9 & LW8_9	LW16_10, LW2_9 & LW9_9	99%
	LW16_10 & LW2_9	LB7_9, LW8_9 & LW9_9	100%
2	LW16_12, LB7_11 & NM4_11	LW2_11, LW8_11 & LW9_11	91%
	LW16_12 & NM4_11	LB7_11, LW2_11, LW8_11 & LW9_11	100%
3	NM4_11, LW9_10 & LW16_10	LB7_10, LW2_10 & LW8_10	95%
	LW9_10 & LW16_10	LB7_10, LW2_10, LW8_10 & NM4_11	97%

among the assemblages. The first of these grouped the newly-founded assemblage with LW9_10 and LW16_10 to the exclusion of LB7_10, LW2_10 and LW8_10. The second division grouped the LW9_10 and LW16_10 to the exclusion of NM4_11, LB7_10, LW2_10 and LW8_10. This result is difficult to interpret. The newly-established assemblage is never grouped with fewer than two of the potential ancestral assemblages, which is in line with the ethnogenesis hypothesis. However, the two assemblages with which the newly-established assemblage is linked, LW9_10 and LW16_10, are themselves grouped together to the exclusion of the newly-established assemblages and the other potential ancestral assemblages. This division is not compatible with the ethnogenesis hypothesis, since it predicts that the relationship between the newly-established assemblages and its ancestors will be multichotomous (Fig. 9.4). It does not predict that, within a clade comprising the newly-founded assemblage and several potential ancestral assemblages, the ancestral assemblages will be grouped to the exclusion of the newly-founded assemblage. The most likely explanation for this unexpected result is that the pits at Langweiler 16 in phase 10 are outliers of the phase 10 settlement at Langweiler 9, and therefore could not have been the ancestor from which the Niedermerz 4 assemblage was derived, even partially. The corollary of this is that there is a well-supported division among the taxa between Niedermerz 4 and Langweiler 9 on the one hand, and Laurenzberg 7, Langweiler 2 and Langweiler 8 on the other. This division is in line with the phylogenesis hypothesis.

In sum, the second set of analyses support the phylogenesis hypothesis rather than the ethnogenesis hypothesis. Two of the analyses offer strong support for the idea that newly-founded assemblages derive from a single ancestral assemblage through descent with modification. The results of the third analysis are more ambiguous. However, the most parsimonious interpretation of these results also supports the notion that the newly-founded assemblages have a single parent among the assemblages in the preceding phase.

The two sets of analyses of the Merzbach data are not compatible with the assertion that cultural assemblages arise predominantly through ethnogenesis. The first set of analyses indicate that ethnogenesis can only account for a minority of the assemblages that are found at the sites which are occupied throughout the 10-phase period, whilst none of the second set of analyses supports the ethnogenetic hypothesis. Thus, if anything, the two sets of analyses suggest that phylogenesis is more important than ethnogenesis in the generation of the patterns observed among the Merzbach pottery assemblages.

One implication of these results is that archaeologists should not simply assume that the assemblages they study are the result of ethnogenesis. Rather, the relative contribution of ethnogenesis and phylogenesis to the generation of the assemblages needs to be determined empirically on a case-by-case basis (see also Bellwood 1996). A second implication of the results is that the processes of colonization and group fission which are usually assumed to lie behind the LBK Early Neolithic expansion into Europe appear to have the cultural consequences we might expect, and were perhaps associated with corresponding linguistic and genetic patterns. It is particularly striking that the cultural consequences of group fission so clearly involve cultural differentiation and branching even at the localized scale of the sites analyzed in this study, given the extensive inter-site interaction and close relationships that can be assumed to have existed. This seems to point to the operation of transmission-isolating mechanisms of the kind discussed by Durham (1992). It also fits with the suggestion made by Frirdich (1994) that the newly-founded Merzbach communities were concerned to establish distinct identities.

Conclusions

This chapter describes two sets of analyses that were designed to evaluate the relative contribution of phylogenesis and ethnogenesis to the formation of a group of pottery assemblages from the European Neolithic. The results of these analyses suggest that phylogenesis played an important role in generating the patterns in the pottery assemblages that have been found at the Merzbach Neolithic sites. Thus, contrary to what some have claimed, ethnogenesis is not the only process responsible for producing the material culture patterns recorded by archaeologists. Phylogenesis should not be dismissed as a factor in the cultural affairs of past human societies.

References

Baum, B.R., 1988. A simple procedure for establishing discrete characters from measurement data, applicable to cladistics. *Taxon* 37, 63–70.

Bellwood, P., 1996. Phylogeny vs reticulation in prehistory. *Antiquity* 70, 881–90.

Boyd, R., M. Borgerhoff Mulder, W.H. Durham & P.J.

Richerson, 1997. Are cultural phylogenies possible? in *Human by Nature*, eds. P. Weingart, S.D. Mitchell, P.J. Richerson & S. Maasen. Mahwah (NJ): Lawrence Erlbaum.

Dewar, R.E., 1995. Of nets and trees: untangling the reticulate and dendritic in Madagascar's prehistory. *World Archaeology* 26, 301–18.

Durham, W.H., 1992. Applications of evolutionary culture theory. *Annual Review of Anthropology* 21, 331–55.

Felsenstein, J., 1985. Confidence limits on phylogenetics: an approach using the bootstrap. *Evolution* 39, 783–91.

Frirdich, C., 1994. Kulturgeschichtliche Betrachtungen zur Bandkeramik im Merzbachtal, in Lüning & Stehli (eds.), 207–394.

Gjessing, H.K. & R.H. Pierce, 1994. A stochastic model for the presence/absence of readings in Niorstigningar saga. *World Archaeology* 26, 268–93.

Guglielmino, C.R., C. Viganotti, B. Hewlett & L.L. Cavalli-Sforza, 1995. Cultural variation in Africa: role of mechanisms of transmission and adaptation. *Proceedings of the National Academy of Sciences of the USA* 92, 7585–9.

Hillis, D.M. & J.J. Bull, 1993. An empirical test of bootstrapping as a method for assessing confidence in phylogenetic analysis. *Systematic Biology* 42, 182–92.

Holden, C. & R. Mace, 1999. Sexual dimorphism in stature and women's work: a phylogenetic cross-cultural analysis. *American Journal of Physical Anthropology* 110, 27–45.

Kirch, P.V. & R.C. Green, 1987. History, phylogeny and evolution in Polynesia. *Current Anthropology* 28, 431–56.

Lüning, J. & P. Stehli (eds.), 1994. *Die Bandkeramik im Merzbachtal auf der Aldenhovener Platte*. Bonn: Habelt.

Mace, R. & M. Pagel, 1994. The comparative method in anthropology. *Current Anthropology* 35, 549–64.

Mattheusser, E., 1994. Eine Entwicklungsgeschichte der Bandkeramik zwischen Rhein und Maas. Unpublished Ph.D. dissertation, Seminar for Pre- and Protohistory, University of Frankfurt, Frankfurt am Main.

Minelli, A., 1993. *Biological Systematics: the State of the Art*. London: Chapman & Hall.

Moore, C.C. & A.K. Romney, 1994. Material culture, geographic propinquity, and linguistic affiliation on the North Coast of New Guinea: a reanalysis of Welsch, Terrell, and Nadolski (1992). *American Anthropologist* 96, 370–96.

Moore., J.H., 1994a. Putting anthropology back together again: the ethnogenetic critique of cladistic theory. *American Anthropologist* 96, 370–96.

Moore, J.H., 1994b. Ethnogenetic theory. *National Geographic Research and Exploration* 10, 10–23.

Ross, M., 1997. Social networks and kinds of speech-community event, in *Archaeology and Language*, vol. 1, eds. R. Blench & M. Spriggs. London: Routledge, 209–61.

Rowlands, M., 1994. Childe and the archaeology of freedom, in *The Archaeology of V. Gordon Childe*, ed. D. Harris. London: UCL Press, 35–54.

Sanderson, M.J., 1995. Objections to bootstrapping phylogenies: a critique. *Systematic Biology* 44, 299–320.

Slowinski, J., 1993. 'Unordered' versus 'ordered' characters. *Systematic Biology* 42, 155–65.

Sober, E., 1988. *Reconstructing the Past: Parsimony, Evolution and Inference*. Cambridge (MA): The MIT Press.

Stehli, P., 1994. Chronologie der Bandkeramik im Merzbachtal, in Lüning & Stehli (eds.), 79–192.

Swofford, D.L., 1991. *Phylogenetic Analysis Using Parsimony, Version 3.0s*. Champaign (IL): Illinois Natural History Survey.

Terrell, J.E., 1988. History as a family tree, history as a tangled bank. *Antiquity* 62, 642–57.

Terrell, J.E., T.L. Hunt & C. Gosden, 1997. The dimensions of social life in the Pacific: human diversity and the myth of the primitive isolate. *Current Anthropology* 38, 155–95.

Welsch, R.L., J. Terrell & J.A. Nadolski, 1992. Language and culture on the North Coast of New Guinea. *American Anthropologist* 94, 568–600.

Part III

Genetic Markers: from the Global to the Continental

Chapter 10

The Demography of Human Populations inferred from Patterns of Mitochondrial DNA Diversity

Laurent Excoffier & Stefan Schneider

Distributions of pairwise differences often called 'mismatch distributions' have been extensively used to estimate the demographic parameters of past population expansions. However, these estimations relied on the assumption that all mutations occurring in the ancestry of a pair of genes lead to observable differences (the infinite-sites model). This mutation model may not be very realistic, especially in the case of the control region of mitochondrial DNA (mtDNA), where this methodology has been mostly applied. A new model that explicitly takes into account a potential heterogeneity of mutation rates is used to estimate the parameters of a stepwise expansion in more than 60 human populations studied for mtDNA first hypervariable segment (HV1). We find that most populations show signals of Pleistocene expansions at approximately similar dates in Africa and Asia, and at more recent dates in Europe, with oldest expansions being found in East Africa. A simple stepwise expansion model cannot explain the genetic diversity of two groups of populations: the first group is made up of most Amerindian populations and the second of most present-day hunter-gatherers from different continents. A multivariate analysis of the genetic distances computed among 61 populations reveals that populations not having gone through demographic expansions show increased genetic distances from other populations, suggesting that the demography of the populations strongly affect their observed genetic affinities. The absence of traces of Pleistocene expansions in present-day hunter-gatherers seems best explained by the occurrence of recent bottlenecks in those populations, which is confirmed by simulations.

Relation between genetic diversity and demography

The amount and the pattern of molecular diversity depend in large part on the past demography of a population. It has long be known that large populations maintain more diversity than small populations (Kimura & Crow 1964; Kimura 1969), and that genetic drift is stronger in small populations, leading to a more rapid change in gene frequencies (e.g. Wright 1931; Kimura 1955). Bottlenecks are also known to reduce transiently the amount of polymorphism in a population (Nei *et al.* 1975). More recently, it has been realized that demographic expansions could also leave specific signatures in the pattern of molecular diversity, like an increased homozygosity level owing to an excess of low frequency alleles (Watterson 1986), phylogenies with star-like shapes (Slatkin & Hudson 1991), unimodal and bell-shaped distributions of pairwise differences (Rogers & Harpending 1992), or an absence of significant linkage disequilibrium between linked loci (Slatkin 1994). The distinction between population expansion and stationarity is best understood within the framework of coalescent theory (conveniently reviewed in Hudson 1990), which aims at reconstructing the history of a sample of genes starting from the present and going backwards in time until the most recent common ancestor of all sampled genes. The gene genealogy resulting from a coalescent process in stationary population is well defined (Kingman 1982a,b): phylogenies have on average long internal branches and small external branches, and very large standard errors associated with each branch length. It follows that the shape and the height of phylogenies inferred from a series of independent loci should be very different. In contrast, in populations having recently expanded from a small number of individuals, such phylogenies have generally short internal branches and long external branches, and the shape of such genealogies is very reproducible across different loci not submitted to selective pressure. The height of the phylogenies will itself mostly depend on the time of the onset of the expansion (see Rogers & Harpending 1992; Slatkin 1994). A simple way to summarize the information embedded in the phylogeny is to plot the distribution of pairwise differences (the mismatch distribution) among all sampled genes. While mismatch distributions are very ragged and multimodal in stationary populations owing to the large variance of the evolutionary process, they are much smoother and unimodal in expanding populations (Harpending et al. 1993; Harpending et al. 1998). Moreover, the mode of the distribution is closely related to the age of the expansion, as it reflects the fact that two genes have generally no common ancestor (i.e. they have not coalesced) until the population was small. This is because the probability of a coalescent event is inversely proportional to the size of the population, and thus extremely small for large populations, while it becomes much larger before the expansion, when the population size was small.

Estimating demographic parameters from mismatch distributions

Such unimodal mismatch distributions have been recognized early while analyzing human mitochondrial DNA data (Di Rienzo & Wilson 1991). Since then, many populations analyzed for the first hypervariable segment of the mtDNA control region (HV1) have shown this pattern (Sherry et al. 1994; Rogers 1995). Assuming a simple model of population growth such as an instantaneous stepwise expansion, the expected mismatch distribution had been initially derived by Li (1977; see also Takahata et al. 1995, for a simpler derivation). It was shown to depend in a complex way on the population sizes be-

fore (N_0) and after (N_1) the expansion, as well as on the age of the expansion expressed in generations (T), all three parameters being scaled by the mutation rate u, as $\theta_0 = 2uN_0$, $\theta_1 = 2uN_1$, and $\tau = 2uT$ (Li 1977). Alan Rogers and Henry Harpending have thus proposed to estimate the three parameters θ_0, θ_1, and τ directly from the observed mismatch distributions, using either least-square (Rogers & Harpending 1992) or moment (Rogers 1995) methods. Their main result was to obtain evidence for Pleistocene and not Neolithic expansions (Sherry et al. 1994; Rogers 1995; Harpending et al. 1998). At this point, note that this result does not mean that no population expansion occurred in the Neolithic, but rather that the methodology only detects the first expansion from a very small to a large population (Rogers & Harpending 1992). Further expansions from an already large population to an even larger one remain undetected, because coalescent events will, in any case, have a very small probability of happening in the large or very large population.

Dealing with heterogeneity of mutation rates

Estimations of demographic parameters from the mismatch distribution have mostly relied on an infinite site model (Kimura 1969) implying that new mutations always occur at different sites. However, evidence has accumulated over the years that there are many mutational hotspots in the control region of the human mtDNA HV1, implying considerable mutation rate heterogeneity among nucleotide sites (Hasegawa et al. 1993; Wakeley 1993; Excoffier & Yang 1999; Meyer et al. 1999). Although it has been claimed that demographic parameter estimation was quite insensitive to the presence of fast sites (Rogers 1992; Rogers et al. 1996), several other authors have shown that it could affect the shape of the mismatch distribution (Lundstrom et al. 1992; Aris-Brosou & Excoffier 1996) and introduce errors in the reconstructed parameters (Bertorelle & Slatkin 1995). In order to address these concerns, we (Schneider & Excoffier 1999) have recently extended Rogers and Harpending's estimation procedure to take into account departures from the infinite-site model that are specific to the mtDNA HV1 region, such as a high transition bias and heterogeneity of mutation rates. We found that the age of the expansion could be underestimated by up to 20 per cent if mutation rate heterogeneity was not accounted for (Schneider & Excoffier 1999). Simulation studies also allowed us to show that the age of the expansion (τ) and the population size before the expansion (θ_0) were quite

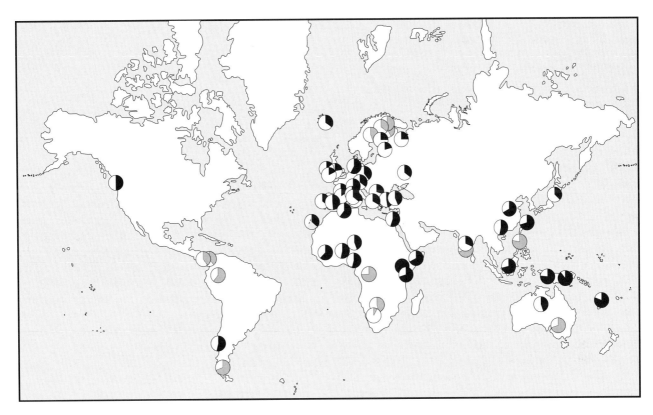

Figure 10.1. *Expansion times of 62 populations analyzed for mtDNA HV1 diversity as reported in Excoffier & Schneider (1999). The Turkana population from Kenya, which shows the oldest expansion time is taken as the reference (black circle). Other populations expansion times are expressed relatively to that time by a black pie. Black pies are indicative of populations showing significant signs of expansions, whereas the molecular diversity of populations with grey pies is not consistent with a mere stepwise expansion.*

well estimated, whereas the size after the expansion (θ_1) was greatly overestimated. Confidence intervals for the three parameters were obtained by a parametric bootstrap approach (Efron & Tibshirani 1993). Using a coalescent approach, samples were simulated according to the estimated demographic parameters, and the estimation procedure was applied each time, in order to get the empirical percentile values for the three estimated parameters. Simulations made for known demographic histories showed that the confidence intervals for τ were quite well estimated, whereas the confidence intervals for θ_0 and θ_1 were much too broad. It follows that the only parameter that can be measured with reasonable confidence is the age of the expansion.

Demographic expansions inferred from mtDNA HV1 polymorphism

In a recent paper (Excoffier & Schneider 1999), we analyzed the pattern of mtDNA HV1 diversity in 62 human populations from all continents. Parameters of demographic expansions were estimated for the observed mismatch distributions. In Figure 10.1, we plot the expansion times of these populations relative to that of the Turkana population in northwest Kenya, which shows the trace of the oldest expansion. In mutation units, the expansion in the Turkana can be dated to $\tau = 2Tu = 0.036$, where u is the average mutation rate per nucleotide. In order to be able to give an absolute date we need to have a good estimation of the mutation rate for mtDNA HV1, which is still under debate. The estimate of the mutation rate varies between about 7 per cent per million years divergence (Hasegawa *et al.* 1993) to more than 150 per cent (Howell *et al.* 1996; Parsons *et al.* 1997; Parsons & Holland 1998). Note that the last estimate has been obtained from pedigree data, and is at least an order of magnitude larger than estimates calibrated by comparing human and chimpanzees (Jazin *et al.* 1998). In a recent review of human and chimpanzee data (Excoffier & Yang 1999), we found that the average number of mutations having occurred during the divergence of a pair of human and chimpanzee sequence was 0.36 (unpublished results, using Tamura & Nei (1993) substitution model with a gamma parameter $\alpha = 0.4$). If humans and chimpanzees diverged 5 million years ago, such a distance means a mutation rate of 3.6×10^{-8} substitutions per site per year or 7.2 per cent divergence per 1 million years. Using this mutation rate for the Turkana expansion leads to an absolute expansion time of 500,000 years ago, with a 95 per cent confidence interval of 332,000–627,000 years assuming no error on the mutation rate. These figures are in clear contrast with our previously published expansion time of 110,000 years for the Turkana (Excoffier & Schneider 1999), obtained using a much faster mutation rate of 33 per cent divergence per million years, published by Ward *et al.* (1991). Some caution thus seems to be needed concerning the absolute dating of human expansions until some better estimates of the mutation rate be available for mtDNA HV1, and we prefer to plot expansion as relative times. Nevertheless, the date of the earliest expansion in eastern Africa seems to be close to or to predate the appearance of the first modern humans in Africa (Mellars & Stringer 1989; Lewin 1993; Rightmire 1998). It would imply that the molecular signature we see in the mismatch distributions could be the trace of the initial rise of the ancestors of modern humans, and not be a mere signal of demographic expansion associated to technology change and geographic expansion (Sherry *et al.* 1994; Rogers 1995).

Average expansion times expressed as a percentage of that of the Turkana were computed for the different continents as follows: Asia and Oceania 65.5 per cent; Sub-Saharan Africa 63.6 per cent; America 51.8 per cent; Europe 38.2 per cent. As can be seen in Figure 10.1, expansion times are generally found to be largest in Asia and Africa and smallest in Europe, in keeping with evidence concerning the settlement of these continents. These early expansion times are also in agreement with recent analyses of nuclear diversity showing quite a large amount of molecular diversity in Africa and Asia, but not in Europe (Harding *et al.* 1997; Hammer *et al.* 1998; Kaessmann *et al.* 1999). However, these relative dates should not necessarily be taken as tracing the settlement history of our species for two reasons. First, human migrations may have occurred in a context of demographically expanding groups, which would preserve the signal of expansions having occurred in the place of origin of the migrants. The signals of expansions we thus see in Asia could therefore be those which occurred elsewhere. Similarly, the old expansions that are attested in a few Amerindian populations could have occurred in Asia, and their signal could have been preserved during migration to the Americas. Second, small migrating groups could have reached a given region and remained small for many generations, before entering a demographic expansion.

Assessing the significance of demographic expansions

Even though the parameters of demographic expansions have been estimated in all samples, it does not mean that these expansions have really taken place. In order to attest the significance of these expansions, we have used two different approaches. We first tested the significance of the demography estimated from the mismatch distribution by a Monte-Carlo coalescent procedure (Schneider & Excoffier 1999). For each population, we used the set of estimated parameters to simulate new samples. A mismatch distribution was obtained in each case, from which an expected mismatch distribution is adjusted by a least-squares method. The sum of squared deviations (SSD) between the expected and the realized mismatch distribution is recorded and compared to the SSD obtained from the original data. We accept the hypothesis of expansion if the proportion of simulated cases where we have a smaller SSD than the observed one is smaller than 5 per cent. The second approach follows from a test of selective neutrality based on Fu's F_S statistic that is especially sensitive to population expansion (Fu 1997). The F_S statistic is the logit of the probability of observing a number of alleles equal to or larger than the observed one given the average number of pairwise differences between DNA sequences (Fu 1997). Large negative values of the statistic are expected in case of demographic expansions. The significance of the statistic was also tested using a coalescent simulation algorithm (Excoffier & Schneider 1999), and a significant statistic is thus taken as evidence for population expansion. In Figure 10.1, we have greyed the pie charts for which at least one of the two tests gave evidence for an absence of expansion. Two main groups of population do not show signs of expansions: several native Amerindian populations (Ngoebe and Kuna samples from Panama, a sample of Colombians, and an Argentinean Mapuche sample) and most present or recent hunter-gatherer populations (!Kung from Botswana, Pygmies from the Democratic Republic of the Congo, Mukhri from India, Australian Aborigines from New South Wales, and four Saami populations from northern Scandinavia).

Recent population bottlenecks in hunter-gatherers

Amerindian and hunter-gatherer populations showed a pattern of molecular diversity that is more compatible with population stationarity than with demographic expansion. However, we have argued (Excoffier & Schneider 1999) that, at least in the case of the hunter-gatherer populations, a more complex scenario involving both an ancient expansion and a recent demographic reduction had to be envisioned. This conclusion was based on the fact that the attested demographic expansions inferred from the mismatch distribution had taken place in the Pleistocene and not in the Neolithic where the distinction between hunter-gatherer and the other populations has occurred. Before the Neolithic, all populations were hunter-gatherers, which implies that the present genetic distinction between food producers and food gatherers has occurred recently, thus during the Neolithic. Using coalescent simulations, we have checked that recent bottlenecks could indeed erase the signals of past expansions (Excoffier & Schneider 1999), so that the raggedness of the observed mismatch distributions in hunter-gatherers could be due to very recent bottlenecks superimposed on older expansions. Note that other complex demographic scenarios such as hidden subdivision (recent admixtures or hybridizations) could lead to ragged distributions. However, most hunter-gatherer and Amerindian populations show a high frequency of pairs of sequences that do not differ from each other. Such a large zero-difference frequency class is not expected in cases of hidden subdivision, where distributions multimodal for the non-zero frequency classes should be often observed (Marjoram & Donnelly 1994).

If hunter-gatherer populations went through a population bottleneck recently, it is likely that it is connected to the rise of Neolithic farmers, who probably disrupted the pre-existing demographic and migratory equilibrium of hunter-gatherer populations, as shown on Figure 10.2. The transition from a stationary population to a meta-population phase with local extinctions and recolonizations can lead to a much more drastic reduction in local and even global effective population size (Barton & Whitlock 1997; Whitlock & Barton 1997) than in the census size. There is thus no need to postulate active participation of farmer populations other than their being sedentary and very numerous to explain the hunter-gatherer demographic decline.

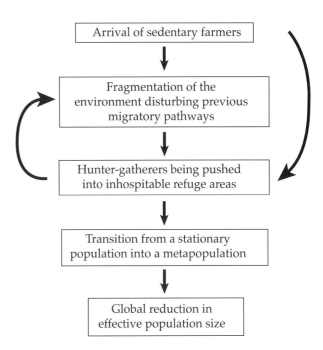

Figure 10.2. *Genetic affinities between 61 populations analyzed for mtDNA HV1 diversity. Populations whose genetic diversity is not compatible with a simple stepwise expansion are shown in boxes. (Adapted from Excoffier & Schneider 1999, fig. 1.)*

Effect of demography on genetic affinities between populations

On Figure 10.3, we have reported the results of a multivariate analysis (multidimensional scaling) of the genetic affinities between 61 human populations analyzed for mtDNA HV1 (one population is missing as compared to the analysis reported in Figure 10.1, because it was analyzed for a segment of HV1 that is not fully overlapping with that of the other populations). Overall we see a good congruence between genetic and geographic groups, as is the case for conventional markers (Cavalli-Sforza et al. 1994). We have, however, highlighted on Figure 10.3 the populations that do not show evidence of population expansions. It is striking that these populations are mostly outliers on the genetic plane. In particular, if one considers only the sub-Saharan populations whose genetic diversity is well explained by demographic expansions, it is clear that they appear genetically quite close to non-Africans. It is only the African populations that have gone through recent bottlenecks that are very divergent. The same observation is valid for the Saami populations from Scandinavia that are extremely different from the other European populations, and for the Amerindian

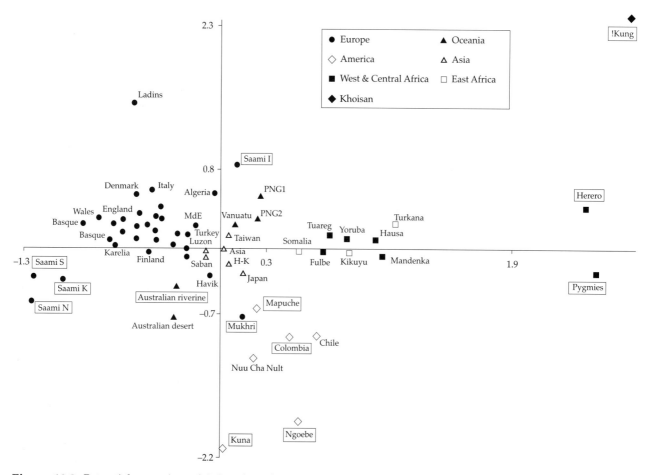

Figure 10.3. *Potential scenario explaining the reduction in effective population size for hunter-gatherer populations.*

populations that show large differences with Asian populations. Founder effects and increased genetic drift in present hunter-gatherers best explain this observation. It confirms that the past demography of the populations can deeply affect our appreciation of their genetic affinities (Nei *et al.* 1975; Chakraborty & Nei 1977; Livshits & Nei 1990). It would thus seem necessary to try to take into account potential differences in effective sizes when inferring genetic distances between populations, a task that begins to be addressed (Wakeley & Hey 1997; Gaggiotti & Excoffier 2000).

Acknowledgements

This work has been made possible thanks to a Swiss National Scientific Foundation No. 31-054059.98. We are grateful to Henry Harpending, Guido Barbujani and Rob Foley for stimulating discussions on the subject. The opinions expressed herein do not necessarily reflect theirs.

References

Aris-Brosou, S. & L. Excoffier, 1996. The impact of population expansion and mutation rate heterogeneity on DNA sequence polymorphism. *Molecular Biology and Evolution* 13, 494–504.

Barton, N.H. & M.C. Whitlock, 1997. The evolution of metapopulations, in *Metapopulation Biology: Ecology, Genetics, and Evolution*, eds. I.A. Hanski & M.E. Gilpin. San Diego (CA): Academic Press, 183–210.

Bertorelle, G. & M. Slatkin, 1995. The number of segregating sites in expanding human populations, with implications for estimates of demographic parameters. *Molecular Biology and Evolution* 12, 887–92.

Cavalli-Sforza, L.L., P. Menozzi & A. Piazza, 1994. *The History and Geography of Human Genes*. Princeton (NJ): Princeton University Press.

Chakraborty, R. & M. Nei, 1977. Bottleneck effects on average heterozygosity and genetic distance with the stepwise mutation model. *Evolution* 31, 347–56.

Di Rienzo, A. & A.C. Wilson, 1991. Branching pattern in the evolutionary tree for human mitochondrial DNA. *Proceedings of the National Academy of Sciences of the USA* 88, 1597–601.

Efron, B. & R.J. Tibshirani, 1993. *An Introduction to the Bootstrap*. London: Chapman & Hall.

Excoffier, L. & S. Schneider, 1999. Why hunter-gatherer populations do not show signs of Pleistocene demographic expansions. *Proceedings of the National Academy of Sciences of the USA* 96, 10,597–602.

Excoffier, L. & Z. Yang, 1999. Substitution rate variation among sites in the mitochondrial hypervariable region I of humans and chimpanzees. *Molecular Biology and Evolution* 16, 1357–68.

Fu, Y-X., 1997. Statistical tests of neutrality of mutations against population growth, hitchhiking and background selection. *Genetics* 147, 915–25.

Gaggiotti, O. & L. Excoffier, 2000. A simple method to remove the effect of unequal population sizes and bottleneck effect in pairwise genetic distances. *Proceedings of the Royal Society of Biological Sciences*.

Hammer, M.F., T. Karafet, A. Rasanayagam, E.T. Wood, T.K. Altheide, T. Jenkins, R.C. Griffiths, A.R. Templeton & S.L. Zegura, 1998. Out of Africa and back again: nested cladistic analysis of human Y chromosome variation. *Molecular Biology and Evolution* 15, 427–41.

Harding, R., S. Fullerton, R. Griffiths, J. Bond, M. Cox, J. Schneider, D. Moulin & J. Clegg, 1997. Archaic African and Asian lineages in the genetic ancestry of modern humans. *American Journal of Human Genetics* 60, 772–89.

Harpending, H., S.T. Sherry, A.R. Rogers & M. Stoneking, 1993. The genetic structure of ancient human populations. *Current Anthropology* 34, 483–96.

Harpending, H., M.A. Batzer, M. Gurven, L.B. Jorde, A.R. Rogers & S.T. Sherry, 1998. Genetic traces of ancient demography. *Proceedings of the National Academy of Sciences of the USA* 95, 1961–7.

Hasegawa, M., A. DiRienzo, T.D. Kocher & A.C. Wilson, 1993. Toward a more accurate time scale for the human mitochondrial DNA tree. *Journal of Molecular Evolution* 37, 347–54.

Howell, N., I. Kubacha & D.A. Mackey, 1996. How rapidly does the human mitochondrial genome evolve? *American Journal of Human Genetics* 59, 501–9.

Hudson, R.R., 1990. Gene genealogies and the coalescent proces, in *Oxford Surveys in Evolutionary Biology*, eds. D.J. Futuyma & J.D. Antonovics. New York (NY): Oxford University Press, 1–44.

Jazin, E., H. Soodyall, P. Jalonen, E. Lindholm, M. Stoneking & U. Gyllensten, 1998. Mitochondrial mutation rate revisited: hot spots and polymorphism. *Nature Genetics* 18, 109–10.

Kaessmann, H., F. Heissig, A. von Haeseler & S. Pääbo, 1999. DNA sequence variation in a non-coding region of low recombination on the human X chromosome. *Nature Genetics* 22, 78–81.

Kimura, M., 1955. Random genetic drift in a multiallelic locus. *Evolution* 9, 419–35.

Kimura, M., 1969. The number of heterozygous nucleotide sites maintained in a finite population due to the steady flux of mutations. *Genetics* 61, 893–903.

Kimura, M. & J.F. Crow, 1964. The number of alleles that can be maintained in a finite population. *Genetics* 49, 725–38.

Kingman, J.F.C., 1982a. The coalescent. *Stochastic Processes and Applications* 13, 235–48.

Kingman, J.F.C., 1982b. On the genealogy of large populations. *Journal of Applied Probability* 19A, 27–43.

Lewin, R., 1993. *Human Evolution: an Illustrated Introduction*. Oxford: Blackwells Scientific Publications.

Li, W.H., 1977. Distribution of nucleotide differences between two randomly chosen cistrons in a finite population. *Genetics* 85, 331–7.

Livshits, G. & M. Nei, 1990. Relationships between intrapopulational and interpopulational genetic diversity in man. *Annual of Human Biology* 17, 501–13.

Lundstrom, R., S. Tavaré & R.H. Ward, 1992. Modeling the evolution of the human mitochondrial genome. *Mathematical Bioscience* 112, 319–35.

Marjoram, P. & P. Donnelly, 1994. Pairwise comparisons of mitochondrial DNA sequences in subdivided populations and implications for early human evolution. *Genetics* 136, 673–83.

Mellars, P. & C. Stringer (eds.), 1989. *The Human Revolution: Biological Perspectives in the Origins of Modern Humans*. Princeton (NJ): Princeton University Press.

Meyer, S., G. Weiss & A. von Haeseler, 1999. Pattern of nucleotide substitution and rate heterogeneity in the hypervariable regions I and II of human mtDNA. *Genetics* 152, 1103–10.

Nei, M., T. Maruyama & R. Chakraborty, 1975. The bottleneck effect and genetic variability in populations. *Evolution* 29, 1–10.

Parsons, T.J. & M.M. Holland, 1998. Reply to Jazin *et al.* *Nature Genetics* 18, 110.

Parsons, T.J., D.S. Muniec, K. Sullivan, N. Woodyatt, R. Alliston-Greiner, M.R. Wilson, D.L. Berry, K.A. Holand, V.W. Weedn, P. Gill & M.M. Holland, 1997. A high observed substitution rate in the human mitochondrial DNA control region. *Nature Genetics* 15, 363–8.

Rightmire, G.P., 1998. The first anatomically advanced humans from South Africa and the Levant, in *The Origins and Past of Modern Humans: Towards Reconciliation*, vol. 3, eds. K. Omoto & P.V. Tobias. Singapore: World Scientific, 126–38.

Rogers, A., 1992. Error introduced by the infinite-sites model. *Molecular Biology and Evolution* 9, 1181–4.

Rogers, A., 1995. Genetic evidence for a Pleistocene population explosion. *Evolution* 49, 608–15.

Rogers, A. & H. Harpending, 1992. Population growth makes waves in the distribution of pairwise genetic differences. *Molecular Biology and Evolution* 9, 552–69.

Rogers, A., A.E. Fraley, M.J. Bamshad, W.S. Watkins & L.B. Jorde, 1996. Mitochondrial mismatch analysis is insensitive to the mutational process. *Molecular Biology and Evolution* 13, 895–902.

Schneider, S. & L. Excoffier, 1999. Estimation of demographic parameters from the distribution of pairwise differences when the mutation rates vary among sites: application to human mitochondrial DNA. *Genetics* 152, 1079–89.

Sherry, S.T., A.R. Rogers, H. Harpending, H. Soodyall, T. Jenkins & M. Stoneking, 1994. Mismatch distributions of mtDNA reveal recent human population expansions. *Human Biology* 66, 761–75.

Slatkin, M., 1994. Linkage disequilibrium in growing and stable populations. *Genetics* 137, 331–6.

Slatkin, M. & R.R. Hudson, 1991. Pairwise comparisons of mitochondrial DNA sequences in stable and exponentially growing populations. *Genetics* 129, 555–62.

Takahata, N., Y. Satta & J. Klein, 1995. Divergence time and population size in the lineage leading to modern humans. *Theoretical Population Biology* 48, 198–221.

Tamura, K. & M. Nei, 1993. Estimation of the number of nucleotide substitutions in the control region of mitochondrial DNA in humans and chimpanzees. *Molecular Biology and Evolution* 10, 512–26.

Wakeley, J., 1993. Substitution rate variation among sites in hypervariable region I of human mitochondrial DNA. *Journal of Molecular Evolution* 37, 613–23.

Wakeley, J. & J. Hey, 1997. Estimating ancestral population parameters. *Genetics* 145, 847–55.

Ward, R.H., B.L. Frazier, K. Dew-Jager & S. Pääbo, 1991. Extensive mitochondrial diversity within a single Amerindian tribe. *Proceedings of the National Academy of Sciences of the USA* 88, 8720–24.

Watterson, G.A., 1986. The homozygosity test after a change in population size. *Genetics* 112, 899–907.

Whitlock, M.C. & N.H. Barton, 1997. The effective size of a subdivided population. *Genetics* 146, 427–41.

Wright, S., 1931. Evolution in Mendelian populations. *Genetics* 16, 97–159.

Chapter 11

Nuclear Genetic Variation of European Populations in a Global Context

Kenneth K. Kidd, Judith R. Kidd, Andrew J. Pakstis, Batsheva Bonné-Tamir & Elena Grigorenko

In collaboration with several colleagues we have assembled two large data sets on populations from most major regions of the world. A data set of 94 STRP loci on 11 populations shows a pattern consistent with the 'Out of Africa' model for modern human diversification with Europeans then East Asians then American Indians showing successively greater divergence from modern Africans and successively lower heterozygosity. A separate data set of 30 populations but typed uniformly for 17 loci, many of which are multi-site haplotypes, shows the same general pattern with the European samples tightly clustered, the East Asian samples tightly clustered and the American Indians sampled clustered, but with greater divergence among the populations. In the larger study, the bootstrap values provide strong (>80 per cent) support for the global structuring by geographic region but little support for most of the specific relationships within each geographic region. Within Europe and the Middle East, the Adygei, from the Krasnodar region along the Black Sea in southern Russia, tend to be the outliers in a tree representation connecting closest to the base of the European branch. Middle Eastern populations generally connect much closer to typical northwestern European samples (such as the Irish and Danes) and the Russians and Finns tend to form a distinct pair. As additional populations are added and additional loci typed, we expect the pattern of relationships among the European populations to stabilize. With respect to linkage disequilibrium we find that Europeans, in general, show intermediate levels between the very low levels and short molecular extent of disequilibrium seen in sub-Saharan Africans and the higher levels and longer molecular extent for disequilibrium seen in East Asian and especially American Indian populations.

One objective of studies of genetic markers is an understanding of the history of our species: how different populations got to their current locations, where they came from, who they are descended from, etc. Different types of genetic markers can shed light on different aspects of this history. Genetic variation in mitochondrial DNA (mtDNA) and Y-chromosome DNA can trace matrilineal and patrilineal ancestry, respectively. Both types of DNA provide extremely important data but each is essentially only one locus.

Since a population loses the mtDNA of a woman who has only sons and the Y-chromosome DNA of a man who has only daughters, these genetic markers may illuminate little about the broad ancestry of most of the genes in a population. A full picture of the histories of populations requires studies of markers in the recombining parts of the nuclear DNA, primarily the autosomes. These markers are inherited through both sexes and passed on to both sexes. Moreover, hundreds of such markers exist and alleles at different marker loci can be inherited completely independently from one another because of recombination across the generations. Thus, studies of genetic variation at multiple autosomal marker loci allow greater statistical accuracy than studies based on any one marker alone; indeed, evidence suggests that dozens of independent markers are required for reasonable accuracy in estimates of genetic similarity of populations.

Just as mtDNA variation and Y-chromosome variation provide different types of historical information, not all autosomal nuclear markers provide equivalent types of data. We are studying multiallelic short tandem repeat polymorphisms (STRPs), biallelic single nucleotide polymorphisms (SNPs), biallelic insertion-deletion polymorphisms (indels), and haplotypes comprised of multiple markers of all types in a short (up to 100 kb) segment of DNA. STRP loci have multiple alleles consisting of the number of tandem repeats of a short (usually 2 to 5 nucleotides) sequence. Their specific mode of mutation results in relatively rapid change (in evolutionary terms) and recurrent generation of alleles of the same size; this may limit their usefulness for very ancient relationships but allows good resolution among closely related populations. In contrast, SNPs and indels usually have very low mutation rates and recurrent mutation is unlikely to be a factor in the time span since modern *Homo sapiens* evolved. Thus, SNPs and indels are generally suitable for tracing back relationships that are ancient in hominid terms. Haplotypes provide an intermediate type of data — multiple alleles because of the multiple combinations that often exist in a population but a higher 'mutation' rate than SNPs because recombination can generate 'new' combinations. Because of recombination haplotypes will tend through time toward completely random combinations in a population. However, at sites over short molecular distances many human populations show remarkably non-random combinations of alleles, referred to as linkage disequilibrium. This non-randomness provides an interesting additional tool for determining the history of a population: similar patterns of non-randomness in different populations suggest the possibility of a common ancestry and different degrees of non-randomness suggest different histories of the populations since older, larger populations will generally show less linkage disequilibrium than younger, smaller populations.

The overall picture of recent human evolution that is emerging from the accumulating genetic evidence is an African origin followed by expansion of anatomically modern humans from Africa into all the rest of the world. We believe a fundamental factor in this expansion is the founder effect coming out of Africa with profound consequences for the genetics and subsequent evolution (including adaptation and disease susceptibility) of all non-African populations. Evidence from multiple sources argues that modern Africans (read as 'sub-Saharan Africans' throughout) have more genetic variation than modern Europeans or modern Asians: mtDNA (starting with Cann *et al.* 1987; Vigilant *et al.* 1991; Horai *et al.* 1993; Chen *et al.* 1995; and more recently Watson *et al.* 1997; among many others); STRPs (Bowcock *et al.* 1994; Calafell *et al.* 1998; in prep.; Deka *et al.* 1995; Jorde *et al.* 1995; Mateu *et al.* 1999); minisatellites (Armour *et al.* 1996); resequencing SNP discovery (Cargill *et al.* 1999; Halushka *et al.* 1999); Y chromosome (Underhill *et al.* in prep.); other nuclear haplotype data (Zietkiewicz *et al.* 1997; 1998; Kidd *et al.* 1998; 2000; Tishkoff *et al.* 1996; 1998). These data, viewed simply as levels of variation, argue that African populations have had a larger effective population size for a longer time than non-African populations. More variation in Africa is certainly compatible with an African origin but is also compatible with other scenarios. Templeton (1999) notes the greater heterozygosity of RFLPs in Europeans relative to sub-Saharan Africans as evidence against a strictly out-of-Africa conclusion, but it has already been shown that the European ascertainment bias can explain that observation (Bowcock *et al.* 1994; Kidd & Kidd 1996; Rogers & Jorde 1996).

Stronger evidence (than just higher levels of variation) of an African origin comes from other analyses. One such is the *pattern* of variation: in general, heterozygosity decreases with geographic distance from Africa, and variation outside of Africa tends to be a subset of what is in Africa. This pattern of variation holds for STRPs (Calafell *et al.* 1998; in prep.), described in more detail later, minisatellites (Armour *et al.* 1996), and haplotypes (Tishkoff *et al.* 1996; 1998; Kidd *et al.* 1998; 2000, and others). Most of these data argue for a significant founder effect associated with migration out of Africa. Also, the imbalance index of Kimmel *et al.* (1998) is consistent with a bottleneck most ancient in Africa, more recent in Europe and even more recent in Asia. Finally, linkage disequilibrium is greater outside Africa and extends molecularly farther along the DNA (Tishkoff *et al.* 1996; 1998; Kidd *et al.* 1998; Pakstis *et al.* 1997; Kidd *et al.* 2000). Our view of the founder effect is represented in our 'Pointillist View of Recent Evolution' and the 'Wisteria Vine View of Recent Evolution' (both on Kidd Lab Web Site 1999).

The requirements that every population needs to be studied for every marker, and that data on multiple marker loci must be studied, pose serious problems for global overviews such as the above summary. Adequate data are rare for a truly global picture. Instead, what is generally developing is for different researchers to study different genetic polymorphisms in different populations, such that the matrix of 'markers studied' by 'populations studied' is largely empty with little ability to compare populations studied by different groups because completely different loci were studied. This empty matrix problem arises in large part because there are now thousands of polymorphic loci known, each of which is a potentially useful genetic marker for population studies, and there is no easy way to learn what loci have already been studied on what populations — the data are scattered through the medical, genetic, forensic, and anthropological literature. One hope is that the Human Genome Diversity Project will result in greater coordination. In the meantime, only a few studies involve multiple markers on a truly global sample of populations; we are working to accumulate such a data set and make it readily available through a data base, ALFRED, accessible through the internet (Kidd Lab Web Site 1999). In the following section, we review some previous results and present some new interim analyses, based on our growing data set, that focus on European populations in a global context.

Analyses and discussion

Many loci on a few populations
We have studied 94 short tandem repeat polymorphisms (STRPs) to determine the frequencies of their alleles in 11 different populations representing all major regions of the world (Calafell *et al.* 1998; in prep.). The study included one European population, Danes, and one southwest Asian population, Druze, plus a sample of individuals of mixed European ancestry. For all but the mixed Europeans two

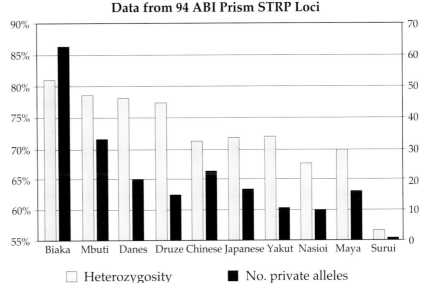

Figure 11.1. *STRP heterozygosity and private alleles. These values are based on 94 STRP loci from the ABI Prism linkage mapping panels from chromosomes 5 through 11. The per cent heterozygous is the average per locus value and is a measure of genetic variation within a population. The number of private alleles (alleles seen only in that single population sample) is a measure of genetic differences between populations. Most private alleles are rare and constitute a small fraction of the alleles in any population. These data are taken from Calafell* et al. *(in prep.); similar data on the first half of these loci can be found in Calafell* et al. *(1998).*

different statistical summaries of those 94 loci are summarized in Figure 11.1, the average per cent heterozygosity and the number of private alleles. The percent heterozygosity, the per cent of the people in a population that are heterozygous at a particular locus, has been estimated from the allele frequencies assuming Hardy-Weinberg ratios; here we present the average of those values for all 94 loci. This average ranges from over 80 per cent in one African population and over 70 per cent through Europe, Asia, and North America to its lowest value of 55 per cent in a very small, isolated tribe in South America. This is essentially a clinal distribution from the largest values in Africa, then decreasing from West to East across Eurasia, then from North America to the lowest value in South America. However, essentially two-thirds of the substantial variation that exists in Africa is present in South America. The Melanesian population we studied, the Nasioi, is similarly a very isolated population and our sample of it very small, but it also retains substantial genetic variation. The difference in heterozygosity between the African samples and the European and southwest Asian samples is not great. We think this is an artefact of a European bias in the selection of these loci: only loci with heterozygosities >70 per cent in European samples were selected as good markers for genetic linkage studies. The other statistic shown in Figure 11.1 gives a different picture of the relationships between European and African populations. The number of private alleles is the count across all 94 loci of alleles that were seen only in that single population. This tally shows quite dramatically that a much larger number of alleles occurs only in African populations and not in other parts of the world. We can see this pattern because the European ascertainment bias at these multiallelic loci is different from that at biallelic loci. At STRPs we are not limited to looking at alleles identified in Europeans and the number of alleles can increase greatly without increasing heterozygosity appreciably, given an already multiallelic locus.

We have used the allele frequencies at these loci to calculate pairwise genetic distances among the ten population samples in Figure 11.1 plus a sample of mixed Europeans and have estimated additive trees with bootstrap values (see '94-locus Neighbor-Joining Tree': Kidd Lab Web Site 1999). This analysis of all 94 loci gives essentially an identical tree to one based on half the data (Calafell et al. 1998). Three points from the analysis are relevant to European genetic history: the tight clustering of the three European populations, the place of this cluster in the tree, and the bootstrap values around this location. The three European populations cluster very tightly with little biological significance to the mixed Europeans being closer to the Danes than to the Druze; the three, as a group, are separated by an appreciable branch from the rest of the tree. The 'European' branch connects to the tree between the African populations and the East Asian, American Indian, and Melanesian populations which share a common branch connecting them to the African and European clusters. Finally, the bootstrap values for the African branch and the European branch are both 100 per cent, indicating the highest possible statistical support for these two clusters. If the Melanesians are omitted from the analysis (they are not placed into the tree with much precision because their allele frequencies are so distinctly different from all other populations studied), the branch extending to the East Asians and American Indians also has a bootstrap value of 100 per cent.

Many populations but fewer loci
The pattern of regional relationships based on the 94 STRP loci is identical to the pattern we see for a much larger number of populations based on our accumulating data on SNPs and haplotypes, including data already published, at least in part, for individual loci: CD4 (Tishkoff et al. 1996), DM (Tishkoff et al. 1998), DRD2 (Kidd et al. 1998), SLC6A3 (Kang et al. 1999), COMT (Palmatier et al. 1999). We can represent gene frequency data graphically in two ways, a principal components analysis (PCA) and a tree analysis. Based on interim analyses of our ever-increasing data set, we have prepared figures focusing on the relationships of the European and southwest Asian populations in a global context. The data used for these interim analyses consist of allele frequencies at 17 independent loci for 30 distinct population samples. Eight of the 17 loci are treated as haplotypes with two to four sites and there are more than 90 statistically independent alleles in the data set. Figure 11.2 is a section of the 30-population tree that has resulted. Figure 11.3 is a plot of the first two dimensions of a PCA for these same populations, showing much the same clustering of these populations relative to African populations in one 'direction' and all others in another. The most relevant aspect of both representations is the clustering of all the European and southwest Asian populations, relative to the rest of the world. While the statistical support for the position of the European cluster is not as strong as for the analysis of the 94 STRP loci, the clustering is the same and the relevant parts of the tree have moderate bootstrap support of 80 to 90 per cent. In contrast, there is not strong statistical support for the interrelationships of populations within the cluster of European and southwest Asian populations.

Within the 'European' cluster of populations the tree is not meant to represent evolutionary relationships or historical branchings, but it is the best representation of overall genetic similarities but with little statistical support to favour other patterns over this one. However, the pattern seen can illuminate history if interpreted in conjunction with other data. In the tree analysis, one outlier of the cluster is the Adygei from the Krasnador region of southern Russia, the northwestern part of the Caucasus. This population connects closest to the base of the 'European' branch while all others connect farther out on the branch. This placement might reflect either that the Adygei are most like the ancestral European population or that there has been gene flow from populations farther east that has 'pulled' this population closer to where the Asian lineage branches off. There may be historical and/or archeological data to support neither, either, or both possibilities; also, the Caucasus mountains are a genetically complex region with

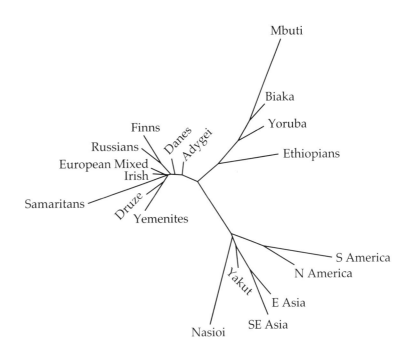

Figure 11.2. *An additive tree of 14 specific populations from Europe and southwest Asia from Africa, and from Melanesia (Nasioi) and Siberia (Yakut) with positions of several Asian and American Indian populations given as regional averages for global reference. This is a section of the best least squares fit found to Tau genetic distances (Kidd & Cavalli-Sforza 1974) based on gene frequencies at 17 independent loci and more than 90 statistically independent alleles. The branch lengths to the Asian and American Indian populations were averaged after the analyses to simplify the figure. A full tree of 30 populations can be found on our web site (Kidd Lab Web Site 1999); it differs slightly in relationships shown in this tree but is not significantly worse statistically. In addition, much of the raw data can be found in ALFRED, the allele frequency data base also accessible through our web site.*

Figure 11.3. *A principal components analysis based on the same genetic distances as the tree in Figure 11.2. Genetic distances to Asian and American Indian populations were averaged to maximize discrimination among the European and southwest Asian populations while preserving the global context. A PCA of all 30 populations based on the same genetic data can be found on our web site (Kidd Lab Web Site 1999). The third principal component accounts for an additional 10.9 per cent of the variance and places the two Asian groups below the plane of the figure and the two American Indian groups about the same distance above the plane; the Melanesians are placed slightly below the plane and the Siberian population stays roughly in the plane. Modified from a figure in Kidd et al. (1999).*

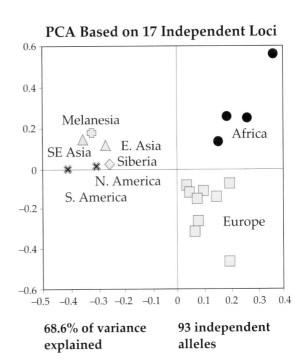

great population diversity based on classical markers (Barbujani *et al.* 1994) and DNA markers are not yet available for other populations in the region. The point is that the genetic data are not definitive at this level. Interestingly, the Russians and Finns do seem to be genetically more similar to each other than to other populations, possibly reflecting a common ancestry between the Finns and this sample of Russians from several hundred kilometres northeast of Moscow.

The terminal branches, those connecting a population at the tip to the rest of the tree, represent the amount of genetic variation that is unique to that sample of that population and includes simple sampling error. However, if there is considerable unique genetic variation, owing to increased genetic drift as a result of historically small population size, that terminal branch will tend to be longer. This appears to be the case of the Samaritans, a population known to have undergone a major reduction in numbers in recent centuries (Bonne 1963). The lack of strong statistical support for any relationships within the cluster is an important point because support is a function of the number of loci and statistically independent alleles studied. With 17 independent loci and over 90 independent alleles, this is one of the largest data sets yet assembled, far exceeding the data available from classical markers. Yet, it is not sufficient to have great confidence in precise relationships of closely related populations.

The PCA (Fig. 11.3) shows some of the same features as the tree in Figure 11.2 but using a different graphic representation. Here the population at the bottom right of the European cluster is the Samaritans. Their displacement from the centre of the cluster reflects the greater random genetic drift they have experienced. Also somewhat displaced from the centre of the cluster is a close pair of populations, the Finns and Russians. In this representation, the Adygei is the population closest to the centre of the entire figure, reflecting that its allele frequencies do not deviate as much from the 'global average' as those of the other Europeans.

Great caution is required in interpreting the results of both the 94-locus and 17-locus studies discussed above, but especially the 17-locus, 30-population study. The major caveat applicable to both the PCA and tree analyses results from our very non-uniform sampling of the world's populations. We have no samples between the Adygei at the east end of the Black Sea and the Cambodians and Chinese near the Pacific coast of Asia. This huge geographic region contains over one-third of the world's population and a very large number of distinct ethnic groups. Nor do we have Canadian, far east Siberian, or North African population samples, populations that might be expected to be genetically intermediate. Consequently, these seemingly distinct clusters cannot be interpreted as distinct 'races'. The representations reflect shared patterns of gene frequencies among the populations sampled, but give no information on relationships to populations not sampled. For example, in the tree representation (Fig. 11.2) the single branch to all European (including southwest Asian) populations indicates they share some gene frequency variation in common. An example of this can be seen at DRD2 where one particular four-site haplotype is present at 30 to 40 per cent in all European and southwest Asian populations studied and much less common to absent elsewhere (Kidd et al. 1998, and unpublished). The clustering in the PCA plot similarly indicates a shared genetic profile. However, this pattern may be altered when other 'European' populations are studied and especially when populations farther east in southwestern Asia are included in an analysis. Indeed, data at classical marker loci show a cline across southern Asia from southwest Asia to southeast Asia (Cavalli-Sforza et al. 1994).

Linkage disequilibrium

Data on linkage disequilibrium in multiple African and non-African populations are now available for several haplotyped loci. The general pattern is seen quite strongly at the four loci for which full data sets have been published — CD4 (Tishkoff et al. 1996), DM (Tishkoff et al. 1998), and DRD2 (Kidd et al. 1998) and PAH (Kidd et al. 2000) — and exists at a locus described so far only in an abstract — HOXB@ (Pakstis et al. 1997). These loci have many haplotypes at common frequencies and little if any disequilibrium in sub-Saharan African populations. All of them show fewer common haplotypes in non-African populations and it is the same small set of haplotypes in all non-African populations with successive subsetting from Europe to East Asia. The linkage disequilibrium also tends to increase from Europe to east Asia and from North America to South America. This pattern provides strong evidence for a significant founder effect associated with the initial expansion out of Africa. This founder event established the linkage disequilibrium that has been preserved and strengthened as modern humans increased in number and spread around the rest of the world.

Discussion and conclusions

Nuclear genes give the overall genetic similarities of populations that are the net result of all historical phenomena. Differences caused by genetic drift are greater when populations suffer bottlenecks, as in the case of the Samaritans, and are smaller when populations have a recent common origin, as may be the case with the Finns and Russians. If those were the only factors, inferring history from present genetic similarities would be easy. However, significant genetic admixture in the ancestry of a population can increase genetic variation in that population and cause it to be 'similar' to all components of its ancestry, complicating interpretation of the genetic data. Another aspect of admixture is that the resulting population is less different from the remote population that was ancestral to the admixing components than it would otherwise be — the admixture counteracts the effects of random genetic drift. In a tree analysis the admixed population will be placed on a short branch that joins the tree somewhere between where the admixing ancestral components connect to the tree. In a PCA representation, the admixed population is placed between the admixing components and displaced toward the centre of the genetic space. If those ancestral components are not represented in the analysis, the interpretation becomes even more difficult. Admixture may be a factor in the placement of the Adygei in the tree analyses. Our mixed European sample, though primarily of western European ancestry, is expected to show the effects of admixture to some extent. Interestingly, it is very close to the Irish sample which should not have a similar degree of admixture. This highlights the problems that can occur from attempting to place a simple historical interpretation on modern genetic differences, especially when based on only a few genetic systems, each of which is a poor indicator of history, and a very incomplete selection of populations. From our analyses it is clear that European populations (geographically defined) share much genetic similarity with populations in southwest Asia and all must be considered 'European' in genetic studies, but we are hesitant to add much finer interpretations until there are considerably more data than just 17 loci, even with over 90 statistically independent alleles. Certainly, the pattern in the tree in Figure 11.2 should not be interpreted as the Danes being most closely related to the population that then radiated out into the rest of Europe and southwest Asia.

The pattern in Figure 11.2 also illustrates the value of having a global context for interpreting genetic similarities among European populations. Without the African and East Asian axis as the 'backbone' of the tree, we might be tempted to place the 'root' of the European populations among the Middle Eastern populations. Though we emphasize the statistical uncertainty of the relationships among the European populations, the actual results are exactly opposite to that expectation. Additional data and populations may change this picture, but that is a question requiring an empirical answer. History and simple observation suggest genetic similarity with populations in northern Africa, supported by various studies of classical markers (Cavalli-Sfroza *et al.* 1994; Kandil *et al.* 1999) and DNA markers (Bertranpetit pers. comm.). Inclusion of those populations may yield two main 'European' branches along either side of the Mediterranean connecting with populations in southwest Asia at the main bifurcation. But that is speculation; we cannot infer from these analyses how inclusion of some of those North African populations will affect the inferred relationships of the populations already studied.

An even greater uncertainty exists for populations in far eastern Europe (e.g. in the Urals) and in central Asia. We do not have any samples from these areas and must be cautious about generating hypotheses based on very partial data. Patterns of genetic similarity need to be interpreted in an appropriate context. A great fallacy often seen in the popular and scientific literature is the implicit assumption that ancient populations were 'pure'. We believe the genetic evidence points to exactly the converse — ancient populations were more varying and loss of variation is responsible for most of the genetic differences seen today. Evidence summarized here argues that as modern humans expanded out of Africa the differences seen across Eurasia gradually accumulated at the front of the expanding population. Thus, peoples in central Asia would have started out as intermediate between European and east Asian peoples. The process of human diversification has been and continues to be a dynamic one with a huge stochastic element and we need to be exceedingly careful about reaching conclusions on too small a sample of populations and/or too small a sample of the genetic material. We believe our study represents the largest data set of autosomal genetic markers studied on a global sample of populations; some elements of the relationships are becoming quite clear but others are still quite uncertain. The spread of modern humans out of Africa accompanied by a significant founder effect is quite clear. The decrease in genetic variation and increase in disequilibrium in east Asian and American populations relative to European populations argues for accumulated genetic drift from West to East after the original migration out of Africa. However, the genetic relationships of individual modern human populations to other closely related populations are not so clear. We need more populations studied for more markers before we can make strong inferences of more recent historical interactions of populations.

Finally, a call for greater cooperation on studies of populations. If there could be greater agreement on which markers to study and greater availability of the raw gene frequency data, the empty matrix problem could be minimized. We think ALFRED, our Web-accessible data base (Kidd Lab Web Site 1999), is important in this regard because it will provide accessibility to our data and added value in terms of links to other data on the same populations as well as other populations studied for the same polymorphisms. We wish to encourage others to study the same markers knowing the data are available on other populations so they can place the populations they study into a global framework. We also hope the availability of the data will stimulate theoreticians to develop new models and methods appropriate to or testable on these data. A comprehensive, complete data set on more populations and better methods of analysis will eventually allow geneticists, archeologists, and historians to realize the promise of genetic data for illuminating our past as a species.

Acknowledgements

These studies would not have been possible without the participation of the people in the various populations who participated by providing samples, the help of many colleagues who collected the various samples, and the work over several years of many technicians, students, and postdocs who typed the samples for the numerous genetic markers. The work was supported in part by NSF grants SBR9408934, SBR9632509, by NIH grants AA09379, GM57672, MH30929, MH39239, and MH50390, and by a grant from the Alfred P. Sloan Foundation.

References

Armour, J.A.L., T. Anttinen, C.A. May, E.E. Vega, A. Sagantila, J.R. Kidd, K.K. Kidd, J. Bertranpetit, S. Paabo & A.J. Jeffreys, 1996. Minisatellite diversity supports a recent African origin for modern humans. *Nature Genetics* 13, 154–60.

Barbujani, G., I.S. Nasidze & G.N. Whitehead, 1994. Genetic diversity in the Caucasus. *Human Biology* 66, 639–68.

Bonne, B., 1963. The Samaritans: a demographic study. *Human Biology* 35, 61–89.

Bowcock, A.M., A. Ruiz-Linares, J. Tomfohrde, E. Minch, J.R. Kidd & L.L. Cavalli-Sforza, 1994. High resolution of human evolutionary trees with polymorphic microsatellites. *Nature* 368, 455–7.

Calafell, F., A. Shuster, W.C. Speed, J.R. Kidd & K.K. Kidd, 1998. Short tandem repeat polymorphism evolution in humans. *European Journal of Human Genetics* 6, 38–49.

Calafell, F., A. Shuster, W.C. Speed, N. Mizuno, J.R. Kidd & K.K. Kidd, in prep. Population genetics of STRPs.

Cann, R.L., M. Stoneking & A.C. Wilson, 1987. Mitochondrial DNA and human evolution. *Nature* 325, 31–6.

Cargill, M., D. Altshuler, J. Ireland, P. Sklar, K. Ardlie, N. Patil, C.R. Lane, E.P. Lim, N. Kalyanaraman, J. Nemesh, L. Ziaugra, L. Friedland, A. Rolfe, J. Warrington, R. Lipshutz, G.Q. Daley & E.S. Lander, 1999. Characterization of single-nucleotide polymorphisms in coding regions of human genes. *Nature Genetics* 22, 231–8.

Cavalli-Sforza, L.L., P. Menozzi & A. Piazza, 1994. *The History and Geography of Human Genes*. Princeton (NJ): Princeton University Press.

Chen, Y-S., A. Torroni, L. Excoffier, A.S. Santachiara-Benerecetti & D.C. Wallace, 1995. Analysis of mtDNA variation in African populations reveals the most ancient of all human continent-specific haplogroups. *American Journal of Human Genetics* 57, 133–49.

Deka, R., L. Jin, M.D. Shriver, L.M. Yu, S. DeCroo, J. Hundrieser, C.H. Bunker, R.E. Ferrell & R. Chakraborty, 1995. Population genetics of dinucleotide $(dC-dA)_n \cdot (dG-dT)_n$ polymorphisms in world populations. *American Journal of Human Genetics* 56, 461–74.

Halushka, M.K., J-B. Fan, K. Bentley, L. Hsie, N. Shen, A. Weder, R. Cooper, R. Lipshutz & A. Chakravarti, 1999. Patterns of single-nucleotide polymorphisms in candidate genes for blood-pressure homeostasis. *Nature Genetics* 22, 239–47.

Horai, S., R. Kondo, Y. Nakagawa-Hattori, S. Hayashi, S. Sonoda & J. Tajima, 1993. Peopling of the Americas, founded by four major lineages of mitochondrial DNA. *Molecular Biology and Evolution* 10, 23–47.

Jorde, L.B., J.J. Bamshad, W.S. Watkins, R. Zenger, A.E. Fraley, P.A. Krakowiak, K.D. Carpenter, H. Soodyall, T. Jenkins & A.R. Rogers, 1995. Origins and affinities of modern humans: a comparison of mitochondrial and nuclear genetic data. *American Journal of Human Genetics* 57, 523–38.

Kandil, M., P. Moral, E. Esteban, L. Autori, G.E. Mameli, D. Zaoui, C. Calo, F. Luna, L. Vacca & G. Vona, 1999. Red cell enzyme polymorphisms in Moroccans and southern Spaniards: new data for the genetic history of the Western Mediterranean. *Human Biology* 71, 791–802.

Kang, A.M., M.A. Palmatier & K.K. Kidd, 1999. Global variation of a 40-bp VNTR in the 3'-untranslated region of the dopamine transporter gene (SLC6A3). *Biological Psychiatry* 46, 151–60.

Kidd, J.R., A.J. Pakstis, H. Zhao, R-B. Lu, F.E. Okonofua, A. Odunsi, E. Grigorenko, B. Bonné-Tamir, J. Friedlaender, L.O. Shulz, J. Parnas & K.K. Kidd, 2000. Haplotypes and linkage disequilibrium at the phenylalanine hydrozylase locus (*PAH*) in a global representation of populations. *American Journal of Human Genetics* 66, 1882–99.

Kidd, K.K. & L.L. Cavalli-Sforza, 1974. The role of genetic drift in the differentiation of Icelandic and Norwegian cattle. *Evolution* 28, 381–95.

Kidd, K.K. & J.R. Kidd, 1996. A nuclear perspective on human evolution, in *Molecular Biology and Human Diversity*, eds. A.J. Boyce & C.G.N. Mascie-Taylor. Cambridge: Cambridge University Press, 242–64.

Kidd, K.K., B. Morar, C.M. Castiglione, H. Zhao, A.J. Pakstis, W.C. Speed, B. Bonné-Tamir, R-B. Lu, D. Goldman, C. Lee, Y.S. Nam, D.K. Grandy, T. Jenkins & J.R. Kidd, 1998. A global survey of haplotype frequencies and linkage disequilibrium at the DRD2 locus. *Human Genetics* 103, 211–27.

Kidd Lab Web Site, 1999. http://info.med.yale.edu/genetics/kkidd. Figures referenced are under 'Illustrations, Teaching Aids, etc.' on the Contents page.

Kimmel, M., R. Chakraborty, J.P. King, M. Bamshad, W.S. Watkins & L.B. Jorde, 1998. Signatures of population expansion in microsatellite repeat data. *Genetics* 148, 1921–30.

Mateu, E., F. Calafell, B. Bonné-Tamir, J.R. Kidd, T. Casals, K.K. Kidd & J. Bertranpetit, 1999. Allele frequencies in a worldwide survey of a CA repeat in the first intron of the CFTR gene. *Human Heredity* 49, 15–20.

Pakstis, A.J., A. Carter, C.M. Castiglione, W.C. Speed, T. Ogura, J.R. Kidd & K.K. Kidd, 1997. At the HOXB@-NGFR loci, the relationship between linkage disequilibrium and physical distance varies across human populations. *American Journal of Human Genetics* 61, A17.

Palmatier, M.A., A. Min Kang & K.K. Kidd, 1999. Global variation in the frequencies of functionally different catechol-O-methyltransferase alleles. *Biological Psychiatry* 46, 557–67.

Rogers, A.R. & L.B. Jorde, 1996. Ascertainment bias in estimates of average heterozygosity. *American Journal of Human Genetics* 58, 1033–41.

Templeton, A.R., 1999. Human races: a genetic and evolutionary perspective. *American Anthropologist* 100, 632–50.

Tishkoff, S.A., E. Dietzsch, W. Speed, A.J. Pakstis, K. Cheung, J.R. Kidd, B. Bonné-Tamir, A.S. Santachiara-Benerecetti, P. Moral, E. Watson, M. Krings, S. Pääbo, N. Risch, T. Jenkins & K.K. Kidd, 1996. Global patterns of linkage disequilibrium at the CD4 locus and modern human origins. *Science* 271, 1380–87.

Tishkoff, S.A., A. Goldman, F. Calafell, W.C. Speed, A.S. Deinard, B. Bonné-Tamir, J.R. Kidd, A.J. Pakstis, T. Jenkins & K.K. Kidd, 1998. A global haplotype analysis of the DM locus: implications for the evolution of modern humans and the origin of myotonic dystrophy mutations. *American Journal of Human Genetics* 62, 1389–402.

Underhill, PA, P. Shen, A.A. Lin, L. Jin, G. Passarino, W.H. Yang, E. Kauffman, F.S. Dietrich, J.R. Kidd, S.Q. Mehdi, T. Jenkins, R.S. Wells, M.T. Seielstad, M. Ibrahim, P. Francalacci, J. Bertranpetit, R.W. Davis, L.L. Cavalli-Sforza & P.J. Oefner, in prep. The architecture of Y-chromosome biallelic haplotype diversity: an emerging portrait of mankind.

Vigilant, L., M. Stoneking, H. Harpending, K. Hawkes & A.C. Wilson, 1991. African populations and the evolution of human mitochondrial DNA. *Science* 253, 1503–7.

Watson, E., P. Forster, M. Richards & H-J. Bandelt, 1997. Mitochondrial footprints of human expansions in Africa. *American Journal of Human Genetics* 61, 691–704.

Zietkiewicz, E., V. Yotova, M. Jarnik, M. Korab-Laskowska, K.K. Kidd, D. Modiano, R. Scozzari, M. Stoneking, S. Tishkoff, M. Batzer & D. Labuda, 1997. Nuclear DNA diversity in worldwide distributed human populations. *Gene* 205, 161–71.

Zietkiewicz, E., V. Yotova, M. Jarnik, M. Korab-Laskowska, K.K. Kidd, D. Modiano, R. Scozzari, M. Stoneking, S. Tishkoff, M. Batzer & D. Labuda, 1998. Genetic structure of the ancestral population of modern humans. *Journal of Molecular Evolution* 47, 146–55.

Chapter 12

Genetic Population Structure of Europeans inferred from Nuclear and Mitochondrial DNA Polymorphisms

Guido Barbujani & Lounès Chikhi

Patterns of genetic diversity in modern populations contain information about past demographic processes. Demographic expansions, admixture and subdivision leave a mark on the entire genome, and so reliable inferences must be based on comparison of as many loci as possible. Spatial autocorrelation analysis of both protein and DNA data shows that broad clines encompassing much of Europe are the rule. The single, most evident exception, is represented by mitochondrial DNA, where a clinal pattern is statistically significant only around the Mediterranean Sea, but not in northern Europe. The similar patterns observed at almost all loci studied suggest that much of the current European diversity has been shaped by some sort of directional expansion from one extreme of the continent, which archaeological evidence places either during the initial Upper Palaeolithic colonization, or during the demic diffusion of Neolithic farmers. That result, and the recent separation of the European gene pools estimated from microsatellite data, do not suggest that phenomena of Mesolithic reexpansion from glacial refugia have played a crucial role in determining the overall population structure of the Europeans.

The earliest human settlement of Europe took place at least 800,000 years ago (Gutin 1995), and several human forms, variously labelled as *Homo erectus*, *Homo antecessor*, archaic *Homo sapiens*, and *Homo sapiens neandertalensis* are documented in the successive millennia in the fossil record. However, it seems highly unlikely that any of them transmitted their genes to contemporary individuals (Krings *et al.* 1997; Ovchinnikov *et al.* 2000). The current Europeans are descended from anatomically modern *Homo sapiens* ancestors, who expanded from Africa some time between 100,000 and 200,000 years ago, and who presumably replaced without admixture the previously existing populations.

Homo sapiens sapiens reached Europe through the Near East. The first anatomically modern Europeans were Upper Palaeolithic hunter-gatherers who are documented around 45,000 years ago (Mellars 1992). They settled over much of the continent, although at low population densities (Bar-Yosef 1992; Kozlowski 1993; Stiner *et al.* 1999). During the Upper Palaeolithic, and in the post-glacial phase which is referred to as Mesolithic period, their populations expanded and contracted, colonized new regions and sometimes underwent local extinctions, especially in the areas where the climate was harshest. A second major immigration process, around 10,000 years ago, was a consequence of the development of the technologies for food production, farming and animal breeding. The first archaeological evidence of farming activities is in the Levant; later specimens demonstrate that farming technologies spread northwards and westwards from there (Clark 1965). Farming societies were established across Europe, all the way to Iberia and the British Isles, between 8000 and 3000 BC (Ammerman & Cavalli-Sforza 1984; Whittle 1996). However, in each locality the transition from hunting and gathering societies to communities based on fully domesticated crops and animals may have taken only a few centuries (Heun *et al.* 1997; Diamond 1997). This set of modifications has been termed the Neolithic Transition or Revolution (see Mannion 1999), and the people who carried into Europe the farming technologies are referred to as Neolithic people.

What is the relative role of the first Palaeolithic colonizers, and of later, Neolithic immigrants, in the ancestry of the European population? This question can be addressed using genetic data, namely, comparing the levels and patterns of genetic polymorphism shown by contemporary Europeans, to the levels and patterns expected under different demographic scenarios. Schematically, one can think that the genetic affinities among today's European populations were determined in the course of the initial, Palaeolithic colonization; successive demographic phenomena, including the Neolithic revolution, did not cause important changes at the genetic level. Alternatively, one can think that the expanding Neolithic communities largely replaced previously settled communities of hunters and gatherers, and therefore that most ancestors of current Europeans arrived in Europe about 10,000 years ago. Between these views (which, for the sake of clarity, have been oversimplified here) there are of course intermediate possibilities. One is that the current European gene pool is largely derived from that of the first Palaeolithic settlers, but that the current pattern of genetic diversity reflects the way people dispersed after the last glaciation (i.e. between 18,000 and 12,000 years ago). Also this model, which we shall here term 'Mesolithic model', implies that the contribution of Neolithic farmers to the current European gene pool was limited. Under a Mesolithic model, the current populations are regarded as largely descended from Palaeolithic ancestors who survived in a few refugia (respectively south of the Pyrenees, the Alps, and the Balkan range) during the last glacial maximum, and expanded north when the glaciers retreated.

In this chapter, the available DNA evidence on the genetic structure of the European populations will be reviewed. Because a crucial point in the debate is the existence and the extent of genetic gradients, one section will be devoted to the description of a statistical method for quantifying geographical patterns of genetic variation, namely spatial autocorrelation analysis. We shall then compare the patterns identified, and some estimates of population divergence based on genetic distances, with the predictions of the main models of evolution of the European gene pool.

Spatial autocorrelation methods

Spatial autocorrelation coefficients may be regarded as correlation coefficients, estimated by comparing values of the same variable at different locations. In genetics, the variables considered are either allele frequencies (Sokal & Oden 1978), or indices of similarity between DNA sequences (Bertorelle & Barbujani 1995). By estimating coefficients of spatial autocorrelation in arbitrary distance classes, the relationships between levels of genetic similarity and geographical distances are quantified. In this way, nonrandom spatial patterns can be recognized, including long-distance differentiation, gradients, partial gradients, and the effects of genetic drift and short-range migration, i.e. isolation by distance.

For estimating spatial autocorrelation coefficients, a matrix of geographical distances between the sampling locations is first constructed. This matrix can be based on the linear distance between samples, but it can also be defined in order to take into account the existence of geographical barriers, or of particular migration routes.

The autocorrelation coefficients are calculated by comparing the values of the variable within predetermined distance classes, e.g. for all pairs of samples separated by distances comprised between 0 and 100 km, 101 and 200 km, and so on. The distance class limits are arbitrary; they are generally fixed at constant intervals, or they are defined so that each class will contain roughly the same number of comparisons. One autocorrelation coefficient is then estimated within each distance class as follows:

$$I = \frac{\{n \, \Sigma_i \neq_j w_{ij} \, (x_i - E[X]) \, (x_j - E[X])\}}{\{\Sigma_i \neq_j w_{ij} \, \Sigma_i \, (x_i - E[X])^2\}} \quad (1)$$

where n is the number of samples, x_i and x_j are the allele frequencies at locations i and j, respectively; w_{ij} is 1 if locations i and j are in the same distance class and 0 if they are not. The expected value of the allele frequency, $E[X]$, is approximated by the observed mean (Sokal & Oden 1978). If the distribution of allele frequencies is random, the expected value of I is close to zero in all distance classes. Genetic resemblance and dissimilarity within one distance class result in I values tending to +1 and −1, respectively.

A series of autocorrelation coefficients can be represented as a correlogram, i.e. a plot of autocorrelation measures against geographic distance. The shape of the correlogram is then used to summarize the spatial patterns, and to draw inferences about the evolutionary processes that have generated those patterns.

With the diffusion of molecular biology techniques, it has become possible not only to detect greater numbers of alleles, but also to quantify the differences between alleles, an unsolved problem in the case of allozymes. In this way, allelic differences may be treated in terms of mutational steps, and population diversity may be described based on both sequence and frequency differences among localities and/or individuals. Autocorrelation statistics that incorporate such an option are termed AIDAs, an acronym for autocorrelation indices for DNA analysis (Bertorelle & Barbujani 1995).

Whereas I can be estimated directly from the allele frequencies, a more complex processing of the data is necessary for the AIDA equivalent of Moran's I, called II. The first step is the construction of a binary data matrix where alleles are coded as a binary vector **p** containing only 0s and 1s. Under reasonable assumptions on the mutational mechanism generating DNA diversity, the three main types of DNA polymorphism, namely restriction fragment length (RFLPs), sequence, and microsatellite polymorphism, can be represented by such vectors.

In the case of restriction haplotypes, a 1 naturally represents the presence of a restriction site whereas a 0 will represent its absence. For sequence data, this way of coding assumes that only two alternative bases can occur at a particular site. Although apparently restrictive, this assumption is consistent both with Watterson's (1975) infinite site model, and with most sequence data. It is however possible to deal with the existence of more than two bases at the same site by coding the four bases as 1000, 0100, 0010, 0001, so that each site will be defined by a four-digit binary code. A plausible mutational process leading to the generation of new alleles in microsatellites and other tandem repeat polymorphisms is the single step mutation model, in which the greater the length difference between two alleles, the more remote their common ancestor (see Chakraborty et al. 1997). As a consequence, alleles can be coded by strings of 1s and 0s, with 1s representing the presence of a repeat and 0s its absence.

Once the data have been coded, an average vector is calculated by taking, for each binary digit, the average across all haplotypes (i.e. all **p** vectors) in the study. This average has no biological meaning, but it is a mathematical object necessary for estimating the autocorrelation indices. From this point on, the calculation of AIDAs will follow the same steps as the classical spatial autocorrelation. In every distance class, II is estimated as follows:

$$II = \frac{[n \, \Sigma_i \Sigma_{j>i} w_{ij} \Sigma_k (p_{ik} - p_k)(p_{jk} - p_k)]}{[W \Sigma_i \Sigma_k (p_{ik} - p_k)^2]} \quad (2)$$

where n is the sample size, W is the number of pairwise comparisons in that distance class, p_{ik} is the value at the kth site of the vector representing haplotype i, p_k is the kth element of the average vector, and w_{ij} is the same as in equation (1). The summation is over all sites and over all haplotypes of the sample.

To test whether the observed coefficients differ significantly from those that might be expected by chance, a randomization approach (Noreen 1989) is followed. A distribution of II pseudovalues is constructed by assigning the haplotypes to random locations, while keeping sample sizes constant. By repeating this procedure many times (usually 1000), and each time recalculating II, an empirical null distribution is obtained. The statistical significance of the observed II values is then assessed by comparison with this distribution.

II shares some properties of Moran's I. It varies between -1 and $+1$, and its expected value is approximately 0 in the absence of geographical patterning ($E[I] = E[II] = -1/(n-1)$, where n is the sample size). But there are differences too. First, the unit of analysis is the haplotype whereas for I it is the allele frequency. As a consequence II can also be calculated between individuals of the same population; in AIDA studies, a zero distance class gives the average autocorrelation within populations. A second difference is that whereas Moran's I is defined for one allele, II summarizes genetic variation for all alleles at one locus. By doing so, the statistical error owing to the existence of rare alleles is reduced, whereas there is an increased chance to detect even limited amounts of spatial patterning.

The main classes of correlograms that can be identified, and the evolutionary processes that may account for each of them, are discussed in Sokal (1979). Schematically, a declining set of autocorrelation coefficients, from positive significant to negative significant, reflects the existence of a cline. Even in the absence of a cline, when highly differentiated populations occupy the extremes of the area studied, the last distance class(es) will show negative autocorrelation, whereas autocorrelation at short-distances will not be significant. Isolation by distance results in positive autocorrelation at short distances, and in declining coefficients beyond a class that depends on the average gene-flow distances (Barbujani 1987). Two clines converging within the area studied often determine a correlogram with a negative peak in one of the intermediate distance classes. Finally, if allele frequencies are distributed at random, the correlogram is expected to show insignificant fluctuations about the expected value.

Two testable predictions

How is DNA variation distributed in Europe, and which of the hypotheses on the origin of the European gene pool accounts best for those distributions? Previous theoretical (Ammerman & Cavalli-Sforza 1984) and simulation work (Rendine et al. 1986; Barbujani et al. 1995) has shown that broad clines encompassing much of the continent can only be generated by a large-scale population expansion occurring from one extreme of the continent. The last demographic process with these characteristics has been the Neolithic demic diffusion. Afterwards, contacts between populations are documented, and their genetic consequences are often evident, but their effects are limited to small regions or single populations (Sokal et al. 1996).

Therefore, two predictions seem reasonable (Table 12.1):
1. The existence of broad clines would indicate that a large-scale, directional expansion, either the initial Palaeolithic colonization or the Neolithic demic diffusion, has been the main phenomenon determining current genetic diversity in Europe; on the contrary, limited gradients radiating from the putative glacial refugia would support a major role of Mesolithic expansions;
2. The expected consequences of a Palaeolithic and of a Neolithic expansion cannot be discriminated in terms of spatial patterns, but the extent of genetic differentiation between populations could tell us whether the current diversity reflects very ancient (Palaeolithic) or more recent (Neolithic) population splits.

Patterns of DNA variation across Europe

Chikhi et al. (1998) collected data on seven DNA loci (four tetranucleotide microsatellites, two minisatellites, and the DQα gene, coding for the α-chain of the HLA-DQ molecule: Table 12.2), and analyzed them by AIDA. For both micro- and minisatellite loci, the allelic differences were expressed as the differences in number of repeats. For DQα, conversely, the six alleles considered differ by known nucleotide substitutions (Gyllensten & Ehrlich 1989), and so evolutionary distances between them were measured in terms of nucleotide substitutions. The overall number of samples was 308, with an average of 44 samples per locus, although the distribution

Table 12.1. *Expected consequences of the three main models of origin of the European gene pool.*

Feature \ Model	Palaeolithic	Mesolithic	Neolithic
Autocorrelation patterns	Mostly clinal	Non clinal	Mostly clinal
Population divergence (years ago)	Around 30–40,000	Around 12–18,000	Less than 10,000

Table 12.2. *Summary of the data analyzed.*

Locus	Type of marker *	Chromosome	No. of samples	No. of chromosomes	No. of alleles †	Reference
FES/FPS	Micro (4)	15	44	22,250	10 (6–15)	Chikhi et al. 1998
FXIIIA	Micro (4)	6	18	8568	18 (2–19)	
HUMTH01	Micro (4)	11	55	28,000	9 (3–10+)	
VWA31A	Micro (4)	12	64	24,610	12 (11–22)	
ApoB$_{3'}$	Mini (15)	2	23	13,412	28 (28–55)	
D1S80	Mini (8)	1	50	17,572	28 (14–41)	
DQα	Sequence	6	54	16,148	6	
P12f2	RFLP	Y	20	1622	2	Semino et al. 1996
49a,f	RFLP	Y	20	2002	2	
DYS19	Micro (4)	Y	59	4665	4	Casalotti et al. 1999
DYS389b	Micro (4)	Y	17	1338	3	
DYS390	Micro (4)	Y	24	1885	4	
DYS391	Micro (4)	Y	22	1265	2	
DYS392	Micro (3)	Y	18	1046	4	
DYS393	Micro (4)	Y	20	1351	3	
DYS19/YAP	Haplotype	Y	19	567	5	
HVR I	Sequence	mtDNA	36	2619	203 **	Simoni et al. 2000

* Length of the repeated motif in parentheses. Micro stands for microsatellites and Mini for minisatellite.
** Based on 22 polymorphic sites.
† Minimum and maximum allele length in parentheses.

available of samples was different for each marker. An average of 424 chromosomes was analyzed for each sample, for a total of 130,560 chromosomes.

The autocorrelation patterns observed at the seven loci show evident similarities (Table 12.3). Because of the limited number of samples available, for two markers, FXIIIA and ApoB$_3$, it was necessary to lump all comparisons beyond 1500 km and 2000 km, respectively. The first clear result is that all correlograms depart from randomness, indicating that there are statistically significant geographical patterns, and confirming the clinal variation that has been descibed in studies of protein markers (Menozzi et al. 1978; Sokal & Menozzi 1982; Cavalli-Sforza et al. 1993; 1994). Second, II is positive and significant within samples (i.e. at distance zero), for all loci. Third, positive values (though not always significant) are observed at short distances for all loci except FXIIIA. Fourth, autocorrelation is virtually zero at intermediate distances. Finally, all loci exhibit negative II values beyond 2000 km, most of them significant.

In brief, geographically close samples are genetically closer than distant samples, with a steady decline towards insignificance between 200 and 500 km (data not shown, see Chikhi et al. 1998, table 3).

At large distances there are comparatively large negative levels of autocorrelation. The rather uniform decrease of autocorrelation at increasing distances within 500 km is an expected consequence of isolation by distance, but the negative long-range autocorrelation cannot possibly be produced by an interaction of local gene flow and genetic drift. Thus, the seven markers studied appear by and large clinally distributed over much of Europe.

Nine markers of the Y chromosome were analyzed by Casalotti et al. (1999) (Table 12.2). Two of them, p12f2 and 49a,f, are essentially biallelic restriction-fragment-length polymorphisms (RFLPs), previously analyzed by Semino et al. (1996). In addition, data were available for six microsatellites. The seventh marker, DYS19/YAP, was obtained by combining information on one microsatellite, DYS19, and on the presence or absence of the only known Alu insertion of the Y chromosome, YAP (Hammer 1994). Sample sizes were smaller than for autosomal markers, ranging from 17 to 59 populations, and from 567 to 4665 chromosomes.

RFLP and microsatellite diversity appear strongly structured in space. II is positive and highly significant in the first two distance classes for all loci, and it declines at increasing distances, reaching

Table 12.3. *Spatial autocorrelation analysis of molecular variation at seven loci. (For autosomic loci and HVR I, II values are multiplied ×10.) D is the upper class limit (km).*

D	0		500		1000		1500		2000		2500		3000		>3000	
ApoB$_3$,	.066	**	.027	**	−.000		−.020		−.015	**	−.033[1]	**				
D1S80	.025	**	.006	**	−.007	**	.004	**	−.003		−.010	**	.019	**	−.035	**
Dqα	.037	**	.010	**	.001		−.006	*	−.013	**	.004		−.007	*	−.010	*
FES/FPS	.042	**	.003		.002		−.003		−.009	**	.002		−.030	**	−.028	**
FXIIIA	.018	**	−.003		−.001		−.003		−.005[2]	**						
HUMTH01	.036	**	.009	**	.004	**	−.011	**	−.007	**	.007	*	−.006	**	−.020	**
VWA31A	.023	**	.006	*	.001		.004		−.003		−.004		−.024	**	−.061	**
DYS19	.035	†	.005	†	.000				−.006	†			−.004	†		
DYS389b	.072	†	.039	†	.009	†			−.016	†			−.030	†		
DYS390	.044	†	.010	†	−.003	†			−.010	†			−.005	†		
DYS391	.059	†	.017	†	−.008	**			−.007	†			−.012	†		
DYS392	.261	†	.114	†	.047	†			−.042	†			−.084	†		
DYS393	.160	†	.083	†	.025	†			−.029	†			−.094	†		
DYS19/YAP	.155	†	.138	†	.022	†			.006	†			−.060	†		
HVR I[3]	.107	†	.006		.024	†	.004		−.008		−.014		−.034	†	−.054	†
HVR I[4]	.767	†	.012	*	−.004	*	.006	*	−.021	**	−.046	**	−.041		−.061	
HVR I[5]	.222	†	.078	†	.044	†	−.001		.017	†			−.040		−.279	†	−.201	†

* $P < 0.05$
** $P < 0.01$
† $P < 0.001$ (Note: for technical reasons, this level of significance could not be tested for the autosomal data of Chikhi et al. 1998.)
[1] this class corresponds to >2000 km
[2] this class corresponds to >1500 km
[3] 34 populations
[4] 17 populations of northern Europe
[5] 14 populations of southern Europe

a highly significant negative peak at extreme distances for all microsatellite markers (Table 12.3). Classical autocorrelation was applied to p12f2 and 49a,f since they are essentially biallelic markers. A steady decline of Moran's I coefficient, from positive within 1000 km through negative significant beyond 2000 km was detected for them as well (Casalotti et al. 1999). Once again, the genetic resemblance between samples separated by less than 500 km is the expected consequence of isolation by distance. However, the overall pattern is clearly a clinal one. Y-chromosome markers are not independently transmitted, and hence these results cannot be considered statistically independent. However, their internal consistency shows that, over Europe, Y-chromosome diversity is distributed in a clinal fashion. Recently Pritchard et al. (1999) analyzed nine markers of the Y chromosome using a likelihood-based method. They conclude that their European sample is consistent with a model of population growth, and that the time at which the population started to grow is consistent with the start of agriculture (with a large 95 per cent confidence interval, though, 4000–32,000 years).

A significant role of Palaeolithic expansions was first advocated by Richards et al. (1996) based on coalescence times of mtDNA haplogroups in Europe. This generated some controversy on the interpretation of mtDNA data (Cavalli-Sforza & Minch 1997; Richards et al. 1997; Barbujani et al. 1998). Afterwards, joint analysis of previous and new mitochondrial genealogies led to the proposal of a less controversial scenario, with Mesolithic (rather than Palaeolithic) expansions playing a major role in shaping European genetic diversity (Richards et al. 1998; Torroni et al. 1998; Sykes 1999). In none of those studies, however, were specific methods for pattern detection applied; hence the importance of the autocorrelation analysis of mitochondrial diversity. Simoni et al. (2000) collected data on sequence polymorphism in the hypervariable region I (HVR I) of the mitochondrial genome. Among more than 2600 individuals typed for a 360-bp DNA segment, 241 sites appeared polymorphic, and 852 distinct haplotypes were detected. That variation reflects in part the effect of highly mutable DNA sites, whose nucleotide content is poorly informative on the genealogical relationships of the alleles. In practice, the data are affected by a high statistical noise. Richards et al. (1998) defined a set of sites whose variation is very informative for the reconstruction of the mitochondrial phylogeny. On the basis of those 22 sites, 203 distinct haplotypes were defined, and subjected to spatial autocorrelation analysis (Simoni et al. 2000).

At a global scale, mitochondrial diversity seems nearly clinal in Europe (data not shown). The values of II decline with distance and reach a negative maximum around 3000 km, although they then increase at greater distances. However, the pattern changes radically when the Saami and Icelandic samples are excluded from the analysis. Most coefficients become insignificant, and no overall pattern is evident (Table 12.2). (On the contrary, removal of the Saami samples does not deeply modify the patterns of nuclear DNA variation described in the previous paragraphs.) However, when northern and southern Europe are separately analyzed, as suggested by previous work (Comas et al. 1997; Malyarchuk 1998), some degree of geographical structuring emerges. North of an imaginary line corresponding roughly to the Alps and the Pyrenees the correlogram is not significant, and from its shape no expansion process can be inferred. On the contrary, the populations living around the Mediterranean Sea show significant structuring also at the mitochondrial level. The peak of maximum differentiation is around 3000 km, but both short-range similarity, and long-range dissimilarity, are significant.

Simoni et al. (2000) also analyzed a southwestern–northeastern transect, from Iberia to northern Scandinavia, which had been suggested as a possible area of Mesolithic expansion (Torroni et al. 1998). No significant autocorrelation pattern was identified there, except for the strong genetic differentiation of Saami with respect to all other western European samples.

The II values observed at autosomic loci, and for mtDNA, are small. To understand this, one has to remember that II can be regarded as the average increase or decrease in the probability to sample the same nucleotide, at a random site, in two sequences separated by a certain spatial distance. The calculation of II involves products of terms much smaller than one (see Equation 2), which increase in number and decrease in value as the number of DNA sites considered increases. Thus, it is to be expected that when the locus studied is highly polymorphic, and when variation is high within samples (as is common in humans: Barbujani et al. 1997), the alleles in two samples will not possibly be much more similar than pairs of random alleles. In Europe, however, II values, small though they might be, differ highly significantly from zero for most loci studied.

Broad genetic gradients have long been known for enzymes and other nuclear gene products (reviewed in Cavalli-Sforza et al. 1994). Because mtDNA variation did not seem clinal in Europe (Richards et al. 1996), it was suggested that protein markers may fail to give a reliable description of the variation existing at the DNA level. But these analyses of seventeen DNA markers prove that clines are the rule, not the exception, in Europe. Because the seven autosomic loci are located on six different chromosomes, linkage disequilibrium is an unlikely cause of that very consistent pattern.

Less-than-optimal sampling, especially in Eastern Europe, and limited sample sizes at certain loci, may have affected the observed patterns, and it would not be surprising if the analysis of larger samples showed details of the spatial patterns that were missed in the studies here reviewed. But the high statistical significance of the clines observed, and the consistent results obtained for several independently-transmitted markers, show that gradients are the main feature of European genetic diversity, both at the protein and at the nuclear DNA level.

To the best of our knowledge, a Mesolithic model has not been formalized yet. It may be that future theoretical efforts may accommodate some level of clinal variation, even in a demographic context dominated by post-glacial reexpansions. At present, models assuming that current genetic variation was largely determined by Mesolithic expansions from Southern and Central Europe are at odds with the spatial patterns observed.

Estimating dates of population splits

What is the best estimate of the age of the directional expansion suggested by spatial autocorrelation analysis? With non-DNA markers, dating is only possible by correlation between genetic and non-genetic information, either archaeological or linguistic (Cavalli-Sforza et al. 1994). However, under a model of population expansion by successive fissions, estimates of the times since separation of populations can be obtained from microsatellite diversity. A measure of genetic distance, $(\delta\mu)^2$ is computed for each locus independently as follows:

$$(\delta\mu)^2 = (\mu_A - \mu_B)^2 \qquad (3)$$

where μ_A and μ_B are the mean allelic lengths of the microsatellite in populations A and B, respectively (Goldstein et al. 1995a). The value of $(\delta\mu)^2$ is expected to increase linearly with time provided that the loci follow a stepwise mutation model and that drift does not overwhelm mutation (see Cooper et al. 1999):

$$(\delta\mu)^2 = 2 \times u \times T \qquad (4)$$

where u is the mutation rate (per locus per generation) and T is time since separation in generations. The mutation rates that were used ($u = 2.8 \times 10^{-4}$ for autosomic microsatellite loci, and $u = 2.0 \times 10^{-4}$ for Y-chromosome microsatellites) are minimum estimates based on studies by Goldstein et al. (1995b), Heyer et al. (1997), and Brinkmann et al. (1998). If anything, these mutation rates should therefore lead to overestimating the time elapsed since population splits. A generation was assumed to correspond to 20 years.

Table 12.4 summarizes the maximum observed divergence times (τ_{max}) estimated by comparing all possible pairs of samples at each locus. Dates calculated from both Y-chromosome and autosome polymorphisms suggest that most European gene pools separated less than 10,000 years ago; all the exceptions include populations which do not speak Indo-European languages. In particular, at the four autosomic loci, in 467 pairwise comparisons, only 15 time estimates were greater than 10,000 years, and all of them involved Turks, Saami, or both, but none of the well-known European genetic outliers, such as Sardinians or Basques (Chikhi et al. 1998).

Different factors may have led to underestimating these separation times, gene flow being the most obvious. The exchange of migrants brings allele frequencies closer to a common, equilibrium value, and reduces the genetic distances between populations. To our knowledge, there is no statistical method to separate the effects of gene flow from those of recent common origin. However, had local gene flow had a major impact upon genetic variation at the continental scale, clear and broad gradients would be uncommon, which is not the case. Even if local gene flow has been important in the evolution of the European

Table 12.4. *Estimated average divergence times between European populations for autosomic and Y-chromosome microsatellites. Note that variation across loci is expected due to the stochasticity of the mutation and demographic processes (Goldstein et al. 1995a).*

Autosomic		Y chromosome	
Chikhi et al. 1998a $\mu = 2.8 \times 10^{-4}$	τ_{max}	Casalotti et al. 1999 $\mu = 2 \times 10^{-3}$	τ_{max}
FES/FPS	4511	DYS19	10,296
FXIIIA	6042	DYS389b	281
TH01	8640	DYS390	3996
VWA31A	6325	DYS391	781
		DYS392	4600
		DYS393	900
Average (autosomic)	6380	Average (Y chromosome)	3476
	Average, 10 loci	4637	

diversity, it is hard to imagine extensive genetic exchange between the populations at the geographical extremes of the European clines, and unless they are Turks or Saami (i.e. non-Indo-European-speakers) their estimated separation times fall in the last 10,000 years.

At any rate, in order to quantify these effects, a computer simulation was run, describing the input of new genes by local gene flow and their extinction due to drift (birth–death process), and estimating the proportion of alleles of foreign origin in a present-day sample (Slatkin & Rannala 1997). The simulation results indicate that that proportion is highly unlikely to exceed 10 per cent, even for large numbers of migrants (15 per generation) exchanged between, say, Iberia and Greece (Chikhi et al. 1998). In other words, to be on the safe side, it is sufficient to increase by ten per cent the estimated divergence times of Table 12.4. After that correction, the maximum time of divergence, averaged across 11 genes, would be 5140, still a value much easier to reconcile with a Neolithic than with a Palaeolithic separation of the European gene pools.

Homoplasy, the creation by mutation of already existing alleles, could also reduce the apparent degree of genetic divergence. However, this factor is taken into account in the mutation model assumed while using the $(\delta\mu)^2$ distance, and therefore it should not be a serious problem. Another factor that may have created gradients and/or reduced the apparent level of genetic differentiation is geographically-variable selection, and selection is notoriously difficult to rule out. However, parallel gradients for many independent loci are the expected consequence of long-range gene flow (Sokal 1979; Ammerman & Cavalli-Sforza 1984). It may well be that one or a few of the observed clines reflect adaptation to (still unspecified) environmental factors, but selective mechanisms do not appear a plausible cause of the overall clinal pattern of nuclear diversity in Europe (see, among others, Cavalli-Sforza et al. 1994).

Elements for a discussion

It seems clear that the analysis of protein data did not misrepresent the basic pattern of genetic variation in Europe, and that this pattern is clinal. Which combination of evolutionary forces led to such a strong geographical patterning over an entire continent is open to discussion. However, no historically or archaeologically documented process after the Neolithic seems to have had the potential for exerting genetic consequences at that scale. The clines that we described are unlikely to have arisen after the Neolithic establishment of large farming communities over much of Europe. Given the number of documented demographic changes in the course of successive history, each of them potentially blurring pre-existing patterns, the degree of genetic structuring observed in Europe must be considered high; and it is that highly significant structure that we have to explain.

What is the best estimate of the age of the clines observed for DNA (including, in southern Europe, mtDNA) as well as for protein, markers? Results based upon the analysis of a relatively limited number of loci (i.e. seven autosomal loci and two non-recombining stretches of DNA) should be taken with caution. The analysis of larger portions of the genome may somewhat modify our views. However, at present these results are in better agreement with a recent (presumably Neolithic) than with a more ancient (Palaeolithic or Mesolithic) separation of the gene pools of even geographically distant populations. The exceptions are groups (such as Saami and Turks) speaking non-Indo-European languages. This fact is in good agreement with the implications of a model proposed by Renfrew (1987), who identified in a single, Neolithic demographic dispersal the origin of the observed archaeological, genetic, and linguistic diversity in Europe. Most populations speaking non-Indo-European languages entered Europe in different times, and are expected to have looser genealogical relationships with Indo-European speakers. The present review of genetic evidence has no bearing on the linguistic aspects of that model, but it shows that, if we focus on Indo-European speakers, we identify a set of populations whose patterns of genetic variation are internally consistent.

The main reason why mitochondrial data were interpreted as suggesting a largely Palaeolithic or Mesolithic origin of the European gene pool was the estimated time depth of mtDNA allele genealogies. Richards et al. (1996) analyzed sequences of the mtDNA control region from 821 individuals sampled in Europe and Turkey. They did not identify clear geographical patterns and, using the method of median networks (Bandelt et al. 1995), they defined several haplogroups. Diversity within haplogroups was estimated, and their age (or coalescence time) was inferred. Because 85 per cent of mtDNA lineages had a Palaeolithic coalescence, Richards et al. (1996) concluded that the people carrying those alleles entered Europe at that time, and estimated the contribution of hunter-gatherers as being approximately 85 per cent.

The relationship between coalescence times and population divergence has been discussed by several authors (see e.g. Pamilo & Nei 1988). In principle, there seems to exist no doubt that the age of a population is not the age of the common molecular ancestor of its set of DNA sequences. Therefore, in the studies here reviewed, the timing of population splits were inferred differently, namely from genetic distances between populations. Also this approach relies on several assumptions, which may or may not be fully met by the data. In addition, these time estimates are affected by large statistical errors. As a consequence, no single value referring to a pair of populations should be taken at its face value, unless it is based on several loci. But a general trend is evident, and it does not suggest an ancient separation between Indo-European speakers, even between those that occupy the extremes of the European gradient.

The patterns shown by mtDNA are similar to those shown by the other markers only in Southern Europe. Human mitochondrial variation is far from being completely understood, although the attempts to identify haplogroups are clarifying the picture (Torroni et al. 1996; 1998; Richards et al. 1998). For the time being, however, it seems that conclusions based on mtDNA data should be generalized only with great caution. Recent selective sweeps (Excoffier 1990; Wise et al. 1997), and a high female mobility (Seielstad et al. 1997; Perez-Lezaun et al. 1999) may account for the comparatively low levels of population differentiation inferred from the maternally-transmitted mitochondrial genes.

By and large, therefore, the overall genetic structure of the European population seems to reflect the consequences of a Neolithic process. This is also supported by a recent mitochondrial study, in which likely traces of Neolithic expansions were recognized in all European groups studied, except Saami (Excoffier & Schneider 1999). But are there alternative explanations? And can we safely assume that every European population was affected to the same extent by the Neolithic transition? The answer to the second question is no. The relative role of Palaeolithic and Neolithic people in the formation of the gene pool of current European populations must have varied. Three main regions where Neolithic technologies developed differently have been identified based on archaeological evidence (Whittle 1996), and differences exist within these regions too. Probably, some Palaeolithic groups (especially in the North) were not deeply affected by the incoming agriculturalists, admixture was the rule in some regions, and Neolithic immigrants totally replaced preexisting groups in some other regions. Methods for inferring complex demographic processes from DNA data are in their infancy, and the first results are still fragmentary (e.g. Beaumont 1999; Pritchard et al. 1999). At present, although no simple, general answer is likely to be accurate, the question whether most genes in the European gene pool were brought into Europe 45,000 or 10,000 years ago is still a crucial one in human population genetics. The results of the studies reviewed in this chapter suggest that most European genes come from ancestors who did not live in Europe, but in the Levant, until 10,000 years ago. When they moved into Europe, they seem to have brought with themselves their genes (as is obvious), the new subsistence technologies, and possibly (Renfrew 1987, but see Renfrew 2000) their Indo-European languages as well.

References

Ammerman A.J. & L.L. Cavalli-Sforza, 1984. *The Neolithic Transition and the Genetics of Populations in Europe.* Princeton (NJ): Princeton University Press.

Bandelt, H-J., P. Forster, B.C. Sykes & M.B. Richards, 1995. Mitochondrial portraits of human populations using median networks. *Genetics* 141, 743–53.

Barbujani, G., 1987. Autocorrelation of gene frequencies under isolation by distance. *Genetics* 117, 777–82.

Barbujani G., R.R. Sokal & N.L. Oden, 1995. Indo-European origins: a computer-simulation test of five hypotheses. *American Journal of Physical Anthropology* 96, 109–32.

Barbujani, G., A. Magagni, E. Minch & L.L. Cavalli-Sforza, 1997. An apportionment of human DNA diversity. *Proceedings of the National Academy of Sciences of the USA* 94, 4516–19.

Barbujani, G., G. Bertorelle & L. Chikhi, 1998. Evidence for Paleolithic and Neolithic gene flow in Europe. *American Journal of Human Genetics* 62, 488–91.

Bar-Yosef, O., 1992. The role of western Asia in modern human origins. *Philisophical Transactions of the Royal Society, London* B 337, 193–200.

Beaumont, M.A., 1999. Detecting population expansion and decline using microsatellites. *Genetics* 153, 2013–29.

Bertorelle, G. & G. Barbujani, 1995. Analysis of DNA diversity by spatial autocorrelation. *Genetics* 139, 811–19.

Brinkmann, B., M. Klintschar, F. Neuhuber, J. Huhne & N. Rolf, 1998. Mutation rates in human microsatellites: influence of the structure and length of the tandem repeat. *American Journal of Human Genetics* 62, 1408–15.

Casalotti, R., L. Simoni, M. Belledi & G. Barbujani, 1999. Y-chromosome polymorphism and the origins of the European gene pool. *Proceedings of the Royal Society* B 266, 1959–65.

Cavalli-Sforza, L.L. & E. Minch, 1997. Palaeolithic and

Neolithic lineages in the European mitochondrial gene pool. *American Journal of Human Genetics* 61, 247–51.

Cavalli-Sforza, L.L., P. Menozzi & A. Piazza, 1993. Demic expansions and human evolution. *Science* 259, 639–46.

Cavalli-Sforza, L.L., P. Menozzi & A. Piazza, 1994. *The History and Geography of Human Genes*. Princeton (NJ): Princeton University Press.

Chakraborty, R., M. Kimmel, D.N. Stivers, L.J. Davison & R. Deka, 1997. Relative mutation rates at di-, tri-, and tetranucleotide microsatellite loci. *Proceedings of the National Academy of Sciences of the USA* 94, 1041–6.

Chikhi, L., G. Destro-Bisol, G. Bertorelle, V. Pascali & G. Barbujani, 1998. Clines of nuclear DNA markers suggest a largely Neolithic ancestry of the European gene pool. *Proceedings of the National Academy of Sciences of the USA* 95, 9053–8.

Clark, J.G.D., 1965. Radiocarbon dating and the expansion of farming culture from the near east over Europe. *Proceedings of the Prehistoric Society* 31, 57–73.

Comas, D., F. Calafell, E. Mateu, A. Pérez-Lezaun, E. Bosch & J. Bertranpetit, 1997. Mitochondrial DNA variation and the origin of the Europeans. *Human Genetics* 99, 443–9.

Cooper, G., W. Amos, R. Bellamy, M.R. Siddiqui, A. Frodsham, A.V.S. Hill & D.C Rubinsztein, 1999. An empirical exploration of the $(\delta\mu)^2$ genetic distance for 213 human microsatellite markers. *American Journal of Human Genetics* 65, 1125–33.

Diamond, J., 1997. Location, location, location: the first farmers. *Science* 278, 1243–4.

Excoffier, L., 1990. Evolution of human mitochondrial DNA: evidence for departure from a pure neutral model of populations at equilibrium. *Journal of Molecular Evolution* 30, 125–39.

Excoffier, L. & S. Schneider, 1999. Why hunter-gatherer populations do not show signs of Pleistocene demographic expansions. *Proceedings of the National Academy of Sciences of the USA* 96, 10,597–602.

Goldstein, D.B., A. Ruiz-Linares, L.L. Cavalli-Sforza & M.W. Feldman, 1995a. An evaluation of genetic distances for use with microsatellite loci. *Genetics* 139, 463–71.

Goldstein, D.B., A. Ruiz-Linares, L.L. Cavalli-Sforza & M.W. Feldman, 1995b. Genetic absolute dating based on microsatellites and the origin of modern humans. *Proceedings of the National Academy of Sciences of the USA* 92, 6723–7.

Gutin, J., 1995. Remains in Spain now reign as oldest Europeans. *Science* 269, 12–13.

Gyllensten, U.B. & H.A. Erhlich, 1989. Generation of single stranded DNA by the polymerase chain reaction and its application to direct sequencing of the HLA-DQA locus. *Proceedings of the National Academy of Sciences of the USA* 85, 7652–6.

Hammer, M., 1994. A recent insertion of an Alu element on the Y chromosome is a useful marker for human population studies. *Molecular and Biological Evolution* 11, 749–61.

Heun, M., R. Schäfer-Pregl, D. Klawan, R. Castagna, M. Acerbi, B. Borghi & F. Salamini, 1997. Site of eikorn wheat domestication identified by DNA fingerprinting. *Science* 278, 1312–14.

Heyer, E., J. Puymirat, P. Dieltjes, E. Bakker & P. de Knijff, 1997. Estimating Y-chromosome specific microsatellite mutation frequencies using deep rooting pedigrees. *Human Molecular Genetics* 6, 799–803.

Kozlowski, J.K., 1993. L'Aurignacien en Europe et au Proche Orient, in *Aurignacien en Europe et au Proche Orient*, eds. L. Banesz & J.K. Kozlowski. Bratislava: Acts of 12th International Congress of Prehistoric and Protostoric Sciences, 283–91.

Krings, M., A. Stone, R.W. Schmitz, H. Krainitzki, M. Stoneking & S. Pääbo, 1997. Neandertal DNA sequences and the origin of modern humans. *Cell* 90, 19–30.

Malyarchuk, B.A., 1998. Mitochondrial DNA markers and genetic demographic processes in Neolithic Europe. *Russian Journal of Genetics* 7, 842–5.

Mannion, A.M., 1999. Domestication and the origins of agriculture: an appraisal. *Progress in Physical Geography* 23, 37–56.

Mellars, P.A., 1992. Archaeology and the population dispersal hypothesis of modern human origins in Europe. *Philisophical Transactions of the Royal Society London* B 337, 225–34.

Menozzi, P., A. Piazza & L.L. Cavalli-Sforza, 1978. Synthetic maps of human gene frequencies in Europeans. *Science* 201, 786–92.

Noreen, E., 1989. *Computer-intensive Methods for Testing Hypotheses: an Introduction*. New York (NY): John Wiley & Sons.

Ovchinnikov, I.V., A. Götherström, G.P. Romanova, V.M. Kharitonov, K. Lidén & W. Goodwin, 2000. Molecular analysis of Neanderthal DNA from the northern Caucasus. *Nature* 404, 490–93.

Pamilo, P. & M. Nei, 1988. Relationships between gene trees and species trees. *Molecular and Biological Evolution* 5, 568–83.

Perez-Lezaun, A., F. Calafell, D. Comas, E. Mateu, E. Bosch, R. Martinez-Arias, J. Clarimon, G. Fiori, D. Luiselli, F. Facchini, D. Pettener & J. Bertranpetit, 1999. Sex-specific migration patterns in Central Asian populations, revealed by analysis of Y-chromosome short tandem repeats and mtDNA. *American Journal of Human Genetics* 65, 208–19.

Pritchard, J.K., M.T. Seielstad, A. Perez-Lezaun & M.W. Feldman, 1999. Population growth of human Y chromosomes: a study of Y chromosome microsatellites. *Molecular Biology and Evolution* 16, 1791–8.

Rendine, S., A. Piazza & L.L. Cavalli-Sforza, 1986. Simulation and separation by principal components of mutiple demic expansions in Europe. *American Naturalist* 128, 681–706.

Renfrew, C., 1987. *Archaeology and Language: the Puzzle of Indo-European Origins*. London: Jonathan Cape.

Renfrew, C., 2000. At the edge of knowability: towards a

prehistory of languages. *Cambridge Archaeological Journal* 10(1), 7–34.

Richards, M., H. Côrte-Real, P. Forster, V. Macaulay, H. Wilkinson-Herbots, A. Demaine, S. Papiha, R. Hedges, H-J. Bandelt & B. Sykes, 1996. Palaeolithic and Neolithic lineages in the European mitochondrial gene pool. *American Journal of Human Genetics* 58, 185–203.

Richards, M., V. Macaulay, B. Sykes, P. Pettit, P. Forster, R. Hedges & H-J. Bandelt, 1997. Palaeolithic and Neolithic lineages in the European mitochondrial gene pool: a response to Cavalli-Sforza and Minch. *American Journal of Human Genetics* 61, 251–4.

Richards, M., V.A. Macaulay, H-J. Bandelt & B.C. Sykes, 1998. Phylogeography of mitochondrial DNA in western Europe. *Annals of Human Genetics* 62, 241–60.

Semino, O., G. Passarino, A. Brega, M. Fellous & A.S. Santachiara-Benerecetti, 1996. A view of the Neolithic demic diffusion in Europe through two Y chromosome-specific markers. *American Journal of Human Genetics* 59, 964–8.

Simoni, L., F. Calafell, D. Pettener, J. Bertranpetit & G. Barbujani, 2000. Geographic patterns of mtDNA diversity in Europe. *American Journal of Human Genetics* 66, 262–78.

Slatkin, M. & B. Rannala, 1997. Estimating the age of alleles by use of intraallelic variability. *American Journal of Human Genetics* 60, 447–58.

Sokal, R.R., 1979. Ecological parameters inferred from spatial correlograms, in *Contemporary Quantitative Ecology and Related Ecometrics*, eds. G.N. Patil & M.L. Rosenzweig. Fairland (MD): International Co-operative Publishing House, 167–96.

Sokal, R.R. & P. Menozzi, 1982. Spatial autocorrelation of HLA frequencies in Europe support demic diffusion of early farmers. *American Naturalist* 119, 1–17.

Sokal, R.R. & N.L. Oden, 1978. Spatial autocorrelation in biology, 1: Methodology. *Biological Journal of the Linnean Society* 10, 199–228.

Sokal, R.R., N.L. Oden, J. Walker, D. di Giovanni & B.A. Thomson, 1996. Historical population movements in Europe influence genetic relationships in modern samples. *Human Biology* 68, 873–98.

Stiner, M.C., N.D. Munro, T.A. Surovell, E. Tchernov & O. Bar-Yosef, 1999. Palaeolithic growth pulses evidenced by small animal exploitation. *Science* 283, 190–94.

Sykes, B., 1999. The molecular genetics of European ancestry. *Philisophical Transactions of the Royal Society London* B 354, 131–9.

Torroni, A., K. Huoponen, P. Francalacci, M. Petrozzi, L. Morelli, R. Scozzari, D. Obinu, M.L. Savontaus & D.C. Wallace, 1996. Classification of the European mtDNAs from an analysis of three European populations. *Genetics* 144, 1835–50.

Torroni, A., H-J. Bandelt, L. D'Urbano, P. Lahermo, P. Moral, D. Sellitto, C. Rengo, P. Forster, M.L. Savontaus, B. Bonné-Tamir & R. Scozzari, 1998. MtDNA analysis reveals a major late Paleolithic population expansion from southwestern to northeastern Europe. *American Journal of Human Genetics* 62, 1137–52.

Watterson, G.A., 1975. On the number of segregating sites in genetical models without recombination. *Theoretical Population Biology* 7, 256–76.

Whittle, A., 1996. *Europe in the Neolithic: the Creation of New Worlds*. Cambridge: Cambridge University Press.

Wise, C.A., M. Sraml & S. Easteal, 1997. Departure from neutrality at the mitochondrial NADH dehydrogenase subunit 2 gene in humans, but not in chimpanzees. *Genetics* 148, 409–21.

Chapter 13

Spatial Variation of mtDNA Hypervariable Region I among European Populations

Lucia Simoni, Francesc Calafell, Jaume Bertranpetit & Guido Barbujani

Genetic diversity in Europe has been interpreted mostly as reflecting phenomena occurring during the Upper Palaeolithic, Mesolithic and Neolithic. A crucial role of the Neolithic demographic transition is supported by the analysis of most nuclear loci, but the interpretation of mtDNA evidence is controversial. To assess whether mtDNA variation is geographically structured within the continent, we collected sequences of the first hypervariable mitochondrial control region in 36 European populations, and analyzed them by means of spatial autocorrelation analysis. In the analysis of the whole continent, limited geographical patterning was observed, which could largely be attributed to a marked difference between the Saami and all other populations. However, an area of significant clinal variation was identified around the Mediterranean Sea (and not in the north), even though the differences between northern and southern populations were insignificant. Both a Palaeolithic expansion, and the Neolithic demic diffusion of farmers, can have determined a longitudinal cline of mtDNA diversity. But the observed patterns do not correspond to the predicted genetic consequences of Mesolithic expansions from glacial refugia. However, there is only a limited correspondence between the patterns inferred from the analysis of mitochondrial and nuclear loci, and neither a model of Palaeolithic expansion nor a model of Neolithic farming dispersal, in their standard versions, can easily account for all of them.

According to the archaeological record, the European gene pool could have been formed during three main periods: the Upper Palaeolithic, Mesolithic and Neolithic (beginning ~10,000 y.a. in the Near East). The first and the third periods are both characterized by population movements from the Near East into much of western and northern Europe. Immigrating groups followed approximately the same routes even if more than 20,000 years had passed between them (see Cavalli-Sforza *et al.* 1994). Between these periods a glaciation took place, at about 20,000 BP, which led to a southward expansion of glaciers. Therefore it is likely that populations of humans, and of other animal and plant species, sought zones where the effect of cooling was milder,

or natural refugia, including areas located south of three mountain chains: the Pyrenees in western Europe, the Alps in the centre and the Balkans in the East. When the climate improved, during the Mesolithic, populations may have re-expanded from these areas into northern Europe. In this context, the term Mesolithic is used loosely to define the period between the last glacial maximum (around 20,000–18,000 years ago) and the origin of agriculture (between 10,000 and 5000 years ago, depending on the area of Europe considered). We are aware that this word may have a slightly different meaning in an archaeological context. Using it, however, simplifies the terminology, because it allows us to distinguish the effects of the initial colonization of Europe (hereafter referred to as the 'Palaeolithic' process), and those of the demographic expansions which may have occurred during the retreat of the glaciers ('Mesolithic' process).

There is no consensus on the relative demographic impact of these migrations, and on their importance in determining current patterns of geographic diversity. However, an effect on the gene pool of the current populations can be hypothesized for each of them: Palaeolithic and Neolithic dispersals were directional processes, whose potential effects include the establishment of southeast–northwest patterns at many loci. Mesolithic gene flow may also have affected the pattern of genetic affinities among European populations (Torroni et al. 1998; Richards et al. 1998), but because it was caused by dispersal from several centres, continent-wide clines are not among its expected consequences. Rather, gradients encompassing part of Europe, and radiating from the putative centres of post-glacial re-expansion should be evident. Because all three prehistoric processes of interest are migratory processes, the consequences of each of them should be apparent over much of the genome, and not only at individual loci.

The genetic structure of current European populations is characterized by broad clines encompassing much of the continent, observed for many classes of genetic markers: blood groups and allozymes (Menozzi et al. 1978; Sokal et al. 1989), histocompatibility alleles (Menozzi et al. 1978; Sokal & Menozzi 1982), and Y-chromosome and autosomal DNA variants (Semino et al. 1996; Casalotti et al. 1999; Chikhi et al. 1998a,b). Until recently, most investigators agreed on the notion that these clines are a consequence of the demic diffusion of the first Neolithic farmers (Ammerman & Cavalli-Sforza 1984; Renfrew 1987). Communities that could produce (as opposed to collect) food tended to grow in size, and to disperse wherever suitable land was available (Pennington 1996). This hypothesis was also supported by comparisons of patterns of linguistic and genetic diversity (Sokal 1988; Barbujani & Pilastro 1993), simulation studies (Rendine et al. 1986; Barbujani et al. 1995) and separation time estimates between eastern and western populations calculated on autosomic DNA variants and Y chromosome, which are all lower than 12,000 y.a. (Chikhi et al. 1998a; Casalotti et al. 1999), and generally much lower than that.

Mitochondrial patterns of variation are not in obvious agreement with the patterns shown by most nuclear markers. European populations appear similar to each other (Pult et al. 1994; Torroni et al. 1994; Sajantila et al. 1996) and only a subset of mitochondrial lineages seems to show any degree of geographical clustering (Richards et al. 1996). Therefore some authors (Richards et al. 1996; Comas et al. 1997) proposed that the Neolithic contribution to the European gene pool would had been overestimated in studies of protein markers. As a consequence, the European population structure would reflect two kinds of effects: an older colonization of the continent, in the Palaeolithic, and successive Mesolithic expansions from glacial refugia, which would account for the absence of continent-wide mitochondrial gradients (Torroni et al. 1998; Sykes 1999). These conclusions were not supported by a quantitative analysis of the geographical patterns of mitochondrial variation.

The principal aims of this study are therefore:
- To define objectively how mtDNA diversity is distributed in Europe, by using spatial autocorrelation analysis;
- To assess if the observed pattern can be easily reconciled with the effects of processes which occurred during the Palaeolithic, Mesolithic or Neolithic periods;
- To compare mitochondrial spatial patterns with those observed using nuclear markers.

Materials and methods

The data base
Our data base was composed of 2619 mtDNA sequences of the hypervariable region I (HVR-I), typed from positions 16024 to 16383 of the Cambridge reference sequence (Anderson et al. 1981). Samples, collected from the literature, are from Europe, the Near East and the Caucasus (Fig. 13.1; Table 13.1). Europe was divided into 36 regions, corresponding generally to entire countries, or, when more detailed information was available, to more restricted areas (such as Cornwall, Sardinia, the eastern Alps) or to well-defined population groups (Adygh, Druzes, Saami, Basques, Catalans).

The number of different sequences observed is 852, and the number of polymorphic sites is 241. There might be high statistical noise due to substitutions which are present in one or two individuals and consequently contain little phylogenetic information. Richards et al. (1998) identified phylogenetically informative polymorphic sites in HVR-I, which allow one to divide the European sequences into various clusters, or haplogroups. Therefore we focused on 22 such variable positions: 16069, 16126, 16129, 16145, 16163, 16172, 16186, 16189, 16193, 16222,

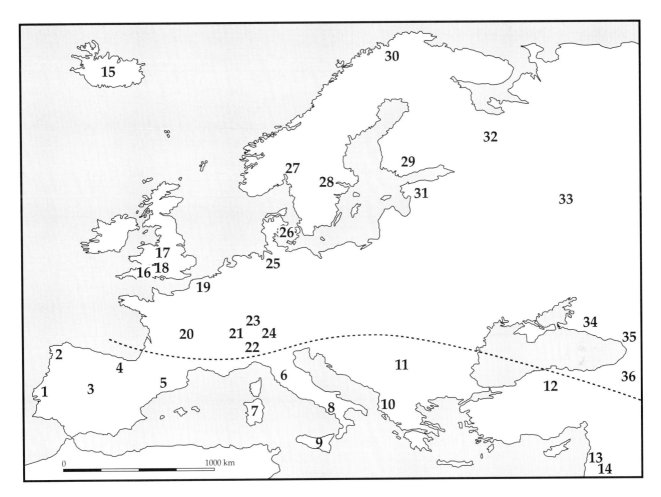

Figure 13.1. *Geographic distribution of sampled populations from Europe, Near East and Caucasus. Numbers represent codes given to populations in Table 13.1. The black dotted line indicates the north–south division used for spatial autocorrelation analysis.*

16223, 16224, 16231, 16261, 16270, 16278, 16292, 16294, 16298, 16311, 16343 and 16356. Variation at other sites was disregarded. In this way, 203 distinct 22-nucleotide haplotypes were obtained from the initial 2619 sequences, each haplotype present in one or more individuals. The sequences were then coded as strings of binary digits, i.e. 0s and 1s, where 0 corresponds to the nucleotide in the Cambridge reference sequence (Anderson *et al.* 1981) and 1 to any observed substitutions.

Spatial autocorrelation analysis
Patterns of mitochondrial variation were summarized by spatial autocorrelation. Measures of overall genetic similarity are evaluated in arbitrary distance classes, and inferences are based on the degree of genetic similarity at various geographical distances. A variable is autocorrelated, positively or negatively, if its value at a given point in space is directly or inversely associated with its values at other locations. Here, we used an approach specifically designed for DNA analysis, AIDA (Bertorelle & Barbujani 1995), which takes into account sequence and frequency information. The autocorrelation index we considered, II, ranges between +1 and −1, and for large sample sizes, its expected value is close to zero. Since II is not normally distributed, its significance was assessed by permutation tests.

The set of spatial autocorrelation coefficients evaluated at various distance classes, or correlogram, can be associated with one or more likely generating processes (Sokal & Oden 1978). A spatially random distribution results in a series of insignificant autocorrelation coefficients at all distances. A decreasing set of coefficients, from positive significant to negative significant, describes a cline, whereas a decreasing correlogram from positive significant through insignificant values at large distances is expected

Table 13.1. *Populations considered in this study.*

Code no.	Population	Size	Reference
1	Portugal	54	Côrte-Real *et al.* 1996
2	Spain-Galicia	92	Salas *et al.* 1998
3	Spain-Mainland	74	Pinto *et al.* 1996; Côrte-Real *et al.* 1996
4	Spain-Basques	106	Bertranpetit *et al.* 1996; Côrte-Real *et al.* 1996
5	Spain-Catalunya	15	Côrte-Real *et al.* 1996
6	Italy-Tuscany	49	Francalacci *et al.* 1996
7	Italy-Sardinia	73	Di Rienzo & Wilson 1991; Rickards (unpublished)
8	Italy-Southern	37	Rickards (unpublished)
9	Italy-Sicily	63	Rickards (unpublished); Nigro *et al.* (unpublished)
10	Albania	42	Belledi *et al.* 2000
11	Bulgaria	30	Calafell *et al.* 1996
12	Turkey	96	Calafell *et al.* 1996; Comas *et al.* 1996; Richards *et al.* 1996
13	Israel-Druze	45	Macaulay *et al.* 1999
14	Near East	42	Di Rienzo & Wilson 1991
15	Iceland	53	Sajantila *et al.* 1995; Richards *et al.* 1996
16	Great Britain-Cornwall	69	Richards *et al.* 1996
17	Great Britain-Mainland	100	Richards *et al.* 1996
18	Great Britain-Wales	92	Piercy *et al.* 1993
19	Belgium	33	De Corte (unpublished)
20	France	111	Le Roux (unpublished)
21	Switzerland	72	Pult *et al.* 1994
22	Italy-Alps	115	Stenico *et al.* 1996
23	Germany-South	249	Richards *et al.* 1996; Lutz *et al.* 1998
24	Austria	117	Handt *et al.* 1994; Parson *et al.* 1998
25	Germany-North	108	Richards *et al.* 1996
26	Denmark	32	Richards *et al.* 1996
27	Norway	30	Dupuy & Olaisen 1996
28	Sweden	32	Sajantila *et al.* 1995
29	Finland	79	Sajantila *et al.* 1995; Richards *et al.* 1996
30	Saami	240	Sajantila *et al.* 1995; Dupuy & Olaisen 1996
31	Estonia	28	Sajantila *et al.* 1995
32	Karelians	83	Sajantila *et al.* 1995
33	Volga Finnic	34	Sajantila *et al.* 1995
34	Caucasus-Adygh	50	Macaulay *et al.* 1999
35	Georgia	45	Comas *et al.* 2000
36	Kurds	29	Comas *et al.* 2000

under isolation by distance, when genetic diversity is the product of the interaction between genetic drift and short-range gene flow (Barbujani 1987). Finally, negative coefficients in the last distance classes reflect some kind of long-range differentiation, i.e. the pattern typical of a subdivided population where geographically extreme samples are also the most differentiated, but there is no overall gradient.

Results

Spatial autocorrelation analysis: AIDA
We initially considered all samples available, then removed some of them, and focused on specific regions of Europe. In the global analysis, *II* is positive and highly significant at distance zero (Fig. 13.2a), i.e. genetic similarity is higher within than between samples. At increasing distances, the level of autocorrelation tends to decrease, in a rather irregular way. Although *II* coefficients are positive and significant for distances shorter than 1500 km, and negative and significant beyond 2000 km, the overall trend is not monotonically decreasing. The greatest divergence is observed between 2500 and 3800 km, corresponding to the comparisons between southeastern Europeans and Saami, whereas sequences sampled at greater distances show a lower number of substitutions. In synthesis, this pattern is not strictly clinal,

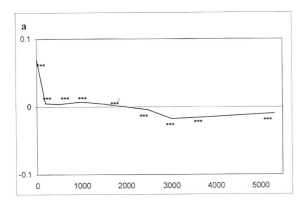
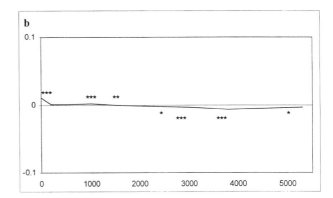
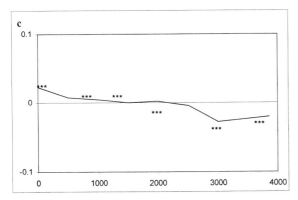
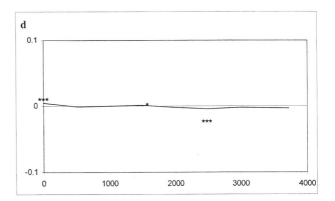

Figure 13.2. *Spatial correlograms on mtDNA HVR-I in Europe: a) all European populations; b) Europe without Saami (i.e. except population coded with number 30 in Table 13.1); c) southern Europe: populations considered in this analysis are coded from 1 to 14 in Table 13.1; d) northern Europe, without Saami: populations considered in this analysis are coded from 15 to 29 in Table 13.1. X-axis: geographic distance between samples; Y-axis: II.*
*** = $P < 0.005$ level; ** = $P < 0.01$; * = $P < 0.05$.

although it shows significant long-distance differentiation.

To assess if the patterns obtained are affected by the presence of genetically isolated populations we excluded from the analysis two samples, known as mitochondrial outliers in Europe: Saami and Ladins, the latter here pooled with other groups from the Alps. Removal of the Alps sample from the database did not change the AIDA profile to any significant degree (data not shown). Conversely, a new pattern emerged when Saami were discarded (Fig. 13.2b). II at distance 0 dropped drastically; the autocorrelation at short distances became insignificant; differences beyond 3000 km, albeit still significant, were reduced to one-fifth or less; and sequences in the extreme distance class no longer differed significantly. This suggests that the previously observed pattern in Europe is largely due to two facts: the differentiation of Saami from all other Europeans (resulting in long-distance negative autocorrelation because of their extreme geographical position, and in short-distance positive autocorrelation owing to the fact that all other samples are comparatively similar to each other); and the internal homogeneity of Saami (resulting in increased autocorrelation at distance zero).

Finally we tested whether different spatial patterns exist within certain regions. In a study of eight European samples, six of them from the Mediterranean area, Comas *et al.* (1997) described an east–west gradient of mean pairwise sequence differences, and suggested that immigration from the Near East could account for it. Also Malyarchuck (1998) described a gradient of mitochondrial RFLPs in southern, but not in northern, Europe. Thus, we independently ran AIDA in the Mediterranean region, and in central-northern Europe. Fourteen and nineteen samples were taken into account, respectively (see legend to Fig. 13.2c, d), whereas the Kurds and two samples from Caucasus were excluded from both analyses.

In the region around the Mediterranean Sea,

AIDA identified a quasi-clinal pattern (Fig. 13.2c). Genetic resemblance appears to decrease smoothly, from positive significant up to 1000 km, to negative significant beyond 2500 km. A different pattern was observed for northern Europe. When the Saami were included, the correlogram was significant; it decreased up to 3000 km, and it increased at greater distances (data not given). But when they were excluded, autocorrelation lost significance in most distance classes, and what remained can be recognized as an insignificant overall pattern (Fig. 13.2d).

Therefore, separate analyses of northern and southern Europe describe two distinct patterns. In northern Europe genetic differences are not related to distance, whereas a cline is apparent around the Mediterranean Sea. We tested whether or not these two regions are genetically differentiated, but the analysis of molecular variance (Excoffier *et al.* 1992) showed that they are not (F_{ST} = 0.0027, P >0.05). Therefore, what we observed is not due to the presence of different alleles, or to different frequencies of these alleles, in northern and southern Europe, but to the different ways these alleles are distributed in space.

Discussion

Our analyses showed that mitochondrial sequences are not clearly patterned over Europe. On a global scale, some degree of short-distance similarity is apparent, as well as negative autocorrelation for samples separated by more than 2000 km. But this picture essentially seems to be determined by the presence of a clear genetic outlier, the Saami. In fact, after their exclusion from the analysis, the pattern does not depart significantly from random expectations. Different results are also obtained when focusing on restricted regions: mitochondrial variation is poorly structured north of an imaginary line corresponding to the latitude of the Pyrenees; whereas a gradient is apparent in southern Europe. Note that the analysis of molecular variance failed to identify any significant differences between these regions, meaning that allele frequencies are roughly the same in the two regions. What changes is their pattern, which is clinal only around the Mediterranean Sea.

The comparison between mitochondrial and nuclear DNA variation shows different features. While nuclear markers tend to show clinal patterns spanning much of the continent, mitochondrial alleles are distributed in clines only in the southern part. A global reanalysis of the existing nuclear evidence is necessary in order to say to what extent nuclear clines may also be steeper around the Mediterranean than elsewhere, in partial agreement with what observed for mitochondrial alleles. We note, however, that clines were detected for nuclear DNA loci, regardless of whether Saami samples were or were not present in the data base (Chikhi *et al.* 1998). A stronger spatial structuring of nuclear than mitochondrial DNA diversity is evident, as has been pointed out by Seielstad *et al.* (1998), using different data and different statistical methods.

It is not easy to envisage a simple model which may reconcile the results of mitochondrial and nuclear studies. Higher female than male mobility (Seielstad *et al.* 1998) may explain why population differences tend to be smaller if estimated at the mtDNA level, but it does not suggest an obvious mechanism leading to the unusual geographical patterns that we described for mtDNA. Further comparisons, and a more thorough sampling of populations, especially in central and eastern Europe, are necessary in order to clarify this point. For the time being, it seems that strong evolutionary inferences can only be based on the consensus among many loci, and the results obtained for most of the loci point to population expansions, probably, in an east–west direction.

The role of the area around the Mediterranean Sea in the evolution of the European genetic diversity deserves further attention. Should nuclear genes also show increased levels of populations structuring in southern Europe, a model of Neolithic expansion could be modified, by incorporating higher migration rates in the South. The central and northern European populations, conversely, would appear to be affected to a more limited extent by the genes carried by the immigrating Neolithic farmers. At any rate, it seems unlikely that Mesolithic re-expansions from glacial refugia may have resulted in clinal patterns, along a southeast–northwest axis. The current European gene pool may be the result of an initial, Upper Palaeolithic colonization of the continent, or the later, Neolithic diffusion of early farmers, or both. More recent demographic processes do not appear to be the most parsimonious explanation for these results. The relationships between pairs of populations have been demonstrably affected by historically-documented migrations (Sokal 1988). However, no population movements occurring after the Neolithic period seems to have had the potential to leave a mark on genetic diversity at a continental scale (Cavalli-Sforza *et al.* 1994). Simulation studies of even complex prehistoric processes also support this view (Barbujani *et al.* 1995). For the future, more modelling, larger sets of data for different genome regions, and computer simulations, seem to be necessary if we are to solve for good the controversy on the origins of the European gene pool.

Acknowledgements

We are grateful to Giorgio Bertorelle and Eduardo Tarazona-Santos for many suggestions to this work; the former also developed for this study an upgraded version of the AIDA software. We also thank Bryan Sykes for giving us access to their unpublished data. This study was supported by grants of the Italian Ministry of University (COFIN 97 and Concerted Actions Italy-Spain), of the Italian CNR (contract 96.01182.PF36 and 97.00683.PF36), and of the Dirección General de Investigación Científico Técnica (DGICT, Spain; grant PB95-0267-C02-01). FC is the recipient of a postdoctoral return contract from DGICT.

References

Ammerman, A.J. & L.L. Cavalli-Sforza, 1984. *The Neolithic Transition and the Genetics of Populations in Europe.* Princeton (NJ): Princeton University Press.

Anderson, S., T. Bankier, B.G. Barrel, M.H.L. De Bruijn, A.R. Coulson, J. Drouin, I.C. Eperon, D.P. Nierlich, B.A. Roe, F. Sanger, P.H. Schreier, A.J.H. Smith, R. Staden & I.G. Young, 1981. Sequence and organization of the human mitochondrial genome. *Nature* 290, 457–65.

Barbujani, G., 1987. Autocorrelation of gene frequencies under isolation by distance. *Genetics* 117, 777–82.

Barbujani, G. & A. Pilastro, 1993. Genetic evidence on origin and dispersal of human populations speaking languages of the Nostratic macrofamily. *Proceedings of the National Academy of Science of the USA* 90, 4670–73.

Barbujani, G., R.R. Sokal & N.L. Oden, 1995. Indo-European origins: a computer simulation test of five hypotheses. *American Journal of Physical Anthropology* 96, 109–32.

Belledi, M., E.S. Poloni, R. Casalotti, F. Conterio, I. Mikerezi, J. Tagliavini & L. Excoffier, 2000. Maternal and paternal lineages in Albania and the genetic structure of Indo-European populations. *European Journal of Human Genetics* 8, in press.

Bertorelle, G. & G. Barbujani, 1995. Analysis of DNA diversity by spatial autocorrelation. *Genetics* 140, 811–19.

Bertranpetit, J., J. Sala, F. Calafell, P.A. Underhill, P. Moral & D. Comas, 1996. Human mitochondrial DNA variation and the origin of Basques. *Annals of Human Genetics* 59, 63–81.

Calafell, F., P. Underhill, A. Tolun, D. Angelicheva & L. Kalaydjieva, 1996. From Asia to Europe: mitochondrial DNA sequence variability in Bulgarians and Turks. *Annals of Human Genetics* 60, 35–49.

Casalotti, R., L. Simoni, M. Belledi & G. Barbujani, 1999. Y-chromosome polymorphism and the origins of the European gene pool. *Proceedings of the Royal Society of London: Biology* 266, 1–7.

Cavalli-Sforza, L.L., P. Menozzi & A. Piazza, 1994. *The History and Geography of Human Genes.* Princeton (NJ): Princeton University Press.

Chikhi, L., G. Destro-Bisol, G. Bertorelle, V. Pascali & G. Barbujani, 1998a. Clines of nuclear DNA markers suggest a largely Neolithic ancestry of the European gene pool. *Proceedings of the National Academy of Science of the USA* 95, 9053-8.

Chikhi, L., G. Destro-Bisol, V. Pascali, V. Baravelli, M. Dobosz & G. Barbujani, 1998b. Clinal variation in the nuclear DNA of Europeans. *Human Biology* 70, 643–57.

Comas, D., F. Calafell, E. Mateu, A. Pérez-Lezaun & J. Bertranpetit, 1996. Geographic variation in human mitochondrial DNA control region sequence: the population history of Turkey and its relationship to the European populations. *Molecular Biology and Evolution* 13, 1067–77.

Comas, D., F. Calafell, E. Mateu, A. Pérez-Lezaun, E. Bosch & J. Bertranpetit, 1997. Mitochondrial DNA variation and the origin of the Europeans. *Human Genetics* 99, 443–9.

Comas, D., F. Calafell, N. Bendukidze, L. Fananas & J. Bertranpetit, 2000. Georgian and Kurd mtDNA sequence analysis shows a lack of correlation between languages and female genetic lineages. *The American Journal of Physical Anthropology* 112, 5–16.

Côrte-Real, H.B., V.A. Macaulay, M.B. Richards, G. Hariti, M.S. Issad, A. Cambon-Thomsen, S. Papiha, J. Bertranpetit & B.C. Sykes, 1996. Genetic diversity in the Iberian Peninsula determined from mitochondrial sequence analysis. *Annals of Human Genetics* 60, 331–50.

Di Rienzo, A. & A.C. Wilson, 1991. Branching pattern in the evolutionary tree for human mitochondrial DNA. *Proceedings of the National Academy of Science of the USA* 88, 597–601.

Dupuy, B.M. & B. Olaisen, 1996. MtDNA sequences in the Norwegian Saami and main populations, in *Advances in Forensic Haemogenetics*, vol. 6, eds. A. Carracedo, B. Brinkmann & W. Bär. Berlin: Springer-Verlag, 23–5.

Excoffier, L., P.E. Smouse & J. Quattro, 1992. Analysis of molecular variance inferred from metric distances among DNA haplotypes: application to human mitochondrial DNA restriction data. *Genetics* 131, 479–91.

Francalacci, P., J. Bertranpetit, F. Calafell & P.A. Underhill, 1996. Sequence diversity of the control region of mitochondrial DNA in Tuscany and its implications for the peopling of Europe. *American Journal of Physical Anthropology* 100, 443–60.

Handt, O., M. Richards, M. Trommsdorff, C. Kilger, J.

Simanainen, O. Georgiev, K. Bauer, A. Stone, R. Hedges, W. Schaffner, G. Utermann, B. Sykes & S. Pääbo, 1994. Molecular genetic analyses of the Tyrolean Ice Man. *Science* 264, 1775–8.

Lutz, S., H.J. Weisser & J.P. Heizmann, 1998. Location and frequency of polymorphic positions in the mtDNA control region of individuals from Germany. *International Journal of Legal Medicine* 111, 67–77.

Macaulay, V., M. Richards, E. Hickey, E. Vega, F. Cruciani, V. Guida, R. Scozzari, B. Bonne-Tamir, B. Sykes & A. Torroni, 1999. The emerging tree of west Eurasian mtDNAs: a synthesis of control-region sequences and RFLPs. *American Journal of Human Genetics* 64, 232–49.

Malyarchuk, B.A., 1998. Mitochondrial DNA markers and genetic demographic processes in Neolithic Europe. *Russian Journal of Genetics* 7, 842–5.

Menozzi, P., A. Piazza & L.L. Cavalli-Sforza, 1978. Synthetic maps of human gene frequencies in Europeans. *Science* 201, 786–92.

Parson, W., T.J. Parsons, R. Scheithauer & M.M. Holland, 1998. Population data for 101 Austrian Caucasian mitochondrial DNA d-loop sequences: application of mtDNA sequence analysis to a forensic case. *International Journal of Legal Medicine* 111, 124–32.

Pennington, R.L., 1996. Causes of early population growth. *American Journal of Physical Anthropology* 99, 259–74.

Piercy, R., K.M. Sullivan, N. Benson & P. Gill, 1993. The application of mitochondrial DNA typing to the study of white Caucasian genetic identification. *International Journal of Legal Medicine* 106, 85–90.

Pinto, F., A.M. Gonzalez, M. Hernandez, J.M. Larruga & V.M. Cabrera, 1996. Genetic relationship between the Canary Islanders and their African and Spanish ancestors inferred from mitochondrial DNA sequences. *Annals of Human Genetics* 60, 321–30.

Pult, I., A. Sajantila, J. Simanainen, O. Georgiev, W. Schaffner & S. Pääbo, 1994. Mitochondrial DNA sequences from Switzerland reveal striking homogeneity of European populations. *Biological Chemistry Hoppe Seyler* 375, 837–40.

Rendine, S., A. Piazza & L.L. Cavalli-Sforza, 1986. Simulation and separation by principal components of multiple demic expansions in Europe. *American Naturalist* 128, 681–706.

Renfrew, C., 1987. *Archaeology and Language: the Puzzle of Indo-European Origins.* London: Jonathan Cape.

Richards, M., H. Côrte-Real, P. Forster, V. Macaulay, H. Wilkinson-Herbots, A. Demaine, S. Papiha, R. Hedges, H-J. Bandelt & B. Sykes, 1996. Paleolithic and Neolithic lineages in the European mitochondrial gene pool. *American Journal of Human Genetics* 59, 185–203.

Richards, M., V.A. Macaulay, H-J. Bandelt & B.C. Sykes, 1998. Phylogeography of mitochondrial DNA in western Europe. *Annals of Human Genetics* 62, 241–60.

Sajantila, A., P. Lahermo, T. Anttinen, M. Lukka, P. Sistonen, M.L. Savontaus, P. Aula, L. Beckman, L. Tranebjaerg, T. Geddedahl, L. Isseltarver, A. Dirienzo & S. Pääbo, 1995. Genes and languages in Europe: an analysis of mitochondrial lineages. *Genome Research* 5, 42–52.

Sajantila, A., A.H. Salem, P. Savolainen, K. Bauer, C. Gierig & S. Pääbo, 1996. Paternal and maternal DNA lineages reveal a bottleneck in the founding of the Finnish population. *Proceedings of the National Academy of Sciences of the USA* 93, 12,035–9.

Salas, A., D. Comas, M.V. Lareu, J. Bertranpetit & A. Carracedo, 1998. MtDNA analysis of the Galician population: a genetic edge of European variation. *European Journal of Human Genetics* 6, 365–75.

Seielstad, M., E. Minch & L.L. Cavalli-Sforza, 1998. Genetic evidence for a higher female migration rate in humans. *Nature Genetics* 20, 278–80.

Semino, O., G. Passarino, A. Brega, M. Fellous & A.S. Santachiara-Benerecetti, 1996. A view of the Neolithic demic diffusion in Europe through two Y chromosome-specific markers. *American Journal of Human Genetics* 59, 964–8.

Sokal, R.R., 1988. Genetic, geographic and linguistic distances in Europe. *Proceedings of the National Academy of Sciences of the USA* 85, 1722–6.

Sokal, R.R. & P. Menozzi, 1982. Spatial autocorrelation of HLA frequencies in Europe supports demic diffusion of early farmers. *American Naturalist* 119, 1–17.

Sokal, R.R. & N.L. Oden, 1978. Spatial autocorrelation in biology. *Biological Journal of the Linnean Society* 10, 199–249.

Sokal, R.R., R.M. Harding & N.L. Oden, 1989. Spatial patterns of human gene frequencies in Europe. *American Journal of Physical Anthropology* 80, 267–94.

Stenico, M., L. Nigro, G. Bertorelle, F. Calafell, M. Capitanio, C. Corrain & G. Barbujani, 1996. High mitochondrial sequence diversity in linguistic isolates of the Alps. *American Journal of Human Genetics* 59, 1363–75.

Sykes, B., 1999. The molecular genetics of European ancestry. *Philosophical Transactions of the Royal Society of London: Biology* 354, 131–9.

Torroni, A., M.T. Lott, M.F. Cabell, Y.S. Chen, L. Lavergne & D.C. Wallace, 1994. MtDNA and the origin of Caucasians: identification of ancient Caucasian-specific haplogroups, one of which is prone to a recurrent somatic duplication in the D-loop region. *American Journal of Human Genetics* 55, 760–76.

Torroni, A., H-J. Bandelt, L. D'Urbano, P. Lahermo, P. Moral, D. Sellitto, C. Rengo, P. Forster, M.L. Savontaus, B. Bonné-Tamir & R. Scozzari, 1998. MtDNA analysis reveals a major late Paleolithic population expansion from southwestern to northeastern Europe. *American Journal of Human Genetics* 62, 1137–52.

Chapter 14

Genetic Data and the Colonization of Europe: Genealogies and Founders

Martin Richards & Vincent Macaulay

DNA variation can be used to infer aspects of the settlement of new territory and of subsequent gene flow. We here clarify issues critical to this task. The first is the question of genealogical resolution. We make the point that fine-grained inference of gene flow requires a sufficiently refined estimate of the genealogy of the locus under study. This depends on the mutational properties of the system, as well as an intelligent choice of markers. The second issue concerns the estimation of the time of settlement. We sample likely source populations in order to identify founder types and use the diversity derived from these types in the descendant population to assess time depth. Here we argue that 'founder analysis', which has been applied in the past to the Americas and the Pacific, can be applied to more complex processes, such as the multiple waves of settlement and expansion in Europe.

Questions of resolution

Most researchers working on mitochondrial DNA (mtDNA) variation in humans have employed one of two techniques: analysis of the entire mitochondrial genome by means of restriction enzymes (usually 14; referred to as high-resolution restriction-fragment length polymorphism, or RFLP, analysis), or direct sequencing of the hypervariable control region (or, more usually, the most informative segment, referred to as HVS-I). Although the former method tends to be rather more informative, the majority of workers have focused on the latter, which has the advantages of being relatively inexpensive and technically less demanding. Unfortunately, the exclusive use of HVS-I sequence data in analyses has led the field into considerable difficulties (e.g. Maddison *et al.* 1992).

Mitochondrial DNA, like the Y chromosome, is potentially a highly informative marker for human prehistoric expansions and migrations, but its potential has often not been fully realized due simply to problems of phylogenetic resolution. Both systems have a lower effective population size than autosomal loci, rendering them more sensitive to founder effects and genetic drift — a useful property for markers which are used to detect geographical substructure in the human population. However, the main reason that they are potentially so informative is because of the possibility of estimating their phylogeny. Because both are non-recombining, there is a single tree associated with each system, the genealogy of the individuals who contained mtDNAs or Y chromosomes ancestral to those in a modern sample. On top of which, there is a sufficient number of informative phylogenetic characters to be able to recover a significant proportion of that genealogy. Although broad demographic models — such as population expansion *versus* stationarity — can often be distinguished without highly resolved genealogies (e.g. Rogers 1995; Reich & Goldstein 1998), fine-scale inferences about gene flow, small-scale dispersals and founder events (so-called 'phylogeographic analysis': Avise *et al.* 1987; Templeton *et al.* 1995) can only be made from a finely resolved (in an ideal case, bifurcating) genealogy. A lack of informative mutations can lead to unrecognized migrations by ancestors of the sample and a concomitant loss in the power to detect gene flow.

The shape of the genealogy of the ancestors of a sample of non-recombining DNA is independent of the mutation process (under neutrality) but strongly dependent on demographic history. For example, a major demographic expansion from a small number of founders creates a 'rake-like' genealogy, in which most of the coalescences of lines of descent (events at which two ancestral DNAs have just one 'parent' in the previous generation) occur close to the start of the expansion (when the effective population size was small and, as a result, genetic drift was high: Rogers & Jorde 1995). If enough mutations have occurred during the time since the expansion, this genealogical pattern can be detected as a star-shaped gene tree. Expansions of slower rate result in less star-like trees, and other demographic histories such as prolonged constant population size result in trees with different shapes. Star-like sub-trees are usually seen with human mtDNA, and are attributed to expansions during the Late Pleistocene, although the possibility of selection, which might also generate such a pattern, has not been systematically explored. This star-like property is fortunate, since the time of the most recent common ancestor (TMRCA) of such trees (or sub-trees) can be estimated better, given a calibration of the mutation rate of the system, than the TMRCA of other trees. For mtDNA, at least, there is some confidence in the mutation rate (Macaulay *et al.* 1997).

The only way to estimate the genealogy is to study the (independent) effect of mutation. The mutational process can be pictured as randomly scattering mutational hits over a pre-existing genealogy, rendering at least some of the branching structure visible. Clearly, the more mutations that are thrown at the genealogy, the more of the structure will be revealed. This is where the great advantage of the non-recombining mtDNA and Y chromosome becomes apparent. Most nuclear loci undergo recombination every generation, so that the length of a continuous sequence that can be traced back through the genealogy becomes progressively shorter as time goes by. Furthermore, the base substitution rate of nuclear loci is low, so that the mutations are scattered much more thinly across the genealogy. The result is a phylogenetic tree with few mutations and relatively little structure, resolving only a small fraction of coalescences in the genealogy, as in the β-globin gene tree (Harding *et al.* 1997), or the tree of the relatively large X-chromosome locus Xq13.3 (Kaessmann *et al.* 1999). Because they are non-recombining, mtDNA and the Y chromosome are not subject to this limitation in the same way. The Y chromosome, like other nuclear loci, has a low base substitution rate but has 60 million base pairs of (non-recombining) DNA for mutations to hit. Recent studies have shown that there is indeed a wealth of base substitution markers (known as single nucleotide polymorphisms, or SNPs) on the Y chromosome (Underhill *et al.* 1997). By contrast, mtDNA is only about 16,500 base pairs in length, but it has the added advantage of a mutation rate ten times as fast as nuclear loci for most of its length and ten times faster again in the control region. Therefore, in principle, both mtDNA and the Y chromosome offer much greater opportunities for genealogical reconstruction and demographic inference via this methodology than nuclear systems.

Work on demographic inference using the Y chromosome has only begun to bear fruit in the last few years as new markers have been discovered, whereas mtDNA variation has now been studied quite intensively for more than ten years. Nevertheless, mtDNA research has been hampered because workers have often settled on a particular subset of markers without a critical assessment of their utility for tree reconstruction. In particular, either the first hypervariable segment of the control region (HVS-I) or, less frequently, restriction-fragment length polymorphisms (RFLPs) from around the whole molecule assayed with a given set of enzymes (initially 5 or 6, and subsequently, in the high-resolution version, 12 or 14) have been used. Needless to say, the greatest problems have arisen when workers have focused on low-resolution RFLPs (in particular, the work of Templeton and his co-workers hits difficulties on this score: Templeton 1997), or on HVS-I. In the case of the latter, problems arose especially as a result of applying inappropriate phylogenetic estimation algorithms (designed for interspecific data) and, when these failed, falling back upon rather crude population summary measures that had been developed for use with allele-frequency data from many loci (Richards *et al.* 1997).

More recently, however, more appropriate tree reconstruction procedures have been employed, and these have further highlighted the poor resolution of the mitochondrial characters being assayed. To alleviate this problem, several groups have begun to combine control-region and coding-region data in order to resolve the trees more adequately. This approach has yielded results in essentially just three parts of the world to date: western Eurasia (Europe and the Near East: Macaulay *et al.* 1999), Siberia (Starikovskaya *et al.* 1998; Schurr *et al.* 1999) and America (Brown *et al.* 1998). Once the major branches in the gene tree have been resolved, a phylogeographic approach to detecting prehistoric expansions and migrations can be taken, which examines the geographic distribution and time depth of individual mtDNA types. In this way, some progress has been made towards defining the ancestral mtDNA pools which contributed to the settlements of the Americas in the Late Pleistocene (Torroni *et al.* 1993; Forster *et al.* 1996), the Remote Pacific in the mid-Holocene (Sykes *et al.* 1995; Richards *et al.* 1998a) and Europe, both in the Upper Palaeolithic and the Neolithic (Richards *et al.* 1996; 1998b).

On the basis of the mitochondrial evidence, at least, it seems indisputable now that anatomically modern humans evolved in sub-Saharan Africa and spread only recently to other parts of the world. This remained controversial for so long essentially because of problems of phylogenetic resolution. Both control-region (Vigilant *et al.* 1991) and 12-enzyme RFLP data (Cann *et al.* 1987) taken alone were poorly resolved in PAUP analyses, the root of the tree being outside sub-Saharan Africa in some resolutions (Maddison *et al.* 1992). RFLP analyses employing only six enzymes were, not surprisingly, even worse and led several workers, who had rooted the tree on the most frequent mitochondrial haplotype (Excoffier 1990; Templeton 1993), to suggest a multiregional origin for modern humans. This haplotype was actually the out-of-Africa type, which explained its ubiquitous presence worldwide. High-resolution RFLP analysis was already indicating that most mitochondrial clades, or haplogroups, were continent-specific (Ballinger *et al.* 1992; Torroni *et al.* 1993). The basic structure of the global genealogy can be briefly summarized.

A single HVS-I sequence type, characterized by a T (thymine) at nucleotide position (np) 16223 (using the numbering of the Cambridge Reference Sequence, CRS: Anderson *et al.* 1981), was the sub-Saharan African founder type for virtually all Eurasians approximately 70,000 years ago (Watson *et al.* 1997; Macaulay *et al.* 1999), encompassing not only east and west Eurasians, but also most north Africans (Macaulay *et al.* 1999), and probably Papuans (Sykes *et al.* 1995) and aboriginal Australians as well (van Holst Pellekaan *et al.* 1998; Redd & Stoneking 1999). There is a major split within most non-African populations between mtDNAs with a C at np 16223 (cluster R) and those with the ancestral T: presumably the T→C mutation occurred soon after the 'out-of-Africa' founder event. Both are present in east and west Eurasians, but further branching in the genealogy soon after the geographical split has resulted in clades which are specific either to east or to west, but which are widespread within each region (Fig. 14.1). Incidentally, this major split was not identified by high-resolution RFLP typing, where the deepest split within east Eurasians is between the eastern super-cluster M and other types, which include both R and a substantial minority of mtDNAs which (like M) have 16223T. This highlights the advantage of combining the two kinds of data — even the 14-enzyme RFLP analysis alone fails to resolve the genealogy adequately, although it has recently been extended by incorporating np 10873 (10871*MnlI*), which appears to distinguish cleanly the non-M Eurasian types (10873T, now referred to as haplogroup N) from African L3 types (10873C) (Quintana-Murci *et al.* 1999). Haplogroup N therefore includes both R and the non-M 16223T types in Eurasia. Haplogroup M occurs at highest frequencies (~75 per cent) in the southern parts of the Indian subcontinent (Passarino *et al.* 1996) and at somewhat lower frequencies in most east Eurasian populations, including New Guinea (Stoneking *et al.* 1990; Ballinger *et al.* 1992; Torroni *et al.* 1994b). Preliminary work indicates that the south Asian and east Asian M clades are distinct (Kivisild *et al.* 1999; Quintana-Murci *et al.* 1999).

We have recently been devoting time to resolving the tree in west Eurasia. Deep rooting clades within the region had been identified separately on the basis of HVS-I sequences (Bertranpetit *et al.* 1996; Richards *et al.* 1996; Wilkinson-Herbots *et al.* 1996) and 14-enzyme analysis (Torroni *et al.* 1994a, 1996). The relationship between the two taxonomies had been partially elucidated by typing samples for HVS-I sequences and certain diagnostic RFLPs (Torroni *et al.* 1996): haplogroups I, J, K, T, W and X could be correlated with HVS-I motifs in this way. Later, a study of 67 partial coding-region plus control-region sequences (Hofmann *et al.* 1997) provided yet more information: in particular confirming the single deep 16223T→C mutation mentioned above, and showing that haplogroup K was a sub-clade of haplogroup U (Macaulay *et al.* 1999). In order to synthesize the various schemes, we have performed combined HVS-I sequencing and 14-enzyme analysis, extended by a few additional RFLPs, including an informative character in HVS-II (np 00073: Wilkinson-Herbots *et al.* 1996) on a sample of individuals from northern Israel and the northwest Caucasus (Macaulay *et al.* 1999). One practical result of the study is a set of diagnostic markers for all the major clusters present in the region. These can be used in a hierarchical typing scheme to establish quickly the position in

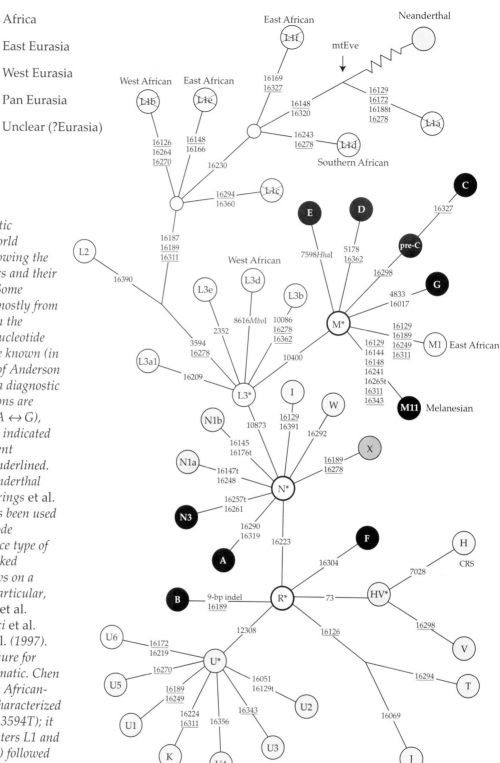

Figure 14.1. *A schematic representation of the world mtDNA phylogeny, showing the principal named clusters and their geographical location. Some diagnostic mutations (mostly from HVS-I) are indicated on the branches; the affected nucleotide position is shown where known (in the numbering system of Anderson et al. 1981), otherwise a diagnostic RFLP, or indel; mutations are transitions (C ↔ T or A ↔ G), unless a transversion is indicated with a 't' suffix; recurrent mutations are shown underlined. The position of the Neanderthal sequence obtained by Krings et al. (1997) is shown and has been used to root the tree at the node representing the sequence type of mitochondrial Eve (marked 'mtEve'). This tree draws on a number of sources, in particular, the papers of Macaulay et al. (1999), Quintana-Murci et al. (1999) and Watson et al. (1997). Note that the nomenclature for African clades is problematic. Chen et al. (1995) defined the African-specific haplogroup L, characterized by +3592HpaI (that is, 3594T); it consisted of two subclusters L1 and L2. Watson et al. (1997) followed this suggestion, referring to −3592HpaI lineages (the remaining African and all Eurasian types) as L3. Unfortunately, the tree of Chen et al. was incorrectly rooted: L1 is not a clade, as the Neanderthal outgroup confirms. Therefore, in the nomenclature of Richards et al. (1998b), the groups labelled here as L1a through to L1f should be redefined (e.g. as L1, L4, L5, etc.). However, pending new data, we leave the earlier system intact, but **sous rature** in the manner of Heidegger (Derrida 1976).*

the tree of a sample. We employed such an approach to characterize a sample of North Algerian Berber speakers, and demonstrated that a cluster which they harbour in high proportion, and which was impossible to classify on the basis of just HVS-I sequencing (Côrte-Real *et al.* 1996), was a sub-clade of haplogroup U, U6, a sister-cluster to the oldest Europe-specific cluster U5, and of a similar age. Hence it is a good candidate for an indigenous North African mtDNA cluster, perhaps carried by the first anatomically modern humans there.

The resolution process is ongoing, with information from newly discovered polymorphisms and newly typed samples shedding light onto old data sets. As an illustration of some of the issues, consider haplogroup H, which is phylogenetically problematic. It is the largest cluster in West Eurasia and is very star-like. In Figure 14.2 is shown the full median network of a set of H mtDNAs (identified on the basis of the two restriction site losses −7025*Alu*I and −14766*Mse*I) that have been typed for HVS-I sequence and high-resolution RFLPs (Macaulay *et al.* 1999; Torroni *et al.* 1999). We also include the recently resequenced Cambridge reference sequence material (Andrews *et al.* 1999). The network necessarily contains all most parsimonious (MP) trees (of length 61 steps). Additional information allows for a tentative reduction of this network almost to a single tree: the links that can be removed are shown dashed. The additional types used are shown as open squares (Antonio Torroni and Martin Richards, unpublished data). They are placed within the network when this is possible; when it is not, they mark the branch that is clarified. For example, in the top left of the diagram, the existence in other samples of the path 16293–16311 (with a pan-European distribution), leading to 16092–16293–16311 (eastern European), in turn leading to 16092–16189–16293–16311 (Caucasian) suggests that we resolve the 16189 *versus* 16311 incompatibility as two 16189s. Similarly, the extant type 16129–16316 (pan-European) suggests a parallelism at the hypervariable site 16093. The existence of the type +4793e (in a Sardinian with the CRS in HVS-I), in conjunction with the 16517*Hae*III status, suggests that (*contra* Macaulay *et al.* 1999, where +4793e is resolved into two events) this is a unique mutation defining a clade.

Although in this case we can weed out most MP trees by drawing on additional information, it is nevertheless unlikely that the reduced network contains the correct gene tree. There are likely to be 'hidden mutations' (Bandelt *et al.* 2000) present. The problem is that in such a star with a large number of branches, mutations are likely to hit the same site (particularly a hypervariable site) on different branches. The lack of intermediates along the branches (especially in small samples) makes these well nigh impossible to resolve. As a result, the parallelism appears as a single deep mutation clustering the two branches, and the tips seem more related than in fact they are. A potential example of this might be the two longer branches sharing (the hypervariable) 16362 mutation in Figure 14.2. This is just what is called 'long-branch attraction' in the tree-reconstruction literature: however here it is a problem even for short branches without intermediates, simply because there are so many of them. An unfortunate result of this is that if one is trying to detect gene flow by placing migration events parsimoniously on the tree, one can be misled, since the artefactual node generated in the process can look like a founder type, if its descendants occur in different populations.

The phylogeny for some parts of the world is still ambiguous. Whilst West Eurasian variation is now quite well understood, and a sketch at least exists for the African phylogeny (where many of the clades are much more distinct at the HVS-I level) few workers have attempted to combine RFLP and HVS-I data for East Asia (Shields *et al.* 1993; Redd *et al.* 1995; Sykes *et al.* 1995; Kolman *et al.* 1996; Brown *et al.* 1998; Starikovskaya *et al.* 1998). Moreover, these studies were in fact generally motivated by studies of Native Americans or Polynesians, rather than focusing on East Asia in its own right. South Asia also remains underexplored at the present time. Nevertheless, some conclusion can be drawn from inspection of the existing data. Two major subclusters of R can be discerned in the HVS-I data, although their boundaries are unclear. They are haplogroup B — characterized by the so-called 'Asian-specific' 9-base-pair deletion and one or more control-region variants — and haplogroup F, characterized by a transition in the control region at np 16304. A further major subcluster of R seems localized to New Guinea (Redd & Stoneking 1999). Several further clusters within the 16223T group can also be identified. However, overall, the genealogy from HVS-I data remains poorly resolved. As an illustration, less than 50 per cent of Japanese (Horai *et al.* 1996) or Korean (Lee *et al.* 1997) sequences can be resolved in the genealogy (*contra* Horai *et al.* 1996). This implies that less than half of the East Eurasian HVS-I sequences currently available for analysis (approximately 1500) are at present informative for phylogeographic studies. As one moves into Southeast Asia, the situation becomes

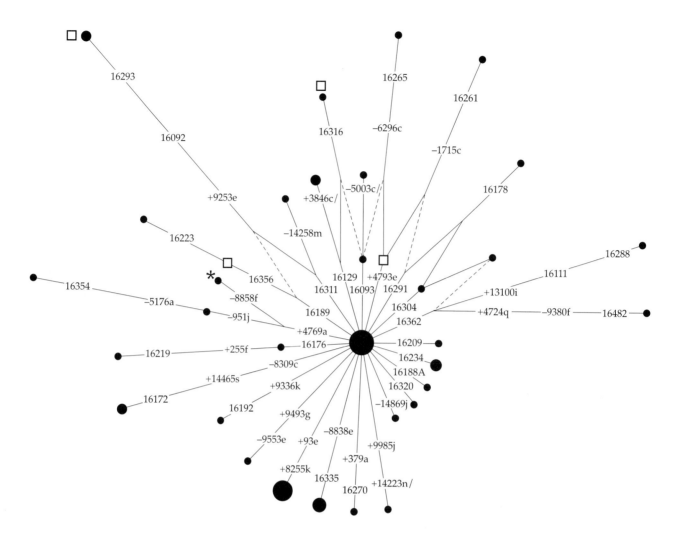

Figure 14.2. *The full median network of a sample of haplogroup H mtDNAs (Macaulay et al. 1999; Torroni et al. 1999). The node marked '*' is the type of the recently resequenced CRS material (Andrews et al. 1999), which is now no longer a chimera, and which incorporates corrections to the sequence. We ignore the hypervariable 16517HaeIII character. RFLP changes are indicated as site gains ('+') or losses ('−') with respect to the node at the centre of the star; the position in the reference sequence of the start of the recognition sequence is shown with a lower case suffix which indicates the enzyme involved (see Macaulay et al. 1999 for details). HVS-I mutations are indicated by numbers in the 16000s. The filled circles represent sampled individuals. A circle's area is proportional to the frequency in the sample: the smallest circles are singletons, the largest represents 13 individuals. Unfilled squares signal additional external information as discussed in the main text.*

even worse: as table 2 of Torroni *et al.* (1994b) shows, only 46 per cent of Vietnamese, 35 per cent of Malays, 19 per cent of Sabah aborigines and only 10 per cent of Malay aborigines fall into well-characterized RFLP haplogroups. Australasia also awaits more complete characterization.

Founder analysis

Three components of the phylogeographic approach that has been developed for the study of demographic history are: i) the gene tree or network; ii) the geographic distribution of the samples; and iii) the time depths of phylogenetic clusters. A well-resolved phylogeny is critical, yet, as we have discussed, HVS-I data rarely yields unambiguous phylogenies, owing to its high rate of recurrent mutation. These ambiguities can be managed using a parsimony-based network method (Bandelt *et al.* 1995; 2000), which summarizes alternative trees in a single diagram. However, as the time depth of the genealogy increases, the difficulties become quite severe. Since

HVS-I is useful for dating, and does include stable motifs for many important clades, while high-resolution RFLP analysis is time-consuming and expensive, and often does not resolve the tree sufficiently by itself, an alternative approach is to work with HVS-I sequences supplemented by additional character data assayed by restriction enzyme analysis at selected diagnostic sites. This approach (which is analogous to an approach adopted for Y-chromosome analysis, where a combination of seemingly rather stable biallelic markers and faster evolving microsatellites is used) is very efficient.

The particular phylogeographic approach that we advocate here has become known as 'founder analysis'. It is based on a model of population history that envisages new populations emerging principally as a result of older populations throwing off occasional small groups of colonists who move away and establish a community in new territory elsewhere. The aim is to identify migration events from a putative source population to a putative descendant population. The source population may be identified, for example, on the basis of archaeological evidence, or from elevated levels of genetic diversity compared to the descendant population. In the case of Europe, it is natural to take the Near East (including Anatolia, the region of the Fertile Crescent and its vicinity and the Arabian peninsula) as a potential source for European DNA, since the archaeological record suggests that agriculture evolved in the Fertile Crescent (in particular in Syrio-Palestine) and spread from there (Henry 1989). Going further back in time, it has also been suggested that anatomically modern humans spread first from the region of the Levant into Europe (Mellars 1992), although this is much less clear. The earliest radiocarbon dates are similar for the Levant and for Southeast Europe (approximately 47,000 years uncalibrated for the Levant and greater than 43,000 years uncalibrated for Southeast Europe: Marks & Kaufman 1983; Gilead 1993; Gamble 1999) and it is possible that there could have been some source to the east which may have given rise to both.

Founder analysis aims to identify the subset of genetic diversity in the descendant population which did not arise in the source and hence does not predate the migration event. It attempts to do this by identifying which sequence types moved to the descendant population (the 'founder types' or 'founders') and to measure only the diversity that has accumulated on top of those founder types, while excluding the variation that separates them. The basic idea is illustrated in Figure 14.3. The pre-existing

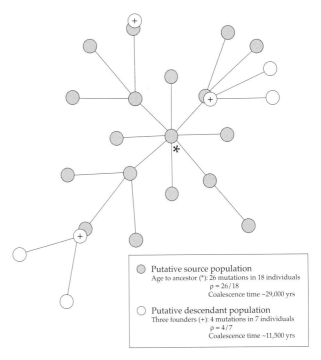

Figure 14.3. *The principles of founder analysis. Putative founder types are identified as sequence matches between source and descendant populations and the diversity that has accumulated* within the descendant population *is estimated using the statistic ρ, the average number of transitions from each founder type to the samples in its respective founder cluster. Scaling by a calibrated mutation rate (20,000 yrs per transition here) yields an estimate of the average coalescence time of the founder clusters, and, assuming rapid expansion of the founders in their new territory, an estimate of the migration time. Here we assume that the three founders were involved in the same migration event; however they could also be treated independently.*

diversity is the reason for the frequently cited doctrine that the coalescence time of sequence types within a population tends to be greater than the age of the population itself (Nei *et al.* 1975; Barbujani *et al.* 1998). It seems likely that this is usually the case, although if the new population suffers a prolonged period of reduced size, drift can eliminate so much diversity that the coalescent might actually postdate the founding of the population.

The obvious way to identify founders is to look for matching types, i.e. those haplotypes that occur in both source and descendant populations. Assuming the populations are otherwise isolated, every instance of a match indicates a single migration event, the movement of an ancestral type from one population to the other. To be slightly more sophisticated,

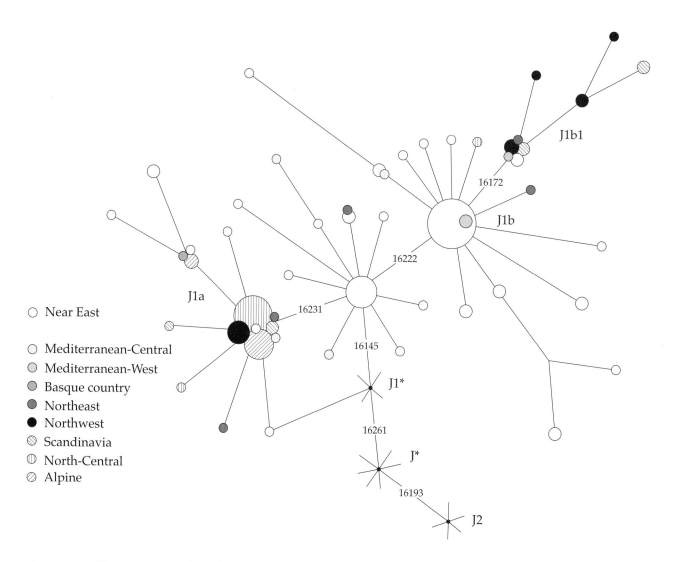

Figure 14.4. *The phylogeography of haplogroup J1. The diagram is a phylogenetic network of samples assigned to J1 on the basis of their HVS-I sequences. Sample origin is indicated by shading, shown in the key. The J* node differs from the CRS by transitions at 16069 and 16126.*

one should check for 'matching' empty nodes in the gene tree or network, as well as for filled extant types; that is, any node with descendants in both populations should be considered a candidate for a founder lineage. This was the approach pioneered by Stoneking and co-workers (Stoneking & Wilson 1989; Stoneking *et al.* 1992) in the Pacific and Torroni *et al.* (1993) for the Americas. The former, applying this analysis to the colonization of Australia, referred to the descendants of each founder as members of a 'maternal clan'.

An actual case is shown in Figure 14.4, which represents part of the phylogeny for haplogroup J in Europe and the Near East. The two central expansion nodes (with motifs 16069T–16126C–16145A–16261T and 16069T–16126C–16145A–16222T–16261T) are largely Near Eastern, suggesting an origin in the Near East with occasional migration of accidentals into Europe (the accidentals occur mostly along the Mediterranean). However, the two main expansion nodes at the tips (with the motifs 16069T–16126C–16145A–16231C–16261T, forming cluster J1a, and 16069T–16126C–16145A–16172C–16222T–16261T, forming cluster J1b1) indicate expansions in Europe (with J1a having expanded earlier, as well, in the Levant). Moreover, the expansions in Europe are geographically specific. Cluster J1a shows a central European pattern, which takes in northwest Europe and Scandinavia, whilst J1b1 indicates an expansion in the Atlantic zone, between Iberia and Britain. Indeed, J1b1 also reaches Norway and Iceland, perhaps as a result of channels opened for the trade of women during the Viking age, whilst male Scandinavian genes were travelling in the opposite direction (Wilson *et al.* submitted).

In practice, the identification of founders can be problematic. Two major factors which are likely to confound the analysis are inadequate sampling and back-migration. As to the first, if there is insufficient sampling of the source population, many founder types will be missed. This was a weakness of our previous work (Richards *et al.* 1996), where we attempted to date European lineages on the basis of a comparison with only 42 individuals from Arabia. Whilst the approach was moderately effective for haplogroup J, which is common in the Arabian peninsula, we underestimated the number of founders in haplogroup H in Europe, since that clade is almost absent from Arabia. The age of founder types can be expected to become lower as the sample size increases until all founders have been identified. (Since founder types can be lost by drift in the source population, an assumption is that the source is sufficiently large that drift is negligible on the time-scale of interest here.)

The second problem — that of back-migration from the descendant into the source population — is particularly acute for the case of Europe and the Near East. In the colonization both of Australia and of the Americas, there seems to have been little back-migration into the Asian source populations. The same can hardly be said to be true of Europe and the Near East. Throughout the historical period, and presumably before, there has been a continual traffic from Europe into the Levant. Therefore, some way of identifying and allowing for back-migration is necessary.

The way we have dealt with this problem is to develop criteria for identifying candidate sequence types that have evolved in the source population. Such types would be expected each to have given rise to derived types in the source, so that the more derived types of each candidate that are sampled in the source, the greater the confidence that that candidate type has evolved *in situ*. A hierarchy of such threshold criteria, demanding more or less derived types in order for a sequence type to qualify as a founder, can be established, so that the effect of varying the stringency can be explored empirically. One approach in the European/Near Eastern context is to explore the effect of such criteria on the distribution of two haplogroups for which there is good evidence of a European origin: U5 and V. Haplogroup

U5 is believed to have originated in Europe approximately 45,000–50,000 years ago, with the arrival of the first ('Cro-Magnon') anatomically modern humans. It occurs in Europe at a frequency of about 10 per cent, but in the Near East at only 1–1.5 per cent. Most of the Near Eastern sequence types are towards the tips of the phylogeny, implying that they are more recent arrivals. This suggests a substantial level of back-migration, of the order of 10–15 per cent. Similarly, haplogroup V is believed to have arisen in southwestern Europe about 13,000 years ago (Torroni *et al.* 1998). However, since V has had less time to evolve and as a result the ancestral type is still by far the most common type, it is less clear-cut whether V in the Near East is ancestral or migrant.

The founder analysis makes a number of other assumptions. Since we are dealing with control-region variation, we need to be able to take into account recurrent mutation. Parallelisms and back-mutations would generate matches between source and descendant populations that were the result of convergence rather than common descent. We attempt to minimize this effect by strongly down-weighting very fast sites (16129, 16189, 16311 and 16362) in the network reduction procedure, so that mutations at these sites are resolved into multiple events whenever possible. However, as the discussion of the haplogroup H network above indicated, parallelisms are unlikely always to be identified, and indeed can generate phantom founders.

Finally, the founder analysis assumes that a founder event has occurred. It is not clear that it would be able to yield a meaningful result if populations had evolved as a result of roughly equal fission events, a model often assumed in classical population genetics (such as when tree models are applied with populations as the taxa). A number of cases in the mitochondrial record support the suggestion that new populations are often formed by quite severe founder events rather than by fission. The most obvious, as discussed above, is the founding of Eurasia from (most probably) a single African mitochondrial type (defined in HVS-I by a type matching the CRS except for a thymine at 16223). However, the settlement of the Americas (where there are less than ten major HVS-I founder types) and of the remote Pacific (where there are less than five, of which only one is very common) provide other examples. The strong effects of drift on mtDNA (and, even more so, on the Y chromosome) seem to act in favour of such an analysis in these cases.

A preliminary founder analysis of European

mtDNAs (Richards *et al.* 1996), while limited by the paucity of available Near Eastern samples, resulted in a novel outcome, suggesting that most of the modern European mtDNA landscape was formed neither in Early Upper Palaeolithic colonization (Sherry *et al.* 1994) nor as a result of Neolithic demic diffusion from the Near East (Ammerman & Cavalli-Sforza 1984), but rather in Lateglacial re-expansions within Europe itself. The suggestion was that the mtDNAs which had arrived and evolved in Europe 20,000–50,000 years ago, during the Early or Middle Upper Palaeolithic, were compressed into refugia in the southern part of the continent with the onset of the Last Glacial Maximum (LGM) — around 20,000 years ago — and subsequently re-expanded dramatically as the climate improved in the Late Upper Palaeolithic — from about 16,000 years ago. This process, which is also seen in other species (Cooper *et al.* 1995; Taberlet *et al.* 1998), has been documented in some detail using radiocarbon evidence from the western part of the continent (Housley *et al.* 1997) and it seems likely that the expansion of the Magdalenian complex was a demic-diffusion-like process from the southwest into either thinly-populated or empty territory. Moreover, Lateglacial population densities are believed to have been much higher than prior to the LGM (Gamble 1993). This picture has received support from an analysis of mtDNA haplogroup V (Torroni *et al.* 1998), which, it has been suggested, evolved in southwestern Europe around the time of the expansion and was dispersed with it, probably along with haplogroup H. We are currently in the final stages of retesting this picture by carrying out a new founder analysis of European mtDNA with a greatly enlarged sample size from the Near East. It remains to be seen whether this picture of a largely Lateglacial ancestry for the majority of European mtDNAs will survive.

Acknowledgements

We thank Hans-Jürgen Bandelt, David Goldstein and Antonio Torroni for stimulating discussions and The Wellcome Trust for financing the work described here through its fellowship programme.

References

Ammerman, A.J. & L.L. Cavalli-Sforza, 1984. *The Neolithic Transition and the Genetics of Populations in Europe.* Princeton (NJ): Princeton University Press.

Anderson, S., A.T. Bankier, B.G. Barrell, M.H.L. de Bruijn, A.R. Coulson, J. Drouin, I.C. Eperon, D.P. Nierlich, B.A. Roe, F. Sanger, P.H. Schreier, A.J.H. Smith, R. Staden & I.G. Young, 1981. Sequence and organization of the human mitochondrial genome. *Nature* 290, 457–65.

Andrews, R.M., I. Kubacka, P.F. Chinnery, R.N. Lightowlers, D.M. Turnbull & N. Howell, 1999. Reanalysis and revision of the Cambridge reference sequence for human mitochondrial DNA. *Nature Genetics* 23, 147.

Avise, J., J. Arnold, R.M. Ball, E. Bermingham, T. Lamb, J.E. Neigel, C.A. Reeb & N.C. Saunders, 1987. Intraspecific phylogeography: the molecular bridge between population genetics and systematics. *Annual Review of Ecology and Systematics* 18, 489–522.

Ballinger, S.W., T.G. Schurr, A. Torroni, Y.Y. Gan, J.A. Hodge, K. Hassan, K-H. Chen & D.C. Wallace, 1992. Southeast Asian mitochondrial DNA analysis reveals genetic continuity of ancient mongoloid migrations. *Genetics* 130(1), 139–52.

Bandelt, H-J., P. Forster, B.C. Sykes & M.B. Richards, 1995. Mitochondrial portraits of human populations using median networks. *Genetics* 141, 743–53.

Bandelt, H-J., V. Macaulay & M. Richards, 2000. Median networks: speedy construction and greedy reduction, one simulation and two case studies from human mtDNA. *Molecular Phylogenetics and Evolution* 16, 8–28.

Barbujani, G., G. Bertorelle & L. Chikhi, 1998. Evidence for Paleolithic and Neolithic gene flow in Europe. *American Journal of Human Genetics* 62, 488–91.

Bertranpetit, J., F. Calafell, D. Comas, A. Pérez-Lezaun & E. Mateu, 1996. Mitochondrial DNA sequences in Europe: an insight into population history, in *Molecular Biology and Human Diversity*, eds. A.J. Boyce & C.G.N. Mascie-Taylor. Cambridge: Cambridge University Press, 112–29.

Brown, M.D., S.H. Hosseini, A. Torroni, H-J. Bandelt, J.C. Allen, T.G. Schurr, R. Scozzari, F. Cruciani & D.C. Wallace, 1998. MtDNA haplogroup X: an ancient link between Europe/Western Asia and North America? *American Journal of Human Genetics* 63, 1852–61.

Cann, R.L., M. Stoneking & A.C. Wilson, 1987. Mitochondrial DNA and human evolution. *Nature* 325, 31–6.

Chen, Y-S., A. Torroni, L. Excoffier, A.S. Santachiara-Benerecetti & D.C. Wallace, 1995. Analysis of mtDNA variation in African populations reveals the most ancient of all human continent-specific haplogroups. *American Journal of Human Genetics* 57, 133–49.

Cooper, S.J., K.M. Ibrahim & G.M. Hewitt, 1995. Postglacial expansion and genome subdivision in the European grasshopper *Chorthippus parallelus*. *Molecular Ecology* 4, 49–60.

Côrte-Real, H.B.S.M., V.A. Macaulay, M.B. Richards, G. Hariti, M.S. Issad, A. Cambon-Thomsen, S. Papiha, J. Bertranpetit & B.C. Sykes, 1996. Genetic diversity in the Iberian peninsula determined from mitochondrial sequence analysis. *Annals of Human Genetics* 60, 331–50.

Derrida, J., 1976. *Of Grammatology*. Baltimore (MD): Johns

Hopkins University Press.

Excoffier, L., 1990. Evolution of human mitochondrial DNA: evidence for departure from a pure neutral model of populations at equilibrium. *Journal of Molecular Evolution* 30, 125–39.

Forster, P., R. Harding, A. Torroni & H-J. Bandelt, 1996. Origin and evolution of Native American mtDNA variation: a reappraisal. *American Journal of Human Genetics* 59, 935–45.

Gamble, C., 1993. *Timewalkers: the Prehistory of Global Colonization*. Stroud: Alan Sutton.

Gamble, C., 1999. *The Palaeolithic Societies of Europe*. Cambridge: Cambridge University Press.

Gilead, I., 1993. The Upper Paleolithic in the Levant. *Journal of World Prehistory* 5, 105–54.

Harding, R.M., S.M. Fullerton, R.C. Griffiths, J. Bond, M.J. Cox, J.A. Schneider, D.S. Moulin & J.B. Clegg, 1997. Archaic African *and* Asian lineages in the genetic ancestry of modern humans. *American Journal of Human Genetics* 60, 772–89.

Henry, D.O., 1989. *From Foraging to Agriculture: the Levant at the End of the Ice Age*. Philadelphia (PA): University of Pennsylvania Press.

Hofmann, S., M. Jaksch, R. Bezold, S. Mertens, S. Aholt, A. Paprotta & K.D. Gerbitz, 1997. Population genetics and disease susceptibility: characterization of central European haplogroups by mtDNA gene mutations, correlations with D loop variants and association with disease. *Human Molecular Genetics* 6, 1835–46.

Horai, S., K. Murayama, K. Hayasaka, S. Matsubayashi, Y. Hattori, G. Fucharoen, S. Harihara, K.S. Park, K. Omoto & I.H. Pan, 1996. MtDNA polymorphism in East Asian populations, with special reference to the peopling of Japan. *American Journal of Human Genetics* 59, 579–90.

Housley, R.A., C.S. Gamble, M. Street & P. Pettitt, 1997. Radiocarbon evidence for the lateglacial human recolonisation of northern Europe. *Proceedings of the Prehistoric Society* 63, 25–54.

Kaessmann, H., F. Heissig, A. von Haeseler & S. Pääbo, 1999. DNA sequence variation in a non-coding region of low recombination on the human X chromosome. *Nature Genetics* 22, 78–81.

Kivisild, T., M.J. Bamshad, K. Kaldma, M. Metspalu, E. Metspalu, M. Reidla, S. Laos, J. Parik, W.S. Watkins, M.E. Dixon, S.S. Papiha, S.S. Mastana, M.R. Mir, V. Ferak & R. Villems, 1999. Deep common ancestry of Indian and western-Eurasian mitochondrial DNA lineages. *Current Biology* 9, 1331–4.

Kolman, C., N. Sambuughin & E. Bermingham, 1996. Mitochondrial DNA analysis of Mongolian populations and implications for the origin of New World founders. *Genetics* 142, 1321–34.

Krings, M., A. Stone, R.W. Schmitz, H. Krainitzki, M. Stoneking & S. Pääbo, 1997. Neandertal DNA sequences and the origin of modern humans. *Cell* 90, 19–30.

Lee, S.D., C.H. Shin, K.B. Kim, Y.S. Lee & J.B. Lee, 1997. Sequence variation of mitochondrial DNA control region in Koreans. *Forensic Science International* 87, 99–116.

Macaulay, V., M.B. Richards, P. Forster, K.A. Bendall, E. Watson, B.C. Sykes & H-J. Bandelt, 1997. MtDNA mutation rates — no need to panic. *American Journal of Human Genetics* 61, 983–6.

Macaulay, V., M. Richards, E. Hickey, E. Vega, F. Cruciani, V. Guida, R. Scozzari, B. Bonné-Tamir, B. Sykes & A. Torroni, 1999. The emerging tree of west Eurasian mtDNAs: a synthesis of control-region sequences and RFLPs. *American Journal of Human Genetics* 64, 232–49.

Maddison, D., M. Ruvolo & D.L. Swofford, 1992. Geographic origins of human mitochondrial DNA: phylogenetic evidence from control region sequences. *Systematic Biology* 41, 111–24.

Marks, A.E. & D. Kaufman, 1983. Boker Tachtit: the artefacts, in *Prehistory and Paleoenvironments in the Central Negev, Israel*, vol. III, ed. A.E. Marks. Dallas (TX): Southern Methodist University Press, 69–125.

Mellars, P., 1992. Archaeology and the population-dispersal hypothesis of modern human origins in Europe. *Philosophical Transactions of the Royal Society of London* series B 337, 225–34.

Nei, M., T. Maruyama & R. Chakraborty, 1975. The bottleneck effect and genetic variability in populations. *Evolution* 29, 1–10.

Passarino, G., O. Semino, L.G. Bernini & S. Santachiara-Benerecetti, 1996. Pre-Caucasoid and Caucasoid genetic features of the Indian population, revealed by mtDNA polymorphisms. *American Journal of Human Genetics* 59, 927–34.

Quintana-Murci, L., O. Semino, H-J. Bandelt, G. Passarino, K. McElreavey & A.S. Santachiara-Benerecetti, 1999. Genetic evidence for an early exit of *Homo sapiens sapiens* from Africa through eastern Africa. *Nature Genetics* 23, 437–41.

Redd, A.J. & M. Stoneking, 1999. Peopling of Sahul: mtDNA variation in aboriginal Australian and Papua New Guinean populations. *American Journal of Human Genetics* 65, 808–28.

Redd, A.J., N. Takezaki, S.T. Sherry, S.T. McGarvey, A.S.M. Sofro & M. Stoneking, 1995. Evolutionary history of the COII/tRNA^{Lys} intergenic 9-base-pair deletion in human mitochondrial DNAs from the Pacific. *Molecular Biology and Evolution* 12, 604–15.

Reich, D.E. & D.B. Goldstein, 1998. Genetic evidence for a Paleolithic human population expansion in Africa. *Proceedings of the National Academy of Sciences of the USA* 95, 8119–23.

Richards, M., H. Côrte-Real, P. Forster, V. Macaulay, H. Wilkinson-Herbots, A. Demaine, S. Papiha, R. Hedges, H-J. Bandelt & B. Sykes, 1996. Paleolithic and Neolithic lineages in the European mitochondrial gene pool. *American Journal of Human Genetics* 59, 185–203.

Richards, M., V. Macaulay, B. Sykes, P. Pettitt, R. Hedges, P. Forster & H-J. Bandelt, 1997. Reply to Cavalli-

Sforza and Minch. *American Journal of Human Genetics* 61, 251–4.

Richards, M., S. Oppenheimer & B. Sykes, 1998a. MtDNA suggests Polynesian origins in eastern Indonesia. *American Journal of Human Genetics* 63, 1234–6.

Richards, M., V.A. Macaulay, H-J. Bandelt & B.C. Sykes, 1998b. Phylogeography of mitochondrial DNA in western Europe. *Annals of Human Genetics* 62, 241–60.

Rogers, A.R., 1995. Genetic evidence for a pleistocene population explosion. *Evolution* 49, 608–15.

Rogers, A.R. & L.B. Jorde, 1995. Genetic evidence on modern human origins. *Human Biology* 67, 1–36.

Schurr, T.G., R.I. Sukernik, Y.B. Starikovskaya & D.C. Wallace, 1999. Mitochondrial DNA variation in Koryaks and Itel'men: population replacement in the Okhotsk Sea-Bering Sea region during the Neolithic. *American Journal of Physical Anthropology* 108, 1–39.

Sherry, S.T., A.R. Rogers, H. Harpending, H. Soodyall, T. Jenkins & M. Stoneking, 1994. Mismatch distributions of mtDNA reveal recent human population expansions. *Human Biology* 66, 761–75.

Shields, G.F., A.M. Schmiechen, B.L. Frazier, A. Redd, M.I. Voevoda, J.K. Reed & R.H. Ward, 1993. MtDNA sequences suggest a recent evolutionary divergence for Beringian and northern North American populations. *American Journal of Human Genetics* 53, 549–62.

Starikovskaya, Y.B., R.I. Sukernik, T.G. Schurr, A.M. Kogelnik & D.C. Wallace, 1998. MtDNA diversity in Chukchi and Siberian Eskimos: implications for the genetic history of ancient Beringia and the peopling of the New World. *American Journal of Human Genetics* 63, 1473–91.

Stoneking, M. & A.C. Wilson, 1989. Mitochondrial DNA, in *The Colonization of the Pacific: a Genetic Trail*, eds. A.V.S. Hill & S. Serjeantson. Oxford: Oxford University Press, 215–45.

Stoneking, M., L.B. Jorde, K. Bhatia & A.C. Wilson, 1990. Geographic variation in human mitochondrial DNA from Papua New Guinea. *Genetics* 124, 717–33.

Stoneking, M., S.T. Sherry, A.J. Redd & L. Vigilant, 1992. New approaches to dating suggest a recent age for the human mtDNA ancestor. *Philosophical Transactions of the Royal Society of London* Series B 337, 167–75.

Sykes, B., A. Leiboff, J. Low-Beer, S. Tetzner & M. Richards, 1995. The origins of the Polynesians — an interpretation from mitochondrial lineage analysis. *American Journal of Human Genetics* 57, 1463–75.

Taberlet, P., L. Fumagalli, A.G. Wust-Saucy & J.F. Cosson, 1998. Comparative phylogeography and postglacial colonization routes in Europe. *Molecular Ecology* 7, 453–64.

Templeton, A.R., 1993. The 'Eve' hypotheses: a genetic critique and reanalysis. *American Anthropologist* 95, 51–72.

Templeton, A.R., 1997. Testing the Out of Africa replacement hypothesis with mitochondrial DNA data, in *Conceptual Issues in Modern Human Origins Research*, eds. G.A. Clark & C.M. Willermet. New York (NY): Aldine de Gruyter, 329–60.

Templeton, A.R., E. Routman & C.A. Phillips, 1995. Separating population structure from population history: a cladistic analysis of the geographical distribution of mitochondrial DNA haplotypes in the tiger salamander, *Ambystoma tigrinum*. *Genetics* 140, 767–82.

Torroni, A., T.G. Schurr, M.F. Cabell, M.D. Brown, J.V. Neel, M. Larsen, D.G. Smith, C.M. Vullo & D.C. Wallace, 1993. Asian affinities and continental radiation of the four founding Native American mtDNAs. *American Journal of Human Genetics* 53, 563–90.

Torroni, A., M.T. Lott, M.F. Cabell, Y-S. Chen, L. Lavergne & D.C. Wallace, 1994a. MtDNA and the origin of Caucasians: identification of ancient Caucasian-specific haplogroups, one of which is prone to a recurrent somatic duplication in the D-loop region. *American Journal of Human Genetics* 55, 760–76.

Torroni, A., J.A. Miller, L.G. Moore, S. Zamudio, J.G. Zhuang, T. Droma & D.C. Wallace, 1994b. Mitochondrial DNA analysis in Tibet: implications for the origin of the Tibetan population and its adaptation to high altitude. *American Journal of Physical Anthropology* 93, 189–99.

Torroni, A., K. Huoponen, P. Francalacci, M. Petrozzi, L. Morelli, R. Scozzari, D. Obinu, M.L. Savontaus & D.C. Wallace, 1996. Classification of European mtDNAs from an analysis of three European populations. *Genetics* 144, 1835–50.

Torroni, A., H-J. Bandelt, L. d'Urbano, P. Lahermo, P. Moral, D. Sellitto, C. Rengo, P. Forster, M.L. Savontaus, B. Bonné-Tamir & R. Scozzari, 1998. MtDNA analysis reveals a major Late Paleolithic population expansion from southwestern to northeastern Europe. *American Journal of Human Genetics* 62, 1137–52.

Torroni, A., F. Cruciani, C. Rengo, D. Sellitto, N. López-Bigas, R. Rabionet, N. Govea, A.L. de Munain, M. Sarduy, L. Romero, M. Villamar, I. del Castillo, F. Moreno, X. Estivill & R. Scozzari, 1999. The A1555G mutation in the 12S rRNA gene of human mtDNA: recurrent origins and founder events in families affected by sensorineural deafness. *American Journal of Human Genetics* 65, 1349–58.

Underhill, P.A., L. Jin, A.A. Lin, S.Q. Mehdi, T. Jenkins, D. Vollrath, R.W. Davis, L.L. Cavalli-Sforza & P.J. Oefner, 1997. Detection of numerous Y chromosome biallelic polymorphisms by denaturing high-performance liquid chromatography. *Genome Research* 7, 996–1005.

van Holst Pellekaan, S.M., M. Frommer, J.A. Sved & B. Boettcher, 1998. Mitochondrial control-region sequence variation in aboriginal Australians. *American Journal of Human Genetics* 62, 435–49.

Vigilant, L., M. Stoneking, H. Harpending, K. Hawkes & A.C. Wilson, 1991. African populations and the evolution of human mitochondrial DNA. *Science* 253, 1503–7.

Watson, E., P. Forster, M. Richards & H-J. Bandelt, 1997. Mitochondrial footprints of human expansions in Africa. *American Journal of Human Genetics* 61, 691–704.

Wilkinson-Herbots, H., M. Richards, P. Forster & B. Sykes, 1996. Site 73 in hypervariable region II of the human mitochondrial genome and the origin of European populations. *Annals of Human Genetics* 60, 499–508.

Wilson, J.F., D.A. Weiss, E.W. Hill, M.G. Thomas, N. Bradman, D.G. Bradley & D.B. Goldstein, submitted. Gene–culture coevolution in the British Isles.

Chapter 15

The Phylogeography of European Y-chromosome Diversity

Zoë H. Rosser, Matthew E. Hurles, Tatiana Zerjal, Chris Tyler-Smith & Mark A. Jobling

We have analyzed the distribution of twelve Y-chromosomal haplogroups in a sample of 3594 males from 49 populations within and around Europe. These haplogroups are highly geographically differentiated, with clines for several haplogroups, reflecting possible population movement from the Near East as well as other migration events. Three clades within the Y-chromosomal tree show high frequencies of derived haplogroups accompanied by low frequencies of ancestral haplogroups; global distributions of the ancestral lineages suggest that this pattern may result from range expansions from Asia into Europe.

In a higher resolution study, we are analyzing the distributions of these and other haplogroups within the Iberian peninsula, and combining this with analyses of the multiallelic markers microsatellites and MSY1. These additional markers will allow us to compare lineage diversities between populations and to estimate ages for population movements and fissions. One lineage, haplogroup 22, has been investigated in detail: it probably has an Iberian origin, and contains a signature of gene flow between the Basques and the Catalans during the last few thousand years.

Introduction: the continental picture

Attempts to interpret the genetic landscape of Europe have concentrated on the two opposing hypotheses for the origins of agriculture, an event which dates back to around 10,000 BP. That agriculture arose in the Near East around this time is not disputed; the argument arises over whether its subsequent spread (evinced by the dating of archaeological sites) was accompanied by a movement of people (demic diffusion: Ammerman & Cavalli-Sforza 1984), or whether the ideas and technologies spread without substantial population dispersal (cultural diffusion: Dennell 1983).

These opposing models are undoubtedly oversimplistic, but have been adopted enthusiastically by geneticists, since they predict patterns of genetic diversity which should be easily recognizable. Principal components analysis of gene frequency data reveals clines within Europe, and the first principal component, which has a focus in the Near East, is taken to support the demic diffusion hypothesis (Cavalli-Sforza *et al.* 1993; Menozzi *et al.* 1978). These conclusions have been supported by independent studies: for example, significant partial correlations were found between classical marker frequencies and the relative dates for the origin of agriculture in different locations (Sokal *et al.* 1991); and spatial autocorrelation analysis of DNA-based polymorphisms including microsatellites identified geographical patterns compatible with a substantial directional demographic expansion affecting much of the continent (Chikhi *et al.* 1998). However, analysis of diversity in European mitochondrial DNA (mtDNA) reveals a strikingly homogeneous landscape (Comas *et al.* 1997), with no detectable clines (Simoni *et al.* Chapter 13). Despite this homogeneity, an east–west gradient of pairwise difference has been discerned, and claimed to be compatible with expansion from the Middle East (Comas *et al.* 1997). In contrast, attempts to identify and date founding lineages (Richards *et al.* 1996) have suggested that Palaeolithic lineages persist in Europe to a degree which is inconsistent with the demic diffusion hypothesis.

Europe is remarkable for its linguistic homogeneity. Languages of the Indo-European family are spoken by most populations from India to Ireland (Renfrew 1989). Demic diffusion from the Near East provides a common explanation for both the spread of agriculture and of Indo-European (Renfrew 1987). Diversity exists within Indo-European and, despite its hegemony, some members of other language families do persist. Various methods for the detection of genetic barriers in autosomal gene frequencies within Europe (Barbujani 1991) show that most of these barriers correlate with linguistic boundaries. Language, rather than geography, may be the primary force shaping genetic differentiation within Europe.

The Y chromosome displays a high degree of geographic differentiation, which has been explained by a greater migration of females than males through the phenomenon of patrilocality (Seielstad *et al.* 1998). The Y may therefore constitute a sensitive system for detecting the population movements which have shaped European genetic diversity. Little published data exist on Y-chromosome diversity within Europe. Using two 'classical' Y markers, the highly polymorphic and complex 49f/*Taq*I system (Ngo *et al.* 1986), and the biallelic marker 12f2 (Casanova *et al.* 1985), patterns of diversity were demonstrated which were claimed to be clinal, and to support the demic diffusion model (Semino *et al.* 1996). Subsequent analysis using Y-specific microsatellites (Quintana-Murci *et al.* 1999), and using a combination of microsatellites and two biallelic markers (Malaspina *et al.* 1998), showed similar east–west gradients. 49f diversity has been exploited more fully to analyze the correlation between language, mtDNA diversity, and Y diversity, and suggested that the Y showed the strongest correlation with language (Poloni *et al.* 1997). This certainly appears to be so for the Uralic-speaking Finns, who share most of their mtDNA lineages with Indo-European speakers (Lahermo *et al.* 1996) but, in contrast, around half of their Y lineages with Central Asian Uralic speakers (Zerjal *et al.* 1997).

This study aims: to assay the diversity of Y-chromosomal lineages in a large sample of European males; to explore the geographical distribution of this diversity in the framework of the conflicting hypotheses for the origins of agriculture; to examine the influence of geographical and linguistic barriers to gene flow; and to compare Y-chromosomal genetical cartography with that of the autosomes and mtDNA, in order to explore the differential contributions of males and females to European gene pools.

We take a 'genealogical' approach to Y-chromosome diversity, defining monophyletic 'haplogroups' (hgs) using biallelic polymorphisms such as base substitutions, and then analyzing the diversity within these using Y-specific microsatellites (Kayser *et al.* 1997) and a minisatellite (MSY1; Jobling *et al.* 1998). In this continental study we concentrate on ten biallelic markers, which define a maximum of twelve haplogroups (Fig. 15.1). We have chosen these particular biallelic markers on the basis of previous work by us and others (Hammer *et al.* 1998; Hurles *et al.* 1999; Santos & Tyler-Smith 1996; Semino *et al.* 1996; Underhill *et al.* 1997; Zerjal *et al.* 1997) which indicates that they may be polymorphic within Europe. Ideally all markers within the tree should be analyzed, but practical considerations preclude this: for some markers no PCR assay is available, and some haplogroups are known from extensive data to be specific to non-European populations, e.g. hg 5 in Eastern Asians, including Japanese (Nakagome *et al.* 1992), hg 18 in Native Americans (Santos *et al.* 1999), and hg 6 in sub-Saharan Africans (Jobling 1994; and unpublished observations). When the analysis of European samples is complete with these biallelic markers, multiallelic analysis will be carried out within the same samples: this will be the subject of a future study. As well as this broad study of European diversity, we are examining one region, the Iberian peninsula, in much greater detail. Here, we will exploit the more rapidly mutating polymorphic Y-chromosomal loci to analyze diversity within lineages delineated by the biallelic markers, and to identify further sublineages. The use of the multiallelic systems will also allow us to attempt to place a timescale on the events responsible for current patterns of diversity.

Introduction: the Iberian picture

Lying on the southwestern fringe of Europe, Iberia has an interesting history of cultural influences from the Mediterranean, Africa and mainland Europe. Evidence of regionalization predates the arrival of agriculture around 6500 years before present (YBP), and, it is claimed, persists in cultural, linguistic and genetic diversity today.

In the Palaeolithic, Iberia may have represented a glacial refugium, from which re-expansion may have occurred approximately 12,000 years ago, after the last glacial maximum; the distribution of mtDNA haplogroup V has been adduced to support this idea (Torroni *et al.* 1998). Haplogroup V has a limited geographical distribution, and is observed at high frequencies within some Iberian populations (Basques and Catalans), and in Berber speakers of North Africa, but at its highest frequency in Scandinavian Saami. It has thus been proposed that this haplogroup may have originated in northern Iberia or southwestern France, and spread following the end of the Ice Age. This is in complete contrast to haplogroup H, the putative ancestral haplogroup to haplogroup V, and the most common haplogroup within European populations: this shows a distribution consistent with expansion from the Near East (Torroni *et al.* 1998). If this picture of post-glacial expansion is correct, then we might expect to find its traces with other markers.

More recently invasions and colonizations, for example from Italy, Greece and North Africa (the Moors), movements from France and the mysterious 'Atlantic influence' (Bertranpetit & Cavalli-Sforza 1991) may have shaped Iberian genetic diversity.

Principal components analysis of gene frequency data (Bertranpetit & Cavalli-Sforza 1991) and mtDNA analysis (Côrte-Real *et al.* 1996) have emphasized the distinctiveness of the Basques, as well as suggesting influences from southwestern France, and North Africa. The second synthetic map of PC analysis shows a gradient centred in Iberia, which, it is suggested, is most likely due to climatic/ecological effects or gene flow from Uralic speakers from northwestern Asia (Cavalli-Sforza *et al.* 1994).

The Basques speak a non-Indo-European language with no close affinities to any other extant language (Ruhlen 1991); this linguistic uniqueness has led to the idea that they may represent a Mesolithic relict population, isolated from cultural and genetic exchange by linguistic and geographic barriers. Although specific archaeological evidence for such a picture is lacking (Collins 1986), genetic analysis certainly lends some support to the view of the Basques as an isolate, in unusual frequencies of alleles in blood groups such as Rhesus and ABO (Mourant 1947; 1983), and of disease alleles in the calpain-3 gene, responsible for limb-girdle muscular dystrophy (Urtasun *et al.* 1998). Mitochondrial DNA haplotype analysis has shown the Basques to be significantly different to other European populations (Côrte-Real *et al.* 1996).

In this study we are analyzing ~1500 samples

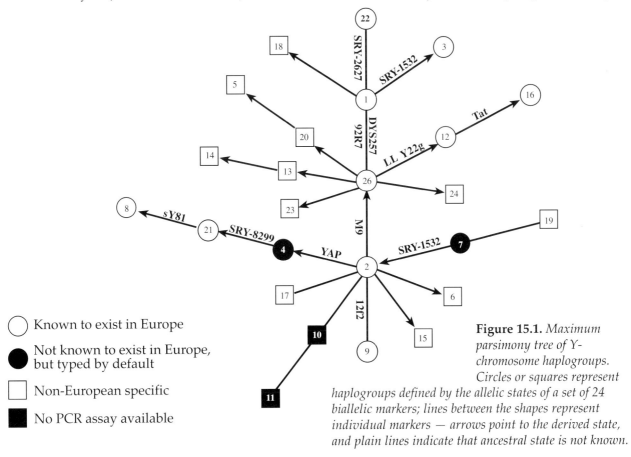

Figure 15.1. *Maximum parsimony tree of Y-chromosome haplogroups. Circles or squares represent haplogroups defined by the allelic states of a set of 24 biallelic markers; lines between the shapes represent individual markers — arrows point to the derived state, and plain lines indicate that ancestral state is not known.*

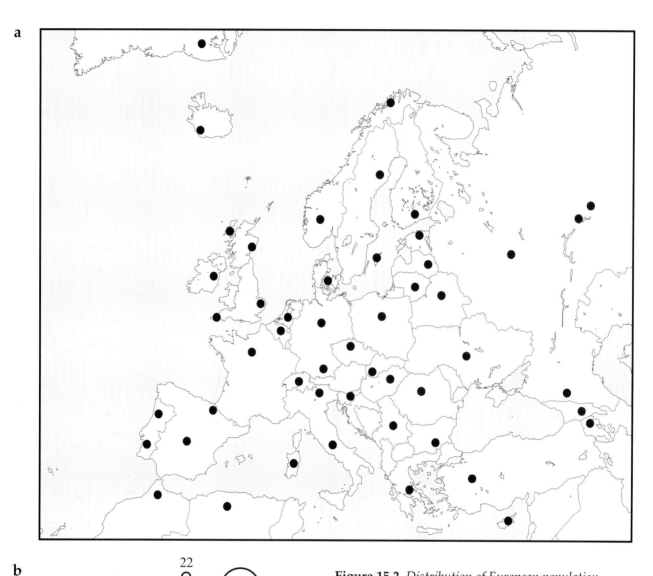

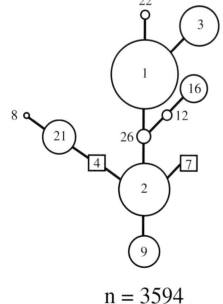

Figure 15.2. *Distribution of European population samples, and summary of haplogroup frequency data. (a) Geographical distribution of samples: the total number of samples is 3594, and most individual samples are of between 50 and 100 individuals. (b) Summary of haplogroup frequency data, presented as an abbreviated version of the tree in Figure 15.1. Circle areas are proportional to haplogroup frequency, and squares (hgs 4, 7) indicate zero frequency in the data set.*

from Iberian and North African populations (classified by grand-paternal birthplace), using 15 biallelic markers (see Fig. 15.1) plus a minimum of seven microsatellites, and, for some sublineages, MSY1. Additional biallelic markers over the set used in the broad European study are being incorporated to add further definition to the study. We aim to describe in detail the landscape of Y diversity with respect to geographical and cultural boundaries within Iberia, to identify lineages which link Iberian to specific non-Iberian populations, and to seek the traces of proposed and known historical events.

Summary of results: Europe

We have analyzed 3594 DNA samples from 49 European and circum-European populations (Fig. 15.2a). These data will be presented and discussed fully elsewhere (Rosser *et al.* in prep.); here we present their general features. The distributions are highly non-random. In the data set as a whole (Fig. 15.2b), frequencies of haplogroups vary from zero (hgs 4, 7) to 39 per cent (hg 1). It is striking that three 'ancestral' haplogroups within the tree (hgs 4, 12, 26) are rare or absent within Europe, while the respective derived haplogroups (hg 21, hg 16, hgs 1 and sub-groups) are common, and all three rare ancestral haplogroups are common in Asia (Hammer *et al.* 1998; Underhill *et al.* 1997; Zerjal *et al.* 1997). This pattern seems consistent with a range expansion out of Asia into Europe. The method of nested cladistic analysis (Templeton *et al.* 1995) has been used to define such expansions more rigorously in other data sets (Hammer *et al.* 1998); we have not used this method ourselves, as we have not collected the appropriate Asian data.

All of the haplogroups which constitute a substantial proportion of European Y chromosomes (hgs 1, 2, 3, 9, 16, 21) show great geographical differentiation. The identification of clines is best done quantitatively, rather than by eye, and this analysis is being carried out using spatial autocorrelation (Bertorelle & Barbujani 1995). Patterns resembling the clines claimed for previous Y data (Semino *et al.* 1996) are recognizable: hg 1 is broadly equivalent to 49f/*Taq*I haplotype 15 (Jobling 1994), and hg 9 is defined by the 12f2/*Taq*I 8kb allele. However, distributions of the remaining haplogroups are very different to these, and cannot be interpreted as a simple reflection of population movement from the Near East. It seems clear that one of the contributions of Y-chromosome diversity to the debate on European origins will be to show that the history was complex.

Summary of results: Iberia

Our study of Y diversity within Iberia is at a preliminary stage. We have already investigated the geographic distribution of hg 22 chromosomes within Iberia (Hurles *et al.* 1999), where the Basques (11 per cent) and Catalans (22 per cent) have the highest frequencies (see Fig. 15.3), with a limited number

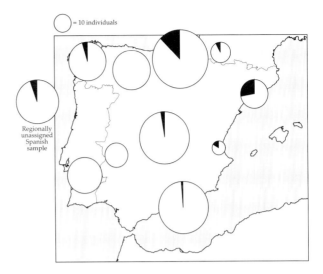

Figure 15.3. *Distribution of haplogroup 22 chromosomes within Iberia. Areas of circles are proportional to population sample sizes, and haplogroup 22 chromosomes are indicated by dark sectors. (See Hurles* et al. *1999.)*

also being found in other Iberian populations; large samples from Aragon and the Pyrenees remain to be typed. Very few examples are detected in the rest of Europe. All hg 22 chromosomes cluster tightly in median joining networks constructed using both microsatellite and minisatellite data, and it is likely that this haplogroup has an Iberian origin, with an age, calculated by different methods from microsatellite and minisatellite haplotype diversity, of approximately 1600–3500 years (with wide 95 per cent confidence intervals). Because of the young age of this haplogroup, it is unlikely to represent a Y-chromosomal signal of the post-glacial out-of-Iberia expansion (Torroni *et al.* 1998). Non-Iberian hg 22 chromosomes in Europe and in South America (Bianchi *et al.* 1997) may reflect the activities of Basque cohorts of the Roman army, and the conquistadores, respectively, and the high sharing of hg 22 at high frequency between Basques and Catalans suggests that, for males at least, the Basque — non-Basque linguistic barrier has not been insurmountable (Hurles *et al.* 1999).

Acknowledgements

Both the European and Iberian Y-chromosome diversity studies involve a large number of collaborators, who are listed below. We thank them for contributing data and DNA samples to these projects. Our work is funded by the Wellcome Trust (including grant no. 057559/Z/99/Z to M.A.J.), the MRC, the BBSRC, and the CRC.

European collaborators: D. Alavantic, A. Amorim, W. Amos, M. Armenteros, E. Arroyo, G. Barbujani, L. Beckman, J. Bertranpetit, E. Bosch, D. Bradley, G. Brede, G.C. Cooper, H.B.S.M. Côrte-Real, P. de Knijff, R. Decorte, Y.E. Dubrova, O. Evgrafov, S. Glisic, E.W. Hill, A. Jeziorowska, L. Kalaydjieva, M. Kayser, S.A. Kravchenco, J. Lavinha, L.A. Livshits, A. López de Munain, S. Maria, K. McElreavey, T.A. Meitinger, B. Melegh, R.J. Mitchell, J. Nicholson, S. Nørby, A. Novelletto, A. Pandya, J. Parik, P.C. Patsalis, L. Pereira, B. Peterlin, G. Pielberg, M.J. Prata, C. Previdere, K. Rajczy, L. Roewer, S. Rootsi, D.C. Rubinsztein, J. Saillard, F.R. Santos, G. Stefanescu, M. Stenico, B. Sykes, A. Tolun, R. Veitia, R. Villems.

Iberian collaborators: M. Aler, M. Armenteros, E. Arroyo, J. Bertranpetit, L. Blazquez, E. Bosch, A. Cambon-Thomsen, J. Carrill, H.B.S.M. Côrte-Real, J. Lavinha, A. López de Munain, B. Martinez-Jaretta.

References

Ammerman, A.J. & L.L. Cavalli-Sforza, 1984. *Neolithic Transition and the Genetics of Populations in Europe.* Princeton (NJ): Princeton University Press.

Barbujani, G., 1991. What do languages tell us about human microevolution? *Trends in Ecology and Evolution* 6, 151–6.

Bertorelle, G. & G. Barbujani, 1995. Analysis of DNA diversity by spatial autocorrelation. *Genetics* 140, 811–19.

Bertranpetit, J. & L.L. Cavalli-Sforza, 1991. A genetic reconstruction of the history of the population of the Iberian Peninsula. *Annals of Human Genetics* 55, 51–67.

Bianchi, N.O., G. Bailliet, C.M. Bravi, R.F. Carnese, F. Rothhammer, V.L. Martínez-Marignac & S.D.J. Pena, 1997. Origin of Amerindian Y chromosomes as inferred by the analysis of six polymorphic markers. *American Journal of Physical Anthropology* 102, 79–89.

Casanova, M., P. Leroy, C. Boucekkine, J. Weissenbach, C. Bishop, M. Fellous, M. Purrello, G. Fiori & M. Siniscalco, 1985. A human Y-linked DNA polymorphism and its potential for estimating genetic and evolutionary distance. *Science* 230, 1403–6.

Cavalli-Sforza, L.L., P. Menozzi & A. Piazza, 1993. Demic expansions and human evolution. *Science* 259, 639–46.

Cavalli-Sforza, L.L., P. Menozzi & A. Piazza, 1994. *The History and Geography of Human Genes.* Princeton (NJ): Princeton University Press.

Chikhi, L., G. Destro-Bisol, G. Bertorelle, V. Pascali & G. Barbujani, 1998. Clines of nuclear DNA markers suggest a largely Neolithic ancestry of the European gene pool. *Proceedings of the National Academy of Sciences of the USA* 95, 9053–8.

Collins, R., 1986. *The Basques.* Oxford: Blackwell.

Comas, D., F. Calafell, E. Mateu, A. Pérez-Lezaun, E. Bosch, & J. Bertranpetit, 1997. Mitochondrial DNA variation and the origin of the Europeans. *Human Genetics* 99, 443–9.

Côrte-Real, H.B.S.M., V.A. Macaulay, M.B. Richards, G. Hariti, M.S. Issad, A. Cambon-Thomsen, S. Papiha, J. Bertranpetit & B.C. Sykes, 1996. Genetic diversity in the Iberian Peninsula determined from mitochondrial sequence analysis. *Annals of Human Genetics* 60, 331–50.

Dennell, R., 1983. *European Economic Prehistory: a New Approach.* London: Academic Press.

Hammer, M.F., T. Karafet, A. Rasanayagam, E.T. Wood, T.K. Altheide, T. Jenkins, R.C. Griffiths, A.R. Templeton & S.L. Zegura, 1998. Out of Africa and back again: nested cladistic analysis of human Y chromosome variation. *Molecular Biology and Evolution* 15, 427–41.

Hurles, M.E., R. Veitia, E. Arroyo, M. Armenteros, J. Bertranpetit, A. Pérez-Lezaun, E. Bosch, M. Shlumukova, A. Cambon-Thomsen, K. McElreavey, A. López de Munain, A. Röhl, I.J. Wilson, L. Singh, A. Pandya, F.R. Santos, C. Tyler-Smith & M.A. Jobling, 1999. Recent male-mediated gene flow over a linguistic barrier in Iberia, suggested by analysis of a Y-chromosomal DNA polymorphism. *American Journal of Human Genetics* 65, 1437–48.

Jobling, M.A., 1994. A survey of long-range DNA polymorphisms on the human Y chromosome. *Human Molecular Genetics* 3, 107–14.

Jobling, M.A., N. Bouzekri & P.G. Taylor, 1998. Hypervariable digital DNA codes for human paternal lineages: MVR-PCR at the Y-specific minisatellite, MSY1 (DYF155S1). *Human Molecular Genetics* 7, 643–53.

Kayser, M., A. Caglià, D. Corach, N. Fretwell, C. Gehrig, G. Graziosi, F. Heidorn, S. Herrmann, B. Herzog, M. Hidding, K. Honda, M. Jobling, M. Krawczak, K. Leim, S. Meuser, E. Meyer, W. Oesterreich, A. Pandya, W. Parson, G. Penacino, A. Pérez-Lezaun, A. Piccinini, M. Prinz, C. Schmitt, P.M. Schneider, R. Szibor, J. Teifel-Greding, G. Weichhold, P. de Knijff & L. Roewer, 1997. Evaluation of Y-chromosomal STRs: a multicenter study. *International Journal of Legal Medicine* 110, 125–33.

Lahermo, P., A. Sajantila, P. Sistonen, M. Lukka, P. Aula, L. Peltonen & M-L. Savontaus, 1996. The genetic relationship between the Finns and the Finnish Saami (Lapps): analysis of nuclear DNA and mtDNA. *American Journal of Human Genetics* 58, 1309–22.

Malaspina, P., F. Cruciani, B.M. Ciminelli, L. Terrenato, P. Santolamazza, A. Alonso, J. Banyko, R. Brdicka, O. Garcia, C. Gaudiano, G. Guanti, K.K. Kidd, J. Lavinha, M. Avila, P. Mandich, P. Moral, R. Qamar, S.Q. Mehdi, G. Ragusa, G. Sefanescu, M. Caraghin, C. Tyler-Smith, R. Scozzari & A. Novelletto, 1998. Network analyses of Y-chromosomal types in Europe, northern Africa, and western Asia reveal specific patterns of geographic distribution. *American Journal of Human Genetics* 63, 847–60.

Menozzi, P., A. Piazza & L.L. Cavalli-Sforza, 1978. Synthetic maps of human gene frequencies in Europeans. *Science* 201, 786–92.

Mourant, A.E., 1947. The blood groups of the Basques. *Nature* 160, 505.

Mourant, A.E., 1983. *Blood Relations*. Oxford: Oxford University Press.

Nakagome, Y., S.R. Young, A. Akane, H. Numabe, D.K. Jin, Y. Yamori, S. Seki, T. Tamura, S. Nagafuchi, H. Shiono & Y. Nakahori, 1992. A Y-associated allele may be characteristic of certain ethnic groups in Asia. *Annals of Human Genetics* 56, 311–14.

Ngo, K.Y., G. Vergnaud, C. Johnsson, G. Lucotte & J. Weissenbach, 1986. A DNA probe detecting multiple haplotypes of the human Y chromosome. *American Journal of Human Genetics* 38, 407–18.

Poloni, E.S., O. Semino, G. Passarino, A.S. Santachiara-Benerecetti, L. Dupanloup, A. Langaney & L. Excoffier, 1997. Human genetic affinities for Y-chromosome P49a,f/*Taq*I haplotypes show strong correspondence with linguistics. *American Journal of Human Genetics* 61, 1015–35.

Quintana-Murci, L., O. Semino, E. Minch, G. Passarino, A. Brega & A.S. Santachiara-Benerecetti, 1999. Further characteristics of proto-European Y chromosomes. *European Journal of Human Genetics* 7, 603–8.

Renfrew, C., 1987. *Archaeology and Language: the Puzzle of Indo-European Origins*. London: Jonathan Cape.

Renfrew, C., 1989. The origins of Indo-European languages. *Scientific American* 261, 106–14.

Richards, M., H. Côrte-Real, P. Forster, V. Macaulay, H. Wilkinson-Herbots, A. Demaine, S. Papiha, R. Hedges, H-J. Bandelt & B. Sykes, 1996. Paleolithic and Neolithic lineages in the European mitochondrial gene pool. *American Journal of Human Genetics* 59, 185–203.

Rosser, Z.H., T. Zerjal, M.E. Hurles, M. Adojaan, D. Alavantic *et al.*, in prep. The genetical cartography of human Y-chromosomal variation in Europe.

Ruhlen, M., 1991. *A Guide to the World's Languages*. London: Edward Arnold.

Santos, F.R. & C. Tyler-Smith, 1996. Reading the human Y chromosome: the emerging DNA markers and human genetic history. *Brazilian Journal of Genetics* 19, 665–70.

Santos, F.R., A. Pandya, C. Tyler-Smith, S.D.J. Pena, M. Schanfield, W.R. Leonard, L. Osipova, M.H. Crawford & R.J. Mitchell, 1999. The central Siberian origin for Native American Y chromosomes. *American Journal of Human Genetics* 64, 619–28.

Seielstad, M.T., E. Minch & L.L. Cavalli-Sforza, 1998. Genetic evidence for a higher female migration rate in humans. *Nature Genetics* 20, 278–80.

Semino, O., G. Passarino, A. Brega, M. Fellous & A.S. Santachiara-Benerecetti, 1996. A view of the Neolithic demic diffusion in Europe through two Y chromosome-specific markers. *American Journal of Human Genetics* 59, 964–8.

Simoni, L., F. Calafell, D. Pettener, J. Bertranpetit & G. Barbujani, 2000. Geographic patterns of mtDNA diversity in Europe. Spatial variation of mtDNA hypervariable region I among European populations. *American Journal of Human Genetics* 66, 262–78.

Sokal, R.R., N.L. Oden & C. Wilson, 1991. Genetic evidence for the spread of agriculture in Europe by demic diffusion. *Nature* 351, 143–5.

Templeton, A.R., E. Routman & C.A. Phillips, 1995. Separating population structure from population history: a cladistic analysis of the geographical distribution of mitochondrial DNA haplotypes in the tiger salamander, *Ambystoma tigrinum*. *Genetics* 140, 767–82.

Torroni, A., H-J. Bandelt, L. D'Urbano, P. Lahermo, P. Moral, D. Sellitto, C. Rengo, P. Forster, M-L. Savontaus, B. Bonné-Tamir & R. Scozzari, 1998. MtDNA analysis reveals a major Late Paleolithic population expansion from Southwestern to Northeastern Europe. *American Journal of Human Genetics* 62, 1137–52.

Underhill, P.A., L. Jin, A.A. Lin, S.Q. Mehdi, T. Jenkins, D. Vollrath, R.W. Davis, L.L. Cavalli-Sforza & P.J. Oefner, 1997. Detection of numerous Y-chromosome biallelic polymorphisms by denaturing high-performance liquid chromatography. *Genome Research* 7, 996–1005.

Urtasun, M., A. Saenz, C. Roudaut, J.J. Poza, J.A. Urtizberea, A.M. Cobo, I. Richard, F. Garcia Bragado, F. Leturcq, J.C. Kaplan, J.F. Marh Masso, J.S. Beckmann & A. López de Munain, 1998. Limb-girdle muscular dystrophy in Guipuzcoa (Basque Country, Spain). *Brain* 121, 1735–47.

Zerjal, T., B. Dashnyam, A. Pandya, M. Kayser, L. Roewer, F.R. Santos, W. Schiefenhövel, N. Fretwell, M.A. Jobling, S. Harihara, K. Shimizu, D. Semjidmaa, A. Sajantila, P. Salo, M.H. Crawford, E.K. Ginter, O.V. Evgrafov & C. Tyler-Smith, 1997. Genetic relationships of Asians and northern Europeans, revealed by Y-chromosomal DNA analysis. *American Journal of Human Genetics* 60, 1174–83.

Part IV

Regional Studies in the Molecular Genetic History of Western Europe

Chapter 16

Human Y-chromosomal Networks and Patterns of Gene Flow in Europe, West Asia and North Africa

Patrizia Malaspina, Fulvio Cruciani, Antonio Torroni, Luciano Terrenato, Andrea Novelletto & Rosaria Scozzari

We examined 1801 Caucasian males sampled in 33 European, 15 West Asian and 3 North African locations, for the following Y-chromosome specific markers: the alphoid HindIII site, the YAP insertion, the A-G SNS at SRY10831, the YCAII and DYS413 (CA)n microsatellite polymorphic systems. In a previous work on 908 subjects and 33 locations we had shown that haplotypes defined by these markers can be grouped into networks which display a high degree of geographic specificity. By doubling the sample size and increasing the locations, the number of networks did not increase proportionally, showing that much of the overall variation has been sampled. The three networks previously identified only by peculiar microsatellite allele states, could not be linked to any other network by assuming a single (CA) change. They show that multirepeat mutations can be considered founder events on new lineages recognizable from the pre-existing background. The geographic distributions of the different networks highlight centres of high frequencies in western Europe, northwestern Africa, northeastern Europe, Sardinia and the eastern Mediterranean. The analysis of the 1275 pairwise F_{ST} values show a much higher dependence of this measure on distance for comparisons within Europe vs comparisons within other continents or between continents. These results indicate much larger similarity between the gene pools of eastern Europeans and western Asians as contrasted to a sharp distinction in the comparisons between eastern and western Europe and between Europe and northern Africa. These data show that the European–Asia continental boundary did not act as a strong barrier to male-specific gene flow.

Markers of the genetic diversity of the human Y chromosome are gaining an ever-increasing value in understanding human microevolution. In previous works we have shown that the combined use of binary (such as SNS and Alu insertions) and microsatellite markers greatly expands the possibilities of detecting different Y-chromosomal lineages in the gene pools of extant populations (Malaspina *et al.* 1998; Scozzari *et al.* 1999).

The rationale of this approach is outlined in Figure 16.1. A putative 7-locus microsatellite founder haplotype (B,B,B,B,B,B,B) is supposed to undergo mutational events that change one locus at a time by the insertion/deletion of a single repeated unit (represented as consecutive letters, lower left part of graph). Several lines of evidence (Di Rienzo *et al.* 1994; Brinkmann *et al.* 1998) show that this is by far the most common type of microsatellite mutation and, in this example, the six related haplotypes can be arranged into a one-step network. Uncommon multirepeat mutational events (e.g. B,B,B,B,B,B,B → B,B,B,B,F,B,B) can be considered as rare events suddenly producing novel haplotypes that cannot be linked to the previous ones by the one-step criterion.

These new haplotypes then become founders of new networks, as the continuing process of one-step mutations generates further derivative haplotypes (upper right part of graph). Thus, multirepeat mutations can specifically identify groups of chromosomes with a common ancestry and, sometimes, lineages that otherwise cannot be recognized with binary markers. In addition, information on repeat variation within well-defined lineages can be used to estimate each lineage antiquity, although with large confidence intervals (Goldstein et al. 1996).

We examined 1801 Caucasian males sampled in 51 locations of Europe, West Asia and North Africa for three Y-specific binary polymorphisms (YAP insertion, alphoid HindIII site, A-G SNS at SRY10831), and the two (CA)n microsatellite systems YCAII and DYS413.

We analyzed the variation of microsatellite markers between chromosomes of the same binary haplotype. The search for all one-step relationships among microsatellite haplotypes resulted in six large, one small and eight very small (no more than three haplotypes and four subjects each) one-step networks, whose main features are listed in Table 16.1.

The spatial variations of overall network frequencies across the 51 locations were then analyzed. Network 1.1 reaches its highest frequency in northeastern Europe with a decreasing cline towards the southwest. The lowest incidence is observed among Basques (0.04). Network 1.2 is confirmed to be present mainly in Mediterranean populations. It reaches frequencies exceeding 0.30 only in Crete and southwestern Turkey (Novelletto et al. Chapter 24). Network 1.3 is identified as a group of chromosomes peculiar to Sardinians. Network 2.1 includes a set of chromosomes highly frequent in Africa, north to the Equator (Scozzari et al. 1999). In this data set it shows the highest frequencies in northern Africa. Within Europe, it is found in southern Italy and the Balkans, decreasing in the south–north direction. In five samples from the Iberian peninsula the frequency of this network is <0.09. Network 3.1G is found at high frequencies (>0.46) in all populations of western Europe but not in northeastern Europe. Network 3.1A reaches frequencies of 0.30–0.40 in northeastern Europe and in Pakistan. The distribution of network 3.1A shows an almost complete complementarity to

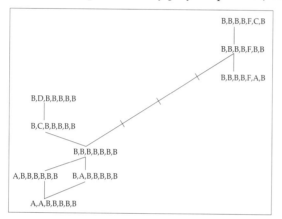

Figure 16.1. *A putative 7-locus microsatellite founder haplotype (B,B,B,B,B,B,B) is supposed to undergo mutational events that change one locus at a time by the insertion/deletion of a single repeated unit (represented as consecutive letters, lower left part of graph).*

Table 16.1. *Main features of 15 Y-chromosomal haplotype networks.*

Binary haplotype[a]	Network	No. of subjects (% of the entire study)	Major haplotype in the network (CA units)[b]	No. of carriers	Estimated antiquity
− + G	1.1	523 (29.0)	22-22-22-21	44	Pre-Neolithic
− + G	1.2	166 (9.2)	22-19-17-17	91	Pre-Neolithic/Neolithic transition
− + G	1.3	91 (5.1)	21-11-21-21	72	Neolithic
+ + G	2.1	216 (12.0)	21-19-24-23	38	Pre-Neolithic
− − G	3.1G	480 (26.7)	23-19-23-23	253	Pre-Neolithic
− − A	3.1A	222 (12.3)	23-19-22-22	136	Pre-Neolithic/Neolithic transition
− + G	1.4	15 (0.8)	24-23-20-20	5	Not applicable
	8 minor networks	22 (1.2)			
	Unclassified	66 (3.7)			
TOTAL		**1801**			

[a] Reported as YAP, Alphoid HindIII, SRY10831
[b] Reported as YCAIIa-YCAIIb-DYS413a-DYS413b

the distribution of its precursor, network 3.1G, and thus contributes to the definition of a sharp frequency change across Central Europe. Network 1.4 is observed only at low frequencies (<0.05) in populations of the central-eastern Mediterranean area.

We also used network frequencies to measure pairwise genetic dissimilarities between populations (F_{ST}). These were related to the geographic distance between populations, thus allowing inferences about the pattern of male-specific gene flow over the area under scrutiny. First, among the conventional geographic borders between the continents here reported, the European–Asian border is not associated with higher F_{ST}-on-distance slopes. Instead, the highest slope is observed within Europe, to which the Basque population contributes to a large extent. In general, all pairwise inter-population comparisons except those involving the Basques produce a slope of F_{ST}-on-distance that is one half or less than that reported by Seielstad et al. (1998). Second, different linguistic affiliations are often associated with reduced levels of genetic similarity. However, this is not an absolute rule. For example, large F_{ST} are observed between Indo-European-speaking populations residing in Europe.

In this work for the first time the frequency and F_{ST}-on-distance patterns are shown to complement each other in revealing the effects of peopling processes that led to the observed frequency distributions. The main conclusions can be summarized as follows:

1. The strongest genetic boundary around Europe is at the Gibraltar Straits, whereas a considerable contribution of African Y chromosomes is found in southeastern Europe.
2. Genetic boundaries within Europe are stronger than between Europe and Asia. This pattern is only partly accounted for by the peculiarities of Basques and still persists between Indo-European-speaking populations.
3. The overall picture is that of recent lineages entering Europe from the East and overlaying the pre-existing background up to approximately a line connecting the Eastern Alps to the Baltic Sea.
4. The extreme model of a demic diffusion spanning the entire European continent from the Levant, associated with the spread of agriculture (Cavalli-Sforza et al. 1994), must be confronted with the sharp genetic discontinuity in Central Europe. This is more easily explained by a two-phase process: a major spread of farmers to Eastern Europe, followed by a further cultural spreading of farming towards the West.

Acknowledgements

We gratefully acknowledge P. Santolamazza for technical assistance. We wish to thank the following colleagues, who contributed to this work with sample collection: N. Akar, A. Alonso, M. Avila, V. Bakalli, J. Banyko, R. Brdicka, M. Caraghin, O. Garcia, C. Gaudiano, G. Guanti, J. Jaruzelska, K. K. Kidd, A. Kozlov, J. Lavinha, B. Malyarchuk, P. Mandich, S. Q. Mehdi, M. M. Memmi, E. Michalodimitrakis, P. Moral, A. Pangrazio, J. Parik, R. Qamar, A. Ragusa, V. Romano, M. Stefan, G. Stefanescu, M. Stenico, C. Tyler-Smith, L. Varesi, R. Villems, G. Vona. Work supported by CNR grants 98.00485.CT04 (AN), 97.00712.PF36 (LT), 97.00702.PF36 (RS) and PRIN MURST 1997 (LT and RS).

References

Brinkmann, B., M. Klintschar, F. Neuhuber, J. Hühne & B. Rolf, 1998. Mutation rate in human microsatellites: influence of the structure and length of the tandem repeat. *American Journal of Human Genetics* 62, 1408–15.

Cavalli-Sforza, L.L., P. Menozzi & A. Piazza, 1994. *The History And Geography of Human Genes*. Princeton (NJ): Princeton University Press.

Di Rienzo, A., A.C. Peterson, J.C. Garza, A.M. Valdes, M. Slatkin & N.B. Freimer, 1994. Mutational processes of simple-sequence repeat loci in human populations. *Proceedings of the National Academy of Sciences of the USA* 91, 3166–70.

Goldstein, D.B., L.A. Zhivotovsky, K. Nayar, A. Ruiz-Linares, L.L. Cavalli-Sforza & M.W. Feldman, 1996. Statistical properties of the variation at linked microsatellite loci: implications for the history of human Y chromosomes. *Molecular Biology and Evolution* 13, 1213–18.

Malaspina, P., F. Cruciani, B.M. Ciminelli, L. Terrenato, P. Santolamazza, A. Alonso, J. Banyko, R. Brdicka, O. Garcia, C. Gaudiano, G. Guanti, K.K. Kidd, J. Lavinha, M. Avila, P. Mandich, P. Moral, R. Qamar, S.Q. Mehdi, A. Ragusa, G. Stefanescu, M. Caraghin, C. Tyler-Smith, R. Scozzari & A. Novelletto, 1998. Network analyses of Y-chromosomal types in Europe, North Africa and West Asia reveal specific patterns of geographical distribution. *American Journal of Human Genetics* 63, 847–60.

Scozzari, R., F. Cruciani, P. Santolamazza, P. Malaspina, A. Torroni, D. Sellitto, B. Arredi, G. Destro-Bisol, G. De Stefano, O. Rickards, C. Martinez-Labarga, D. Modiano, G. Biondi, P. Moral, A. Olckers, D.C. Wallace & A. Novelletto, 1999. Combined use of biallelic and microsatellite Y chromosome polymorphisms to infer affinities among African populations. *American Journal of Human Genetics* 65, 829–46.

Seielstad, M.T., E. Minch & L.L. Cavalli-Sforza, 1998. Genetic evidence for a higher female migration rate in humans. *Nature Genetics* 20, 278–80.

Chapter 17

Autosomal and Mitochondrial Genetic Diversity in Sicily

Valentino Romano, Francesco Calì, Peter Forster, Rosaldo P. D'Anna, Anna Flugy, Giacomo De Leo, Alfredo Salerno, Ornella Giambalvo, Giuseppe Matullo & Alberto Piazza

Two questions relevant for a better understanding of the genetic structure of the extant population of Sicily concern: i) the genetic relationship between Sicilians and the surrounding populations of continental Europe, the Near East and Africa; and ii) the possible existence of a genetic differentiation within the island. A third, and more general question, involves the choice of genetic markers in such studies for verifying whether the information gained with different *markers allows the reconstruction of a coherent picture of the genetic structure and history of the Sicilian population. In the present study we have collected a large number of Sicilian samples, according to well-defined anthropological criteria, and have tested them with the following molecular markers: i) neutral and pathological mutations of the phenylalanine hydroxylase gene analyzed in families with phenylketonuria; ii) autosomal 'microsatellites' (short tandem repeat loci); and iii) mutations in the mitochondrial genome analyzed in samples of the general population. In this chapter we present a preliminary account of the results obtained with these three markers.*

The first evidence of human presence in Sicily, the largest of the Mediterranean islands (25,708 sq. km), can be dated back to the Palaeolithic; since then the island was settled by Neolithic farmers from Anatolia and the Near East, the Italic peoples from the Italian peninsula, Phoenicians, Greeks, Romans, Byzantines, Arabs, and Normans (Finley 1979; Mack Smith 1994). Whether these invasions and settlements had a real demographic impact able to modify permanently the structure of the population has only been a speculative matter, mostly based on the study of the material culture and literary sources. Early Neolithic settlements have been found predominantly in the fertile plains and coastal areas of southeastern Sicily (e.g. 'Stentinello') (Tusa 1983). Thucidides, the Greek historian, spoke of the existence of three populations inhabiting the island just before the Greek colonization of Sicily: the *Elimi* in the extreme western triangle, the *Sicani* in the central part and the *Siculi* in the eastern side of the island. Archaeological research seems to substantiate the literary sources in terms of material culture (La Rosa 1989). In the Early Bronze Age another source of possible ethnic stress was probably caused by the arrival of the so-called *Bell Beaker people* whose origins have been traced to the Atlantic coast of the Iberian peninsula (Tusa 1997). It is interesting to note that most of the sites that have been ascribed to this culture are found in western Sicily. After 1000 BC a west–east differentiation was maintained when Phoenicians and Greeks colonized the island: Phoenicians settled in the northwestern triangle of Sicily, Greeks settled in the remaining coastal belt of the island (Pace 1958). In the Middle Ages two important events that may have left genetic traces along the north–south axis of the island consisted in the arrival of Arabs (from the South) and Normans (from the North) respectively.

Because of the complex history and prehistory of Sicily, it is therefore not surprising that it has often attracted the interest of population geneticists (e.g. Piazza *et al.* 1988; Rickards *et al.* 1998). Nevertheless, we still know very little about its genetic structure.

The present study aims to analyze several samples of the Sicilian population by the use of molecular markers of the nuclear and mitochondrial genomes and to compare molecular data with pre-existing data obtained by the analysis of non-DNA polymorphisms. Some of these molecular markers (autosomal microsatellites, mutations of the mtDNA D-loop region), because of their higher mutation rate, should indeed be more reliable for comparing genetically close populations, than 'classical' markers. Moreover, we shall use genetic data to challenge current archeological models on the prehistoric and historic demographic dynamics of the Sicilian population, taken both in isolation and in relation to the genetic and cultural history of other Mediterranean and European populations.

Population samples and methods

Samples and sampling strategy

Our strategy involves the collection of a large and realistic number of samples so as to have a sufficiently dense grid of points from all over Sicily, each point being a sampled town. Once available, all these samples will be grouped according to various anthropological and geographical criteria and different groups, within a classified category, tested for genetic heterogeneity. Examples of planned comparisons include: geographical (east vs west, north vs south, highland vs lowland, coastal vs hinterland); archaeological (e.g. 'Greek' vs 'Phoenician' sites); linguistic (e.g. analysis of surnames and dialects from different areas). In order to comply with the scope of this multi-purpose sampling protocol, we have adopted a diocese-based sub-division of Sicily. Within each diocese two sites of some archaeological relevance have been selected. The sample sizes usually comprise between 50 and 100 unrelated blood donors. To avoid sampling of individuals of recent immigration, the assignment of blood donors to a specific site (town, village) is made according to the birthplace of their maternal and paternal grandparents. Urban sites are excluded from sampling. To test for possible sampling biases the frequency of each surname in the sample is compared to the frequency of the same surname in the general population.

The samples analyzed in this study refer to: i) a group of 70 PKU families representing the nine provinces of Sicily; ii) a group of 331 healthy blood donors, of both sexes, randomly selected from the general population of the following towns: Troina (diocese of Nicosia, Enna province: 111 individuals), Piazza Armerina (diocese of Piazza Armerina, Enna province: 44 individuals), Sciacca (diocese of Agrigento, Agrigento province: 89 individuals), and Castellammare del Golfo (diocese of Trapani, Trapani province: 87 individuals).

Historical background of the towns of Troina, Sciacca, Castellamare del Golfo and Piazza Armerina

Troina is inhabited by about 10,000 people, and is located in the northeastern triangle of the Sicilian island at 1120 m above sea level. Historically, this town is best known as the first settlement of *Normans* ('The French Vikings') in Sicily (eleventh century AD). The Troina area appears to have been relatively unpopulated until the fourth millennium BC when changes in agricultural practice led to higher demographic pressure: a recent survey provides more detail concerning the demographic changes in this region (S. Stoddart pers. comm.). Troina is located in an area attributed by literary sources to be settled by *Siculi* (likely an Italic group from the central part of the Italian peninsula) in pre-Greek times (La Rosa 1989).

Sciacca is a coastal town located in southwestern Sicily and belongs to the province of *Agrigento*. The population size is close to 40,000. The Romans called it *Thermae Selinuntinae* to indicate it was founded by people from Selinunte (an important Greek archaeological site located nearby to the northwest), and the presence of a hot spring still active today. The name 'Sciacca' derives from the Arab as *saqha*, after Arabs conquered it in AD 814. The presence of Arab culture can still be recognized today in its urban structure (narrow streets, etc.).

Castellammare del Golfo, with 15,000 people, is located on the coast in the extreme northeastern triangle of the island, in an area that, before and after Greeks settled in Sicily, was inhabited by ethnic groups considered to be distinct from those occupying the remaining part of the island (e.g. Elymes, Phoenicians). Before the arrival of the Greeks it was the port of the important Elyme city of Segesta.

Piazza Armerina, with 22,000 people, is located at 697 m above the sea level in the province of Enna. The area surrounding Piazza Armerina was certainly inhabited since the eighth–seventh century BC by Greeks, Romans and Byzantines until the Middle Ages. In the twelfth century AD strong conflicts are reported between Arabs, still present in the island, and the Latin conquerors (*Lombardi*). After the town was destroyed by Arabs (AD 1161), it was re-peopled by a colony of Lombardi coming from Piacenza (in northern Italy). A typical *Gallo-Italic* dialect of northern Italian origin is still spoken by the inhabitants of

Piazza Armerina (the same dialect is also spoken in other towns of eastern Sicily, i.e. *Nicosia, San Fratello*). Less than 10 km from Piazza Armerina there are two important archaeological sites: the Roman *Villa del Casale* (built between the end of third century BC and the beginning of fourth century BC) and *Morgantina*, founded by an Italic tribe (*Morgeti*) coming from central Italy. The city of Morgantina was destroyed due to conflicts between Greeks and Siculi. Around 300 BC Morgantina flourished again under the influence of the Greek city of Syracuse. Morgantina was destroyed again by Romans during the 2nd Punic war (211 BC). After 30 BC the city was abandoned by the few survivors still living there.

The geographical locations of the four towns are shown in Figure 17.1.

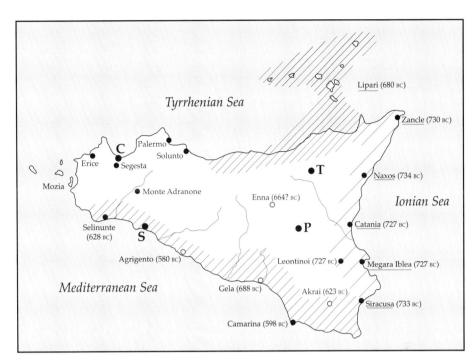

Figure 17.1. *Map of Sicily (modified from Carlotta 1998) showing the position of the four towns (T= Troina; C = Castellammare; S = Sciacca; P = Piazza Armerina) whose population samples were analyzed for autosomal microsatellites and mtDNA polymorphisms. Greek sites (Magna Graecia) are predominantly located along the central and eastern coastal belt (the shaded area) of the island. Among the four populations analyzed with DNA markers, Sciacca is the only one included in the area of Greek influence (see also Fig. 17.2). The dates of foundation of Greek colonies in Sicily are indicated within brackets.*

Surname distribution and assessment of sampling bias
Surname analysis was limited to the towns of Troina, Sciacca, Castellamare del Golfo and Piazza Armerina. Surname proportions in the samples were compared with those of telephone users listed in the telephone directories by using the z-statistics of the normal distribution. This analysis was performed to assess the possible occurrence of sampling biases. Surnames not corresponding to individuals (e.g. companies, shops, etc.) were obviously excluded from the computation.

Analysis of PAH gene mutations and minihaplotypes in PKU families
Extensive data on the molecular basis of PKU in Sicily can be found in Romano *et al.* (1993); Guldberg *et al.* (1993) and Bosco *et al.* (1998). The analysis of one Short Tandem Repeat (STR) and one Variable Number of Tandem Repeat (VNTR) locus, located near the 3' and 5' ends of the phenylalanine hydroxylase gene respectively, was performed as previously described (Goltsov *et al.* 1992; 1993). A segregation analysis of both STR and VNTR alleles was performed in each family with phenylketonuria to identify the minihaplotype (a specific association of VNTR and STR alleles) present in normal and mutant chromosomes. As a rule, a pathological PAH gene mutation linked to the same minihaplotype in two or more populations was considered identical by descent.

Analysis of other autosomal microsatellite loci in samples from the general population
The following four microsatellite polymorphisms were analyzed in the population samples from Troina, Sciacca, Castellamare del Golfo and Piazza Armerina: TPOX (Anker *et al.* 1992), TH01 (M.H. Polymeropoulos *et al.* 1991), vWA/31A (Kimpton *et al.* 1992) and FES/FPS (D.S. Polymeropoulos *et al.* 1991).

HUMTPOX is a microsatellite sequence located on chromosome 2p13. The repeat consists of $[AATG]_n$ motifs. Eight alleles have been identified.

HUMTH01. This short tandem repeat (STR) system is located in intron 1 of the human tyrosine hydroxylase gene on chromosome 11p15.5-p15 and consists of tandemly repeated [AATG]$_n$ motifs. Nine alleles have been identified for this marker.

HUMFES/FPS is a hypervariable microsatellite sequence located near the human C-FES/FPS proto-oncogene and it maps to chromosome 15 (15p25-qter). The HUMFES/FPS sequence consist of variable numbers of tetranucleotide repeat units: [ATTT]$_n$. Eight alleles have been identified for this polymorphism.

HUMvWA31. This is a microsatellite sequence located between nucleotides 1640 and 1794 in intron 40 of the human von Willebrand factor gene on chromosome 12 (12p12-12pter). The repeat is formed by [AGAT]$_n$ motifs. Ten alleles have been identified.

The four loci were analyzed by two different multiplex (vWA31 and FES/FPS) or (TH01 and TPOX) polymerase chain reactions (PCR), performed in 50 μl containing: 5 ng of genomic DNA; 1 U Taq DNA-polymerase (Perkin Elmer, USA); 5 μl reaction buffer 10X (20 mM Tris-HCl pH 8, 100 mM KCl, 0.1 mM EDTA, 1 mM DTT, 50 per cent glycerol, 0.5 per cent Tween 20, 0.5 per cent Nonidet P40); 1.5 mM MgCl$_2$; 0.2 mM of each dNTP; 0.2 μM of each primer. TH01 and TPOX primers were modified by the addition of a dye label (TAMRA: N,N,N',N'-tetramethyl-6-carboxyrhodamine) at the 5' end of one of the two primers used for PCR; vWA31 primer is modified by the addition of a dye label (JOE: 6-carboxy-2', 7-dimethoxy-4',5'-dichlorofluorescein) at the 5' end of one of the two primers used for PCR; FES/FPS primer is modified by the addition of a dye label (6-FAM: 6-carboxyfluorescein) at the 5' end of one of the two primers used for PCR. Primer sequences used for PCR are as described in the ABI PRISM™ STR Primer Set protocol (Perkin Elmer). Conditions used for multiplex PCR were as follows: vWA31and FES/FPS, 93° C for 3'. Each of the 30 cycles was then performed as follows; 94°C for 1', 64° C for 1', 72°C for 1'. At the end of the 30 cycles, samples were kept at 60°C for 30'. Conditions used for the TH01 and TPOX PCRs were as follows: 93°C for 3' followed by 30 cycles, each performed at 94°C for 45", 54°C for 1', 72°C for 1'. Samples were then kept at 60°C for 30'. 1 μl of the PCR products was diluted in 12 μl of deionized formamide plus 1 μl of GeneScan 350 Rox (molecular weight DNA marker), denatured at 95°C for 3 minutes, cooled on melting ice, and loaded on a ABI PRISM 310 Genetic Analyzer (Perkin Elmer, USA). Fragment sizes were estimated by the GeneScan software (Perkin Elmer, USA). To simplify allele typing and to establish the exact number of tetranucleotide repeat units, at least two allele lengths for each locus were sequenced by conventional techniques.

Greek data for the same four microsatellite loci used in the Sicilian samples were obtained by Kondopoulou *et al.* (1999).

Analysis of mtDNA in samples from the general population
The variability of mtDNA in Sicilians was analyzed both in the coding sequence and in the hypervariable segment 1 of the D-loop control region. In the coding sequence, several restriction fragment length polymorphisms (RFLP) were analyzed to reconstruct the most prevalent mtDNA haplogroups. The presence/absence of a restriction site was ascertained by digestion of PCR-amplified segments of mtDNA according to the procedure of Torroni *et al.* (1996). The following sites and enzymes were analyzed: 7025 *Alu*I, 10394 *Dde*I, 4577 *Nla*III, 4216 *Nla*III and 12308 *Hinf*I.

The sequence of HVR1 at positions 16024–16383 was amplified by PCR using the following primers:
5'-CCCTTACTACACAATCAAAG-3' and
5'-CGTGTGGGCTATTTAGGC-3'.
PCR was performed in 50 μl containing: 50 ng of genomic DNA; 1 U Taq DNA-polymerase (Perkin Elmer, USA); 5 μl reaction buffer 10X (20 mM Tris-HCl pH 8, 100 mM KCl, 0.1 mM EDTA, 1 mM DTT, 50 per cent glycerol, 0.5 per cent Tween 20, 0.5 per cent Nonidet P40) 1.5 mM MgCl$_2$ 0.2 mM of each dNTP 0.2 μM of each primer. Cycling conditions used for the PCR were as follows: 96°C for 1 min., 50° C for 1 min., 72°C for 1 min. 30 sec., 30 cycles.

The following primers were used to sequence both strands of DNA:
5'-CACCATTAGCACCCAAAGCT-3' and
5'-CGTGTGGGCTATTTAGGC-3',
using ABI PRISM BigDye Terminator Cycle Sequencing Ready Reaction Kit (PE, Applied Biosystems, Milano, Italy)

Statistical methods
Allele frequencies for each microsatellite locus were estimated by direct gene counting. In order to test whether the populations are in Hardy-Weinberg equilibrium the chi-square test of goodness of fit was used. Pairwise comparisons among samples were examined by estimating the F_{ST} coancestry coefficient, as proposed by Reynolds *et al.* (1983). Principal Components (PCs) Analysis was performed to summarize the information given by 39 distinct alleles for the five population samples of Greece, Piazza Armerina, Castellammare del Golfo, Troina and Sciacca.

Results

PAH gene mutations

In recent years, our work on the molecular basis of PKU in Sicilians has uncovered a remarkable molecular heterogeneity of this disorder in this population with the identification of 45 mutations affecting the gene encoding for the phenylalanine hydroxylase enzyme (Romano *et al.* 1993; Guldberg *et al.* 1993; Bosco *et al.* 1998). The value of heterozygosity (H) computed from the relative frequency of PKU mutations was compared between the Sicilian population and other populations. This analysis revealed that Sicily has one of the highest H values (0.96) in the world, similar to the H value of the (very heterogeneous) population of USA (Bosco *et al.* 1998). This result is consistent with similar findings obtained for thalassemia (Giambona *et al.* 1995) and cystic fibrosis (Rendine *et al.* 1997), thus supporting the notion that these results are not merely reflecting a disease locus-specific feature, but altogether they are likely a hallmark of the strong genetic heterogeneity that characterize the population of Sicily as a whole. Many PAH gene mutations are rare in Sicily and therefore they are not easily amenable to statistical comparison both within the island and between Sicily and other populations. However, the analysis of intragenic minihaplotypes has revealed that a large proportion of these mutations (>50 per cent) are common (and probably identical by descent) to peninsular Italy and continental Europe, suggesting that these have been introduced in Sicily from these areas. Other mutations are also found in North Africa but, unfortunately, no comprehensive studies have been done for these latter populations thus preventing a reliable comparison with Sicily. The best case we have studied concerns instead the most common PKU gene defect in southern Europe and other Mediterranean countries, including Sicily, i.e. the IVS10nt546 mutation. This mutation occurs in Sicilian PKU cases at a frequency of nearly 15 per cent. Specifically, in 10 per cent of Sicilian PKU cases the IVS10nt546 mutation is associated with two intragenic polymorphic alleles (the minihaplotype VNTR 7-STR 252). This mutant minihaplotype shows a gradient of gene frequency from the eastern to the western Mediterranean (Calì *et al.* 1997) similar to the 1st Principal Component of non-DNA markers (Cavalli Sforza *et al.* 1994). We have recently acquired strong evidence that the mutation did not originate in Sicily. This conclusion relies on the absence in Sicily of minihaplotype '7-252' in the normal PAH gene, whereas the disease-causing mutation is frequently linked to the '7-252' minihaplotype in Sicilian (as in Mediterranean) populations. This means that if the mutation had arisen in Sicily, it should today be linked to a frequent minihaplotype present in normal Sicilian PAH alleles, which is not the case.

The diversity of all the minihaplotypes linked to the splicing mutation IVS10nt546 is twice as high in Spain as in the other populations further east, possibly implying an ancient European presence of the disease; however this conclusion (or the alternative one presented in Calì *et al.* 1997) is tentative because sample sizes in this study were rather low (n = 21 to 42 alleles). Unlike the case of the most common mutation causing cystic fibrosis (CF, delta508 mutation) (Morral *et al.* 1994) it is not possible, at present, to date the age of IVS10nt546 mutation. This is because we lack a sufficient number of intragenic multiallelic polymorphisms linked to the mutation that would provide a dating system based on the mutation rates of microsatellite loci, as was performed for the most common CF mutation.

Autosomal microsatellites

We have tested various autosomal microsatellite polymorphisms selected from those commonly used in forensic medicine to analyze several samples of the Sicilian population. The analysis has uncovered genetic affinities between Sicilians (the four pooled samples) and other European and Mediterranean populations. This work is still in progress and we plan to publish the results in full length elsewhere. A result of particular interest, obtained by Principal Components Analysis, consisted of the strong genetic affinity detected beween the samples from 'Sicily' and 'Greece'. PCA was repeated to test for possible heterogeneity existing among the Sicilian samples (Troina, Sciacca, Piazza Armerina and Castellamare del Golfo) in their comparison with the Greek sample. The results of this analysis, summarized in the PC biplot of Figure 17.2, show that, as expected, amongst the four Sicilian samples Sciacca shows the highest affinity with Greece. Compared to the other three Sicilian samples, Sciacca is certainly the best candidate for the presence of 'Greek genes'. Indeed, Sciacca was founded by Selinunte, located at the most extreme southwestern coast of Sicily. West of this point, the island was characterized, in the same period of Greek colonization, by a strong Phoenician presence (Castellamare del Golfo is located just in this area). Phoenicians resisted any attempt at Greek expansion westward. Selinunte was in turn founded in the late period of Greek colonization of Sicily (628 BC) by people coming from Megara

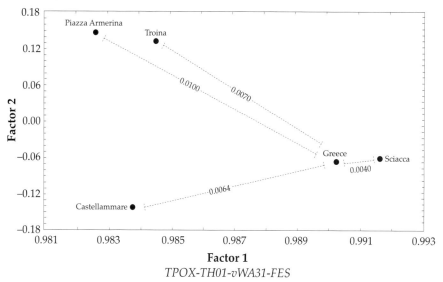

Figure 17.2. *Principal Components Analysis performed using the allele frequencies of four autosomal microsatellites (loci designation is reported below the graph) analyzed in the four Sicilian samples and in one sample from Greece. The highest genetic affinity is obtained between Greece and Sciacca as is also confirmed by the F_{ST} (genetic distance) values superimposed along the dotted lines to the PC biplot. (See also Fig. 17.1 and text for the interpretation of these data.)*

Iblea, a Greek site near Siracusa, founded in 728 BC by Megaresi coming from continental Greece. No such detailed and strong evidence of a Greek connection is available for the other two Sicilian towns analyzed for microsatellite polymorphisms. No relevant Greek sites are indeed present near Troina or Piazza Armerina. These are inland towns, whereas the Greeks mainly occupied coastal sites.

Analysis of mtDNA
We are currently analyzing several samples of the Sicilian population for the variability of mtDNA. The analysis involves both RFLP typing within the mtDNA coding region and DNA sequencing of the hypervariable segment 1 of the D-Loop control region of mitochondrial genome. In Table 17.1 we present the frequency of the most frequent mtDNA RFLP haplogroups in Sicily (Troina + Sciacca) with haplogroup H being the most frequent one. Haplogroup H is also the most frequent mtDNA haplogroup in Europe (Torroni *et al.* 1998). Perhaps the high frequency of haplogroup H in the two Sicilian samples may reflect the presence of a predominant genetic substratum of European origin in Sicily. mtDNA data would thus be consistent with similar conclusions which we have drawn analyzing both microsatellite and PKU data (see above). Moreover, we have thus far detected 94 different HVR1 sequence types in Sicily and have compared them with a world data base of nearly 3500 HVR1 sequences from 31 populations The comparison relied on the estimation of a newly developed Vicinal Sequence Distance (VSD) parameter which measures the sequence similarity between pairs of populations. In short this distance weights for the composition and the frequency of all DNA sequences by ranking those which are similar and not only identical to a given sequence. This new method of analysis seems to suggest a sequence affinity of Sicilians from Troina and Sciacca also with populations of the Atlantic coast of Europe. However this preliminary finding needs to be validated by additional analyses to establish whether we are dealing with a significant genetic relationship or rather with an effect simply caused by the high frequency of haplogroup H shared by Sicily and all European samples collected in the data base.

Concluding remarks and future research perspectives
Our investigation provides new data that may help elucidate the genetic structure of Sicily in a European context. The analysis of the autosomal variability supports the hypothesis of the existence in Sicily of different genetic strata. The results obtained with the three markers used in this study point to the predominance of 'European genes' in Sicily. Another important, though minor, component of the genetic structure of extant Sicilians, as revealed by autosomal data, appears to be of eastern Mediterranean origin: of particular interest is the genetic heterogeneity displayed by the four Sicilian samples in their comparison with the Greek sample (see data from microsatellite polymorphisms). These preliminary findings deserve further investigation for their potential interest in the elucidation of the complex network of movements and colonizations that characterized the rise of *Magna Graecia* in Sicily. However we need to emphasize the tentative nature of these results. One major limitation in the interpretation of these results is the impossibility of dating

Table 17.1. *MtDNA RFLP haplogroups in Sicily.*

MtDNA haplogroups	Sicily (Troina + Sciacca) N*	%
H	98	49.5
U	17	8.6
T	16	8.1
J	13	6.6
K	11	5.5
V	39	19.7
Others	39	19.7
Total	198	100

*N = Number of analyzed individuals

recombining autosomal data and determining direction of migration. For the future, our short-term plan is to collect and analyze additional population samples in Sicily near Greek sites to verify whether it is possible to extend to them the observation made with microsatellites polymorphisms for the population of Sciacca. Furthermore, the reconstruction of the mtDNA tree (Bandelt *et al.* 1995), may lead to the identification of clusters of mutations in the HVR1 region that are specific for different areas of Sicily and/or between Sicilians and other populations.

The value of many molecular markers (recombining and non-recombining ones) for analyzing the genetic variability between populations of distant geographic areas has now been firmly established (Cavalli-Sforza *et al.* 1994; Torroni *et al.* 1998). In contrast it is less clear whether the same is true when studying populations living in small geographic areas (e.g. Sicily). The present study provides encouraging evidence for the view that this is indeed the case, at least for the set of molecular markers and populations samples we have analyzed in Sicily.

Acknowledgements

The expert technical assistance of Valeria Chiavetta, Rosanna Galati and Pietro Schinocca is acknowledged, and also the assistance in mtDNA RFLP analysis by Antonio Torroni. This work was supported by Progetto Finalizzato of Ministry of 'Genetica di popolazione degli alleli PAH in Sicilia: paragone con altri polimorfismi'; and Progetto Finalizzato C.N.R.; Sottoprogetto 4 Beni Culturali: 'Cultural Heritage'; and Progetto MURST 40 per cent and 60 per cent (Italy), and by CNR Projects 'Biological Archive' and 'Biotechnology and Bioinstrumentation'.

References

Anker, R., T. Steinbrueck & H. Donis-Keller, 1992. Tetranucleotide repeat polymorphism at the human thyroid peroxidase (hTPO) locus. *Human Molecular Genetics* 1, 137.

Bandelt, H-J., P. Forster, B.C. Sykes & M. Richards, 1995. Mitochondrial protraits of human populations using median networks. *Genetics* 141, 743–53.

Bosco, P., F. Calì, C. Meli, F. Mollica, E. Zammarchi, R. Cerone, C. Vanni, L. Palillo, D. Greco & V. Romano, 1998. Eight new mutations of the phenylalanine hydroxylase gene in Italian patients with hyperphenylalaninemia. *Human Mutation* 11, 240–43.

Calì, F., I. Dianzani, L.R. Desviat, B. Perez, M. Ugarte, M. Ozguc, Y. Shiloh, S. Giannattasio, C. Carducci, P. Bosco, G. De Leo, A. Piazza & V. Romano, 1997. The STR 252 - IVS10nt546 - VNTR 7 phenylalanine hydroxylase minihaplotype in five Mediterranean samples. *Human Genetics* 100, 350–55.

Carlotta, F., 1998. *Breve storia della Sicilia Antica*. Palermo: Flaccovio Editore.

Cavalli-Sforza, L.L., P. Menozzi & A. Piazza, 1994. *The History and Geography of Human Genes*. Princeton (NJ): Princeton University Press.

Finley, M.I., 1979. *Storia della Sicilia antica*. Bari: Laterza.

Giambona, A., P. Lo Gioco, M. Marino, I. Abate, R. Di Marzo, M. Renda, F. Di Trapani, F. Messana, S. Siciliano, P. Rigano, F.F. Chehab, H.H. Kazazian & A. Maggio, 1995. The great heterogeneity of thalassemia molecular defects in Sicily. *Human Genetics* 95, 526–30.

Goltsov, A.A., R.C. Eisensmith, D.S. Konecki, U. Lichter-Konecki & S.L.C. Woo, 1992. Association between mutations and a VNTR in the human phenylalanine hydroxylase gene. *American Journal of Human Genetics* 51, 627–36.

Goltsov, A.A., R.C. Eisensmith, E.R. Naughton, L. Jin, R. Chakraborty & S.L.C. Woo, 1993. A single polymorphic STR system in the human phenylalanine hydroxylase gene permits rapid prenatal diagnosis and carrier screening for phenylketonuria. *Human Molecular Genetics* 2, 577–81.

Guldberg, P., V. Romano, N. Ceratto, P. Bosco, M. Ciuna, A. Indelicato, F. Mollica, C. Meli, M. Giovannini, E. Riva, G. Biasucci, F. Henriksen & F. Guttler, 1993. Mutational spectrum of phenylalanine hydroxylase deficiency in Sicily: implications for diagnosis of hyperphenylalaninemia in southern Europe. *Human Molecular Genetics* 2, 1703–7.

Kimpton, C.P., A. Walton & P. Gill, 1992. A further tetranucleotide repeat polymorphism in the vWF gene. *Human Molecular Genetics* 1, 287.

Kondopoulou, H., R. Loftus, A. Kouvatsi & C. Triantaphyllidis, 1999. Genetic studies in 5 Greek population samples using 12 highly polymorphic DNA loci. *Human Biology* 71(1), 27–42.

La Rosa, V., 1989. Le popolazioni della Sicilia: Sicani, Siculi, Elimi, in *Italia Omnium Terrarum Parens*, eds. Pugliese

Carratelli. Milan: Instituto Veneto Arte Graficlo, 50–54.

Mack Smith, D., 1994. *Storia della Sicilia medievale e moderna.* Roma-Bari: Laterza.

Morral, N., J. Bertranpetit, X. Estivall, V. Nunes, T. Casals, J. Gimènez, A. Reis, R. Varon-Mateeva, M. Macek Jr, L. Kalaydjieva, D. Angelicheva, R. Dancheva, G. Romeo, M.P. Russo, S. Garnerone, G. Restagno, M. Ferrari, C. Magnani, M. Claustres, M. Desgeorges, M. Schwartz, M. Schwarz, B. Dallapiccola, G. Novelli, C. Ferec, M. Dearce, M. Nemeti, T. Kere, M. Anvret, N. Dahl & L. Kadasi, 1994. The origin of the major cystic fibrosis mutation (deltaF508) in European populations. *Nature Genetics* 7, 169.

Pace, B., 1958. Arte e civiltà della Sicilia antica, in *Società Editrice Dante Alighieri*, ed. B. Pace. 2nd edition. Milan, Rome & Naples: Città di Castello, 177–235.

Piazza, A., N. Cappello, E. Olivetti & S. Rendine, 1988. A genetic history of Italy. *Annals of Human Genetics* 52, 203–13.

Polymeropoulos, D.S., M.H. Rath, H. Xiao & C.R. Merrill, 1991. Tetranucleotide repeat polymorphism at the human c-fes/fps proto-oncogene (FES). *Nucleic Acids Research* 19, 3753.

Polymeropoulos, M.H., H. Xiao, D.S. Rath & C.R. Merrill, 1991. Tetranucleotide repeat polymorphism at the human tyrosine hydroxylase gene. *Nucleic Acids Research* 19, 4018.

Rendine, S., F. Calafell, N. Cappello, R. Gagliardini, G. Caramia, N. Rigillo, M. Silvetti, M. Zanda, A. Miano, F. Battistini, L. Marianelli, G. Taccetti, M.C. Diana, L. Romano, C. Romano, A. Giunta, R. Padoan, A. Pianaroli, V. Raia, G. De Ritis, A. Battistini, G. Grzincich, L. Japichino, F. Pardo, A. Piazza, *et al.* 1997. Genetic history of cystic fibrosis mutations in Italy. I. Regional distribution. *Annals of Human Genetics* 61, 411–24.

Reynolds, J., B.S. Weir & C.C. Cockerham, 1983. Estimation of the coancestry coefficient: basis for a short-term genetic distance. *Genetics* 105, 767–79.

Rickards, O., C. Martinez-Labarga, G. Scano, G.F. De Stefano, G. Biondi, M. Pacaci & H. Walter, 1998. Genetic history of the population of Sicily. *Human Biology* 70, 699–714.

Romano, V., P. Bosco, V. Chiavetta, G. Fasulo, L. Pitronaci, F. Mollica, C. Meli, M. Giovannini, E. Riva, B. Giuffrè, R. Eisensmith, S.L.C. Woo, C. Romano, A. Ponzone, I. Dianzani, C. Camaschella, C. Di Pietro & N. Ceratto, 1993. Geographical distribution of phenylalanine hydroxylase alleles in Sicily. *Developmental Brain Dysfunction* 6, 83–91.

Torroni, A., K. Huoponen, P. Francalacci, M. Petrozzi, L. Morelli, R. Scozzari, D. Obinu, M.L. Savontaus & D. Wallace, 1996. Classification of European mtDNAs from an analysis of three European populations. *Genetics* 144, 1835–50.

Torroni, A., H-J. Bandelt, L. D'Urbano, P. Lahermo, P. Moral, D. Sellitto, C. Rengo, P. Forster, M.L. Savontaus, B. Bonnè-Tamir & R. Scozzari, 1998. MtDNA reveals a major Paleolithic population expansion from southwestern to northwestern Europe. *American Journal of Human Genetics* 62, 1137–52.

Tusa, S., 1983. *La Sicilia nella preistoria.* Palermo: Edizioni Sellerio, 53–111; 121–81.

Tusa, S., 1997. *Prima Sicilia: Alle origini della società siciliana,* ed. S. Tusa. Palermo: Edipoint, 317–32.

Chapter 18

MtDNA Variability in Extinct and Extant Populations of Sicily and Southern Italy

Olga Rickards, Cristina Martínez-Labarga, Rosa Casalotti, Giuseppe Castellana, Anna M. Tunzi Sisto & Francesco Mallegni

A new approach to human evolutionary studies has come about in the last ten years through the development of highly sophisticated molecular biological techniques that directly compare ancestral populations to their present-day counterparts. The molecular approach to human evolutionary studies is potentially useful for the reconstruction of the peopling of the Mediterranean basin, and, in particular, of southern Italy. This area, because of its central geographical position, was crossed by a complex network of migration of people from Phoenicia, Greece, Arabia, the Balkans, the Near and Middle East, and North Africa, resulting in a heterogeneous pattern of both cultural and genetic interactions. To test the extent of the genetic impact these various populations had upon the peoples of southern Italy, a population-level survey of mitochondrial DNA (mtDNA) variation on both ancient and present Mediterranean people was made. DNAs were extracted from bones and teeth obtained from different sites of southern Italy. Endogenous mtDNA sequences from the samples processed up to now were tentatively classified, on the basis of HV1 nucleotide motifs and 00073 status at HV2 as T, J and HV. The findings indicate that haplogroups T and J have been present in southern Italy since the Bronze Age and HV since at least the sixteenth century AD. The other ancient samples turned out to belong to haplogroup H, the most frequent in Europe and one that is also common in the Near East. In addition, several extant populations from various Mediterranean countries were analyzed to permit the diachronic comparison and to reconstruct the phylogenies of the mtDNA lineages. The maternal genealogy of mtDNA haplotytpes identified in Sicily and southern Italy was obtained by network analysis using the median algorithm of Bandelt et al. (1995). Phylogenetic analysis of haplotypes through genetic distances was used to estimate relationships between the Mediterranean populations.

Human diversity in Sicily and southern Italy has largely been shaped by migration processes that have occurred since the initial occupation of the region in Palaeolithic times, followed by the diffusion of Neolithic farming, which entailed population movements from the Near East into Europe, the consistency of which is still under debate.

Historically, a complex network of migrations can be traced from Phoenicia, Greece, Arabia, the Balkans, the Near and Middle East, and North Africa into the central Mediterranean area. With this recurrent colonization, a heterogeneous pattern of cultural interactions (Pallottino 1984; Finley 1985; Moscati 1994) of unknown genetic consequences has emerged.

Much is already known about the main demographic transitions from the patterns of population diversity observed by studies of genetic diversity in extant populations and research into historical, cultural and classic anthropological sources. But the extent of the impact which various population movements had upon the people of southern Italy could not be detected. Investigations into the distribution of traditional marker allele frequencies in the Sicilian and Apulian extant populations (Rickards et al. 1992; 1995; 1998; Walter et al. 1997) clearly highlighted the presence of a more recent gene flow from North Africa and the Middle East superimposed on a predominant Greek genetic substratum. Even so, the studies found no genetic traces of the ancient groups who peopled Sicily and southern Italy in pre-Roman times. These observations disagreed with the results of studies that used other genetic markers (HLA-A, HLA-B, ABO, RH, MN, KEL, HP: Piazza et al. 1988) and rare surnames (Guglielmino et al. 1991) to support the claim that the existence of a more ancient trace in this area can indeed be determined.

Such conflicting results underline the inherent difficulties of attempting to reconstruct molecular human microevolutionary history on the basis of the present gene structure alone.

With the application of the polymerase chain reaction (PCR) (Saiki et al. 1985; Mullis & Faloona 1987) in molecular biology, not only can large sets of molecular data be collected from present populations but from ancient remains as well. The comparison of extinct populations with their extant counterparts permits the direct reconstruction of the evolutionary history of the populations through time, providing a temporal dimension for the study of molecular evolution and human population biology. Thus, ancient DNA (aDNA) analysis constitutes a unique and highly valuable contribution to the debate on the origin of genetic diversity of human populations. But the power of aDNA is at its greatest when it can answer historical questions posed by other disciplines and, in so doing, better explain the reconstruction of human populations of the past, their migration and dispersal (Stone & Stoneking 1993; Lalueza et al. 1997; Izagirre & de la Rúa 1999).

Our case study reports on the peopling of Sicily and southern Italy (Rickards et al. 1998; in press a). We analyzed mitochondrial DNA (mtDNA) variation in contemporary Italian populations and in several ancient samples from Sicily and Puglia. There are many advantages in using mtDNA. Because it is present in high copy numbers in mammalian cells, mtDNA can most likely be detected even in highly degraded samples, such as the ancient ones we studied. Two properties in particular make mtDNA an ideal marker for studying ancestor/descendant relationships. The first concerns its fast evolution rate which enhances the attractiveness of mtDNA in human population studies because of the short evolutionary times characterizing our evolution. The second concern is its maternal mode of inheritance: in fact the lack of segregation and recombination allows the reconstruction of maternal genealogies without the drawbacks connected to reconstruction based on nuclear genes. Moreover, the wealth of available data on mtDNA variability in human populations allows the comparison and the attribution of extinct and extant Italian sequences to geographically or ethnically specific lineages from which migrational patterns can then be reconstructed.

DNA source and methods

DNA was extracted from bone and tooth samples obtained at two sites: the 'Castellucciano' sepulchral cave of Ticchiara (Agrigento, Sicily) and the Bronze Age hypogeum of Madonna di Loreto (Trinitapoli, Foggia, Puglia) (Castellana 1997; Tunzi Sisto 1999). Moreover, the remains of the family of Prince Branciforte, a Sicilian maecenas and benefactor who lived between the sixteenth and seventeenth centuries, were processed for molecular personal identification and mtDNA lineage assignment (Rickards et al. in press a,b). Three stretches of mtDNA were exploited: region V, between the cytochrome oxidase II and the lysine transfer RNA genes (Wrischnik et al. 1987), and the two hypervariable segments, HV1 and HV2, of the control region (Vigilant et al. 1991).

To perform diachronic comparisons with extant populations, we analyzed the mtDNA polymorphisms in 69 individual samples from Sicily and southern Italy.

Matching genetic data from the literature were collected for various European, North African and Middle Eastern populations (Fig. 18.1). Populations which appeared to have undergone drastic founder effects and drift were not included in the data set.

All the ancient samples had been previously studied using archaeological and osteological approaches. Microradiography cross-sections of ancient tissues were obtained and analyzed by TEM, before proceeding with the DNA extractions (Hagelberg et al. 1991; Herrmann & Hummel 1994). All the bones appeared in good condition, with perfectly preserved microstructures.

Procedures in sample preparation and DNA

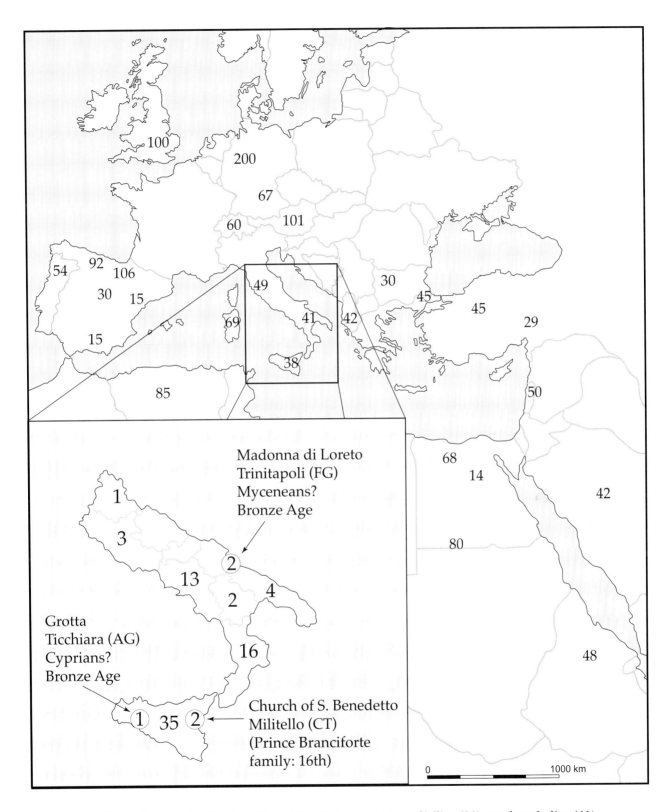

Figure 18.1. *Size and geographical distribution of the study populations. Sicilian (38), southern Italian (41), Ethiopian (48) and Egyptian (14) data refer to this survey. References for the other populations are given in the legend of Table 18.1. The enlarged area shows details of the samples from Sicily and southern Italian regions. The location of the ancient specimens analyzed here is also highlighted.*

extraction, purification, amplification and cloning were performed following all precautionary measures and criteria recommended in aDNA analyses (Rickards et al. in press b; http://www.uniroma2.it/biologia/centri/rickards/). To authenticate the results, we performed at least two independent extractions for each ancient sample, four amplifications for each extraction, and screened 20 clones for each amplification. Extant DNA processing methods and sequencing protocols for both present and ancient DNAs are reported elsewhere (Rickards et al. 1999; in press a,b). None of the ancient mtDNA types matched the DNA sequences of the archaeologists, palaeontologists or molecular anthropologists involved in the study. This was done to exclude the possibility of contemporary human DNA contamination.

MtDNA genealogy within southern Italy was reconstructed by network analysis using the median algorithm of Bandelt et al. (1995). In this kind of analysis alternative evolutionary pathways are retained and represented in the form of reticulations.

The statistical analyses were carried out using the Arlequin software package (Schneider et al. 1997). The selective neutrality of sequences and the demographic equilibrium of the sample from southern Italy were tested using Tajima's D statistics (Tajima 1989).

Results and discussion

Eight ancient specimens and 69 modern individuals (31 Sicilians and 38 southern Italians) were analyzed for both HV1 (16024–16390 bp) and HV2 (00030–00408 bp) sequence polymorphisms. Because some of the archaeological samples proved to belong to the same maternal lineage, only five independent sequences were used for further analyses.

Table 18.1. Table of Sicilian and southern Italian HV1 haplotypes shared with the European, North African and Middle Eastern populations. Five individuals not characterized for HV2 sequence polymorphisms were included in this comparison. Ancient specimens are underlined. Substitutions relative to the Cambridge Reference Sequence (Anderson et al. 1981) are reported (minus 16000). Population samples are coded as follows: Sic = Sicilians and Sit = southern Italians (this survey); Ibe = Iberians (Côrte-Real et al. 1996; Pinto et al. 1996; Salas et al. 1998); Bas = Basques (Bertranpetit et al. 1995; Côrte-Real et al. 1996); NCEu = North Central Europeans (Piercy et al. 1993; Parson et al. 1998; Hofmann et al. 1997; Lutz et al. 1998; Pult et al. 1994); Sar = Sardinians (Di Rienzo & Wilson 1991); Tus = Tuscans (Francalacci et al. 1996); Alb = Albanians (Belledi et al. 2000); Bul = Bulgarians (Calafell et al. 1996); Dru = Druses and Ady = Adygey (Macaulay et al. 1999); Tur = Turques (Calafell et al. 1996; Comas et al. 1996); Mea = Middle Easterns (Di Rienzo & Wilson 1991); Egy = Egyptians (Krings et al. 1999, this survey); Nub = Nubians (Krings et al. 1999); Eth = Ethiopians (this survey); Alg = Algerians (Côrte-Real et al. 1996).

Pop S. Size Hapl. N	Sic 38 32	Sit 41 34	HV1 motif	Ibe 224 127	Bas 106 53	NCEu 528 275	Sar 69 45	Tus 49 40	Alb 42 31	Bul 30 22	Dru 45 26	Tur 74 64	Ady 50 31	MEa 42 39	Egy 82 71	Eth 48 45	Nub 80 53	Alg 85 30
Haplotypes			HV1 motif															
BRAN1	5	7	0	49	25	85	17	9	7	3	5	5	8		5		1	7
SIC31	1		311	6	2	15	3		2	1	1	1	1	1		2		7
ML2		1	69, 126	5		15		2	2	2	1	3	1		1			1
SIC103, CAL106	1	1	224, 311	6	2	8	1	1			3	2			1			
CAL897, LAZ18		2	126, 294, 296		2	8							1	1	1			
BAS9A		1	291			3	2		2			1						
SIC64, SIC95	2		298	3	4	10												4
CAMP49		1	192, 256, 270	1		5			2									
LAZ3397		1	224	1	1			1										
MAR1097		1	362		3	5										5		
CAMP1229		1	278, 311		2								4					
TICH	1		294			1	1											
CAL109		1	126, 294				1							1				
CAL9297		1	69, 126, 278	1		1												
SIC30	1		146, 342		1	1												
CAMP312		1	145, 176, 223, 367												1			
n. of shared haplotypes	16			8	10	12	5	4	5	4	4	5	6	2	4	1	2	4
Per cent of shared hapl.	0.25			0.06	0.19	0.04	0.11	0.10	0.16	0.18	0.15	0.08	0.19	0.05	0.06	0.02	0.04	0.13
Per cent of individuals with shared hapl.	0.38			0.32	0.42	0.30	0.33	0.27	0.36	0.30	0.20	0.15	0.32	0.05	0.10	0.04	0.08	0.22

Four different haplotypes identified in Sicily, Molise and Apulia were associated with the region V 9 bp deletion or triplication. This finding supports the hypothesis of a polyphiletic origin of these length polymorphisms in Europe. All the ancient haplotypes showed the two 9-bp tandem repeats (Rickards *et al.* in prep.).

Sixteen out of the 64 haplotypes identified on the basis of HV1 motif (range of comparison: 16069–16360 bp) also appeared in the other European and Mediterranean populations studied (Fig. 18.1; Table 18.1).

Two of the ancient sequences, BRAN1 (identified in the Branciforte family and identical to the CRS) and ML2 (detected among the Madonna di Loreto remains characterized by two transitions in position 16069 and 16126) were identified in most of the data set populations. The sequence found in the Grotta Ticchiara (TICH) sample was shared only with one Sardinian and one German sample. Among the extant sequences, the two most represented haplotypes were detected in Sicily and Calabria; they were characterised by transitions at 16311, and 16311 and 16224, respectively. The remaining shared haplotypes are also present at low frequencies in northwestern Europe or in both European and eastern Mediterranean populations.

Phylogenetic relationships among southern Italian haplotypes were analyzed by constructing a reduced median network (Bandelt *et al.* 1995) with information on HV1 from 16024 to 16390 bp, combined with the status of both sites 00072 and 00073 of HV2 (Fig. 18.2).

MtDNA lineages cluster into distinct groups, each defined by a set of associated polymorphisms. Cluster designation, attributed on the basis of HV1 and HV2 motifs, follows the classification scheme recently reviewed by Macaulay *et al.* (1999). The resulting topology is characterized by a 'star-like' pattern. Similar topologies have already been reported for other European populations (Richards *et al.* 1998; Macaulay *et al.* 1999). The largest cluster of the sequences obtained (51 per cent) is cluster H, which includes the Cambridge Reference Sequence (CRS) (Anderson *et al.* 1981). This cluster, characterized by the 00073A status, includes fairly short branch derivatives, suggesting a recent expansion (Richards *et al.* 1996). It is the most frequent in Europe and is also common in the Near East (Torroni *et al.* 1998). It can be subdivided into distinct sub-clusters on the basis of motifs consisting of mutations in both HV1 and HV2 (Macaulay *et al.* 1999; Rickards *et al.* in press a; in prep.), and labelled as HB, HC, HD or V and HV. Three of the ancient samples belonged to this main cluster. In particular, the other type identified in the Branciforte family (BRAN2) was characterized as HV on the basis of the 16067 transition. The same motif, recently reported for the Druse population, an ethnic minority living in Israel, Libya and Syria (Macaulay *et al.* 1999), was shared with an extant Sicilian sequence. Turkish sequences, one Sardinian, one Bulgarian, two Nubian, two Ethiopian and two Egyptian types, matched this haplogroup which seems to be specific to eastern Mediterranean areas. This lineage is also present in some central Asian populations (Comas *et al.* 1998). V haplogroup, which reaches high frequencies in northwestern Europe (Torroni *et al.* 1998), was identified in five individuals.

The cluster characterized by transition at 16289 (4 sequences), which we labelled as HB, seems to be unique to southern Italy. The distinguishing motif (16278, 16291) of the other subcluster identified in this study (HC) is also detectable at low frequency in other European populations (Rickards *et al.* in prep.).

That the clusters characterized by G at bp 00073 have longer branches than cluster H is probably an indication of the ancient population expansion of our species out of Africa and its subsequent dispersal into Eurasia. The archaeological sample labelled as ML2 belongs to cluster J; this cluster originated from the Near East (Richards *et al.* 1998) and is characterized by transitions at 16069 and 16126. TICH haplotype, which displays only the 16294 mutation, was tentatively attributed to T haplotype (identified by the 16126 and 16294 motif) on the basis of observations which other authors have made. In fact, in European and European-derived sequences this mutation is rarely seen outside of haplogroup T (Côrte-Real *et al.* 1996; Torroni *et al.* 1996). Moreover, there was the lack of the 16126 mutation (Torroni *et al.* 1996) in two Tuscan sequences, classified as T and T1 by RFLP analysis. The presence of L1b mtDNA lineage, which is typical of sub-Saharan Africa (see Macaulay *et al.* 1999), in one Apulian individual of ascertained Italian maternal ancestry, indicates population movements from this area to southern Italy (Rickards *et al.* in prep.).

MtDNA sequences were also used to further describe the genetic history of southern Italians and to compare the distribution of nucleotide differences between pairs of sequences (Rickards *et al.* in prep.). Under a set of restrictive assumptions, the distribution shape of the pairwise differences (Slatkin & Hudson 1991; Rogers & Harpending 1992) contains information on the demographic history of the popu-

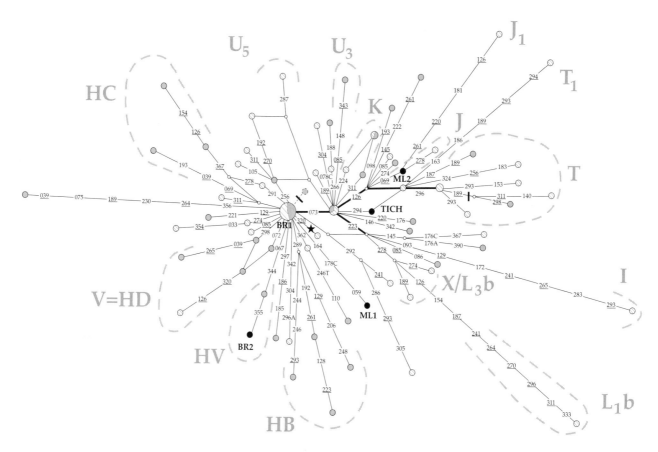

Figure 18.2. *Reduced median network showing the relationship among the different mtDNA haplotypes defined by HV1 motif and the 00073 status. Eight sequences not considered in the previous analysis were included in the network because the shorter stretch sequenced (from 16180 to 16390 bp) could be classified unambiguously. Circles correspond to observed haplotypes, with area proportional to frequency. Solid grey and striped circles denote the extant Sicilian and the continental southern Italian haplotypes, respectively; solid black circles show the ancient sequences. Lines linking nodes represent single substitution referred to CRS (Anderson et al. 1981). Nucleotide positions in HV1 (minus 16000) label the branches and indicate transitions, while transversions are marked by capital letters. Underlined positions state parallel mutational events solved during the reduction procedure. For cluster designation, see the text (Result and discussion section). Region V 9 bp deleted and triplicated haplotypes are marked by transversal black bars and a black star, respectively.*

lations. Simulation tests suggest that a population undergoing rapid expansion is characterized by a single peak in the mismatch distribution, although exceptions have been envisaged (Bertorelle & Slatkin 1995). Rogers (1995) also showed that the peak, i.e. the modal mismatch within a population, tends to travel to the right with time, implying that the higher the average pairwise difference, the more remote the population expansion. Conversely, populations whose size was essentially constant tended to have a more irregular mismatch distribution. These two different patterns relating to specific life conditions lead to different mtDNA phylogenies: clustered trees for populations with constant resources such as hunter-gatherers, for which no changes in population size are expected, and star-like trees for populations with increasing resources, for which either sudden expansion or exponential growth is likely. Southern Italian HV1 sequences showed a unimodal shaped mismatch distribution (mismatch observed mean 4.825) associated with a significant negative value (Tajima's D –2.373; $P = 0.001$) of Tajima neutrality test (Tajima 1989). These results, also confirmed by HV1 and HV2 merged sequence information (mismatch observed mean 7.773; Tajima's D –2.346, $P = 0.002$), are in agreement with the hypothesis of a recent demographic expansion (Rogers & Harpending 1992; Aris-Brosou & Excoffier 1996). The comparison between the HV1 pairwise difference distribution with others observed in Mediterranean populations indicates that the suggested growth in size would have occurred, as in the case of Anatolians (Comas *et al.* 1996), some time between the ancient expansion of the Middle Eastern populations and the recent expansion of central and western Europeans (Rickards *et al.* in prep.).

Conclusions

Our study identified five distinct haplotypes in the archaeological specimens sampled in Sicily and southern Italy which fitted easily into the genetic variation of extant European and Middle Eastern populations. It follows that mtDNA lineages J and T have been present in the human populations living in the Mediterranean area at least since the Bronze Age, and that the HV haplogroup most probably reached Sicily from Middle Eastern countries at least by the sixteenth century AD.

Clusters T and J, both of Middle Eastern origin (Richards *et al.* 1998), have close phylogenetic affinities. T is the more ancient of the two, having a divergence time of at least 46,500 years, while J was very likely first introduced into Europe about 10,000 years ago by Neolithic farmers (Richards *et al.* 1998). The geographic distribution of these haplotypes in extant populations seems to support the claim inferred from the analysis of archaeological and classical anthropological evidence that there is an ancient relationship between Ticchiara and Madonna di Loreto peoples and the Greeks (Castellana 1997; Tunzi Sisto 1999). Furthermore, the genetic distance analysis comparing the ancient samples with extant European and other Mediterranean populations confirmed these results (data not shown).

The use of other ancient mtDNA sequences collected from Late Palaeolithic and Neolithic samples and their comparison with those from the present-day population could, because they shed new light on mtDNA variability, reveal insightful patterns in the population history of southern Italy.

Acknowledgements

This project was supported by CNR, Progetto Finalizzato Beni Culturali, contributions N. 96.01168.PF36 and 97.00689.PF36, allotted to O. Rickards and by CNR fellowships that provided support for C. Martínez Labarga and R. Casalotti.

References

Anderson, S., A.T. Bankier, B.G. Barrell, M.H.L. de Bruijn, A.R. Coulson, J. Drouin, I.C. Eperon, D.P. Nierlich, B.A. Roe, F. Sanger, P.H. Schreier, A.J.H. Smith, R. Staden & I.G. Young, 1981. Sequence and organization of the human mitochondrial genome. *Nature* 290, 457–65.

Aris-Brosou, S. & L. Excoffier, 1996. The impact of population expansion and mutation rate heterogeneity on DNA sequence polymorphism. *Molecular Biology and Evolution* 13, 494–504.

Bandelt, H-J., P. Forster, B. Sykes & M.B. Richards, 1995. Mitochondrial portraits of human populations using median networks. *Genetics* 141, 743–53.

Belledi, M., E.S. Poloni, R. Casalotti, F. Conterio, I. Mikerezi, J. Tagliavini & L. Excoffier, in press. Maternal and paternal lineages in Albania and the genetic structure of Indo-European populations. *European Journal of Human Genetics*.

Bertorelle, G. & M. Slatkin, 1995. The number of segregating sites in expanding human populations, with implications for estimates of demographic parameters. *Molecular Biology and Evolution* 12, 887–92.

Bertranpetit, J., J. Sala, F. Calafell, P.A. Underhill, P. Moral & D. Comas, 1995. Human mitochondrial DNA variation and the origin of Basques. *Annals of Human Genetics* 59, 63–81.

Calafell, F., P.A. Underhill, A. Tolun, D. Angelicheva & L. Kalaydjieva, 1996. From Asia to Europe: mitochondrial DNA sequence variability in Bulgarians and Turks. *Annals of Human Genetics* 60, 35–49.

Castellana, G. (ed.), 1997. *La grotta Ticchiara ed il Castellucciano agrigentino*. Palermo: Regione Sicilia.

Comas, D., F. Calafell, E. Mateu, A. Pérez-Lezaun & J. Bertranpetit, 1996. Geographic variation in human mitochondrial DNA control region sequence: the population history of Turkey and its relationship to the European populations. *Molecular Biology and Evolution* 13, 1067–77.

Comas, D., F. Calafell, E. Mateu, A. Pérez-Lezaun, E. Bosh, R. Martinez-Arias, J. Clarimon, F. Facchini, G. Fiori, D. Luiselli, D. Pettener & J. Bertranpetit, 1998. Trading genes along the silk road: mtDNA sequences and the origin of central Asian populations. *American Journal of Human Genetics* 63, 1824–38.

Côrte-Real, H.B.S.M., V.A. Macaulay, M.B. Richards, G. Hariti, M.S. Issad, A. Cambon-Thomsen, S. Papiha, J. Bertranpetit & B.C. Sykes, 1996. Genetic diversity in the Iberian Peninsula determined from mitochondrial sequence analysis. *Annals of Human Genetics* 60, 331–50.

Di Rienzo, A. & A.C. Wilson, 1991. Branching pattern in the evolutionary tree for human mitochondrial DNA. *Proceedings of the National Academy of Sciences of the United States of America* 88, 1597–601.

Finley, M.I., 1985. *Storia della Sicilia antica*. Bari: Laterza.

Francalacci, P., J. Bertranpetit, F. Calafell & P.A. Underhill, 1996. Sequence diversity in the control region of mitochondrial DNA in Tuscany and its implications for the peopling of Europe. *American Journal of Physical Anthropology* 100, 443–60.

Guglielmino, C.R., G. Zei & L.L. Cavalli-Sforza, 1991. Genetic and cultural transmission in Sicily as revealed by names and surnames. *Human Biology* 63, 607–28.

Hagelberg, E., L.S. Bell, T. Allen, S.J. Jones & J.B. Clegg, 1991. Ancient bone DNA: techniques and applications. *Philosophical Transactions of the Royal Society of London B* 333, 399–407.

Herrmann, B. & S. Hummel, 1994. Introduction, in *Ancient DNA: Recovery and Analysis of Genetic Material from Paleontological, Archaeological, Museum, Medical, and Forensic Specimens*, eds. B. Herrmann & S. Hummel. New York (NY): Springer-Verlag, 1–12.

Hofmann, S., M. Jaksch, R. Bezold, S. Mertens, S. Aholt, A. Paprotta & K.D. Gerbitz, 1997. Population genetics and disease susceptibility: characterization of central European haplogroups by mtDNA gene mutations, correlation with D loop variants and association with disease. *Human Molecular Genetics* 6, 1835–46.

Izagirre, N. & C. de la Rúa, 1999. An mtDNA analysis in ancient Basque populations: implications for haplogroup V as a marker for major Paleolithic expansion from southwestern Europe. *American Journal of Human Genetics* 65, 199–207.

Krings, M., A.E. Salem, K. Bauer, H. Geisert, A.K. Malek, L. Chaix, C. Simon, D. Welsby, A. Di Rienzo, G. Utermann, A. Sajantila, S. Pääbo & M. Stoneking, 1999. MtDNA analysis of Nile river valley populations: a genetic corridor or a barrier to migration? *American Journal of Human Genetics* 64, 1166–76.

Lalueza, C., A. Perez-Perez, E. Prats, L. Cornudella & D. Turbon, 1997. Lack of founding Amerindian mitochondrial DNA lineages in extinct aborigenes from Tierra del Fuego-Patagonia. *Human Molecular Genetics* 6, 41–6.

Lutz, S., H-J. Weisser, J. Heizmann & S. Pollak, 1998. Location and frequency of polymorphic positions in the mtDNA control region of individuals from Germany. *International Journal of Legal Medicine* 111, 67–77.

Macaulay, V., M.B. Richards, E. Hickey, E. Vega, F. Cruciani, V. Guida, R. Scozzari, B. Bonné-Tamir, B. Sykes & A. Torroni, 1999. The emerging tree of west eurasian mtDNAs: a synthesis of control-region sequences and RFLPs. *American Journal of Human Genetics* 64, 232–49.

Moscati, S., 1994. Cosí nacque l'Italia, in *Antiche genti d'Italia*, eds. P.G. Guzzo, S. Moscati & G. Susini. Rome: De Luca, 11–18.

Mullis, K.B. & F.A. Faloona, 1987. Specific synthesis of DNA *in vitro* via a polymerase-catalyzed chain reaction. *Methods in Enzymology* 155, 335–50.

Pallottino, M., 1984. *Storia della prima Italia*. Milan: Rusconi Libri.

Parson, W., T.J. Parsons, R. Scheithauer & M.M. Holland, 1998. Population data for 101 Austrian Caucasian mitochondrial DNA d-loop sequences: application of mtDNA sequence analysis to a forensic case. *International Journal of Legal Medicine* 111, 124–32.

Piazza, A., N. Cappello, E. Olivetti & S. Rendine, 1988. A genetic history of Italy. *Annals of Human Genetics* 52, 203–13.

Piercy, R., K.M. Sullivan, N. Benson & P. Gill, 1993. The application of mitochondrial DNA typing to the study of white Caucasian genetic identification. *International Journal of Legal Medicine* 106, 85–90.

Pinto, F., A.M. Gonzalez, M. Hernandez, J.M. Larruga & V.M. Cabrera, 1996. Genetic relationships between the Canary Islanders and their African and Spanish ancestors inferred from mitochondrial DNA sequences. *Annals of Human Genetics* 60, 321–30.

Pult, I., A. Sajantila., J. Simanainen, O. Georgiev, W. Schaffner & S. Pääbo, 1994. Mitochondrial DNA sequences from Switzerland reveal striking homogeneity of European populations. *Biological Chemistry Hoppe-Seyler* 375, 837–40.

Richards, M.B., H. Côrte-Real, P. Forster, V. Macaulay, H. Wilkinson-Herbots, H. Demaine, S. Papihar, R. Hedges, H-J. Bandelt & B.C. Sykes, 1996. Paleolithic and Neolithic lineages in the European mitochondrial gene pool. *American Journal of Human Genetics* 59, 185–203.

Richards, M.B., V.A. Macaulay, H-J. Bandelt & B.C. Sykes, 1998. Phylogeography of mitochondrial DNA in western Europe. *Annals of Human Genetics* 62, 241–60.

Rickards, O., G. Biondi, G.F. De Stefano, F. Vecchi & H. Walter, 1992. Genetic structure of the population of Sicily. *American Journal of Physical Anthropology* 87, 395–406.

Rickards, O., G. Scano, C. Martínez-Labarga, T. Taraborelli, G. Gruppioni & G.F. De Stefano, 1995. Genetic history of the population of Puglia (southern Italy). *Gene Geography* 9, 25–40.

Rickards, O., C. Martínez-Labarga, G. Scano, G.F. De Stefano, G. Biondi, M. Pacaci & H. Walter, 1998. Genetic history of the population of Sicily. *Human Biology* 70, 699–714.

Rickards, O., C. Martínez-Labarga, J.K. Lum, G.F. De Stefano & R.L. Cann, 1999. Mitochondrial DNA history of the Cayapa Amerinds of Ecuador: detection of additional founding lineages for Native American populations. *American Journal of Human Genetics* 65, 519–30.

Rickards, O., C. Martínez-Labarga, R. Casalotti, G. Castellana, A.M. Tunzi Sisto & F. Mallegni, in press a. Molecular anthropology in the reconstruction of the peopling of southern Italy, in *Proceedings of the 2nd International Congress on: 'Science and Technology for the Safeguard of Cultural Heritage in the Mediterranean Basin, July 5–9 1999, Paris, France*.

Rickards, O., C. Martínez-Labarga, M. Favaro, D. Frezza & F. Mallegni, in press b. DNA analyses of the remains of the family of Prince Branciforte Barresi (16th–17th centuries). *International Journal of Legal Medicine*.

Rickards, O., C. Martínez-Labarga, R. Casalotti, G. Castellana, A.M. Tunzi Sisto & F. Mallegni, in prep. Pattern of mtDNA variation in ancient samples and present populations of southern Italy.

Rogers, A., 1995. Genetic evidence for a Pleistocene population explosion. *Evolution* 49, 608–15.

Rogers, A. & H. Harpending, 1992. Population growth makes waves in the distribution of pairwise genetic differences. *Molecular Biology and Evolution* 9, 552–69.

Saiki, R.K., S. Scharf, F. Faloona, K.B. Mullis, G.T. Horn, H.A. Erlich & N. Arnheim, 1985. Enzymatic amplification of beta-globin genomic sequences and restriction site analysis for diagnosis of sickle cell anemia. *Science* 230, 1350–54.

Salas, A., D. Comas, M.V. Lareu, J. Bertranpetit & A. Carracedo, 1998. MtDNA analysis of the Galician population: a genetic edge of European variation. *European Journal of Human Genetics* 6, 365–75.

Schneider, S., J.M. Kueffer, D. Roessli & L. Excoffier, 1997. *Arlequin ver1.1: a Software for Population Genetic Data Analysis*. Geneva: Genetics and Biometry Laboratory, University of Geneva.

Slatkin, M. & R.R. Hudson, 1991. Pairwise comparisons of mitochondrial DNA sequences in stable and exponentially growing populations. *Genetics* 129, 555–62.

Stone, A. & M. Stoneking, 1993. Ancient DNA from a pre-Columbian Amerindian population. *American Journal of Physical Anthropology* 92, 463–71.

Tajima, F., 1989. Statistical method for testing the neutral mutation hypothesis by DNA polymorphism. *Genetics* 123, 585–95.

Torroni, A., K. Huoponen, P. Francalacci, M. Petrozzi, L. Morelli, R. Scozzari, D. Obinu, M.L. Savontaus & D.C. Wallace, 1996. Classification of European mtDNAs from an analysis of three European populations. *Genetics* 144, 1835–50.

Torroni, A., H-J. Bandelt, L. D'Urbano, P. Lahermo, P. Moral, D. Sellitto, C. Rengo, P. Forster, M.L. Savontaus, B. Bonné-Tamir & R. Scozzari, 1998. MtDNA analysis reveal a major Late Paleolithic population expansion from south-western to north-eastern Europe. *American Journal of Human Genetics* 62, 1137–52.

Tunzi Sisto, A.M., 1999. *Ipogei della Daunia: Preistoria di un territorio*. Foggia: Grenzi Editore.

Vigilant, L., M. Stoneking, H. Harpending, K. Hawkes & A.C. Wilson, 1991. African populations and the evolution of human mitochondrial DNA. *Science* 253, 1503–7.

Walter, H., H. Matsumoto, H. Danker-Hopfe, G.F. De Stefano & O. Rickards, 1997. GM and KM allotypes in nine population samples of Sicily. *Annals of Human Biology* 24, 419–26.

Wrischnik, L.A., R.G. Higuchi, M. Stoneking, H. Erlich, N. Arnheim & A.C. Wilson, 1987. Length mutations in human mitochondrial DNA: direct sequencing of enzymatically amplified DNA. *Nucleic Acids Research* 15, 529–42.

Chapter 19

The Population History of Corsica and Sardinia: the Contribution of Archaeology and Genetics

Laura Morelli & Paolo Francalacci

Mitochondrial DNA polymorphisms are analyzed by PCR and low-resolution restriction analysis in 152 individuals from Corsica and Sardinia. This screening revealed that about 95 per cent of mtDNAs can be grouped in eight of the nine Caucasian haplogroups, labelled H, I, J, K, T, U, V and X. Our results confirm that these haplogroups encompass virtually all the mitochondrial lineages present in Europe and can be detected in both northern and southern European populations. We describe two restriction sites (−73 Alw44I and +75 Sph1) that allow the detection of informative nucleotide changes in HVS-II without further sequencing. Haplogroup H was the most common in this sample, reaching frequencies of about 65 per cent in Central Sardinia. Haplogroup V, originating in the Iberian peninsula, was found only in the Central Sardinian sample. Four of the five Corsican mtDNAs belonging to the haplogroup T show a restriction pattern found only in this population. The sample from Central Sardinia shows a remarkable discontinuity with those from the northern part of the island and from Corsica. The latter areas seem to have undergone a more recent peopling event, possibly related to the arrival of new mitochondrial variability from Continental Italy, while Central Sardinia keeps more archaic features.

Corsica and Sardinia are the most important islands of the western Mediterranean whose environment is mainly characterized by highlands, with little in the way of flat plains. Hominids arrived here before 500,000 BP as proved by the Lower Palaeolithic tools discovered in Perfugas (northern Sardinia) (Martini 1992; Klein Hofmeijer 1997). However, the population density seems to be very low before the Neolithic, with possible gaps in the human presence on the islands.

During the last glaciation (about 18,000–15,000 BP), Sardinia and Corsica were connected to the Italian peninsula through the Tuscan archipelago. During this period the endemic Pleistocene fauna does not show any significant variation as a consequence of the lack of predators including Man (Vigne 1987). However, sporadic human presence in Sardinia has been found at Corbeddu Cave (Sondaar *et al.* 1995) where a human phalanx (20,000 bp) has been found together with pollens relating to glacial flora.

Even though the local evolution of human populations can be taken into account, the arrival of people from outside of the two islands seems to explain the significant increase of human population density observed after 9000 BP. In this period the coastline was identical to that of the present day, separating the northern island of Corsica and the neighbouring Tuscan islands by about 30 km, and the colonization was necessarily made by sea. A maritime arrival can be considered possible since the evidence for navigation in the eastern Mediterranean dates back to this time (Camps 1988; Lanfranchi & Weiss 1997).

The impact of human occupation is also shown by clear faunal change, with the disappearance of Pleistocene fauna, such as the deer *Megaloceros cazioti*, and the replacement with Middle–Upper Pleistocene animals (Klein Hofmeijer 1997). Another important food source for humans was an endemic large hare

(*Prolagus sardous*) that became extinct in the Nuragic period.

Among the human remains of this period it is worth noting the right temporal and left maxilla bones from Corbeddu Cave in central Sardinia (radiocarbon dated to 8750±140 BP) (Sondaar & Spoor 1989; Spoor & Sondaar 1986) and the human female skeleton of Araguina-Sennola (Corsica) dated to 8300±130 bp (Lanfranchi & Weiss 1997).

The cultural features of this human occupation are related to the pre-Neolithic tradition of the western Mediterranean area (Tozzi 1997).

An autochthonous civilization flourished in the Bronze Age with the Megalithic and Nuragic culture (around the second millennium BC) while in the historic period Corsica and Sardinia were in the political orbit of the Etruscans and of the Phoenician-Punics respectively.

The subsequent Roman conquest was mainly directed towards the coastal area, causing the refuge of the Nuragic people in the mountainous central region of Sardinia called by the Romans 'Barbagia' (meaning the country of Barbarians). Toponymic evidence shows the coexistence in Barbagia of northern and southern dialectal variants of the Nuragic language (Paulis 1995) indicating the presence of a mixed population, resulting from the concentration of native people in the mountains. The linguistic features of the Palaeo-Sardinian language points out ancient contacts between Sardinian prehistoric cultures and the Iberian peninsula, and can be framed in a larger Mediterranean context (Paulis 1995). The Latin was super-imposed on a pre-Indo-European Nuragic substratum and evolved into the modern Sardinian language.

Over the Nuragic period up to the Roman age the population size in Sardinia is supposed to have been rather high, with an estimation of about 300,000 inhabitants.

The medieval crisis led to a general demographic decrease, due to the effect of declining economic conditions and to the incidence of epidemic illness such as malaria. The plagues of the fourteenth century reduced the population size to about one third (80–100,000 inhabitants: Vivoli 1982). Among the different historical regions of Sardinia, Gallura was the one that suffered the most significant decrease in population size. Similar events affected the Corsican population in the same period.

Subsequently the population size returned to the previous values. In particular significant Tuscan immigration into Corsica was observed after 1200 because of the Pisan rule on the island. Historical sources attest to the population of Corsican people in the northern part of Sardinian (Gallura). At the present time Sardinia and Corsica total about 1,500,000 and 250,000 inhabitants respectively.

The modern Corsican language is considered part of the Tuscan dialectal system, while the Gallurese dialect is similar to Corsican and presents remarkable differences with other Sardinian languages (Devoto 1977). The Sardinian language is an outlier in the Romance tongues because of the strong conservative features directly derived from Latin.

Sardinia has been widely studied by geneticists and represents one of the major genetic outliers in the European context as in the case of Basques and Lapps. On the basis of classical genetic markers, as blood groups and serum protein polymorphisms, a north–south gradient was observed on the island, reflecting this linguistic and historical partitions (Capello *et al.* 1996; Cavalli-Sforza *et al.* 1994).

The genetics of Corsica is relatively poorly known, especially in comparison with the neighbouring island. Previous studies using classical markers yielded discordant results. A paper by Calafell *et al.* (1996), showed high genetic distance with Sardinia and Tuscany, concluding that the historically known relationships between Corsica and these two regions were mainly cultural, excluding a significant gene flow. On the other hand, a study on 14 genetic markers pointed to a similarity with Sardinia, and larger differences with Tuscanians and Ligurians, which apparently left virtually no trace in the genes of the Corsicans (Vona *et al.* 1995; Varesi *et al.* 1996). However, these results can be explained in terms of genetic drift, particularly significant in island environments.

Other studies focusing on the two towns of Corte and Bastia showed a similarity between Corsica and Italy and a discontinuity between Sardinia and Catalunya (Moral *et al.* 1996).

Mitochondrial DNA (mtDNA) has several unique features that make it ideal for studying the genetic structure of populations: the mitochondrial genome, completely sequenced in humans by Anderson *et al.* (1981), is small and simple to study and available in many copies per cell. Moreover, it is maternally inherited and never recombines. Finally mtDNA evolves five to ten times more quickly than the average of nuclear DNA by the sequential accumulation of mutations along radiating female lineages (Giles *et al.* 1980; Wallace *et al.* 1987). MtDNA has been studied in human population genetics by means of different approaches, such as control region sequencing and coding region restriction analysis.

Parsimony dendrogram of Swedish and Finnish mtDNA examined by PCR amplification and high-resolution restriction analysis, suggested that all these European mtDNA are grouped into ten families of mitochondrial lineages called haplogroups H, I, J, K, M, T, U, V, W and X (Torroni *et al.* 1994; 1996). The restriction analysis of a sample of Tuscan mtDNA (Torroni *et al.* 1996) of which both hypervariable segments of control region were previously determined (Francalacci *et al.* 1996) showed that nearly all Tuscan mtDNA fell into the above-mentioned haplogroups and allowed comprehension of relations between control region nucleotide variability and restriction sites polymorphisms of the coding region.

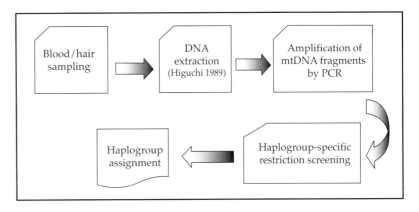

Figure 19.1. *Protocol used for mtDNA haplogroup assignment.*

According to Simoni *et al.* (2000) there are no evidence of clear clinal distribution of mtDNA variability in Europe even though a general southeast–northwest gradient can be observed. However, previous studies showed differences in the distribution of some haplogroups such as H and V, whose frequency is higher in Western Europe (Torroni *et al.* 1996; 1998), and J, whose higher incidence in Eastern European populations is consistent with the Neolithic demic diffusion model (Ammerman & Cavalli-Sforza 1984; Cavalli-Sforza & Minch 1997).

On the basis of control region sequence distribution, there is evidence of early human settlement, probably related to the Palaeolithic occupation, and a subsequent enrichment of the mitochondrial variability coming from the Middle East, following the Neolithic expansion (Comas *et al.* 1996; Francalacci *et al.* 1996; Richards *et al.* 1996).

We analyzed mtDNA variation by PCR and haplogroup-specific restriction screening (see Fig. 19.1) in three population samples not yet examined coming from the centre of Corsica (56 individuals) in the neighbourhood of Corte; from the northeast Sardinian region called Gallura (51 individuals), and from Barbagia (45) in Central Sardinia. These data were compared with Tuscan (49 individuals) and Catalonian (95 individuals) samples (Francalacci *et al.* 1999).

The results confirmed that the nine typically European haplogroups previously identified encompass almost all the mitochondrial variability present in Europe. This observation underlines the remarkable homogeneity in Europe for mtDNA, already noted with sequencing analysis of the control region (Francalacci *et al.* 1996).

Some of the nine haplogroups were not represented in one or more of the Corsican and Sardinian populations. It should be noted, however, that the missing haplogroups are always those which appear at low frequencies in other European groups, and also could reflect the sample sizes of the mtDNA for the studied populations. Results of mtDNA haplogroup assigment are showed in Figure 19.2.

Among the most common haplogroups, H reached the highest values in Barbagia (64.4 per cent) and in Catalonia (45.6 per cent). This result is similar to others observed in western Europe as in the Galician and Basque samples (estimated from the incidence of HVS-I sequences similar or identical to the CRS) (Bertranpetit *et al.* 1995; Salas *et al.* 1998). Tuscan, Corsican and Gallurese samples are closed to the average European percentage (about 41.0 per cent).

The haplogroup V is well represented only in Barbagia (8.9 per cent) and it has also been found in Catalonia (3.3 per cent) while it is completely absent in northern Sardinia, Corsica and Tuscany. According to Torroni *et al.* (1998) this haplogroup originated in southwest Europe about 12,000–16,000 years ago from haplogroup H, and later spread to central-northern Europe and North Africa.

On the other hand, J and T haplogroups evolved more recently in the Neolithic period (Richards *et al.* 1998) and are more widely distributed in Eastern Europe. In fact we found only 2.2 per cent of J in Barbagia and 3.3 per cent in Catalonia, reaching higher frequencies in Tuscany (14.6 per cent), Corsica (10.7 per cent) and Gallura (9.8 per cent).

In addition, we observed the occurrence of a rare recurrent mutation in four of five Corsican individuals belonging to the T haplogroup showing the mutation –8894 HaeIII. This finding allowed the identification of a locally important subgroup that could

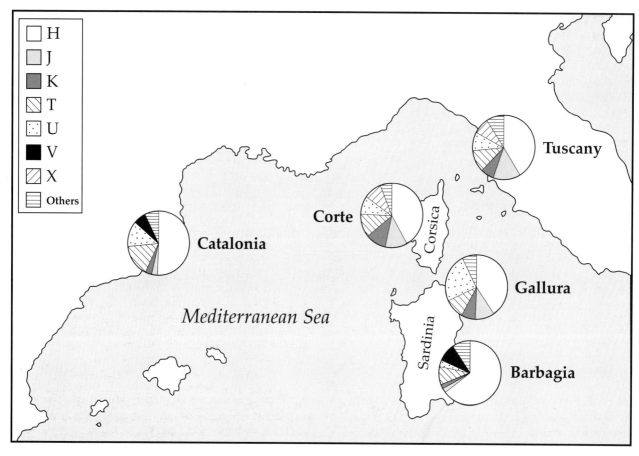

Figure 19.2. *Haplogroup frequency distribution in the samples studied and in control populations.*

be used both for estimating the migration patterns and the time of colonization. Obviously, the newcomers carried a pre-existent variability, but a mutation to the occupation arisen subsequently (thus absent in the source population) can give origin to a cluster of lineages (or sub-haplogroup) whose internal variability can be used for dating the minimum time of the arrival in the area, since the population should have been in place at least for a period of time sufficient to develop the observed variability within the cluster.

Haplogroup U is very important for showing the origins of European populations as it has a very complex phylogenetic history, since, on the basis of HVSI sequence polymorphisms, it is possible to distinguish different sub-haplogroups (from U1 to U6) which presumably arrived in Europe with different population waves. However, since our methodological approach is based on RFLPs it is not possible to separate in our sample the various groups of lineages.

The data here reported point to an apparent discontinuity of the sample from Central Sardinia with the one from the northern part of the island, because of a much higher frequency of haplogroup H, the presence of haplogroup V and a reduced incidence of other haplogroups. In spite of the geographical proximity of the two samples, Gallura seems to have undergone a more recent gene flow, possibly related to the arrival of new mitochondrial variability from continental Italy through Corsica. It should be noted that Barbagia presents more similarities with the Iberian region (for the presence of haplogroup V and the high H incidence) and these two areas seem to share a higher incidence of the pre-Neolithic mitochondrial background.

In conclusion, the genetic structure reflects the historical background of the two islands: while Corsica and northern Sardinia were strongly repopulated by people coming from continental Italy, the hilly region of Central Sardinia was a refugee homeland for the indigenous Sardinian populations escaping the Roman and Phoenician pressure on the coast.

The molecular approach proved to be very useful in understanding the population history of the two islands and the impact of migration, but only an integrated approach, with archaeological, historic, linguistic and genetic data can provide a clearer picture of the complex phenomenon of the human occupation of the islands.

Acknowledgements

This work was supported by CNR 97.00712.PF36 (L. Terrenato) and MURST 60 per cent and ex-40 per cent grants and by a Sassari University Grant for Young Researchers to L.M.

References

Ammerman, A.J. & L.L. Cavalli-Sforza, 1984. *The Neolithic Transition and the Genetics of Populations in Europe.* Princeton (NJ): Princeton University Press.

Anderson, S., A.T. Bankier, B.G. Barrel, M.H.L. de Bruijn, A.R. Coulson, J. Drouin, I.C. Eperon, D.P. Nierlich, B.A. Roe, F. Sanger, P.H. Schreier, A.J.H. Smith, R. Staden & I.G. Yung, 1981. Sequence and organization of the human mitochondrial genome. *Nature* 290, 457–65.

Bertranpetit, J., J. Sala, F. Calafell, P.A. Underhill, P. Moral & D. Comas, 1995. Human mitochondrial DNA variation and the origin of Basques. *Annual of Human Genetics* 59, 63–81.

Calafell, F., J. Bertranpetit, S. Rendine, N. Cappello, P. Mercier, J.P. Amoros & A. Piazza, 1996. Population history of Corsica: a linguistic and genetic analysis. *Annual of Human Biology* 23(3), 237–51.

Camps, G., 1988. *Les mammifères post-glaciaires de Corse.* Paris: C.N.R.S. suppl. 28.

Capello N., S. Rendine, R. Griffo, G.E. Mameli, V. Succa, G. Vona & A. Piazza, 1996. Genetic analysis of Sardinia: I. Data on 12 polymorphisms in 21 linguistic domains. *American Journal of Human Genetics* 60, 125–41.

Cavalli-Sforza, L.L. & E. Minch, 1997. Paleolithic and Neolithic lineages in the European mitochondrial gene pool. *American Journal of Human Genetics* 61, 247–51.

Cavalli Sforza, L.L., P. Menozzi & A. Piazza, 1994. *The History and Geography of Human Genes.* Princeton (NJ): Princeton University Press.

Comas, D., F. Calafell, E. Mateu, A. Pèrez Lezaun & J. Bertranpetit, 1996. Geographic variation in a human mitochondrial DNA control region sequence: the population history of Turkey and its relationship to the European populations. *Molecular Biology and Evolution* 13, 1067–77.

Devoto, G., 1977. *Il linguaggio d'Italia.* Milano: Rizzoli Editore.

Francalacci, P., J. Bertranpetit, F. Calafell & P.A. Underhill, 1996. Sequence diversity of the control region of mitochondrial DNA in Tuscany and its implications for the peopling of Europe. *American Journal of Physical Anthropology* 100, 443–60.

Francalacci, P., R. Montiel & A. Malgosa, 1999. A mitochondrial DNA database: application to problems of nomenclature and to population genetics, in *Genomic Diversity: Applications in Human Population Genetics*, eds. S.S. Papiha, R. Deka & R. Chakraborty. New York (NY): Kluwer Academic/Plenum Publishing Corp, 103–19.

Giles, R.E., H. Blanc, H.M. Cann & D.C. Wallace, 1980. Maternal inheritance of human mitochondrial DNA. *Proceedings of the National Academy of Sciences of the USA* 77, 6715–19.

Higuchi, R., 1989. Simple and rapid preparation of samples for PCR, in *PCR Technology: Principles and Applications for DNA Amplification*, ed. H.A. Erlich. New York (NY): Stockton Press, 31–8.

Klein Hofmeijer, G., 1997. Late Pleistocene deer fossils from Corbeddu Cave, implications for human colonization of the island of Sardinia. *BAR IS* 663, 7–12, 412–14.

Lanfranchi, F. de & M.C. Weiss, 1997. *L'aventure humaine prehistorique en Corse.* Ajaccio: Albiana.

Martini, F., 1992. Early human settlement in Sardinia: the Paleolithic industries, in *Sardinia in the Mediterranean: a Footprint in the Sea*, eds. R.H. Tykott & T.K. Andrews. Sheffield: Sheffield Academic Press, 4048.

Moral, P., M. Memmi, L. Varesi, G.E. Mameli, V. Succa, B. Gutierrez, N. Lutken & G. Vona, 1996. Study on the variability of seven genetic serum protein markers in Corsica (France). *Anthropologischer Anzeiger* 54(2), 97–107.

Paulis, G., 1995. Lingue e popoli della Sardegna preistorica e protostorica. *Antropologia Contemporanea* 18(3–4), 37–40.

Richards, M., H. Côrte-Real, P. Forster, V. Macaulay, H. Wilkinson-Herbots, A. Demaine, S. Papiha, R. Hedges, H-J. Bandelt & B.C. Sykes. 1996. Paleolithic and Neolithic lineages in the European mitochondrial gene pool. *American Journal of Human Genetics* 59, 185–203.

Richards, M., V.A. Macaulay, H-J. Bandelt & B.C. Sykes, 1998. Phylogeography of mitochondrial DNA in western Europe. *Annual of Human Genetics* 62, 241–60.

Salas, A., D. Comas, M.V. Lareu, J. Bertranpetit & A. Carracedo, 1998. MtDNA analysis of the Galician population: a genetic edge of European variation. *European Journal of Human Genetics* 6, 365–75.

Simoni, L., F. Calafell, D. Pettener, J. Bertranpetit & G. Barbujani, 2000. Geographic patterns of mtDNA diversity in Europe. *American Journal of Human Genetics* 66, 262–78.

Sondaar, P.Y. & C.F. Spoor, 1989. Man and the Pleistocene endemic fauna of Sardinia. *Proceedings of the 2nd International Congres of Human Paleontology*, 501–5.

Sondaar, P.Y., R. Elburg, G. Klein Hofmeijer, F. Martini, M. Sanges, A. Spaan & J.A. de Visser, 1995. The human colonization of Sardinia: a Late Pleistocene human fossil from Corbeddu Cave. *Comptes Rendus*

de l'Académie des Sciences de Paris 320(2), 145–50.
Spoor, C.F. & P.Y. Sondaar, 1986. Human fossils from the endemic island fauna of Sardinia. *Journal of Human Evolution* 15, 399–408.
Torroni, A., M.T. Lott, M.F. Cabell, Y.S. Chen, L. Lavergne & D.C. Wallace, 1994. MtDNA and the origin of Caucasian: identification of ancient Caucasian-specific haplogroups, one of which is prone to a recurrent somatic duplication in the D-loop region. *American Journal of Human Genetics* 55, 760–76.
Torroni, A., K. Huoponen, P. Francalacci, M. Petrozzi, L. Morelli, R. Scozzari, D. Obinu, M.L. Savontaus & D.C. Wallace, 1996. Classification of European mtDNAs from an analysis of three European populations. *Genetics* 144, 1835–50.
Torroni, A., H-J. Bandelt, L. D'Urbano, P. Lahermo, P. Moral. D. Sellitto, C. Rengo, P. Forster, M-L. Savontaus, B. Bonné-Tamir & R. Scozzari, 1998. MtDNA analysis reveals a major Late Paleolithic population expansion from Southwestern to Northeastern Europe. *American Journal of Human Genetics* 62, 1137–52.
Tozzi, C., 1997. Le peuplement pléistocéne et de l'Holocéne ancien de la Sardaigne, in Lanfranchi & Weiss (eds.), 72–8.
Varesi, L., M. Memmi, P. Moral, G.E. Mameli, V. Succa & G. Vona, 1996. La distribution de quatorze marqueurs génétiques dans la population de l'île de Corse (France). *Bulletins et Mémoires de la Société d'Anthropologie de Paris* 8(1–2), 5–14.
Vigne, J.D., 1987. Le possibilités d'immigration mammalienne sur les îles de Méditerranée occidentale à la lumière de l'exemple corso-sarde. *IGAL* 11, 259–61.
Vivoli, C., 1982. Villaggi, 'fuochi' e abitanti, in *La Sardegna*, ed. M. Brigaglia. Cagliari: La Torre, 154–8.
Vona, G., M.R. Memmi, L. Varesi, G.E. Mameli & V. Succa, 1995. A study of several genetic markers in the Corsican population (France). *Anthropologischer Anzeiger* 53(2), 125–32.
Wallace, D.C., J.H. Ye, S.N. Neckelmann, G. Singh, K.A. Webster & B.D. Greenberg, 1987. Sequence analysis of cDNAs for the human and boving ATP synthase beta subunit: mitochondrial DNA genes substain seventeen times more mutations. *Current Genetics* 12, 81–90.

Chapter 20

Analysis of the Y-chromosome and Mitochondrial DNA Pools in Portugal

Luísa Pereira, Maria João Prata, Mark A. Jobling & António Amorim

Portugal is the most western country in Iberian Peninsula and, just focusing on historical times, it has been subject to more or less substantial inputs of foreign populations such as Romans, Barbarians and North Africans. The aim of this study was to characterize the Portuguese population relatively to Y-chromosome and the mtDNA variability. We have considered three main regions in the Portuguese population: North, Central and South. 10 Y-chromosome biallelic markers were studied as well as hypervariable regions I and II of the mtDNA control region. This allowed the construction of Y-chromosome and mtDNA haplogroups, both characterized by being highly geographically clustered, specially the first. Results on Y-chromosome biallelic markers revealed a low level of inter-regional diversity, although South Portugal was found to be significantly different from North Portugal. An interesting observation was a clear gradient of increasing frequency from North to South concerning haplogroup 21. As this haplogroup reaches the highest frequencies in Berber-speaking populations, the clinal variation registered might be interpreted as a sign of North African influence in the Portuguese genetic background. Levels of mtDNA diversity at HVRI in the Portuguese sample analyzed were higher than the values published for some neighbouring Iberian populations (Galicia, Basque country). This finding can be interpreted as a result of influx of new sequences from other populations. Sixteen different mtDNA haplogroups were found, including haplogroup U6 that reached the frequency of seven per cent in North Portugal. This haplogroup, which occurs in high frequency in Berber-speaking populations, has been reported to be absent in Europe except in Iberia.

Y-chromosome and mtDNA polymorphisms are particularly suitable for population genetics since they share the special features of haploidy, lack of recombination, uniparental inheritance and a four-fold reduction in effective population size relative to autosomes. Important inferences concerning population origins and movements are made possible by the simultaneous study of Y chromosome and mtDNA, since they give complementary information: the first about male and the second about female lineages. These inferences are relatively easy to derive because the Y and mtDNA haplogroups, especially the former, display a highly geographical clustering and, in some cases, haplotype frequency gradients can be followed through large or restricted geographic regions.

Male and female histories are not necessarily coincident and some results (Seielstad *et al.* 1998), based on Y chromosome and mtDNA SNPs pointed to very different patterns of dispersion of the sexes through the World.

In this work we have analyzed 10 Y-chromosome biallelic markers and HVRI and HVRII regions of mtDNA in Portugal. In the population screening we have considered three main regions in Portugal: North, Central and South.

Table 20.1. *Number of individuals screened in the three Portuguese samples.*

	Y chromosome	mtDNA
North Portugal	329	100
Central Portugal	118	82
South Portugal	49	59

Table 20.3. *Population pairwise F_{ST} P values between samples from North, Central and South Portugal. Values marked by an asterisk are significant at the 5 per cent level.*

	Central Portugal	North Portugal
North Portugal	0.12871 ± 0.0260	
South Portugal	0.16832 ± 0.0473	0.00000* ± 0.0000

Table 20.2. *Haplogroups identified with the ten Y-chromosome biallelic markers studied.*

	YAP	SRY-8299	92R7	SRY-1532	SRY-2627	Tat	SY81	M9	LLY 22g	12f2
2	0	0	0	1	0	0	0	0	0	0
16	0	0	0	1	0	1	0	1	1	0
21	1	1	0	1	0	0	0	0	0	0
8	1	1	0	1	0	0	1	0	0	0
3	0	0	1	0	0	0	0	1	0	0
1	0	0	1	1	0	0	0	1	0	0
22	0	0	1	1	1	0	0	1	0	0
9	0	0	0	1	0	0	0	0	0	1
4	1	0	0	1	0	0	0	0	0	0
7	0	0	0	0	0	0	0	0	0	0
12	0	0	0	1	0	0	0	1	1	0
26	0	0	0	1	0	0	0	1	0	0

The aims of the work were to characterize this Iberian country relative to Y-chromosome and mtDNA variability and to detect any possible genetic record of the several populations that crossed this southwestern region of Europe in prehistoric and historic times.

Materials and methods

Populations and sample sizes
The Portuguese populations analyzed were: North, Central and South, the country be divided up according to the major river basins: the Douro and Tejo. Sample sizes are presented in Table 20.1.

Y-chromosome biallelic markers
The ten biallelic markers analyzed are listed in Table 20.2. The table also displays the compound haplogroups defined by these markers which have already been observed (Jobling, unpublished results). Analytical conditions will be published by M.A. Jobling.

MtDNA
The two hypervariable regions, HVRI and HVRII, were analyzed. Primers for PCR amplification were: L15997 and H16401 for HVRI; L48 and H408 for HVRII. PCR amplification conditions were according to Wilson *et al.* (1995).

The amplified samples were purified with Microspin™ S-300 HR columns (Pharmacia Biotech). Sequencing reaction was performed using the Kit Big-Dye™ Terminator Cycle Sequencing Ready Reaction (Perkin-Elmer) with the primers above described, in the forward and reverse directions. Post cycle sequencing reaction sample purification was undertaken using a $MgCl_2$/ethanol-based protocol. Sequence run and analysis were performed in an ABI 377 sequencer.

Statistical analysis
Sequence diversity parameters and population pairwise differentiation test based upon F_{ST} were obtained with the software ARLEQUIN (Schneider *et al.* 1997).

Results and discussion

I - Y-chromosome biallelic markers
Analysis of the referred ten biallelic markers led to the detection of seven different haplogroups.

Figure 20.1 shows the haplogroup frequency distribution observed in the three Portuguese regions and Table 20.3 presents values of the popula-

tion pairwise differentiation test.
   The major findings were:
- an increasing frequency gradient for haplogroup 21 from North to South Portugal;
- absence of haplogroup 8 (characteristic of Sub-Saharan populations) in the three regions;
- statistically significant difference between South and North Portugal.

II - MtDNA - HVRI and HVRII

Figure 20.2 represents the nucleotide pairwise difference distributions whereas some mtDNA diversity parameters are displayed in Table 20.4.

The different sequences were classified in haplogroups according to Richards *et al.* (1998) and the corresponding distributions are represented in Table 20.5.

For mtDNA, the main results were:
- the nucleotide pairwise difference distributions were very similar in all the Portuguese samples for HVRI and HVRII;
- concerning HVRI, absence of clear bell-shaped distributions of the number of nucleotide pairwise differences and high sequence diversity; both findings differentiate the Portuguese samples from other neighbouring Iberian populations (Côrte-Real *et al.* 1996; Salas *et al.* 1998), suggesting an older or mixed origin for Portugal;
- all major European clusters were detected in the Portuguese samples: about 80 per cent of the sequences belong to the so called Palaeolithic expanded clusters (Richards *et al.* 1998) (V, K, U, U3, U4, U5, W, X, T, I & H) and about 10 per cent to Neolithic expanded ones (J, J1, J2 & T1);
- about 10 per cent of the sequences may represent introductions from Africa;
- North African haplogroup U6 was only detected

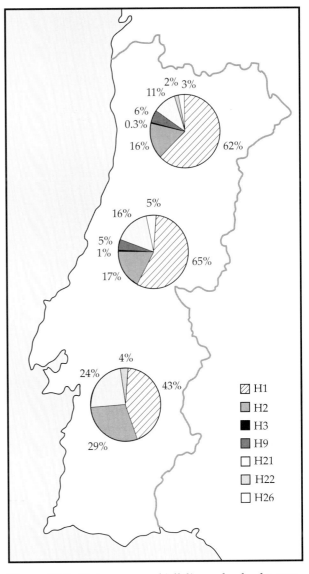

Figure 20.1. *Y-chromosome biallelic marker haplogroup distributions in Portugal.*

Table 20.4. *MtDNA diversity parameters in North, Central and South Portugal, considering HVRI and/or HVRII.*

		% of different haplotypes	Mean no. of nucleotide pairwise differences	Nucleotide diversity
HVRI	North Portugal	67.00	4.784243	0.013290
	Central Portugal	75.60	4.845529	0.013460
	South Portugal	69.49	4.509644	0.012527
HVRII	North Portugal	47.00	3.089091	0.011399
	Central Portugal	50.00	3.247516	0.011983
	South Portugal	61.02	3.552309	0.013060
HVRI + HVRII	North Portugal	84.00	7.873333	0.012478
	Central Portugal	92.68	8.093044	0.012826
	South Portugal	91.53	8.040912	0.012723

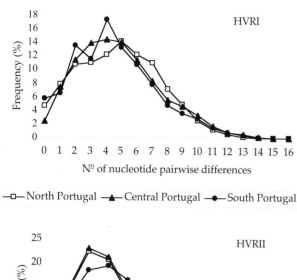

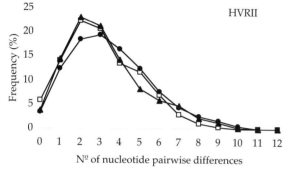

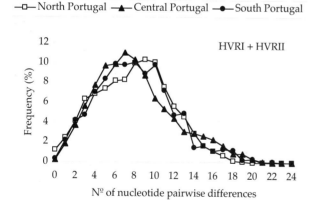

Figure 20.2. *Nucleotide pairwise difference distributions observed in North, Central and South Portugal, considering HVRI and/or HVRII.*

Table 20.5. *MtDNA haplogroup distributions (values in %) in North (NP), Central (CP) and South (SP) Portugal.*

	NP	CP	SP		NP	CP	SP
H	41.00	36.59	38.98	J1	2.00	–	3.39
V	8.00	3.66	6.78	J2	2.00	1.22	–
K	3.00	7.32	6.78	X	–	3.66	5.08
U	6.00	9.76	10.17	W	2.00	1.22	–
U3	1.00	–	–	T	3.00	9.76	10.17
U4	2.00	3.66	–	T1	7.00	1.22	–
U5	7.00	6.10	6.78	I	1.00	1.22	1.70
U6	7.00	–	–	L1b	–	1.22	1.70
JT*	1.00	–	–	L2	3.00	2.44	1.70
J	2.00	4.88	5.08	L3a*	2.00	6.10	1.70

Conclusions

In the Y-chromosome pool, the increasing frequency gradient of haplogroup 21 from north to south has also been described for other European regions (Hammer *et al.* 1998) and has been related to the Neolithic diffusion through the continent. However, since this haplogroup presents high frequencies in North African populations and it is known that multiple historic and pre-historic contacts between North Africans and Iberians have occurred, this cline could at least be enhanced (if not originated) by interchanges between these populations, besides those determined by Neolithic diffusion.

The importance of the Islamic influence in Western Iberia, which started at the beginning of the eighth century AD, was markedly heterogeneous (Saraiva 1993). While northern regions remained totally or practically untouched, in the central and especially the southern regions, Islamic administration lasted up to the thirteenth century. This pattern clearly mimics the clinal variation registered for haplogroup 21.

The absence of the characteristic Sub-Saharan haplogroup 8 suggests that interbreeding between African male slaves who entered Europe from the fifteenth to the last century, and Portuguese females must has been very restricted, with a minor impact in the paternally inherited Portuguese gene pool.

In the maternal gene pool, the distribution of the U6 haplogroup suggests a North African influence restricted, surprisingly, to North Portugal. This pattern is not consistent with the above-described chronology of Islamic administration in Portugal and may represent either another population movement not yet identified, or the effects of drift.

in North Portugal, where it reached the frequency of 7 per cent;
- other African haplogroups (Rando *et al.* 1998) were detected in all Portuguese regions analyzed, ranging from 5 per cent in North and South Portugal to 10 per cent in Central Portugal: L1b (West African), L2 (Pan African) and L3a (Sub-Saharan African).

With respect to the detection of L haplogroups all over the country, it is possible that this reflects the recent presence of Black African slaves in Portugal. If this association is valid, then the African presence has led to a significant contribution to the present day mtDNA Portuguese gene pool.

The differential influence of the Black African genes in the Y chromosome and mtDNA seems to point to a much more frequent interbreeding between autochthonous males with Black African female slaves than the opposite, a conclusion which is in accordance with the socio-cultural relationships between Portuguese and African slaves.

Acknowledgements

This work was partially supported through a grant (PRAXIS BD/13632/97) and a project (PRAXIS/2/2.1/BIA/196/94) from Fundação para a Ciência e a Tecnologia. M.A.J. is a Wellcome Senior Research Fellow in Basic Biomedical Science (grant no. 057559/Z/99/Z).

References

Côrte-Real, H., V.A. Macaulay, M.B. Richards, G. Hariti, M.S. Issad, A. Cambon-Thomsen, S. Papiha, S. Bertranpetit & B.C. Sykes, 1996. Genetic diversity in the Iberian Peninsula determined from mitochondrial sequence analysis. *Annals of Human Genetics* 60, 331–50.

Hammer, M.F., T. Karafet, A. Rasanayagam, E.T. Wood, T.K. Altheide, T. Jenkins, R.C. Griffiths, A.R. Templeton & S.L. Zegura, 1998. Out of Africa and back again: nested cladistic analysis of human Y chromosome variation. *Molecular Biology and Evolution* 15(4), 427–41.

Rando, J.C., F. Pinto, A.M. González, M. Hernández, J.M. Larruga, V.M. Cabrera & H-J. Bandelt, 1998. Mitochondrial DNA analysis of Northwest African populations reveals genetic exchanges with European, Near-Eastern and sub-Saharan populations. *Annals of Human Genetics* 62, 531–50.

Richards, M.B., V.A. Macaulay, H-J. Bandelt & B.C. Sykes, 1998. Phylogeography of mitochondrial DNA in western Europe. *Annals of Human Genetics* 62, 241–60.

Salas, A., D. Comas, M.V. Lareu, J. Bertranpetit & A. Carracedo, 1998. MtDNA analysis of the Galician population: a genetic edge of European variation. *European Journal of Human Genetics* 6, 365–75.

Saraiva, J.H., 1993. *História de Portugal*. 4th edition. Lisbon: Publicações Europa-América.

Schneider, S., J.M. Kueffer, D. Roessli & L. Excoffier, 1997. *Arlequin ver 1.1: a Software for Population Genetic Data Analysis*. Genetic and Biometry Laboratory, University of Geneva, Switzerland.

Seielstad, M.T., E. Minch & L. Cavalli-Sforza, 1998. Genetic evidence for a higher female migration rate in humans. *Nature Genetics* 20, 278–80.

Wilson, M.R., J.A. Di Zinnow, D. Polanskey, J. Replogle & B. Budowle, 1995. Validation of mitochondrial DNA sequencing for forensic casework analysis. *International Journal of Legal Medicine* 108, 68–74.

Chapter 21

Morphological and Molecular Evolution in Prehistoric Skeletal Populations

Neskuts Izagirre, Patricia Artiach & Concepion de la Rua

For a long time, morphometric traits of the skeletal remains had been the only data available to discuss the biological relationships among past populations. In this work we analyze, by means of the RFLPs, the mtDNA recovered from the skeletal remains excavated in three prehistoric sites in the Basque Country (SJAPL, Longar and Pico Ramos). The frequency of the mtDNA haplogroups show a similar distribution in the samples of SJAPL and Pico Ramos, which present the lowest distance values and clustering together in the trees. Longar, shows higher distances with each of them, although lower than with modern samples. On the other hand, the analysis of the frequency of the cranial non-metric traits shows that SJAPL (Araba) and Longar (Nafarroa) are not statistically different populations, clustering together in the UPGMA tree. We have estimated the relative advantages of both data (genetic and morphologic) and we have also discussed the discrepant results obtained in this work, as well as their use in determining the biological relationships between different human groups.

For a long time the morphometric characteristics of the skeleton provided the only method available for determining the biological relationships between human past groups. However, these characteristics are polygenic in nature and their expression is controlled by an unknown combination of environmental and genetic factors (Cavalli-Sforza & Bodmer 1971). This means that it is hard to interpret the results and difficult to establish whether the similarities observed between populations are due to genetic relationships between them or to environmental factors of various kinds.

The development of molecular biology techniques (i.e. PCR: Saiki *et al.* 1985; Mullis & Faloona 1987) has enabled us now to obtain genetic data on extinct populations. In this way we can characterize prehistoric human groups from both a morphological and a genetic perspective.

In this study we analyzed 62 non-metric cranial traits, and the variability of mtDNA by means of RFLP analyses, performed on human remains recovered from three prehistoric sites in the Basque Country: SJAPL, Longar and Pico Ramos. Our intention was to compare the information obtained with morphological and molecular markers.

Materials and methods

The remains analyzed came from the following sites:

* SJAPL, located in the upper Ebro valley in the province of Araba. The remains belong to the late Neolithic period (5070±150 to 5020±140 years BP, according to ^{14}C dating of two bone samples: Etxebarria & Vegas 1988). The minimum number of individuals (MNI) recovered is 289, and genetic analyses were performed on 63 individuals.

* The Pico Ramos cave on the Atlantic coast of the Basque Country (Bizkaia province). The remains belong to the Chalcolithic period based on the stratigraphic sequence (4110±110 to 4790±110 years BP)

(Zapata 1995). At least 104 individuals were recovered, and 24 individuals were analyzed using RFLPs.

* The Longar burial chamber located in the province of Nafarroa, dated at 4445±70 to 4580±90 years BP, corresponding to the Late Neolithic–ancient Chalcolithic (Armendariz & Irigarai 1995). The MNI for this site is 69, and RFLP analyses were performed on 27 individuals.

Analysis of non-metric cranial traits

Tests were performed to check for the presence/absence of a total of 62 non-metric cranial traits, as described in Hauser & De Stefano (1989), Castro & Quevedo (1983–84) and Klug & Wittwer-Backofen (1983). Of all these characteristics, only those showing significant heredability (Sjøvold 1984) were used to calculate the MMD (Mean Measure Divergence) distance matrix (Sjøvold 1973).

The results for the non-metric cranial traits obtained here were compared with data on other present-day, historical and prehistoric populations. The prehistoric populations used were Turks, Syrians, Lebanese (Klug & Wittwer-Backofen 1983), Ukrainians (Cesnys & Konduktorova 1982; in Hauser & De Stefano 1989) and Siberians (Kozintsev 1972; in Hauser & De Stefano 1989); medieval populations from Hito in Cantabria (Galera 1989) and Garai in the Basque Country (unpublished data) were used, along with the population from Aldaieta (Basque country, sixth century) (unpublished data) and present-day population samples of Spaniards (Gil 1985), Romans (Vecchi 1968), Sardinians (Cosseddu *et al.* 1979), Dutch (Perizonius 1979), Bedouins (Henke & Disi 1981; in Klug & Wittwer-Backofen 1983) and Basques (unpublished data).

Analysis of mtDNA

MtDNA variability was studied by means of RFLP analysis (Torroni *et al.* 1996). The DNA for this was extracted from intact teeth using the phenol-chloroform method. These teeth were decontaminated with acid to eliminate any contaminant DNA from their surfaces. The use of intact teeth reduces the risk of contamination of the dental pulp by modern DNA and prevents the destruction of the endogenous DNA by acids.

To analyze the restriction polymorphisms a set of primers was designed to amplify fragments less than 120 pbs long, located all over the mitochondrial genome (Izagirre & de la Rua 1999).

The analyses included two types of control: a blank control on the extraction to check for contamination during the extraction process at the laboratory, and a negative control of amplification to check for contamination during the amplification stage.

The series used to compare genetic data were Tuscans, Finns, Swedes, and Basques from the present day, as published by Torroni *et al.* (1996).

Results

Non-metric cranial traits

The frequency of each non-metric trait was checked, and comparisons between populations were made using a χ^2 test.

The prehistoric sample from Pico Ramos was not included in the χ^2 tests because the remains were highly fragmented, which limited the number of observations for each characteristic. A comparison of all the Basque populations taken as a whole (prehistoric, medieval and present day) showed statistically significant differences. ($\chi^2 = 702.777$, g.f. = 207,

Table 21.1. *Comparison via the χ^2 test between the Basque populations analyzed in this study and a set of Eurasian populations.*

	SJAPL	Longar	Pais Vasco	Aldaieta	Garai	Hito	Spaniards	Romans
SJAPL		49,364	119.825**	103.039**	194.87**	79.845**	717.738**	328.068**
Longar			67.284*	37,908	90.548**	16,627	205.584**	42.990**
Pais Vasco				63,153	153.094**	49.59**	689.548**	159.538**
Aldaieta					144.213**	13,919	229.966**	103.158**

	Sardinians	Dutch	Turks	Syrians	Lebanese	Bedouins	Ukrainians	Siberians
SJAPL	244.456**	796.336**	170.581**	215.229**	253.253**	270.170**	412.487**	296.620**
Longar	24,458	209.883**	69.139**	98.997**	89.888**	140.861***	96.679**	159.538**
Pais Vasco	119.28**	413.867**	92.245**	102.042**	146.777**	337.476**	245.438**	179.199**
Aldaieta	88.223**	393.903**	83.647**	126.909**	102.835**	138.888**	125.146**	180.386**

* $p < 0.05$ ** $p < 0.01$

$p < 0.0001$), from which we can deduce that there is heterogeneity among the Basque populations analyzed for this set of characteristics.

The χ^2 test comparison of the populations analyzed in this study together with the other populations indicated in the bibliography (Table 21.1) shows statistically significant differences between the prehistoric population of SJAPL and all the others, except the prehistoric population from Longar. Similar results were found on comparing the prehistoric sample from Longar with the remaining populations, except that there were no significant differences with the populations from Hito (Cantabria), Sardinia and Aldaieta (Basque Country).

We then calculated the MMD biological distance matrix (Sjøvold 1973), using the data on the 14 characteristics showing significant heredability. Figure 21.1a shows the neighbour-joining tree constructed on the basis of this MMD matrix, where all the Basque populations can be seen to form a group to which the two populations from the Iberian Peninsula (medieval from Hito and present-day Spaniards) are linked. The Western Asia populations are split into two different groups and the European are together in another group.

Analysis of mtDNA haplogroup diversity
Genetic analysis was performed only on those samples which presented negative amplification and extraction controls. To double check the results obtained with dental samples, we carried out a parallel analysis on 92 bone samples (left femur) recovered from the SJAPL site. Although it was not possible to associate the bone samples with any particular dental remains, we analyzed 4577 *Nla*III restriction site on both types of sample (bone and teeth). The same digestion pattern was found in the bone samples as in the teeth, i.e. presence of cleavage for 4577 *Nla*III. A more detailed discussion of the mtDNA analysis method can be found in Izagirre & de la Rua (1999).

Table 21.2 shows the haplogroup distribution found in the remains recovered from the three prehistoric Basque sites. The distribution of haplogroups H, K and U is similar at all three sites, with haplogroup H having the highest frequency in all areas. It is also noteworthy that haplogroups I, W and V were not found at any of the prehistoric sites.

Two peculiarities were observed at the Longar site: the absence of haplogroup J and the presence of

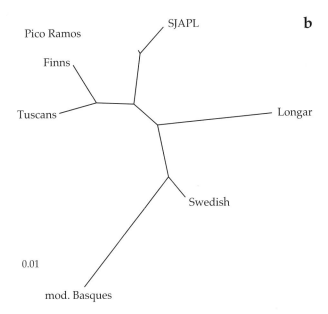

Figure 21.1. *Neighbour-joining trees: a) Based on the MMD matrix for the non-metric skull characteristics; and b) constructed on the basis of the distance matrix from Cavalli-Sforza & Edwards' (1967) cord measurement and mtDNA data.*

individuals which we were unable to classify in any of the haplogroups described.

Diversity analysis shows similar diversity values in ancient and modern populations (Table 21.3). This, and the fact that three different sites are involved, suggests that few if any individuals from the same family were sampled.

The neighbour-joining tree constructed on the basis of Cavalli-Sforza & Edwards' (1967) cord dis-

Table 21.2. *MtDNA haplogroup frequencies at the three prehistoric sites analyzed in this study and their corresponding standard deviations (se).*

	Prehistoric sites		
Haplogroup	SJAPL Frequency ± *se*	Longar Frequency ± *se*	Pico Ramos Frequency ± *se*
H	0.377 ± 0.060	0.408 ± 0.094	0.374 ± 0.091
J	0.164 ± 0.047	–	0.167 ± 0.070
K	0.230 ± 0.054	0.222 ± 0.080	0.167 ± 0.070
U	0.180 ± 0.049	0.148 ± 0.068	0.125 ± 0.062
T, X	0.049 ± 0.027	0.148 ± 0.070	0.167 ± 0.070
W	–	–	–
I	–	–	–
V	–	–	–
Others	–	0.074 ± 0.050	–

SJAPL (Araba); Longar (Nafarroa) y Pico Ramos (Bizkaia).

Table 21.3. *MtDNA diversity based on mtDNA haplogroup frequencies in modern European populations and on three prehistoric Basque samples.*

Populations	h[a]
MODERN POPULATIONS	
Tuscans (Italy)	0.769
Finns (Finland)	0.785
Basques (Gipuzkoa)	0.713
Swedes (Sweden)	0.762
PREHISTORIC BASQUE SAMPLES	
SJAPL[b] (Araba)	0.756
Pico Ramos (Bizkaia)	0.793
Longar (Nafarroa)	0.762

[a] genetic diversity (Nei 1987)
[b] SJAPL: San Juan Ante Portam Latinam

tance shows that the prehistoric populations from SJAPL and Pico Ramos have the least distance. The present-day Basque sample shows the greatest distance as regards all the populations included in our analysis. The prehistoric sample from Longar is least distant from the Swedish population (0.036), owing to the absence or low frequency of haplogroup J in both populations. However, Longar shows lesser distances from the other two prehistoric Basque populations than from the remaining populations analyzed.

Discussion

In this study we observe that in the neighbour-joining trees constructed on the basis of the non-metric trait data (Fig. 21.1a), all the Basque populations (both prehistoric and modern) are grouped together, regardless of their temporal and spatial origin. Although the expression of non-metric cranial traits is due to an unknown combination of genetic and environmental factors, these results seem to support the hypothesis that the influence of the genetic component in this case is highly important.

Discrepancies appear in the results of analyses of the relationships between the prehistoric populations from the Basque Country depending on whether non-metric cranial traits or genetic data are considered. The least distance based on non-metric traits is between the prehistoric populations of SJAPL and Longar, but on the basis of genetic distance it is between SJAPL and Pico Ramos. This discrepancy is hard to evaluate, because for the Pico Ramos prehistoric population it was only possible to collect non-metric cranial trait data from a very small sample, because of the fragmented condition of the bones recovered. We believe that these data are not significant, so Pico Ramos is not considered in the comparisons between populations by means of χ^2 testing (Table 21.1). However the differentiation of Longar on the basis of genetic data is explained by the absence of individuals from haplogroup J and the presence of individuals classified under the haplogroup 'others'. Furthermore, from an archaeological viewpoint, the burial structure of Longar is peculiar, and unlike the Chalcolithic structures typically found in the Basque Country (the dolmens in the Rioja Alavesa area). Burial structures similar to that at Longar have been found in the south of the Iberian Peninsula and in the Paris Basin. However the personal effects found at Longar are similar to those described for nearby sites (Armendariz & Irigarai 1995).

In this study we have arrived at results as regards relationships between populations when using morphological data which are different from those obtained with genetic data. The differences are difficult to resolve with the data available at present. Among other factors, we could point to the scant mtDNA data available on extinct populations and the fact that since mtDNA is maternally inherited it reflects only the history of women. Differences in migration patterns between men and women, which have begun to show up in various studies of present-day populations (Cavalli-Sforza & Minch 1997; Seielstad *et al.* 1998;

Simoni *et al.* 1999) are also a factor which must be taken into account in interpreting genetic data on the prehistoric populations analyzed here.

Acknowledgements

This work was supported by the project GN 154.310-0001-94 (Gobierno de Navarra and U.P.V./E.H.U.) and UPV 154.310-EA008/98 (U.P.V./E.H.U.).

References

Armendariz, J. & S. Irigarai, 1995. *La Arquitectura de la Muerte: el Hipogeo de Longar (Viana, Navarra), un Sepulcro Colectivo de 2.500 A.C.* Estella: Centro de Estudios Tierra-Estella, Lizarraldeko Ikastetxea.

Castro, M. & S. Quevedo, 1983–84. Proposiciones metodológicas para el estudio de los rasgos no-métricos en el cráneo humano. *Boletin del Museo Nacional de Historia Natural de Chile* 40, 173–210.

Cavalli-Sforza, L.L. & W.F. Bodmer, 1971. *The Genetics of Human Populations.* San Francisco (CA): W.H. Freeman.

Cavalli-Sforza, L.L. & A.W.F. Edwards, 1967. Phylogenetic analysis: models and estimation procedures. *Evolution* 32, 550–70. (Also *American Journal of Human Genetics* 19, 233–57.)

Cavalli-Sforza, L.L. & E. Minch, 1997. Paleolithic and Neolithic lineages in the European mitochondrial gene pool. *American Journal of Human Genetics* 61, 247–51.

Cesnys, G. & T.S. Konduktorova, 1982. Non-metric features of the skull in people of the Chernyakhovskaya culture. *Voprosy Antropologii* 70, 62–76.

Cosseddu, G.G., G. Floris & G. Vona, 1979. Sex and side differences in the minor non-metrical cranial variants. *Journal of Human Evolution* 8, 685–92.

Etxebarria, F. & J.I. Vegas, 1988. ¿Agresividad social o guerra durante el Neo-eneolítico en la cuenca media del Valle del Ebro? A propósito de San Juan ante Portam Latinam. *Munibe* 6, 105–12.

Galera, V., 1989. La Población Medieval Cántabra de Santa Maria de Hito. Aspectos Paleobiodemográficos, Morfológicos, Paleopatológicos, Paleoepidemiológicos y de Etnogénesis. Unpublished Ph.D. thesis, Universidad de Alcalá, 82–6.

Gil, C., 1985. Análisis de la simetría de los rasgos epigenéticos craneales bilaterales en una población esquelética española. *Actas IV Congreso Español de Antropologia Biológica* (Barcelona), 473–82.

Hauser, G. & G.F. De Stefano, 1989. *Epigenetic Variants of the Human Skull.* Stuttgart: E. Schweizerbart'sche Verlagsbuchhanlung.

Henke, W. & A. Disi, 1981. The craniology of recent Jordanian Bedouins. *Journal of Mediterranean Anthropology and Archaeology* 1, 217–48.

Izagirre, N. & C. de la Rua, 1999. An mtDNA analysis in ancient Basque populations: implications for haplogroup V as a marker for a major Paleolithic expansion from southwestern Europe. *American Journal of Human Genetics* 65, 199–207.

Klug, S. & U. Wittwer-Backofen, 1983. Diskreta im populationsvergleich. *Homo* 34, 153–68.

Kozintsev, A.G., 1972. Non-metrical traits on crania of 1 thousand BC from Minusinsic hollow. *Arch. Anat. Histol. Embryol.* (Leningrad) 62, 53–9.

Mullis, K.B. & F.A. Faloona, 1987. Specific synthesis of DNA *in vitro* via a polymerase-catalysed chain reaction. *Methods in Enzymology* 155, 335–50.

Nei, M., 1987. *Molecular Evolutionary Genetics.* New York (NY): Columbia University Press.

Perizonius, W.R.K., 1979. Non-metric cranial traits: symmetry and side difference. *Proceedings van de Koninklijke Nederlandse Akademie van Wetenschappen*, 82, 91–112.

Saiki, R.K., S. Scharf, F. Faloona, K.B. Mullis, G.T. Horn, H.A. Erlich & N. Arnheim, 1985. Enzymatic amplification of β-globulin genomic sequences and restriction site analysis for diagnosis of sickle cell anemia. *Science* 230, 1850–54.

Seielstad, M.T., E. Minch & L.L. Cavalli-Sforza, 1998. Genetic evidence for a higher female migration rate in humans. *Nature Genetics* 20, 278–80.

Simoni, L., F. Calafell, J. Bertranpetit & G. Barbujani, 1999. Geographical patterns of mtDNA diversity in Europe. Euroconference: *'Human Diversity in Europe and Beyond: Retrospect and Prospect'*. Cambridge, 1999.

Sjøvold, T., 1973. The occurrence of minor non-metrical variants in the skeleton and their quantitative treatment for population comparisons. *Homo* 24, 204–33.

Sjøvold, T., 1984. A report on the heritability of some cranial measurements and non-metric traits, in *Multivariate Statistical Methods in Physical Anthropology*, eds. GN. Van Vark & W.W. Howels. Boston (MA): D. Reidel, 289–321.

Torroni, A., K. Huoponen, P. Francalacci, M. Petrozzi, L. Morelli, R. Scozzari, D. Obinu, M-L. Savontaus & D.C. Wallace, 1996. Classification of European mtDNAs from an analysis of three European populations. *Genetics* 144, 1835–50.

Vecchi, F., 1968. Sesso e variazione di caratteri discontinui del cranio. *Rivista di Antropologia* 55, 283–90.

Zapata, L., 1995. La excavación del depósito sepulcral calcolítico de la cueva Pico Ramos (Muskiz, Bizkaia). La industria ósea y los elementos de adorno. *Munibe* 47, 35–90.

Chapter 22

Y-chromosome Variation and Irish Origins

Emmeline W. Hill, Mark A. Jobling & Daniel G. Bradley

The origins of the people of Ireland are enigmatic and controversial (Waddell 1998; Mallory & McNeill 1991) and have been little investigated from a genetic perspective. Ireland's population, inhabiting an island on the western edge of Europe, may possess a genetic heritage that has been relatively undisturbed by the major demographic movements that have shaped mainland European diversity (Cavalli-Sforza et al. 1994). We have analyzed Irish Y-chromosome variation, using surnames to subdivide the sample into components of known historic and prehistoric (Gaelic) origin, to which 1000-year-old geographical information may also be assigned. Strikingly, western Gaelic samples are almost fixed (98 per cent) for the putatively ancestral European haplogroup 1, as defined by biallelic markers of low mutation rate (Hill et al. 2000). Analysis of STR variation within this predominant Irish haplogroup suggests a Late Neolithic/Early Bronze Age origin (4200 BP) possibly in a population expansion facilitated by the introduction of farming. The presence of other haplogroups (27 per cent) in eastern Gaelic samples yields the first genetic evidence for a substantial secondary, prehistoric contribution to the gene pool of the island.

Variation on the non-recombining portion of the human Y chromosome contains a history of paternal lineages (Jobling & Tyler-Smith 1995); the presence of slow-mutating biallelic markers allows the definition of monophyletic haplogroups, which can then be analyzed and dated using more rapidly mutating loci such as simple tandem repeats (STRs). We typed 221 Y chromosomes from Irish males for seven bi-allelic markers and six STR loci.

Surnames have been used in Ireland from *c.* AD 950 and originate as markers of complex local kinship systems (Smyth 1997). Since both surnames and Y chromosomes are paternally inherited we divided our Irish sample into seven surname cohorts: four are of prehistoric, Gaelic, origin corresponding to the four ancient provinces of Ireland (Ulster, Munster, Leinster and Connaught: MacLysaght 1997), and three are diagnostic of historical (non-Gaelic) influxes (Scottish, Norman/Norse, English), in particular the conquests and/or settlements by the Vikings (ninth–tenth century), the Anglo-Normans (twelfth century) and the English Tudor, Cromwellian and Williamite armies (sixteenth–seventeenth century) (Smyth 1997). Figure 22.1 shows the approximate locative origins of the surnames sampled for this work. The geographic distribution of the original locations of the surnames, identified from the Heraldic scroll and map of family names and origins of Ireland (published by Mullins of Dublin, Ireland), suggests that a representative sample from all regions of the country has been used.

Methods

Population samples
Sampling was with informed consent, and was anonymous apart from the Irish samples where surname was recorded. The numbers of chromosomes in each sample were: Ulster (22), Munster (37), Leinster (30), Connaught (57), England (40), Scotland (17), Norman/Norse (18), Denmark (12) and Turkey (50).

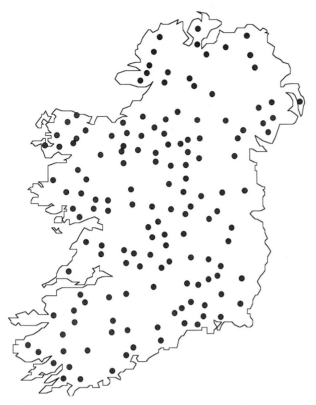

Figure 22.1. *Approximate locative origins of surnames sampled in Ireland. Surnames used in Ireland since c. AD 950, are markers of complex local kinship systems.*

DNA extraction
Buccal cells were collected on cytology brushes, and DNA extracted using the manufacturer's protocol for the Biorad Instagene™ Matrix.

Haplotyping
The seven biallelic markers were: SRY-1532 (Kwok et al. 1996), M9 (Underhill et al. 1997; Hurles et al. 1998), YAP (Hammer & Horai 1995), SRY-8299 (Whitfield et al. 1995), sY81 (Seielstad et al. 1994), 92R7 (Mathias et al. 1994; Hurles & Tyler-Smith, unpublished), SRY-2627 (Veitia et al. 1997). Haplogroups identified in this study have the following haplotypes ('1' referring to derived, and '0' ancestral state; markers in the order just given): hg 1: 1100010; hg 2: 1000000; hg 3: 0100010; hg 21: 1011000; hg 26: 1100000. The six STR loci were DYS19, DYS389I, DYS390, DYS391, DYS392, and DYS393 27. Amplifications were in a 20 μl volume in a Hybaid OmniGene™ 96-well thermal cycler; with the exception of YAP, biallelic markers were typed by PCR-RFLP assays (Hammer & Horai 1995), and STRs were typed by polyacrylamide gel electrophoresis after radioactive end-labelling.

Data analysis
Analysis of Molecular Variance (AMOVA) was performed using Arlequin (Schneider et al. 1997). Maximum consensus parsimony networks were constructed using the Phylip package (Felsenstein 1993).

Results and discussion

Results of two analyses of molecular variance, using biallelic marker and STR variation separately in these cohorts are given in Table 22.1. Both calculations indicate an among-group partitioning of molecular variance which, in a permutation test, shows a significant genetic difference between Gaelic and non-Gaelic surname samples. This validates our use of surnames to distinguish historic and prehistoric influences in patterns of genetic variation.

The biallelic markers assayed here define nine haplogroups (hg) found to be highly non-randomly distributed among human populations (Jobling & Tyler-Smith 1995; Santos & Tyler-Smith 1996), including our samples (Table 22.2). The distribution of hg 1 is of particular interest, being found at a very high frequency in Ireland (78.1 per cent in the island as a whole). When the surname subdivision is imposed, a cline of hg 1 frequency within Irish samples is revealed, with samples of exogenous origin showing the lowest frequencies (e.g. English: 62.5 per cent, Scottish: 52.9 per cent), and western Gaelic samples the highest (near fixation in Connaught: 98.3 per cent). The differences in the frequency of hg 1 between Irish Gaelic and non-Gaelic Y chromosomes ($P < 0.001$) and between eastern and western Gaelic Y chromosomes ($P < 0.001$) are highly significant and persist when any duplicated surnames are removed. Importantly, when we performed the analysis using just one sample chosen randomly from each of duplicated surnames, the divisions in the data (Gaelic vs non-Gaelic and East vs West Gaelic) remain highly significant ($P < 0.001$). The relative proportions of hg 1 in each group also remain highly similar to those for the whole sample (e.g. 97.7 per cent compared to 98.3 per cent in Connaught; 93.9–94.6 per cent in Munster). The biggest change is detected in the smallest sample, Norman/Norse, which decreases from 83.0 per cent to 76.9 per cent but which does not alter conclusions or inference.

An additional level of security in our assertions comes from dividing the data in an alternative way. We recorded the male lineage grandparental origin of each subject which permitted us to partition Y chromosomes by recent geography. When this was done we found a similar trend of Gaelic/non-Gaelic and East vs West Gaelic division in the data. However, the differences in frequency were less pronounced and less significant ($P < 0.05$). Overall, the surname division of the data, despite the undoubted inherent error, is successful. It results in a significant partitioning of the genetic variation which is sharper and more significant for this 1000-year-old tracing than when grandparental origin is used (100-year-old geographical tracing).

Eighty per cent ($n = 26$: Jobling 1994) of European hg 1 Y chromosomes belong to 'haplotype 15' (ht 15), defined using the complex 49f/*Taq*I polymorphic system (Ngo *et al.* 1986). Using this relationship we can draw upon published information for 49f haplotypes to provide a richer context for the Irish data. 49f/ht 15 frequencies follow a genetic cline within Europe (Semino *et al.* 1996), extending from the Near East (1.4 per cent in Turkey; equivalent to ~1.8 per cent hg 1) to a peak in the Spanish Basque country (72.2 per cent: Lucotte & Hazout 1996); ~89 per cent hg 1) in the west (Fig. 22.1: Hill *et al.* 2000). This cline mirrors other genetic gradients in Europe, particularly that discerned in the first principal component of variation of a large classical data set (Cavalli-Sforza *et al.* 1994), and is most easily explained through the demic diffusion of Neolithic farmers from the Near East into Europe over a 5000 year period starting approximately 8000 BC (Ammerman & Cavalli-Sforza 1984; Sokal *et al.* 1991; Barbujani *et al.* 1994). When the surname-divided Irish data are appended to this cline, it continues to the western edge of Europe, with a high frequency extreme of 98.3 per cent in the Connaught sample. This near-maximal difference between the extremes of the gradient is consistent with the observation that human populations show higher between-group variability (F_{ST}) for paternally inherited variation than for other genetic systems (Seielstad *et al.* 1998).

In a maximum parsimony phylogenetic analysis of both biallelic and STR variation between Irish Gaelic haplotypes (Fig. 22.2a–d), the hg 1 chromosomes consistently cluster together with the highest frequency haplotypes occupying central positions, suggesting a coherent common ancestry. The smaller number of hg 2 and other haplotypes do not show such coherence, which is consistent with their being immigrants from a separate ancestral source.

The distribution of hg 1 within Ireland (and within a European context), together with its coherent STR phylogeny, suggests it as the earlier, indigenous Irish variant. By taking the ancestral haplotype as that with the modal allele for each STR and calculating the average squared distance (Goldstein *et al.* 1995) between it and other variants we estimated a date for Irish hg 1 coalescence of 4200 BP (95 per cent C.I. 1800–14,800 BP). This assumes a generation time of 27 years (Weiss 1973), and an STR mutation rate of 0.21 per cent (Heyer *et al.* 1997) estimated from direct observation of Y STR loci (although the most applicable, this is the highest published estimate for Y STR loci and the true coalescence may conceivably be older. For example, by using a mutation rate of 0.12 per cent (Bianchi *et al.* 1998) a coalescence date of 7300 BP is estimated). This date of 4200 BP falls close to the Irish Neolithic/Bronze Age transition, by which stage there is clear archaeological evidence for a substantial settled population on the island

Table 22.1. *AMOVA test of population substructure.*

	SNPs		STRs	
Source of variation	% variation	*p* value =	% variation	*p* value =
Among groups	10.42	<0.0099	12.31	0.0302
Among populations within groups	5.29	<0.0099	1.13	0.1522
Within populations	84.29	0.9901	86.56	0.0000

Population groupings were made according to surname origins. Groups 1: Ulster, Munster, Leinster, Connaught; Group 2: English, Scottish, Norman and Norse.

Table 22.2. *SNP haplogroup frequencies by population.*

hg	Ireland (All) (221)	Ireland (Gaelic) (146)	Ulster (22)	Munster (37)	Leinster (30)	Connaught (57)	England (40)	Scotland (17)	Norman and Norse (18)
1	78.1	89.7	81.8	94.6	73.3	98.3	63.0	52.9	83.0
2	17.0	8.9	18.2	5.4	20.0	1.8	33.0	35.3	5.6
3	0.8	–	–	–	–	–	–	5.9	–
21	2.8	0.7	–	–	3.3	–	5.0	5.9	5.6
26	1.3	0.7	–	–	3.3	–	–	–	5.6

a. Ulster (*n* = 22)

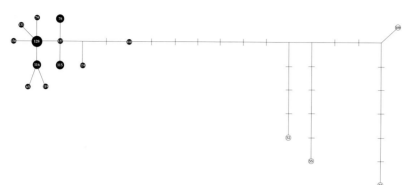

b. Leinster (*n* =30)

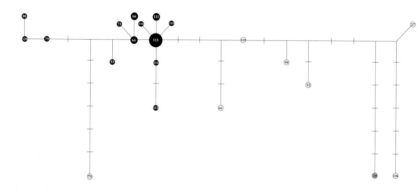

c. Connaught (*n* = 57)

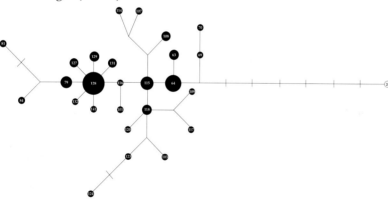

d. Munster (*n* = 37)

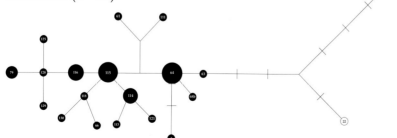

(Waddell 1998; Mallory & McNeill 1991). Estimates of coalescence dates between hg 1 and the other haplogroups are significantly older, and, with the exception of the hg 1–hg 26 comparison (8300 BP) all predate, by at least 3000 years, the 9000-year history of human habitation in Ireland (Waddell 1998; Mallory & McNeill 1991). These results are consistent with an origin of the bulk of Irish Y-chromosome variation in an agriculturally facilitated expansion of variants which themselves were of the pre-Neolithic European type. This tentatively indicates that agriculture may have been adopted by a Western European Mesolithic population, rather than resulting from significant population replacement at the fringe of Europe (Waddell 1998; Mallory & McNeill 1991).

Dramatic archaeological shifts within Ireland's prehistory are well established but the question whether these were accompanied by any significant demographic intrusion remains controversial

Figure 22.2. *Maximum Consensus parsimony networks of Irish Gaelic haplotypes (a. Ulster; b. Leinster; c. Connaught; d. Munster) summarizing both STR and SNP variation. Branch lengths are proportional to the number of mutational steps and node areas are proportional to haplotype frequencies. SNP haplogroups are coded as follows: hg 1, black; hg 2, white; hg 21, grey; hg 26, diagonal stripes (no. 75 in b). The major features of the phylogenies are the separate clustering of haplogroups, and the tight clustering of the Irish hg 1 haplotypes around a few numerous central variants, which were present in most parsimonious trees and which were stable in repetitions of the analysis with randomized input.*

(Waddell 1998; Mallory & McNeill 1991). A comparison of western and eastern Irish Gaelic cohorts strongly suggests a substantial influx marked by Leinster and Ulster non-hg 1 haplotypes. The persistence of this geographical pattern suggests that its origins are after the initial hg 1 population expansion. If one assumes a predominantly intact Y-chromosome surname linkage then the non-hg 1 introgression is suggested as having its roots in migrations prior to the first historical colonizations over a millennium ago.

Acknowledgements

This work was funded by Forbairt/Enterprise Ireland grant. M.A.J. is a Wellcome Trust Career Development Fellow. We are indebted to R.T. Loftus, O. Ertugrul, B. Simonsen, and many volunteers for sample provision. This work was funded by Forbairt/Enterprise Ireland grand no. SC/95/128.

References

Ammerman, A.J. & L.L. Cavalli-Sforza, 1984. *Neolithic Transition and the Genetics of Populations in Europe.* Princeton (NJ): Princeton University Press.

Barbujani, G., A. Pilastro, S. de Domenico & C. Renfrew, 1994. Genetic variation in North Africa and Eurasia: Neolithic demic diffusion vs. Paleolithic colonisation. *American Journal of Physical Anthropology* 95, 137–54.

Bianchi, N.O., C.I. Cartanesi, G. Bailliet, V.L. Martinez-Marignac, C.M. Bravi, L.B. Vidal-Rioja, R.J. Herrera & J.S. Lopez-Camelo, 1998. Characterization of ancestral and derived Y chromosome haplotypes of New World native populations. *American Journal of Human Genetics* 63, 1862.

Cavalli-Sforza, L.L., P. Menozzi & A. Piazza, 1994. *The History and Geography of Human Genes.* Princeton (NJ): Princeton University Press.

Felsenstein, J., 1993. *PHYLIP.* Seattle (WA): University of Washington.

Goldstein, D.B., A.R. Linares, L.L. Cavalli-Sforza & M.W. Feldman, 1995. Genetic absolute dating based on microsatellites and the origin of modern humans. *Proceedings of the National Academy of Sciences of the USA* 92, 6723–7.

Hammer, M.F. & S. Horai, 1995. Y chromosomal DNA variation and the peopling of Japan. *American Journal of Human Genetics* 56, 951–62.

Heyer, E., J. Puymirat, P. Dieltjes, E. Bakker & P. de Knijff, 1997. Estimating Y chromosome specific microsatellite mutation frequencies using deep rooting pedigrees. *Human Molecular Genetics* 6, 799–803.

Hill, E.W., M.A. Jobling & D.G. Bradley, 2000. Y chromosomes and Irish origins. *Nature* 404, 351–2.

Hurles, M.E., C. Irven, J. Nicholson, P.G. Taylor, F.R. Santos, J. Loughlin, M.A. Jobling & B.C. Sykes, 1998. European Y-chromosomal lineages in Polynesians: a contrast to the population structure revealed by mtDNA. *American Journal of Human Genetics* 63, 1793–806.

Jobling, M.A., 1994 A survey of long-range DNA polymorphisms on the human Y chromosome. *Human Molecular Genetics* 3, 107–14.

Jobling, M.A. & C. Tyler-Smith, 1995. Fathers and sons: the Y chromosome and human evolution. *Trends in Genetics* 11, 449–56.

Kayser, M., A. Caglia, D. Corach, N. Fretwell, C. Gehrig, G. Graziosi, F. Heidorn, S. Herrmann, B. Herzog, M. Hidding, K. Honda, M. Jobling, M. Krawczak, K. Leim, S. Meuser, E. Meyer, W. Oesterreich, A. Pandya, W. Parson, G. Penacino et al. 1997. Evaluation of Y-chromosomal STRs: a multicenter study. *International Journal of Legal Medicine* 110, 125–33.

Kwok, C., C. Tyler-Smith, B.B. Mendonca, I. Hughes, G.D. Berkovitz, P.N. Goodfellow & J.R. Hawkins, 1996. Mutational analysis of the 2kb 5' to SRY in XY females and XY intersex subjects. *Journal of Medical Genetics* 33(6), 465–8.

Lucotte, G. & S. Hazout, 1996. Y-chromosome DNA haplotypes in Basques. *Journal of Molecular Evolution* 42, 472–5.

Ngo, K.Y., G. Vergnaud, C. Johnsson, G. Lucotte & J. Weissenbach, 1986. A DNA probe detecting multiple haplotypes of the human Y chromosome. *American Journal of Human Genetics* 38, 407–18.

MacLysaght, E., 1997. *The Surnames of Ireland.* Dublin: Irish Academic Press.

Mallory, J.P. & T.E. McNeill, 1991. *The Archaeology of Ulster from Colonization to Plantation.* Belfast: The Institute of Irish Studies, The Queen's University of Belfast.

Mathias, N., M. Bayés & C. Tyler-Smith, 1994. Highly informative compound haplotypes for the human Y chromosome. *Human Molecular Genetics* 3, 115–23.

Santos, F.R. & C. Tyler-Smith, 1996. Reading the human Y chromosome: the emerging DNA markers and human genetic history. *Brazilian Journal of Genetics* 19, 665-70

Schneider, S., J-M. Kueffer, D. Roessli & L. Excoffier, 1997. *Arlequin.* Geneva: Genetics and Biometry Laboratory, University of Geneva.

Seielstad, M.T., J.M. Hebert, A.A. Lin, P.A. Underhill, M. Ibrahim, D. Vollrath & L.L. Cavalli-Sforza, 1994. Construction of human Y-chromosomal haplotypes using a new polymorphic A to G transition. *Human Molecular Genetics* 3, 2159–61.

Seielstad, M.T., E. Minch & L.L. Cavalli-Sforza, 1998. Genetic evidence for a higher female migration rate in humans. *Nature Genetics* 20, 278–80.

Semino, O., G. Passarino, A. Brega, M. Fellous & A.S. Santachiara-Benerecetti, 1996. A view of the Neolithic demic diffusion in Europe through two Y chromosome-specific markers. *American Journal of Human Genetics* 59, 964–8.

Smyth, W.J., 1997. The making of Ireland: agendas and perspectives in cultural geography, in *In Search of Ireland: a Cultural Geography*, ed. B. Graham. London: Routledge.

Sokal, R.R., N.L. Oden & C. Wilson, 1991. Genetic evidence for the spread of agriculture in Europe by demic diffusion. *Nature* 351, 143–5.

Underhill, P.A., L. Jin, A.A. Lin, S.Q. Mehdi, T. Jenkins, D. Vollrath, R.W. Davis, L.L. Cavalli-Sforza & P.J. Oefner, 1997. Detection of numerous Y chromosome biallelic polymorphisms by denaturing high-performance liquid chromatography. *Genome Research* 7, 996–1005.

Veitia, R., A. Ion, S. Barbaux, M.A. Jobling, N. Souleyreau, K. Ennis, H. Ostrer, M. Tosi, T. Meo, J. Chibani, M. Fellous & K. McElreavey, 1997. Mutations and sequence variants in the testis-determining region of the Y chromosome in individuals with a 46,XY female phenotype. *Human Genetics* 99, 648–52.

Waddell, J., 1998. *The Prehistoric Archaeology of Ireland*. Galway: Galway University Press.

Weiss, K.M., 1973. Demographic models for anthropology. *American Antiquity* 38, 1–186.

Whitfield, L.S., J.E. Sulston & P.N. Goodfellow, 1995. Sequence variation of the human Y chromosome. *Nature* 378, 379–80.

Chapter 23

The 'Travellers': an Isolate within the Irish Population

David T. Croke, Orna Tighe, Charles O'Neill & Philip D. Mayne

The Travellers are an endogamous community (numbering approximately 21,000) who maintain a distinct identity within Irish society and are commercial/industrial nomads. The origins of the Travellers as a distinct group are the subject of some debate but two hypotheses are extant: an 'endogenous' origin by some isolating mechanism and a 'migration' hypothesis. We are studying two metabolic defects common among the Travellers, transferase-deficient galactosaemia [MIM No. 230400] and glutaric acidaemia type I [MIM No. 231670]. The Travellers exhibit the highest incidence of transferase-deficient galactosaemia yet reported (1 in 480) and have a single causative mutation (Q188R) in the Galactose-1-phosphate uridyltransferase locus with a carrier frequency of 1 in 11 (as compared to 1 in 107 in the non-Traveller population). In the case of glutaric acidaemia type I, the Traveller incidence is approximately 1 in 1000 with a carrier frequency of approximately 1 in 16 (vs 1 in 200). We propose that founder effect coupled with rapid population expansion has produced the high mutant allele frequencies observed for these disorders among the Travellers and our preliminary data may favour an endogenous origin for the group. We are now beginning a study of Y-chromosome and mtDNA lineages within the Traveller and non-Traveller population groups to further investigate this question.

In recent years considerable advances have been made in our understanding of the history of the human population by means of the study of biological markers of population diversity. In particular, studies have been carried out in 'mainstream' European populations in addition to isolates such as the Saami, Basques and Sardinians. One isolate remains poorly characterized, the Irish 'Travellers'.

The Travellers as a distinct ethnic group

The Travellers are a community who maintain a distinct cultural identity within Irish society, having a common language known as 'Gammon' or 'Shelta', the origin of which is unclear. Travellers are commercial/industrial rather than pastoral nomads, who tend to live in groups of two to three families. Large family size and endogamy are among a number of distinct cultural characteristics which tend to distinguish them from the non-Traveller community (Ni Shuinear 1994). The origins of the Travellers as a distinct group within the Irish population are the subject of some debate but two main views are extant which may be summarized as the 'migration' and 'endogenous' hypotheses (Murphy *et al.* 1999). The 'migration hypothesis' suggests that the Travellers are derived from a group of nomadic craftspeople who migrated into Ireland from Europe at some point in the past, and who have subsequently maintained genetic isolation and have retained many distinct cultural characteristics. The 'endogenous hypothesis' suggests that the Travellers are descended from a genetic isolate derived from the original population of the island which, over time, has evolved an ethnic identity distinct from that of the general population. Traveller groups move from place to place quite frequently and can cover a large geographic range within and often outside of Ireland (Wales & England). Traveller mobility has in the past facilitated a wide choice of marriage partners within their community; anecdotal evidence suggests that this may now be decreasing because of objections from nearby non-Traveller communities to transient settlement and to the establishment of local-authority funded 'halting sites'.

Traveller population size and demographics

Based upon 1996 data, the Traveller population numbers approximately 21,000 individuals, or approximately 0.58 per cent of the total Republic of Ireland population of approximately 3.6 million (Central Statistics Office, Ireland 1998). A survey of census findings for the community from the 1940s to the present shows that, in recent decades, the Traveller population has been expanding at a rate of approximately 7 per cent per annum (Central Statistics Office, Ireland 1998; Barry & Daly 1988; Government of Ireland 1963). Current demographic data on the Travellers is scarce. The most comprehensive study carried out thus far is the Travellers' Health Status Study of 1986–1987 (Barry & Daly 1988; Barry et al. 1989). This study detailed for the first time the disadvantaged situation of the Traveller community relative to the remainder of the Irish population. For example, birth rate and infant mortality were found to be significantly elevated while average life expectancy was decreased by over ten years. Efforts have been made to improve Traveller living conditions and the delivery of health care to the community. This may in part account for the rate of Traveller population expansion recorded since 1970 and for the decrease in infant mortality which has reduced from 110 per thousand in 1961 to 18.1 per thousand in 1986 (Government of Ireland 1963; Barry et al. 1989).

Inherited metabolic disease in the Traveller community

Several recessive metabolic defects, including transferase-deficient galactosaemia (MIM 230400) and glutaric acidaemia type I (MIM 231670) have a significantly higher incidence among the Travellers as compared to the non-Traveller population. A survey of newborn screening records for the Irish Republic covering a 25 year period (1972–1996) has allowed accurate estimates of the incidence of these conditions to be made for the first time (Murphy et al. 1999). The incidence of transferase-deficient galactosaemia in the Traveller community at 1 in 480 (1 in 30,000 non-Traveller) is one of the highest yet reported. The incidence of glutaric acidaemia type I has not been studied extensively world-wide but data from a number of studies suggest it to be approximately 1 in 30,000; the value estimated for the Traveller community is significantly higher at 1 in 1000 (unpublished observations). Such incidence values obviously reflect high mutant allele frequencies.

DNA samples from a transferase-deficient galactosaemia patient cohort consisting of 56 individuals were screened for mutation in the Galactose-1-phosphate Uridyltransferase (GALT) gene by standard methods (Murphy et al. 1996; 1999). The predominant mutant allele was found to be Q188R, being the sole mutant allele among the Travellers and the majority mutant allele among the non-Travellers (89.1 per cent). We have also demonstrated that the Q188R mutation is in linkage disequilibrium with a *Sac* I RFLP in the Traveller and non-Traveller groups (data not shown). The predicted carrier frequency of 1 in 11 among the Travellers was confirmed by anonymous population screening for Q188R.

We propose that the increased mutant allele frequencies among the Travellers may be the result of founder effect. That founder effect coupled with rapid expansion can produce substantial increases in allele frequency has been elegantly demonstrated by genetic and genealogic studies of the modern Saguenay population in Quebec, Canada (Heyer 1995; Heyer & Tremblay 1995). Family sizes among the Travellers are consistently large and there is evidence from recent surveys that the Traveller population has undergone considerable expansion (see above). Our *Sac* I linkage disequilibrium findings may suggest that the Q188R galactosaemia mutation, on a *Sac* I (+) chromosomal 'background', was present in the indigenous population before the Travellers separated and was carried into the Traveller population by its founders thus favouring an endogenous origin for the Travellers.

Population genetic studies of the Traveller community

The only data which would support a population genetic study of the Travellers derives from field work carried out in Ireland in 1970–1972 by a group from the University of Kansas (Crawford & Gmelch 1974) and, unfortunately, only a subset of this is in the public domain. We have performed a limited phylogenetic analysis of this data for a number of

The 'Travellers'

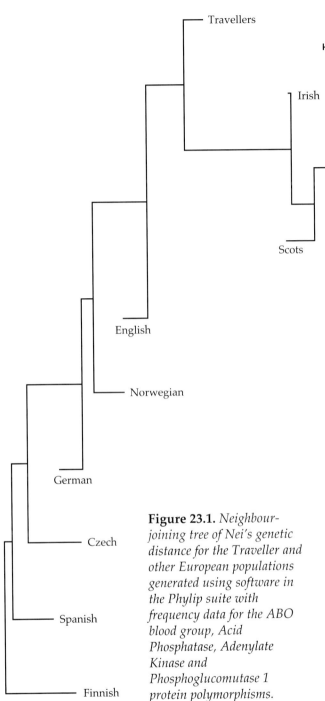

Figure 23.1. *Neighbour-joining tree of Nei's genetic distance for the Traveller and other European populations generated using software in the Phylip suite with frequency data for the ABO blood group, Acid Phosphatase, Adenylate Kinase and Phosphoglucomutase 1 protein polymorphisms.*

blood group and erythrocyte protein markers in the Travellers, together with data on neighbouring populations (Cavalli-Sforza *et al.* 1994). Allele frequencies for ABO blood groups, Acid Phosphatase, Adenylate Kinase & Phosphoglucomutase 1 were analyzed using the Gendist and Neighbour programs in the Phylip suite (Felsenstein 1993) and a neighbour-joining tree was generated (Fig. 23.1). This shows the Travellers clustering with the other populations of the British Isles and may support our hypothesis of an endogenous origin for this community. However, the limited nature of the data set and of the analysis must be borne in mind in interpreting the result. Clearly, further consideration of these issues will only be possible on the basis of detailed demographic studies of the Traveller population and of molecular genetic studies of the Irish population.

Acknowledgements

We gratefully acknowledge financial assistance from The Child Health Foundation (The Children's Hospital), The Health Research Board of Ireland and the Royal College of Surgeons in Ireland.

References

Barry, J. & L. Daly, 1988. *The Travellers' Health Status Study: Census of Travelling People, November 1986.* Dublin: The Health Research Board of Ireland.

Barry, J., B. Herity & J. Solan, 1989. *The Travellers' Health Status Study: Vital Statistics of Travelling People, 1987.* Dublin: The Health Research Board of Ireland.

Cavalli-Sforza, L.L., P. Menozzi & A. Piazza, 1994. *The History and Geography of Human Genes.* Princeton (NJ): Princeton University Press.

Central Statistics Office, Ireland, 1998. *The Demographic Situation in the Traveller Community in April 1996.* Dublin: Government Publications Office.

Crawford, M.H. & G. Gmelch, 1974. Human biology of the Irish tinkers: demography, ethnohistory and genetics. *Social Biology* 21, 321–31.

Felsenstein, J., 1993. *Phylip v3.5c.* Seattle (WA): University of Washington.

Government of Ireland, 1963. *Report of the Commission on Itinerancy.* Dublin: Government Publications Office.

Heyer, E., 1995. Genetic consequences of differential demographic behaviour in the Saguenay region, Quebec. *American Journal of Physical Anthropology* 98, 1–11.

Heyer, E. & M. Tremblay, 1995. Variability of the genetic contribution of Quebec population founders associated to some deleterious genes. *American Journal of Human Genetics* 56, 970–78.

Murphy, M., D. Sexton, C. O'Neill, D.T. Croke, P.D. Mayne & E.R. Naughten, 1996. Frequency distribution of the Q188R mutation in the Irish galactosaemic population. *Journal of Inherited Metababolic Disease* 19, 217–19.

Murphy, M., B. McHugh, O. Tighe, P.D. Mayne, C. O'Neill, E.R. Naughten & D.T. Croke, 1999. The genetic basis of transferase-deficient galactosaemia in Ireland and the population history of the Irish Travellers. *European Journal of Human Genetics* 7, 549–54.

Ni Shuinear, S., 1994. Irish Travellers, ethnicity and the origins question, in *Irish Travellers: Culture and Ethnicity*, eds. M. McCann, S. O'Siochain & J. Ruane. Belfast: The Institute of Irish Studies, The Queen's University of Belfast, 54–77.

Part V

Regional Studies in the Molecular Genetic Prehistory of Eastern Europe and Beyond

Chapter 24

Y-chromosome Diversity in the Eastern Mediterranean Area

Andrea Novelletto, Marla Tsopanomichalou, Aphrodite Loutradis,
Eleni Plata, Marilena Giparaki, Emmanuel N. Michalodimitrakis,
Nejat Akar, Nicholas Anagnou, Luciano Terrenato &
Patrizia Malaspina

In a previous work we described a human Y-chromosomal lineage defined by short alleles at the dinucleotide microsatellite system DYS413 on YAP–, alphoid HindIII+ chromosomes. Assuming a stepwise mutation model, related chromosomes of this lineage (network 1.2) were found at high frequency in Crete. They were also found in southern Italy, southeastern Europe and Turkey. By analyzing the microsatellite CA variance, the origin of the mutation defining this lineage was estimated at approx. 300 generations ago, making this an excellent marker for inferences on the peopling of eastern Mediterranean. We refine here the description of the occurrence of this lineage in the eastern Mediterranean and its molecular radiation. The results show that continental Turkey and Crete are characterized by low frequencies of YAP+ chromosomes. The island of Crete and southwestern Turkey share frequencies of network 1.2 chromosomes above 30 per cent, with a decreasing cline in the southwest-to-northeast direction within Turkey. Molecular radiation within Crete is revealed by peculiar microsatellite variants. This may be the result of local mutations that underwent a rapid increase in frequency because of drift.

Over the past few years several publications have agreed upon the idea that DNA polymorphisms of the human Y chromosome are selectively neutral and are the most powerful markers in discriminating between populations (Jobling & Tyler-Smith 1995; Hammer *et al.* 1997; Scozzari *et al.* 1997). In fact, they reveal higher quotas of variation among populations as compared to mitochondrial and autosomal markers (Seielstad *et al.* 1998). This makes them the best markers for a detailed analysis of neighbouring populations. The overall variation is due to the contribution of polymorphisms with only two possible allelic states (binary polymorphisms such as Single Nucleotide Subtitutions or Alu insertions) as well as multiallelic polymorphisms (such as those generated by length differences at the microsatellite loci). Binary polymorphisms, in which mutational events can be considered very rare or even unique, have been used to reconstruct univocally Y-chromosome phylogenies. The joint use of binary markers with fast evolving mini- and microsatellites increases the possibility of analyzing recent lineages and defining population processes on smaller geographic and time scales. This is based on the observation that microsatellite mutations most often involve the insertion/deletion of a single repeated unit. However, uncommon multirepeat mutational events can be considered rare events which specifically identify groups of chromosomes with a common ancestry and, sometimes, lineages that cannot otherwise be recognized with binary markers. We have recently shown that the distribution of Y chromosomes, as defined by a combination of binary and microsatellite markers, follows specific geographic patterns across Europe (Malaspina *et al.* 1998). Three ancient lineages (whose origin can be estimated approximately in the pre-

Neolithic period) are found throughout Europe albeit at variable frequencies. On the other hand, three additional and more recent lineages are found only in specific regions of the territory and display focal distributions. One of these, called network 1.2, is centred on the Aegean area and is characterized by a multirepeat microsatellite mutation at DYS413 estimated to have occurred 300–500 generations ago. This is, therefore, a potential marker for appropriately investigating the movements across the Aegean Islands, Crete, Anatolia and mainland Greece of populations whose gene pools can be distinguished from the pre-existing background.

We have analyzed 497 males collected in different locations in Turkey, Greece and the islands of Crete, Cyprus, Mitilini and Chios for four binary markers of the Y chromosome: the YAP insertion, DYS257, the alphoid *Hind*III site and SRY10831. A subset of these (333) was also analyzed for the CA dinucleotide microsatellites YCAII and DYS413. The DYS257 A allele was highly concordant with the absence of the alphoid *Hind*III site, and the SRY10831 A allele was found only on *Hind*III-/DYS257A chromosomes, denoting that SRY10831 A is the derived allele state at this site. Thus, all chromosomes could be fitted into a portion of the larger tree proposed for the Y-chromosome phylogeny by several authors (Hammer *et al.* 1998; Zerjal *et al.* 1999). The background chromosomes lack YAP, have the alphoid site and carry G at SRY10831 (− + G). From this background, two independent mutation events produced chromosomes with the insertion of the YAP element (+ + G) and chromosomes that lost the HindIII site (− − G), respectively. The G to A reversion at SRY10831 occurred on a chromosome of this latter type (− − A).

Table 24.1 reports absolute and relative frequencies of the chromosome types in each location. A highly significant heterogeneity is revealed by χ^2 (37.1, 15 d.f., $p < 0.001$). Crete and Turkey have by far the highest frequency of (− + G) chromosomes and the lowest frequency of (+ + G) chromosomes. The (− − G) chromosomes seem to have decreasing frequencies in eastern locations, and the (− − A) chromosomes have the lowest frequency in Crete.

We integrated these data with the analysis of dinucleotide microsatellites of the CA(n) type, namely YCAII and DYS413. Each of these systems consists of two duplicated loci. We grouped chromosomes that fitted the pure stepwise mutation model (i.e. that differ by no more than a single CA unit at one locus) within each lineage defined by binary markers. Each group constitutes, therefore, a one-step network.

The data refer to 333 out of the 500 samples for which typing was completed. The entire sample was divided into seven subsamples for mainland Greece, three samples for the districts of Crete and two Turkish subsamples representative of Central Anatolia and the southwestern coast of Turkey.

Based on microsatellites, (− + G) chromosomes could be divided into two one-step networks. The first is network 1.1. The frequencies of this background are more or less evenly distributed and range from 47 to 25 per cent. The second one-step network (network 1.2) groups chromosomes characterized by very small CA sizes at DYS413. The frequencies of this network are significantly heterogeneous across locations ($\chi^2 = 26$, 12 d.f., $p < 0.01$). In Crete this network accounts for 25 per cent of all chromosomes in Rethimnon, 39 per cent in Chania and 34 per cent in Heraklion. The only other sample with a frequency over 30 per cent is from southwestern Turkey. In mainland Greece it is 18 per cent in Serres and not higher than 14 per cent in the rest of the samples. Interestingly, the three westernmost samples have the lowest frequencies. This correlates well with the distribution on Neolithic archaeological sites in mainland Greece, which overlap with the concentration

Table 24.1. *Absolute and relative frequencies of three main Y-chromosomal lineages in six eastern Mediterranean regions.*

Sample	Chromosomes* (%)								TOTAL
	− + G		+ + G		− − G		− − A		
Continental Turkey	54	(76.1)	4	(5.6)	4	(5.6)	9	(12.7)	71
Cyprus (Turkish)	13	(59.1)	5	(22.7)	1	(4.5)	3	(13.6)	22
Mitilene	17	(60.7)	6	(21.4)	2	(7.1)	3	(10.7)	28
Chios	16	(41.0)	10	(25.6)	9	(23.1)	4	(10.3)	39
Crete	110	(69.2)	18	(11.3)	20	(12.6)	11	(6.9)	159
Continental Greece	90	(50.6)	42	(23.6)	23	(12.9)	23	(12.9)	178

* Reported as YAP, Alphoid HindIII, SRY10831.

of cultivable lands in the eastern part of the country. This adds to the concept that network 1.2 is indeed a marker of the Neolithic peopling of mainland Greece. One hypothesis is that its penetration in the western mountain area is still low today. As far as Crete is concerned, samples from this island show greater similarity to southwestern Turkey than to mainland Greece.

We then analyzed the molecular structure of haplotypes in more detail by partitioning the overall network 1.2 into three graphs, each including only chromosomes sampled in a specific location (Fig. 24.1). One can see that the major haplotype 22-19-17-17 is the commonest in Greece, Crete and Turkey. Some minor haplotypes are found in one or two locations only. However, an entire branch of the network (the left part of panel C) is found only in Crete. This is a very peculiar example of molecular radiation within the island with almost half of the chromosomes of this network characterized by 18 and 16 repeats at DYS413.

This work shows that, as far as the Y chromosome is concerned, the Cretean population has some peculiar characteristics. In particular, we found a remarkable instance of a local increase in the frequency of a single or a group of related haplotypes. This can be attained by joining the data from binary markers and the fast evolving microsatellites. The data show that a detailed microgeographic analysis of Greece, Turkey and Aegean islands is feasible. This would be of great value in understanding the peopling of this part of the Mediterranean basin. However, this will require samples twice the size of those presented here in order to reach statistical significance for more than one network. At present, the highest frequencies of network 1.2 chromosomes in southwestern Turkey and Crete do not allow us to make inferences on the direction of population movements between these areas. In this context, a straightforward interpretation is that of a strong contribution to the Cretean gene pool from Anatolia associated with the first settlements into the island (Renfrew 1998). However, this needs to be confronted with the alternative hypothesis of a local increase of network 1.2 types

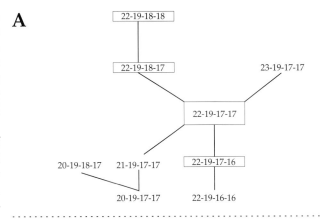
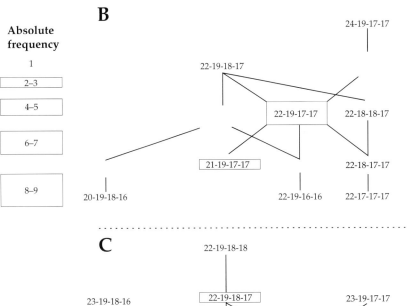

Figure 24.1. *Occurrence of network 1.2 haplotypes in three sampling locations: mainland Greece (panel A), Turkey (panel B), Crete (panel C). All possible one-step relationships are indicated by connecting lines. Haplotypes with the same combined (CA)n length are placed on the same horizontal level. Discrete sizes of rectangles are used to indicate the abundance of each haplotype. The same haplotype occupies the same position in all panels. Note missing haplotypes in some panels and the high frequency of haplotype 22-19-18-16 in Crete.*

within the island (documented in our data), followed by an outward movement associated with the cultural dominance exerted by Crete in the eastern Mediterranean area. The finding of network 1.2 chromosomes in Greek-dominated areas of southern Italy (Magna Graecia: Malaspina *et al.* Chapter 16), denotes the effects of population movements in the first millennium BC.

Acknowledgements

Work supported by CNR grants 98.00485.CT04 (AN), 97.00712.PF36 (LT), and PRIN MURST 1997 (LT).

References

Hammer, M.F., A.B. Spurdle, T. Karafet, M.R. Bonner, E.T. Wood, A. Novelletto, P. Malaspina, R.J. Mitchell, S. Horai, T. Jenkins & S.L. Zegura, 1997. The geographic distribution of human Y chromosome variation. *Genetics* 145, 787–805.

Hammer, M.F., T. Karafet, A. Rasanayagam, E.T. Wood, T.K. Altheide, T. Jenkins, R.C. Griffiths, A.R. Templeton & S.L. Zegura, 1998. Out of Africa and back again: nested cladistic analysis of human Y chromosome variation. *Molecular Biology and Evolution* 15, 427–41.

Jobling, M.A. & C. Tyler-Smith, 1995. Fathers and sons: the Y chromosome and human evolution. *Trends in Genetics* 11, 449–56.

Malaspina, P., F. Cruciani, B.M. Ciminelli, L. Terrenato, P. Santolamazza, A. Alonso, J. Banyko, R. Brdicka, O. Garcia, C. Gaudiano, G. Guanti, K.K. Kidd, J. Lavinha, M. Avila, P. Mandich, P. Moral, R. Qamar, S.Q. Mehdi, A. Ragusa, G. Stefanescu, M. Caraghin, C. Tyler-Smith, R. Scozzari & A. Novelletto, 1998. Network analyses of Y-chromosomal types in Europe, North Africa and West Asia reveal specific patterns of geographical distribution. *American Journal of Human Genetics* 63, 847–60.

Renfrew, C., 1998. Word of Minos: the Minoan contribution to Mycenean Greek and the linguistic geography of the Bronze Age Aegean. *Cambridge Archaeological Journal* 8(2), 239–64.

Scozzari, R., F. Cruciani, P. Malaspina, P. Santolamazza, B.M. Ciminelli, A. Torroni, D. Modiano, D.C. Wallace, K.K. Kidd, A. Olckers, P. Moral, L. Terrenato, N. Akar, R. Qamar, A. Mansoor, S.Q. Mehdi, G. Meloni, G. Vona, D.E.C. Cole, W. Cai & A. Novelletto, 1997. Differential structuring of human populations for homologous X and Y microsatellite loci. *American Journal of Human Genetics* 61, 719–33.

Seielstad, M.T., E. Minch & L.L. Cavalli-Sforza, 1998. Genetic evidence for a higher female migration rate in humans. *Nature Genetics* 20, 278–80.

Zerjal, T., A. Pandya, F.R. Santos, R. Adhikari, E. Tarazona, M. Kayser, O. Evgrafov, L. Singh, K. Thangaraj, G. Destro-Bisol, M.G. Thomas, R. Qamar, S.Q. Mehdi, Z.H. Rosser, M.E. Hurles, M.A. Jobling & C. Tyler-Smith, 1999. The use of Y-chromosomal DNA variation to investigate population history: recent male spread in Asia and Europe, in *Genomic Diversity: Applications in Human Population Genetics*, eds. S.S. Pahiha, R. Deka & R. Chakraborty. New York (NY): Kluwer Academic/Plenum Publishers, 91–102.

Chapter 25

The Topology of the Maternal Lineages of the Anatolian and Trans-Caucasus Populations and the Peopling of Europe: Some Preliminary Considerations

Kristiina Tambets, Toomas Kivisild, Ene Metspalu, Jüri Parik, Katrin Kaldma, Sirle Laos, Helle-Viivi Tolk, Mukaddes Gölge, Halil Demirtas, Tarekegn Geberhiwot, Surinder S. Papiha, Gian Franco de Stefano & Richard Villems

Here we discuss how our understanding of the peopling of Europe by modern humans may be improved by results which can be obtained in the investigation of genetic lineages of populations living in Anatolia and the Trans-Caucasus: Turks, Armenians, Georgians and Ossetes (Fig. 25.1). These four populations occupy a geographic area commonly believed to have great importance for the peopling of Europe. The present paper is directly related to our other paper in this volume (Kivisild et al.) which primarily addresses the genetics of Indian populations and our understanding of the first waves of migration of modern humans out of Africa and the peopling of Eurasia in general.

The historical background

Anatolia and the Trans-Caucasus are famous for their extraordinarily deep and rich history. They are close to and partially overlap with the Fertile Crescent; they include the upper parts of the Tigris and Euphrates basins and Lake Urmia. With the oldest signs of food production anywhere and with the oldest urban settlements which even seem to predate these, the region continued to play an important role for many millennia. Somewhat surprisingly, Anatolia itself is so far very poor in Palaeolithic finds. In contrast, the western Caucasus, in particular the Black Sea coastal area, Colchis and the Upper Kura and Arax River basins have yielded findings that cover a long time span starting from the Upper Palaeolithic and extending well into the Lower Palaeolithic (Rybakov *et al.* 1984). There are numerous (uncorrected) radiocarbon dates for the region, starting from about 35,000 before present that presumably belong to modern humans. The same area is equally rich in Mesolithic findings (Rybakov *et al.* 1989). However, this surprisingly different picture between Anatolia and Georgia may have a trivial explanation: lack of Lower Palaeolithic finds in Anatolia may be the result of the limited fieldwork. (P. Dolukhanov, pers. comm.).

The Neolithic and Chalcolithic periods of western Asia, including Anatolia and the Trans-Caucasus area, are much better documented, with numerous findings throughout southern Anatolia and Upper Mesopotamia as well as the surroundings of Lakes Van, Urmia and Sevan. The Bronze and Iron Ages of Anatolia and the Trans-Caucasus are again extraordinarily rich in important and often culturally overlapping civilizations (Hattian, Hittite, Hurrian and others).

The Hittite language was one of the earliest written Indo-European languages. It is believed that migration starting from Thrace around 1300–1200 BC, bringing to Anatolia yet another wave of Indo-European-speaking populations. This invasion has

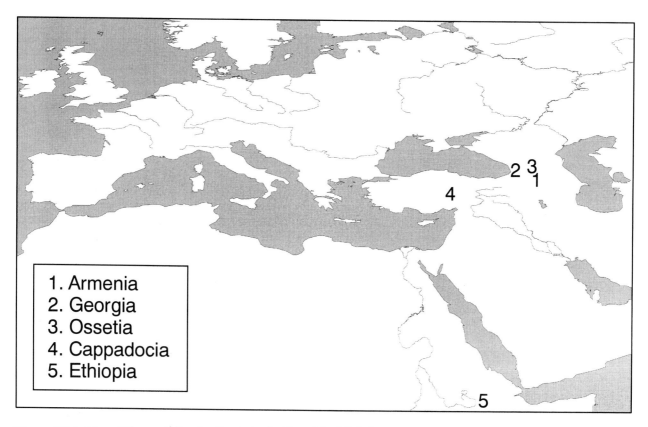

Figure 25.1. *Map of the area. Ossetes live in both sides of the High Caucasus; our sample was collected in Ossetian villages in Georgia.*

been correlated with the destruction of Troy VIIb and the end of Hattusas (the latter dated around 1180 BC because of an abrupt end of written sources), but sophisticated neo-Hittite and Hurrian-Urartian cultures remained and developed in southeastern and eastern Anatolia. Meanwhile, Ionian Greeks started to build their civilization in western Anatolia and these, together with new Indo-European-speaking states further east — Lydians and Medians — became, for a while, part of Achemenid Persia and were taken over by Alexander's empire and its successor states. The following centuries saw the presence of Rome and Byzantium in western Anatolia as far as Armenia and successively Seleucid and Parthian empires in the east. Slow westward advancement of various Turkic nomads, their conversion to Islam in Central Asia by AD 970 and, finally, the fall of Constantinople in AD 1453 brought this period to an end.

The populations which we have studied are linguistically diverse. Turkish, a novelty in Anatolia, belongs to the Altaic group of languages, while Ossetian and Armenian are branches of the Indo-European phylum of languages. Kartvelian, spoken by Georgians (self-name Kartvels), belongs to the southern branch of the Caucasian group of languages, considered to be as distant from the Indo-European phylum as, for example, the Afro-Asiatic (Hamito-Semitic) group. Furthermore, Georgians are often considered as an autochthonous population of the region. The Nostratic school of linguists suggests that the Caucasian languages may have had, until the Neolithic, a much wider geographic distribution, including southern Europe. Apart from this, they also argue that the homeland of the Indo-European languages is in eastern Anatolia — Armenian Highlands (e.g. Gamkrelidze & Ivanov 1984).

Despite their language, maternal lineages of present-day Turks have little in common genetically with the Siberian Altaic populations; instead Anatolian genetic continuity is likely. However, this continuity is itself diverse and complex as well as deep. Linguistically, Hurrians and Urartians did not speak Indo-European languages and the extent of genetic change during the replacement of Urartian (itself perhaps a branch of Hurrian) by Indo-European Armenian, which occurred at about 700–600 BC, is unclear. The nature of the Hurrian language is obscure, since being of the agglutinative type, it was neither Semitic nor Indo-European. By analogy with the Turkish invasion of at that time largely Greek-speaking Anatolia, the much earlier linguistic change from Urartian to Armenian might have had only a minor effect on the genes of the population living in this area, particularly so for maternally inherited mtDNA. There seems to be an agreement now that out of the languages extant today, Greek is the closest to Armenian. The subsequent colourful history of Armenians is much better documented: here it is worthwhile noting that Greater Armenia covered a much wider area than the present-day Republic of Armenia and that at some point in Armenian history, about 800 years ago, 'Lesser Armenia' functioned even as a state largely within the borders of historic Cilicia, situated south and east of Cappadocia.

Finally, the Indo-Iranian speaking Ossetes are considered to be direct descendants of the historic Alans. Relative to Sarmatians, it is believed (depending on the interpretation of names in Chinese chronicles), that at some time they lived much further east, being a threat to China's northern borders. Defeated by the Huns in AD 375, most Alans fled the northern Pontic-Caspian steppes and took refuge in the northern Caucasus, pushed further south by subsequent attacks by Mongols and other invaders. At present, they live both north and south of the High Caucasus, in the Russian Federation and in Georgia, respectively.

Samples

Our sample of Ossetes was collected in Ossetian settlements in Georgia, south of the High Caucasus, Armenian samples in Armenia and Georgian samples from Tbilisi and the Megrelian part of Georgia. The Turkish samples are from the historic Cappadocian area. Ethiopian samples were collected from four different localities in Ethiopia. Population and sub-population (tribal, where appropriate) affiliation and birthplaces of the four grandparents/two parents of a donor were established by questioning. It turned out that a substantial fraction of the grandparents of Armenians in our sample were from the Lake Van area, from the neighbourhood of Erzurum and even from the shores of Lake Urmia. Samples were analyzed as explained in the accompanying paper (Kivisild *et al.* Chapter 31).

Anatolia and particularly the Caucasus area, are ethnically and linguistically most diverse and we are well aware that our analysis covers only a fraction of it. On the other hand, all populations under the study here are numerous; except for the Ossetes, their numbers are in millions and they constitute an absolute majority of the people living in the area.

Formulation of the questions

Out of a potentially long list of questions, we concentrate below on only a few. The first is, in the most general terms, the topology of the phylogenetic tree of the maternal lineages of the Anatolian and the Trans-Caucasus populations. The second is the comparison of these results with data available for other major geographic areas/groups of populations such as Europeans, Indians and Mongoloids. However, our main thrust is an attempt to contribute to the understanding of the place of these populations in the formation of genetic diversity we see among the extant Europeans. To do so, we use two approaches. Firstly, we compare European and Anatolian-Trans-Caucasian populations with respect to the topology of the mtDNA tree. Secondly, we seek to identify the sub-founders within the reconstructed lineage clusters and to compare the corresponding inferred coalescence ages for the populations, living in Europe and in the Anatolia–Trans-Caucasus. Here, we limit our analysis to some case studies. Finally, we bring in the 'sub-Saharan African angle' in order to discuss a few specific aspects of the mtDNA tree, first of all related to the varieties of haplogroup M, found in Anatolia.

Previous results

Although studies using classical genetic markers to compare the four populations investigated here are limited (Cavalli-Sforza *et al.* 1994), they do provide a clear general picture. Among western Asian populations, Turks, Georgians and Armenians all belong to separate branches of one of the two major genetic clusters (e.g. fig. 4.15.1 in Cavalli-Sforza *et al.* 1994). In a more general tree, covering 39 Asian populations (Cavalli-Sforza *et al.* 1994, fig. 4.10.1), they group together with Indian populations and Arabs in a cluster that is well separated from Mongoloids and other Asians. Synthetic genetic maps of the world, based on the first, second and third principal components, all show that the corresponding gene frequencies in Anatolia and the Trans-Caucasus area map together with those characteristic for western Eurasia and northern Africa (Cavalli-Sforza *et al.* 1994).

Three previously published papers deal with mtDNA hypervariable region (HVR) I (or I and II) sequences of Turks and cover, together, about 100 individuals (Richards *et al.* 1996; Calafell *et al.* 1996; Comas *et al.* 1996). All of them lack RFLP identification of mtDNA haplogroups: that was one of the reasons why we investigated a further 400 Turks, using a combined HVR I and RFLP analysis. The Adygeis, a small population living in the northern Caucasus, are well studied by extensive RFLP and HVR I analysis (Macaulay et al. 1999). Partial RFLP analysis of the Ethiopian mtDNA haplogroups revealed the presence of haplogroups U and M among Ethiopians (Passarino et al. 1998) and a recent paper (Quintana-Munci et al. 1999) links Ethiopia and Eurasia via 'the South Route'. Our three previous papers (Metspalu et al. 1999; Kivisild et al. 1999a,b) cover selected general and specific aspects of the distribution of mtDNA haplogroups in western Eurasia and India.

Empirical data

Frequencies of mtDNA haplogroups in Anatolian–Trans-Caucasus populations

Table 25.1 compares mtDNA haplogroup frequencies of Armenians, Georgians, Ossetes and Turks among each other and with Europeans and Indians as 'outgroups'. The overall result is clear: the distribution of haplogroups in all of the four populations studied is much closer to that typical of European rather than to Indian populations. Yet the differences between mtDNA haplogroups in Europeans and Anatolian–Trans-Caucasians are both quantitative and qualitative. Unfortunately, this table lacks data on other Turkish language group populations. However, putative assignment of approximately 200 mtDNA HVR-I sequences of different Central Asian Turkish-speaking populations (Comas et al. 1998) into a median network tree (see reconstruction in Kivisild et al. 1999a) shows that significantly more than a half of their mtDNA variants belong to haplogroups M, A, F and B, infrequent in all four Anatolian — Trans-Caucasus populations (Table 25.1), but typical for Mongoloids (Torroni et al. 1993). The major European haplogroup H in Anatolians–Trans-Caucasians is somewhat less frequent than in European Caucasoids (Torroni et al. 1994), specifically so among Georgians and Ossetes (Table 25.1). The other major haplogroup, U, is nearly equally frequent in Europeans (Torroni et al. 1996; Kivisild et al. 1999b), Anatolians–Trans-Caucasians and in Indians (Kivisild et al. 1999b). However, the three geographic areas differ substantially when sub-clusters of U are compared (Table 25.2). Some of these differences will be discussed below; here we indicate one aspect of these differences which seems to be interesting. Namely, excluding a few exceptions such as the lack of sub-cluster U4 mtDNAs in our Ossetian sample, all basic western Eurasian varieties of haplogroup U are much more evenly represented among Anatolians–Trans-Caucasians than in Europeans. This finding can be understood in the context

Table 25.1. *MtDNA haplogroups in Anatolian–Trans-Caucasus populations (%).*

Haplogroup	Armenians 192	Georgians 139	Ossetes 187	Turks 388
A	0	0	0	0.5
Ä	2.6	2.2	11.8	0.8
B	0.5	0	0	0
F	0	0	0	0.3
H	30.9	17.3	18.7	25.0
I	1.6	2.2	4.3	2.3
J	8.9	3.6	18.7	10.9
K	7.9	10.1	1.1	5.9
L	0.5	0	1.1	0.3
M	0	2.2	2.1	4.1
M1	0	0.7	0	0.3
Õ	1.0	0	0	1.3
O*	7.3	7.2	13.4	3.6
P*	0	1.4	0	0.5
V+pV	0	0.7	0	0.3
R*	1.0	5.0	0	1.7
pJT/pHV	0.5	0.7	1.1	2.3
T	11.5	12.9	6.9	11.9
U	20.4	21.6	17.9	19.1
W	1.0	1.4	2.1	3.9
X	2.1	10.1	0.5	4.4
Ü	2.1	0	0	0.3

Notes:
1. Ä is defined by transitions at nps 16,223, 16,145 and a transversion at np 16,176G or A and +*Ava*II 8252; +*Hph*I 10,237; –*Nla*III 12,501; –*Dde*I 1719.
2. Õ is defined by transitions at nps 15,043, 16,223, 16,172, 16,248, a transversion at np 16,147A and +*Hph*I 10,237; –*Nla*III 12,501; –*Dde*I 1719.
3. pJT/pHV is defined by transitions at 16,126, but without +*Nla*III 4126 and 00073A (see also text).
4. O* is defined as an internal node with 00073A -*Mse*I 14,766 and +*Alu*I 7025 (see Fig. 25.4).
5. R* is defined as an internal node with –*Mbo*II 12,704; +*Mse*I 14,766 and 16,223C; here are included haplotypes deriving from R* which do not belong to any of the described so far haplogroups (e.g. U, T, J etc.).
6. M1 is defined as M with an array of transitions at nps 16,223, 16,129, 16,189 and 16,249 (see Fig. 25.2).
7. P* is defined by +*Mse*II 14,766; 00073G and 16,217C (see Fig. 25.3).
8. Ü is defined by transitions at nps 16,201, 16,223 and 16,265 and –*Dde*I 10,398; +*Hph*I 10,238; –*Nla*III 12,501; –*Dde*I 1719; *Hph*I 10,237.
9. Haplogroup K is a sub-cluster of haplogroup U.

Table 25.2. *Distribution of haplogroup U varieties in Anatolians–Trans-Caucasians and in some other populations*

Population	N	n	%	Per cent from haplogroup U								
				U⁰	U1	U2	U3	U4	U5	U6	U7	K
Armenians	192	54	28.1	1.8	16.7	7.4	16.7	13.0	13.0	0	3.7	27.8
Georgians	139	44	31.7	0	11.4	4.5	13.6	25.0	11.4	0	2.3	31.7
Ossetes	187	36	19.3	0	11.1	25.0	16.7	0	16.7	0	25.0	18.9
Turks	388	92	23.7	5.4	15.2	4.3	22.8	4.3	21.5	0	6.5	24.7
A+G+T	719	190	26.6	3.1	14.1	5.2	18.3	11.5	22.0	0	4.7	27.2
Europe-1	509	118	23.2	5.1	2.5	10.2	6.8	15.3	41.5	0	0	18.6
Europe-2	484	109	22.5	2.8	8.3	4.6	3.7	10.1	33.0	1.8	0.9	35.8
Europe-3	382	100	26.2	3.0	1.0	2.0	1.0	20.0	62.0	0	0	11.0
Indians	550	72	13.1	0	2.3	77.9	0	4.7	1.2	0	12.7	1.2
Ethiop.	270	15	5.6	6.7	6.7	13.3	13.3	0	0	40.0	0	20.0
Nile Valley	255	12	4.7	0	8.3	0	25.0	0	16.7	16.7	8.3	25.0

Notes:
1. U⁰ is defined as +12,308 *HinfI* and –11,465 *Tru1I* mtDNAs that do not belong to any of the indicated above subclusters of haplogroup U.
2. A+G+T: Armenians, Georgians and Turks: this paper.
3. Europe-1: Russians, Poles, Czechs, Slovaks: our unpublished data and Orekhov *et al.* 1999.
4. Europe-2: Sienans, Tuscans, Sardinians, French, Albanians: our unpublished data and Di Rienzo & Wilson 1991; Torroni *et al.* 1996.
5. Europe-3: Estonians, Finns, Karelians: our unpublished data and Sajantila *et al.* 1995; Richards *et al.* 1996; Villems *et al.* 1998.
6. Indians: Kivisild *et al.* 1999b.
7. Nile Valley populations: deduced from HVR I sequence information (Krings *et al.* 1999).

of early and subsequent migrations of modern humans to Europe with respect to the coalescence ages of individual sub-clusters of this complex mtDNA haplogroup in Europe and in western and southern Eurasia.

Comments on haplogroup M

Haplogroup M is, somewhat surprisingly, infrequent in Anatolian–Trans-Caucasus populations, not exceeding a few per cent in Turks, Ossetes and Georgians and was absent among 200 Armenians studied by us (Table 25.1). Although the overall frequency of haplogroup M lineages is low among Turks, our large sample size makes their total number sufficient for phylogenetic analysis (Fig. 25.2). Most of the lineages are typically Mongoloid-specific varieties of this cluster, found equally frequently among Central Asian Turkish-speaking peoples but not in Indians. It is likely that their presence indicates a trace of the Altaic Turkish maternal lineages among contemporary Turks. Interestingly, some of the haplogroup M lineages found among Turks and Georgians belong to the northeastern African specific sub-cluster of haplogroup M (M1), characterized by an HVR I motif of four transitions at nps 16,189, 16,223, 16,249 and 16,311 (Rando *et al.* 1998). We have also found the same sub-cluster of haplogroup M among southern Sicilians and the characteristic motif of four transitions (without haplogroup identification) is present in the published Iberian and Sardinian mtDNA HVR-I sequences (Côrte-Real *et al.* 1996; Di Rienzo & Wilson 1991). This variety of haplogroup M is found in northern and northwestern Africans (Rando *et al.* 1998), but at considerably lower frequency and — what is more important — at lower diversity, than can be observed among Ethiopians (Fig. 25.3), where its frequency in our sample was 15.5 per cent ($N = 270$). Further search of the literature revealed that the same motif is also frequent in populations living in the Nile Valley from Mediterranean to southern Sudan (Krings *et al.* 1999). However, diversity of haplogroup M among Africans and in the European Mediterranean islands is practically restricted to a few founder lineages, and, in contrast to Indian and Mongoloid populations, we have not found in Ethiopian populations haplotypes, corresponding to M* node, frequent among southern, eastern and central Asian

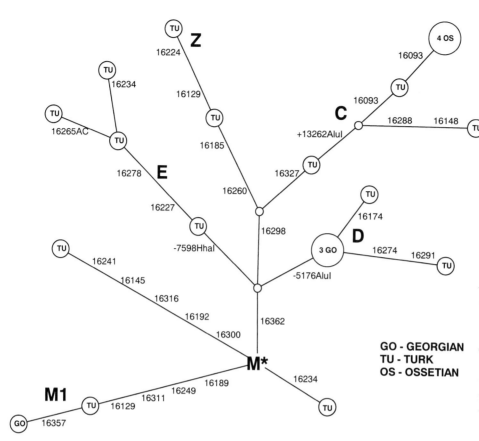

Figure 25.2. *22 haplogroup M mtDNAs found among 906 Turks, Ossetes, Georgians and Armenians. 'African' M1 is present in one Turkish and one Georgian individual. The nature of mutations in mtDNA hypervariable region I is specified only for transitions.*

GO - GEORGIAN
TU - TURK
OS - OSSETIAN

populations.

We consider it likely that haplogroup M arose some 50,000–70,000 years ago (Chen *et al.* 1995) and suggest that its carriers migrated from southern to southeastern Asia before its further diversification and underwent the first expansion/diversification event simultaneously in southern and southeastern Asia around 50,000–60,000 years ago (see Kivisild *et al.* Chapter 31). Time depth of sub-haplogroup M1 is uncertain to us because of the topology of this lineage cluster. However, diversification at two putative internal sub-founders: after transitions at np 16,129 and at np 16,359 (Fig. 25.3), exhibit very similar expansion times around 12,000–13,000 BP (12,800±3400 and 11,900±2300 BP, with n = 22 and 44, respectively). Interestingly, a similar value (about 12,800 BP) can be obtained for a yet another subfounder of a branch characterized by nine base-pair deletion in the intergenic tRNALys/COII region (Fig. 25.3). On the other hand, M1 is probably much older than these apparent expansion phases: accumulation of an array of transitions takes time. In a recent paper by Quintana-Murci *et al.* (1999) the authors argue that the presence of haplogroup M1 in northern Africa/Ethiopia is best interpreted as an evidence for an African origin of haplogroup M. We feel that the question is still unsettled: a widespread presence of the nodal variant for haplogroup M in southern, eastern and central Asia and its lack in Africa plus a vast diversity of this haplogroup in the former regions compared to a single 'ripe' branch in Africa and in Mediterranean basin, sub-haplogroup M1, offer several different scenarios.

However, as far as Anatolians are concerned, the important point is that the frequency of haplogroup M in this Turkish-speaking population is an order of magnitude lower than among the Central Asian Turkish speakers. Here, it is interesting to add that Turkish-speaking Tartars in Russia, living predominantly in the Volga basin, possess this haplogroup at frequencies around 13 per cent: intermediate between the low in Turks and the high in Central Asians (V. Orekhov pers. comm.).

From R* to O*: a tiny but phylogeographically interesting internal node P*

As far as Caucasians are concerned, R* and O* are the two important internal nodes of the mtDNA phylogenetic tree. At the level of the present resolution of the topology of the tree, these two nodes are separated by two mutations. Looking from the direction of the African root, the first is the loss of the *Mse*I restriction site at np 14,766 and the next is a G to A transition at np 00073 (see Fig. 25.2a in the accompanying paper by Kivisild *et al.*). We have now found a number of haplotypes that radiate from this node which we propose to identify as P* (Fig. 25.4), all of them harbouring an additional transition at np 16,217. Their phylogenetic position linked to geographic distribution is interesting and, we think, important: it extends from the western Mediterranean to India, including the Trans-Caucasus.

The internal node O, ancestral to about a half of Europeans*

Metspalu *et al.* (1999) have shown that the mtDNA pools in Armenians, Georgians and Ossetes are particularly rich in haplotypes that belong to neither haplogroup H nor V, but are nevertheless derivatives of (or coincide with) the internal node O* (HV* in Richards *et al.* 1998). With new data on Turks available, this picture is even richer and proves that this internal node, important for the understanding of the history of maternal lineages of the present-day Europeans, is particularly diverse and frequent in Anatolians–Trans-Caucasians (Fig. 25.5). Maternal lineages of four populations studied by us — Turks, Armenians, Georgians and Ossetes — are represented in all lineage clusters, deriving from internal node O*, while Indians, although represented, are distributed less widely and an overall frequency of such lineages in Indian populations studied by us is around 1.5 per cent (Kivisild *et al.* 1999b) compared to 7–13 per cent in Trans-Caucasus area populations and around 4 per cent in Turks.

One of the several lineage clusters additional to H and V, deriving from node O*, is defined by a transition at np 16,067 (Fig. 25.5). The coalescence age for this cluster is approximately 30,000±4000 BP ($n = 38$; $\rho = 1.5$), suggesting that its expansion has started before the Last Glacial Maximum (LGM). This time estimate is somewhat older than similarly calculated coalescence age for haplogroup H (see below). The spread of this cluster is of particular

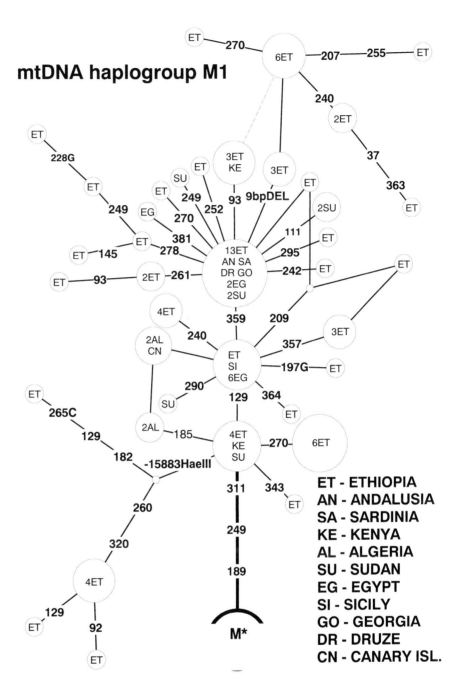

Figure 25.3. *Topology and phylogeography of sub-haplogroup M1 in Africa and western Eurasia.*

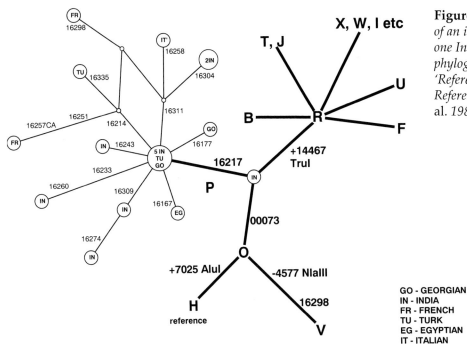

Figure 25.4. *Phylogenetic position of an internal node P*, occupied by one Indian, and topology and phylogeography of haplogroup P. 'Reference' indicates Cambridge Reference Sequence (Anderson et al. 1981).*

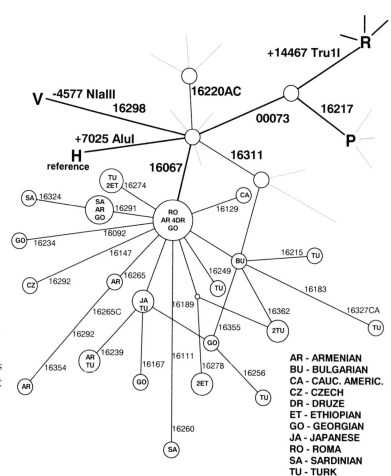

Figure 25.5. *Phylogenetic position of an internal node O* (HV* in Richards et al. 1998) and topology and phylogeography of a lineage cluster with a characteristic founder transition at np 16,067. 'Reference' indicates Cambridge Reference Sequence (Anderson et al. 1981).*

interest: it is most frequent in the Trans-Caucasus populations, seems to be absent in Indians (none out of about 1200 Indian mtDNAs studied) and is rare in Europe, where it seems to be relatively frequent only among eastern-central Mediterranean islands. Its presence in northeastern African populations is also evident. However, in western, northern and northeastern Europe its frequency is below 0.5 per cent. Thus, we see here a pre-LGM, Upper Palaeolithic mtDNA lineage cluster that has an epicentre of frequency and diversity in the Trans-Caucasus area populations, but it is virtually absent in most of Europe as well as in India, while present in northwestern Africa and central and eastern Mediterranean (Druzes, Sardinians, Cretans). Although one can always blame random genetic drift, it nevertheless raises a question about selectivity in possible westward migrations: not all of the mtDNA pool which one would be expecting to find in Anatolia at the beginning of Neolithic, can be sampled among the present-day Europeans, except in a restricted Mediterranean area.

Sub-haplogroups U1, U3 and U7 are more frequent in Anatolians–Trans-Caucasians than in Europeans

Table 25.2 shows that three sub-clusters of haplogroup U (U1, U3 and U7), are more frequent in Anatolians–Trans-Caucasians than in Europeans. Sub-cluster U7 is present among Turks, Armenians, Georgians and Ossetes (Table 25.2; Fig. 25.6), as well as in Druzes (Macaulay *et al.* 1999). However, this sub-cluster has hardly penetrated the European mtDNA pool: we did not find U7 among French, Russian, Polish, Czech, Croatian, Cretan, and Estonian populations (*N* ~1400), and we could infer only a single U7 from the data published by others on German, Swiss, Austrian, Iberian, British, Finnish, Karelian and Saami HVR-I sequences (*N* ~ 2300). Interestingly enough, like O*–16,067, U7 seems to be present in Sardinians (Di Rienzo & Wilson 1991; identification by HVR-I data only). Further-

more, while Indian and western Eurasian (including Europe, Trans-Caucasus and Anatolia) varieties of U2 differ profoundly (Kivisild *et al.* 1999b), with U7 we see a different picture: Indian and Anatolian–Trans-Caucasus haplotypes even coincide partially (Fig. 25.6). It suggests that the cluster has not only a common founder but the carriers of this particular variety of mtDNA seem not to have been isolated from each other until relatively recently. Nevertheless, a nodal haplotype of a sub-founder with motif 16,318AT — 16,309 is abundant only in Indians (Fig. 25.6).

On the other hand, we have not found sub-haplogroup U3 in Indians, whereas it forms a major

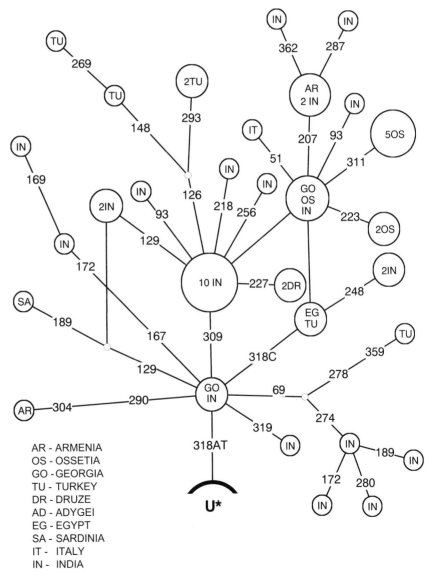

Figure 25.6. *Topology and phylogeography of mtDNA sub-haplogroup U7. Mutations in mtDNA hypervariable region I sequence are shown less 16,000.*

branch of the Armenian, Georgian, Ossetian and Turkish haplogroup U sub-clusters (Table 25.2). This sub-cluster is less frequent in European populations, except in some eastern Mediterranean islands (Metspalu et al. in prep.). It is also present in Ethiopians and, presumably, in Nile Valley populations outside Ethiopia.

U1, in turn, although most frequent among Anatolians–Trans-Caucasians, can be found at lower frequencies also in Indians and Europeans as well as in Ethiopians (Table 25.2).

Hence, a different east–west polarity of sub-haplogroups U1, U3 and U7, as well as of 'non-canonical' derivatives of O* (see, e.g. Metspalu et al. 1999), shows that populations living in Anatolia and the Trans-Caucasus area share maternal lineages with southern Asians, northeastern Africans and with Europeans in a complex, not yet fully understood manner.

Several sub-clusters of typically western Eurasian mtDNA haplogroups display an Early Neolithic beginning of expansion in Europe compared to 20,000 BP or earlier in Anatolians–Trans-Caucasians.

In our previous paper (Metspalu et al. 1999) we described several sub-clusters of western Eurasian mtDNA haplogroups which are shared between Europeans and Anatolians–Trans-Caucasians, but exhibit profoundly different expansion times. Examples for such sub-clusters can be found in many haplogroups, most evidently among T, J and U. Here we bring just one example among haplogroup T (Fig. 25.7). As a rule, the inferred coalescence times of such sub-cluster in Anatolians–Trans-Caucasians lie around 20,000–25,000 BP, while their equivalents in Europe around 8000–9000 BP.

16,126 — 00073A variants of mtDNA in Anatolians–Trans-Caucasians, which are virtually absent in Europeans but more frequent in Ethiopians and likely also in the Nile Valley populations

Haplogroups T and J derive from an internal node R* (Fig. 25.4). Both these haplogroups are common in Europeans (Torroni et al. 1994; 1996; Richards et al. 1998; Kivisild et al. 1999b) and also among Anatolians–Trans-Caucasians (Table 25.1). They are sister groups because they share a number of common mutations compared to the Cambridge Reference Sequence, not shared by other derivatives of the internal node R* (Macaulay et al. 1999). The combination of two mutations: transition at np 16,126 and the gain of *Nla*III site at np 4216, are most informative. We have found in Anatolians–Trans-Caucasians, albeit at low frequencies, mtDNA variants which possess a characteristic for T and J transition at np 16,126 but lack the indicated RFLP site at np 4216 and we classified them as pre-JT (Table 25.1). In fact, they are not 'pre-JT' and the variants found by us are also not monophyletic. There are several such sub-clusters, some of them specific to Indians only; the most frequent among them is a sub-cluster possessing an additional transition at np 16,362 and A at np 00073, characteristic for O* derivatives. Such mtDNA variants were earlier described among Druze population as pre JT*/preHV* (Macaulay et al. 1999). We found such mtDNA lineages in Georgians and Turks (Table 25.1). These lineages are rare in Europe: for example, out of more than 1100 Germans (Pfeiffer et al. 1999; Richards et al. 1996; Lutz et al. 1998) only a single HVR I haplotype can be tentatively (no RFLP data available) assigned to this lineage cluster, suggesting frequencies in northern Europe of around 0.1 per cent. However, we found that this type of lineage is rather frequent in Ethiopian populations. Furthermore, the same types of HVR I motifs are also present in the Nile Valley populations (Krings et al. 1999). In our Ethiopian sample their frequency is 10 per cent (N = 270), exceeding significantly that in Anatolians–Trans-Caucasians (Table 25.1) as well as the frequency of haplogroups T and J in Ethiopians, which in our sample is 2.3 and 1.1 per cent, respectively. Inferred from HVR I data of Krings et al. (1999), the frequency of the 16,126–16,362 lineage in the Nile Valley populations is approximately 4 per cent: somewhat higher than in Anatolians but still more than twice as low as in Ethiopians.

We feel that this finding might be important in further understanding ancient migrations from and to eastern–northeastern Africa. Lack of similar data on many crucial in this context populations does not allow further speculations but we remind here that similar phylogeographic spread is true for mtDNA sub-cluster M1, discussed above: found in Turks and Georgians, very rare in Europeans and frequent in Ethiopians and likely present also in the Nile Valley (defined as Egypt and Sudan: see Krings et al. 1999) populations.

Sub-cluster of haplogroup H with a motif of transitions at nps 16,293 and 16,311 is present in Europeans but absent in Anatolians–Trans-Caucasians

We started to pay attention to this motif because we found it at a relatively high frequency among Estonians and did not find it at all among nearly a thousand Turks, Armenians, Ossetes and Georgians. Searches among further populations showed that this sub-lineage and its derivatives are indeed spread quite unevenly among the Caucasoids of western

Anatolian and Trans-Caucasus Populations

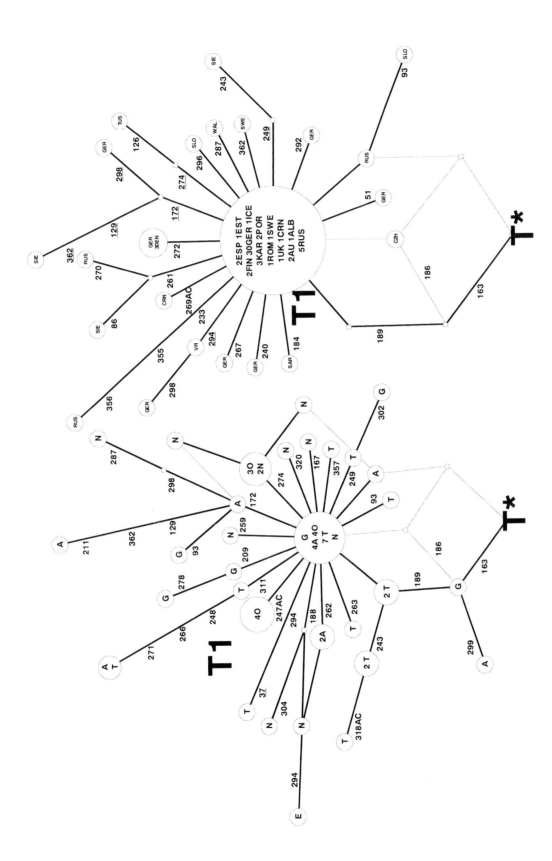

Figure 25.7. Sub-haplogroup T1 in Anatolians–Trans-Caucasians, Nile Valley populations, Ethiopians and in Europeans. Mutations in mtDNA hypervariable region I sequence are shown less 16,000 and T* differs from the CRS by transitions at nps 16,126 and 16,294. Abbreviations: A = Armenians; G = Georgians; O = Ossetes; T = Turks; E = Ethiopians; N = Nile Valley populations; ALB = Albanians; AU = Austrians; CRN = Croatians; DEN = Danish; ESP = Spanish; EST = Estonians; FIN = Finnish; GER = Germans; ICE = Icelanders; KAR = Karelians; POR = Portugese; ROM = Gypsies; RUS = Russians; SAR = Italians (Sardinia); SIE = Italians (Siena); SLO = Slovaks; SWE = Swedes; SZH = Czechs; TUS = Italians (Tuscany); UK = British; VFI = Volga-Finnish; WAL = Welsh-Finnish. References to HVR 1 sequences published elsewhere (HVR 1 and RFLP) are given in the text.

Eurasia (Table 25.3). Although frequent among Estonians, it cannot be considered specific for Finno-Ugric-speaking populations since it is low among Finns and undetected so far in Saamis and Karelians (Sajantila *et al.* 1995). Table 25.3 indicates that this motif has not reached either western or eastern Mediterranean, Anatolian or the Trans-Caucasian populations. Elsewhere in Europe its frequency is generally low; an interesting exception is the eastern Mediterranean Albanian population, where the sub-lineage is well represented. Because of its topology, the time depth of the entire sub-cluster is unclear. However, we have calculated approximate ages of the beginning of expansion for the two very similar sub-founders: 16,293, 16,311 and 16,293, 16,311, 16,278, respectively. The result suggests that this geographically localized expansion started about 5200±1700 BP, i.e. at the end of Neolithic and was almost simultaneous with an expansion of several cultures all over Europe, such as Linear Pottery, Impressed Ware and early pottery-bearing sites in the eastern Baltic area and in central and southern Russia. However, the reason why we have chosen to present this case-study is to show an example of how sub-founders inside haplogroup H may possibly be informative for determining demographic expansion ages that differ substantially from the estimate of the coalescence time of the haplogroup as a whole. And not only in haplogroup H. For example, haplogroup U contains yet another additional minor branch, characterized by a transition at np 16,146. This branch can be found among populations from Iberia to Karelia, but we have not sampled it in Anatolians–Trans-Caucasians or in Africans.

General discussion

The first massive investigation of the European mtDNA pool (Richards *et al.* 1996) re-initiated an intensive scientific debate about the peopling of Europe. Here, the central issue became the age and the place of origin of the extant European gene lineages, particularly with respect to mtDNA. Coalescence age calculations in that paper were used to challange the demic diffusion hypothesis, which stresses a Neolithic gene flow from the Middle East and/or Anatolia to Europe, as a result of food production, i.e. the spread of agriculture. The demic diffusion hypothesis, in its genetic form, is supported by the east-west gradient of the distribution of gene frequencies

Table 25.3. *Distribution of the 16,293–16,311 motif of haplogroup H.*

Population	N	n	Frequency	Reference
Estonians	148	9	6.1	this work
Finns	73	2	2.7	Sajantila *et al.* 1995; Richards *et al.* 1996
Karelians	82	0	0	Sajantila *et al.* 1995
Saami	373	0	0	Sajantila *et al.* 1995; Dupuy & Olaisen 1996; Delghandi *et al.* 1998
Slovaks	130	5	3.8	this work
Poles	93	2	2.1	this work
Czechs	94	2	2.1	this work
Russians	192	4	2.1	this work and Orekhov *et al.* 1999
Croats	400	2	0.5	this work
Albanians	129	6	4.7	this work
Swiss	74	1	1.3	Pult *et al.* 1994
UK	161	2	1.2	Richards *et al.* 1996
Germans	854	9	1.1	Pfeiffer *et al.* 1999; Richards *et al.* 1996
Sardinians	69	1	1.4	Di Rienzo & Wilson 1991
Sienese	150	1	0.7	this work
French	88	0	0	this work
Turks	388	0	0	this work
Iberians	269	0	0	Côrte-Real *et al.* 1996; Bertranpetit *et al.* 1995; Richards *et al.* 1996; Comas *et al.* 1998
Armenians	192	0	0	this work
Ossetes	201	0	0	this work
Georgians	132	0	0	this work
Uigurs	50	1	2	Comas *et al.* 1998
Indians	1008	0	0	this work
Ethiopians	270	0	0	this work

explaining the first principal component (for a review see Cavalli-Sforza *et al.* 1994). In its strong form, the demic diffusion hypothesis suggests that the 'original European' hunter-gatherers perhaps survived only in geographically isolated areas (in the Pyrenees, in Sardinia, the Saami etc.). The critique points out that: a) most mtDNA lineage clusters in Europe seem to be older than 10,000 years (more precisely: display signs of expansion before the Neolithic period); b) gene frequency gradients do not indicate the direction of the movement; c) several anthropological findings suggest continuity of morphological markers throughout the critical period. In contrast, recent autosomal DNA marker studies prompted proponents of the Neolithic migration to suggest that at best a small fraction of the present-day genetic lineages of Europeans is older than the Neolithic (e.g. Chikhi *et al.* 1998). The debate is ongoing and a forthcoming paper (Simoni *et al.* 2000) argues against a Late Upper Palaeolithic re-expansion hypothesis as suggested by Torroni *et al.* (1998).

Discussing the impact of the Neolithic demic diffusion from Anatolia/Middle East, we first need to remind ourselves that the first principal component and the synthetic map derived from it explains about a quarter of classical genetic markers out of those included in the analysis (Cavalli-Sforza *et al.* 1994). That, of course, is far from any nearly total replacement scenario.

It is clear now that from the point of view of mtDNA haplogroups, European and western Asian populations can be viewed as a largely homogeneous phylogeographic entity of Palaeolithic origin. Even the extremes like Saami mtDNA, are only seemingly different from typical for the western Eurasian populations mtDNA pool: the apparent differences are perhaps caused by very strongly pronounced random genetic drift, which has drastically reduced the number of mtDNA haplogroups and lineages in this population. However, only a few Mongoloid-specific mtDNA varieties are present in Saami, in average around 5 per cent of their total mtDNA pool. Hence, the Saami mtDNA pool is not an outlier relative to that of the European Caucasoids, as it has been suggested (Sajantila *et al.* 1995; Simoni *et al.* 2000) but a set of a subset: we think that this distinction is very important for the understanding of biological history of their maternal lineages and of the Saami population in general (Villems *et al.* 1998).

The second conclusion derives from the same often stressed homogeneity of the western Eurasian mtDNA pool: any contemporary analysis should, to be informative, make at least an attempt to analyze mtDNA data phylogenetically. Much has been done already (e.g. Torroni *et al.* 1994; 1996; 1997; Richards *et al.* 1996; Macaulay *et al.* 1999), but much is still left to be done. The wealth of the data makes it evident that some mtDNA lineage clusters and sub-clusters, for example haplogroup V, were 'born' in Europe after the LGM, but before the Neolithic (Torroni *et al.* 1998). However, there is no definite answer available at present about the proportion of such lineage clusters/sub-clusters in the European mtDNA pool. Above, we used one sub-cluster of haplogroup H as an example of a 'European-born' varieties. There are likely a number of other sub-clusters like that.

We have, however, found that several sub-clusters of the common Eurasian mtDNA haplogroups, which are present both in the Anatolian–Trans-Caucasus populations studied by us, as well in the eastern and western European populations, show systematically much later signs of expansion in Europe than in western Asia (Metspalu *et al.* 1999). This difference is large: in average 18,000–22,000 BP for Anatolians compared to 7000–9000 BP for Europeans. This significant time difference can be interpreted as supporting an Early Neolithic migration of the carriers of these lineages to Europe, coinciding with their expansion in new surroundings. Such a dissection of composite lineage clusters is not always straightforward and here we have limited our analysis to sub-clusters which exhibit star-like expansion patterns. Meanwhile, it is interesting to note that these sub-clusters, taken together, may well explain 25 per cent of the existing mtDNA variety in the extant Europeans. However, this is a summary quantitative estimate and the analogy with the above-mentioned gradient seen in the frequency of the first principal component is superficial. On the other hand, if we believe (and we do) that the identical (coinciding in their characteristic HVR polymorphic sites and diagnostic RFLP sites) sub-clusters in western Asian and in European mtDNA pools are indeed monophyletic and have not arisen independently, then we must accept that identical sub-clusters have expanded at widely different times in different localities. If this conclusion were true, one would have to be careful in the interpretation of the coalescence age calculations based on haplotype diversity of lineage clusters covering large geographic areas: obtained estimates may result from many discrete expansion events and the net outcome would be significantly less informative than it might seem at the first glance.

Since haplogroup H is the most frequent mtDNA lineage cluster in Europe, we feel that further progress in the reconstruction of the demographic history of the Caucasoid maternal lineages depends on progress in better understanding the topology of this haplogroup. So far, however, we can say that the coalescence age of haplogroup H appears to be slightly older for Anatolian–Trans-Caucasus than for western European populations. For Georgians, the estimate is 22,500±4500 BP (with $n = 23$; $\rho = 1.1$); for Armenians 23,500±3000 BP ($n = 53$; $\rho = 1.17$) and for Turks 25,000±2.300 BP ($n = 97$; $\rho = 1.24$). Similar calculations gave an even earlier date for Ossetes — about 29,000 BP — but since the topology of the phylogenetic tree of haplogroup H among Ossetes is not star-like, its coalescence age might be overestimated. One may also consider the different demographic history of Ossetes, discussed above. Otherwise, we see that coalescence ages for haplogroup H lineages for the three major Anatolian–Trans-Caucasus populations coincide within the limits of error and are close to coalescence ages for several discussed above mtDNA sub-haplogroups. The last finding is reasonable: one expects that demographic expansions during the Palaeolithic took place in a given geographic region for various reasons, expected to affect simultaneously a population as a whole. The time estimates here are slightly younger than those suggested by other authors (Torroni *et al.* 1998) for the Near Eastern populations (28,000–30,000 BP) and somewhat older than for the western European populations (~13,000–15,000 BP). For Estonians (a Finno-Ugric language-speaking population) we obtained the corresponding value as 30,000 BP, but this higher estimate is partially caused by a relatively high frequency among Estonians of derivatives of the discussed above haplogroup H sub-lineage with a motif of transitions at nps 16,293 and 16,311.

Summary

1. MtDNA lineages of all four Anatolian–Trans-Caucasus populations of this study contain Mongoloid-specific varieties at surprisingly low frequencies and an absolute majority of mtDNA haplogroups found are those present also in Europe. Furthermore, the frequencies of the major haplogroups are largely similar to those observed for Europeans;
2. MtDNA lineages which derive from an internal node O (HV in Richards *et al.* 1998) are significantly more divergent in Armenians, Georgians and Ossetes than they are among European populations;
3. Contrary to the former, haplogroup V, a sister group to haplogroup H, is virtually absent in Anatolian–Trans-Caucasus populations. Taking the last two observations together and bearing in mind that haplogroup H itself is frequent and diverse among Turks, Armenians, Georgians and Ossetes, one can assume that haplogroup V arose in Europe and has never migrated back to western Asia, while the latter area and Europe not only share its sister-group H, but H is also a dominant variety of mtDNA in both these regions;
4. A substantial fraction of the shared lineages between European and Anatolian–Trans-Caucasus populations exhibit much earlier expansion times in the latter area. The differences are large: around LGM for Anatolians and Early Neolithic for Europeans. Since we assume that these sub-clusters are monophyletic, this finding may be explained in terms of demic diffusion, but this hypothesis leaves several questions to be answered;
5. Anatolian–Trans-Caucasus populations share sub-haplogroup U7 with Indians. This lineage cluster, easily identified by characteristic transversion at np 16,318, is virtually absent in Europeans, but relatively frequent in Indians (Kivisild *et al.* 1999b). Furthermore, the Trans-Caucasus and Anatolian populations possess mtDNA lineages deriving from the internal node R*, but do not classify as any of the common haplogroups such as B, F, H, J, K, T, U or V. Frequencies of such lineages are low, but the lineages overlap with those which are frequent in Indian populations, suggesting some limited westward migration of people from southeastern Asia;
6. The presence of only about 4 per cent predominantly Mongoloid-type haplogroup M lineages in the Anatolian mtDNA pool and their absence among Armenians suggests a limited maternal lineage gene flow from the original Altaic language group people to present-day Turkey. This finding is in sharp contrast to the Central-Asian Turkish-speaking populations, where such variants make up about a half of all mtDNA lineages.

In conclusion one may say that except haplogroup V, the European mtDNA haplogroups are a representative subset of the set of haplogroups present in Anatolian and the Trans-Caucasus area populations. In addition, the mtDNA pool of the latter possesses lineages shared by Indians and Central Asians. Occasional additional lineages (e.g. varieties of L) are present, in this geographic area, in frequencies much below polymorphic frequencies. Some other region-specific sub-clusters such as predominantly northern African U6 and M1 — 'an African M' — rare in Mediterranean Europe, are equally absent or rare in Anatolian populations. We think that this excludes models which predict recent substantial flow of maternal lineages from northern Africa and/or Ethiopia to western Eurasia. On the other hand, low frequency of haplogroup H in Ethiopians tells us that any backwards migration must have been either very selective or predating the spread of this major haplogroup in western Asia.

Unless new and significantly earlier archaeological dates for the presence of modern humans in Anatolia, Mesopotamia and Iran emerge, it seems to us that the occupation of western Eurasia by these modern humans took place 40,000–45,000 years ago, covering most of the area, including possibly Scandinavia and other parts of northern Europe, which were subsequently under ice during the LGM. Perhaps the glaciers pushed these early Nordic humans south because south became north: much of the faunal material from the southwest French refugium is reindeer. It is quite likely that during global warming after the end of the LGM reindeer gradually returned north as did the people who hunted them. It is not migration in its typical later sense, creating gene frequency gradients because of admixture with locals: there were no northern locals present. Random drift (including bottlenecks) was probably the main force adjusting mtDNA haplogroup frequencies, but it worked only by reducing possible initial diversity. Proto-Saamis carried their maternal lineages back to the north and this is why we can see that more than 90 per cent of Saami mtDNA lineages are typically European, consisting predominantly of one very old branch, U5, and of another, much more recent branch, V, which is nevertheless a phylogenetic sister group to that most common among western Eurasians-haplogroup H. And because of that, the identity rather than relative frequencies of particular haplogroups in the Saami mtDNA pool is an important issue to consider.

Molecular genetics is only starting to contribute towards the understanding of the reconstruction of demographic history of modern human populations. As far as maternal lineages are concerned, further priorities as we can see them at present include more refined analysis of the topology of the mtDNA tree (additional markers in the coding region) and, of course, covering of hitherto un- and under-sampled regions, specifically Afghanistan, Iran, Iraq and Syria in the 'southern front' and the Ukraine and the steppe belt between Black Sea and Altai Mountains. Parallel studies of paternal lineages and autosomal markers are essential since the information obtained from the mtDNA alone, however valuable it may be, is only a part of the full genetic picture that must be digested together with archaeological, paleoclimatic and linguistic data.

Acknowledgements

We are grateful to Dr A. Torosjan for his help in collecting the Trans-Caucasian samples. We thank Dr P. Rudan for providing Croatian samples, Dr I. Mikerezi for collecting Albanian samples, Dr V. Ferak for providing Czech and Slovakian samples, Dr M. Claustres for the French samples and Dr S. Koziel for Polish samples; analyses of these data will be published in full length elsewhere. We thank Dr M. Stoneking and Dr S. Meyer for sending the Nile Valley sequences. We thank I. Hilpus and J. Lind for technical assistance.

This work was supported by Citrina Foundation UK and by Estonian Science Fund grants. R.V. wishes to thank Wellcome Trust for an award to participate in Human Genome Diversity meeting, September 1999.

References

Anderson, S., A.T. Bankier, B.G. Barrell, M.H. de Bruijn, A.R. Coulson, J. Drouin, I.C. Eperon, D.P. Nierlich, B.A. Roe, F. Sanger, P.H. Schreier, A.J.H. Smith, R. Staden & I.G. Young, 1981. Sequence and organization of the human mitochondrial genome. *Nature* 290, 457–65.

Bertranpetit, J., J. Sala, F. Calafell, P.A. Underhill, P. Moral & D. Comas, 1995. Human mitochondrial DNA variation and the orgin of Basques. *Annals of Human Genetics* 59, 63–81.

Calafell, F., P. Underhill, A. Tolun, D. Angelicheva & L. Kalaydieva, 1996. From Asia to Europe: mitochondrial DNA variability in Bulgarians and Turks. *Annals of Human Genetics* 60, 35–49.

Cavalli-Sforza, L.L., P. Menozzi & A. Piazza, 1994. *The History and Geography of Human Genes*. Princeton (NJ): Princeton University Press.

Chen, Y.S., A. Torroni, L. Excoffier, A.S. Santachiara-Benerecetti & D.C. Wallace, 1995. Analysis of mtDNA variation in African populations reveals the most ancient of all human continent-specific haplogroups. *American Journal of Human Genetics* 57, 133–49.

Chikhi, L., G. Destro-Bisol, G. Bertorelle, V. Pascali & G. Barbujani, 1998. Clines of nuclear DNA markers suggest a largely Neolithic ancestry of the European gene pool. *Proceedings of the National Academy of Sciences of the USA* 95, 9053–8.

Comas D., F. Calafell, E. Mateu, A. Pérez-Lezuan & J.

Bertranpetit, 1996. Geographic variation of human mitochondrial DNA control region sequence: the population history of Turkey and its relationship to the European populations. *Molecular Biology and Evolution* 13, 1067–77.

Comas, D., F. Calafell, E. Mateu, A. Pérez-Lezuan, E. Bosch, R. Martinez-Arias, J. Clarimon, F. Facchini, G. Fiori, D. Luiselli, D. Pettener & J. Bertranpetit, 1998. Trading genes along the Silk Road: mitochondrial DNA sequences and the origin of Central Asian populations. *American Journal of Human Genetics* 63, 1824–38.

Côrte-Real, H.B.S.M., V.A. Macaulay, M.B. Richards, G. Hariti, M.S. Issad, A. Chambon-Thomsen, S. Papiha, J. Bertranpetit & B.C. Sykes, 1996. Genetic diversity in the Iberian Peninsula determined from mitochondrial sequence analysis. *Annals of Human Genetics* 60, 331–50.

Deka, R. & S.S. Papiha (eds.), 1999. *Genome Diversity*. New York (NY): Kluwer Academic/Plenum Publishers.

Delghandi, M., E. Utsi & S. Krauss, 1998. Saami mitochondrial DNA reveals deep maternal lineage clusters. *Human Heredity* 48, 108–14.

Di Rienzo, A. & A.C. Wilson, 1991. Branching pattern in the evolutionary tree for human mitochondrial DNA. *Proceedings of the National Academy of Sciences of the USA* 88, 1567–601.

Dupuy, B.M. & B. Olaisen, 1996. MtDNA sequences in the Norwegian Saami and main populations, in *Advances in Forensic Haematogenetics*, vol. 6, eds. A. Carracedo, B. Brinkmann & W. Bär. Berlin: Springer-Verlag, 23–5.

Gamkrelidze T.V. & V.V. Ivanov, 1984. *Indo-European Language and Indo-Europeans*. Tbilisi: Tbilisi University Press.

Kivisild, T., M.J. Bamshad, K. Kaldma, M. Metspalu, E. Metspalu, M. Reidla, S. Laos, J. Parik, W.S. Watkins, M.E. Dixon, S.S. Papiha, S.S. Mastana, M.R. Mir, V. Ferak & R. Villems, 1999a. Deep common ancestry of Indian and western-Eurasian mitochondrial DNA lineages. *Current Biology* 9, 1331–4.

Kivisild T., K. Kaldma, M. Metspalu, J. Parik, S.S. Papiha & R. Villems, 1999b. The place of the Indian mitochondrial DNA variants in the global network of maternal lineages and the peopling of the old world, in Deka & Papiha (eds.), 135–52.

Krings, M., A.H. Salem, K. Bauer, H. Geisert, A.K. Malek, L. Chaix, C. Simon, D. Welsby, A. Di Rienzo, G. Utermann, A. Sajantila, S. Pääbo & M. Stoneking, 1999. MtDNA analysis of Nile Valley populations: a genetic corridor or a barrier to migration. *American Journal of Human Genetics* 64, 1166–76.

Lahermo, P., A. Sajantila, P. Sistonen, M. Lukka, P. Aula, L. Peltonen & M-L. Savontaus, 1996. The genetic relationship between the Finns and Finnish Saami (Lapps): analysis of nuclear DNA and mtDNA. *American Journal of Human Genetics* 58, 1309–22.

Lutz, S., H.J. Weisser & J.P. Heizmann, 1998. Location and frequency of polymorphic positions in the mtDNA control region of individuals from Germany. *International Journal of Legal Medicine* 111, 67–77.

Macaulay, V.A., M.B. Richards, E. Hickey, E. Vega, F. Cruciani, V. Guida, R. Scozzari, B. Bonné-Tamir, B. Sykes & A. Torroni, 1999. The emerging tree of the West Eurasian mtDNAs: a synthesis of control region sequences and RFLPs. *American Journal of Human Genetics* 64, 232–49.

Metspalu, E., T. Kivisild, K. Kaldma, J. Parik, M. Reidla, K. Tambets & R. Villems, 1999. The Trans-Caucasus and the expansion of the Causasoid-specific human mtDNA kineages, in Deka & Papiha (eds.), 121–34.

Orekhov, V., A. Poltoraus, L.A. Zhivotovsky, P. Ivanov & N. Yankovsky, 1999. Mitochondrial DNA sequence diversity in Russian. *FEBS Letters* 19, 197–201.

Passarino, G., O. Semino, L. Quintana-Murci, L. Excoffier, M. Hammer & A.S. Santachiara-Benerecetti, 1998. Different genetic components in the Ethiopian population, identified by mtDNA and Y-chromosome polymorphisms. *American Journal of Human Genetics* 62, 420–34.

Pfeiffer, H., B. Brinkman, J. Hühne, B. Rolf, A.A. Morris, R. Steigner, M.M. Holland & P. Forster, 1999. Expanding the forensic German mitochondrial DNA control region database: genetic diversity as a function of sample size and microgeography. *International Journal of Legal Medicine* 112, 291–8.

Pult, I., A. Sajantila., J. Simanainen, O. Georgiev, W. Schaffner & S. Pääbo, 1994. Mitochondrial DNA sequences from Switzerland reveal striking homogeneity of European populations. *Biological Chemistry Hoppe-Seyler* 375, 837–40.

Quintana-Murci, L., O. Semino, H-J. Bandelt, G. Passarino, K. McElreavey & A.S. Santachiara-Benerecetti, 1999. Genetic evidence for an exit of *Homo sapiens* from Africa vie East-Africa. *Nature Genetics* 23, 437–41.

Rando, J.C., F. Pinto, A.M. Gonzalez, M. Hernandez, J.M. Larruga, V.M. Cabrera & H-J. Bandelt, 1998. Mitochondrial DNA analysis of northwest African populations reveals genetic exchanges with European, near-eastern and sub-Saharan populations. *Annals of Human Genetics* 62, 531–50.

Rando, J.C., V.M. Cabrera, J.M. Larruga, M. Hernandez, A.M. Gonzalez, F. Pinto & H-J. Bandelt, in press. Phylogeographic patterns of mtDNA reflecting the colonisation of the Canary Islands.

Richards, M.B., H. Côrte-Real, P. Forster, V. Macaulay, H. Wilkinson-Herbots, A. Demaine, S. Papiha, R. Hedges, H-J. Bandelt & B. Sykes, 1996. Paleolithic and Neolithic lineages in the European mitochondrial gene pool. *American Journal of Human Genetics* 59, 185–203.

Richards, M.B., V.A. Macaulay, H-J. Bandelt & B.C. Sykes, 1998. Phylogeography of mitochondrial DNA in western Europe. *Annals of Human Genetics* 325, 241–60.

Rybakov, B.A., R.M. Munchayev, V.A. Bashilov & P.G. Gaidukov (eds.), 1984. *Archaeology of the USSR: Paleolithic of USSR*. Moscow: Nauka. [In Russian.]

Rybakov, B.A., R.M. Munchayev, V.A. Bashilov & P.G. Gaidukov (eds.), 1989. *Archaeology of the USSR:*

Mesolithic of USSR. Moscow: Nauka. [In Russian.]

Sajantila, A., P. Lahermo, T. Lukka, P. Sistonen, M-L. Savontaus, P. Aula, L. Beckman, L. Tranebjaerg, T. Gedde-Dahl, L. Issel-Tarvel, A. DiRienzo & S. Pääbo, 1995. Genes and languages in Europe: an analysis of mitochondrial lineages. *Genome Research* 5, 42–52.

Simoni, L., F. Calafell, D. Pettener, J. Bertranpetit & G. Barbujani, 2000. Geographic patterns of mtDNA diversity in Europe. *American Journal of Human Genetics* 66, 262–78. [Electronically published in *American Journal of Human Genetics* December 1, 1999.]

Torroni, A., T.G. Schurr, M.F. Cabell, M.D. Brown, J.V. Neel, M. Larsen, C.M. Vullo & D.C. Wallace, 1993. Asian affinities and continental radiation of the four founding Native American mtDNAs. *American Journal of Human Genetics* 53, 563–90.

Torroni, A., M.T. Lott, M.F. Cabell, Y-S. Chen, L. Lavergne & D.C. Wallace, 1994. MtDNA and the origin of Caucasians: identification of ancient Caucasian-specific haplogroups, one of which is prone to a recurrent somatic duplication in the D-loop region. *American Journal of Human Genetics* 55, 760–76.

Torroni, A., K. Huoponen, P. Francalacci, M. Petrozzi, L. Morelli, R. Scozzari, D. Obidu, M-L. Savontaus & D.C. Wallace, 1996. Classification of European mtDNAs from an analysis of three European populations. *Genetics* 144, 1835–50.

Torroni, A., M. Petrozzi, L. D'Urbino, D. Sellitto, M. Zaviani, F. Carrara, C. Carducci, V. Leuzzi, V. Carelli, P. Barboni, A. De Negri & R. Scozzari, 1997. Haplotype and phylogenetic analysis suggest that one European-specific mtDNA background plays a role in the expression of Leber hereditary optic neuropathy by increasing the penetrance of the primary mutations 11778 and 14484. *American Journal of Human Genetics* 60, 1107–21.

Torroni, A., H-J. Bandelt, L. D'Urbino, P. Lahermo, P. Moral, D. Sellitto, C. Rengo, P. Forster, M-L. Savontaus, B. Bonné-Tamir & R. Scozzari, 1998. mtDNA analysis reveals a major Late Paleolithic population expansion from southwestern to northeastern Europe. *American Journal of Human Genetics* 62, 1137–52.

Villems, R., M. Adojaan, T. Kivisild, E. Metspalu, J. Parik, G. Pielberg, S. Rootsi, K. Tambets & H-V. Tolk, 1998. Reconstruction of maternal lineages of Finno-Ugric speaking people and some remarks on their paternal inheritance, in *The Roots of Peoples and Languages of Northern Eurasia*, vol. I, eds. K. Julku & K. Wiik. Jyväskylä: Gummerus Kirjapaino Oy, 180–200.

Chapter 26

Network Analysis of Y-chromosome Compound Haplotypes in Three Bulgarian Population Groups

Boriana Zaharova, Anja Gilissen, Silvia Andonova-Baklova, Jean-Jacques Cassiman, Ronny Decorte & Ivo Kremensky

Five Y-chromosome STRs (DYS19, DYS390, DYS391, DYS392 and DYS393) and the biallelic polymorphism DYS287 (YAP) were analyzed in a sample of 172 unrelated males from the three largest ethnic groups in Bulgaria: Bulgarians, Bulgarian Turks and Gypsies. The performed network analysis of six-locus compound haplotypes defined at least two male lineages (YAP+ and YAP−) in each population and demonstrated variability within them. High heterogeneity in the groups of Bulgarian Turks and Gypsies as well as a great extent of admixture between Bulgarian Turks and Bulgarians were revealed. Strong founder effect was observed in the main male Gypsy lineage and an isolated founder haplotype was also detected. Relatively recent immigration of the Bulgarian founder YAP+ haplotype was supposed.

Recently, a number of polymorphisms on the Y chromosome have been described and tested in different population samples (Underhill *et al.* 1997; Roewer *et al.* 1996; Jobling *et al.* 1996; Santos *et al.* 1996). Some of them, such as biallelic polymorphisms, proved to be quite useful for distinguishing between major populations and groups of related chromosomes (Jobling & Tyler-Smith 1995). Others, such as the highly variable tandem repeats, were more informative for studying diversity within particular populations or male lineages (de Knijff *et al.* 1997; Kayser *et al.* 1997). Therefore, the investigation of compound haplotypes, based on both sets of polymorphisms, seems to be the most appropriate approach when studying the history and evolution of human male lineages (Jobling & Tyler-Smith 1995; Jobling *et al.* 1997).

The aim of this survey was to define the main male lineages in the three largest Bulgarian population groups: Bulgarians, Bulgarian Turks and Gypsies. For this purpose six-locus compound chromosome Y haplotypes, based on the biallelic DYS287 (YAP) polymorphism and five microsatellites (DYS19, DYS390, DYS391, DYS392, DYS393), were studied among 172 unrelated males (70 Bulgarians, 48 Bulgarian Turks and 54 Gypsies) from Bulgaria. A network analysis of compound haplotypes was performed to better understand the structure of the investigated groups and relationships between them.

Materials and methods

Population samples
The major population group in Bulgaria (Bulgarians) accounts for 86 per cent of the population. Bulgarian Turks and Gypsies are the two largest ethnic minorities in Bulgaria (10 and 4 per cent, respectively).

Y-chromosome haplotype analysis was performed using three population samples consisting of 70 unrelated Bulgarian, 48 Turkish and 54 Gypsy males from various geographical regions of Bulgaria. The DNA samples were selected from the Bulgarian DNA bank including unaffected members of families with common Mendelian disorders. The selection was based on the following criteria: the individuals were of known ethnic origin, had different surnames and their birthplaces covered the whole territory of the country proportionally to the population density.

Analysis

Microsatellite analysis

A multiplex PCR of microsatellites DYS390, DYS391 and DYS393 and standard single locus amplifications of markers DYS19 and DYS392 were performed using primer sequences described in the Genome Database (GDB, The Johns Hopkins University, Baltimore). The forward primers were 5'-FITC-labelled. Reactions were performed in a total volume of 25 μl containing respectively: 50 ng of DNA, 0.3–0.8 μM of each primer, 1 × Gelatine Buffer (Perkin Elmer), 1.5 mM $MgCl_2$, 200 μM dNTPs and 1–1.5 U TaqGold (Perkin Elmer). Tween 20 to a final concentration of one per cent and BSA were added respectively to the multiplex amplification mix and the singleplex reactions. The cycling conditions were: initial denaturation for 10 min. at 94°C, followed by 30 cycles of 30 sec. at 94°C, 30 sec. at 54°C, 40 sec. at 72°C (for multiplex DYS390, DYS391, DYS393); 30 cycles of 20 sec. at 94°C, 20 sec. at 55°C, 20 sec. at 72°C (for DYS19 locus); 35 cycles of 20 sec. at 94°C, 45 sec. at 53 °C, 30 sec. at 72°C (for DYS392); final extension for 5 min. at 72°C.

The PCR products were diluted up to 1:3 and mixed with two appropriate internal standards (Decorte & Cassiman 1996). Fragment analysis was performed on an ALF DNA sequencer (Amersham Pharmacia Biotech), in denaturing Hydrolink Long Ranger (six per cent) gels for 90 min. at 2000 V, 70mA, 45W, 50°C. Determination of alleles in bp was automatically performed by Fragment Manager Version 1.2 software (Amersham Pharmacia Biotech) in comparison with allelic ladders, consisting of sequenced alleles for each locus.

Biallelic polymorphism DYS287 (YAP)

The DYS287 (YAP) locus was analyzed by standard PCR using flanking primers (Hammer & Horai 1995) that amplify either a 150 bp product (YAP⁻) or a 455 bp product (YAP⁺), which were resolved on two per cent ethidium-bromide-stained agarose gels.

Haplotype analysis

After genotyping, five-locus microsatellite haplotypes were constructed for each individual, with the five loci in the following order: DYS19-DYS390-DYS391-DYS392-DYS393. The YAP state of each microsatellite haplotype was determined. For ease of haplotype interpretation, the allele designation for each locus was re-coded into a simple code where number one indicates the shortest observed allele for a given locus, number two — the allele being one repeat longer, etc. For the analyzed loci, the shortest alleles contained the following number of repeats: 13 for DYS19, 8 for DYS390; 9 for DYS391; 10 for DYS392; 12 for DYS393.

Statistical analysis

Haplotype diversities were calculated according to Nei (1987). Networks of 'adjacent' (one repeat difference) haplotypes were constructed according to Cooper et al. (1996).

Results

Cooper et al. (1996), reported that maximum parsimony methods are not always informative and able to produce meaningful phylogenetic trees. The authors modified an approach, based on the concept of parsimonious networks and successfully applied it for interpretation of compound Y-chromosome haplotype data. The proposed network analysis of 'adjacent' (one repeat difference) haplotypes seems to be an informative tool for studying the evolution of the human male lineages.

Considering single-step mutational changes at one single locus, networks of 'adjacent' haplotypes were constructed for the three Bulgarian population groups studied (Fig. 26.1a,b,c).

Among the 70 Bulgarians investigated 40 different haplotypes were found. The observed haplotype diversity in the group was 0.966. Twenty-five haplotypes can be placed on three separate networks (Fig. 26.1a). One of them is very small and consists of only two haplotypes. The second one (surrounded by a circle on Fig. 26.1a) connects all YAP⁺ chromosomes (20 per cent) in the group. The third, largest network, unites 20 different YAP⁻ haplotypes and demonstrates a well-connected core structure of common haplotypes. The biggest branches of the network (horizontal levels 14 and 15) comprise the most common YAP⁻ haplotypes, which have many adjacent neighbours and form the centre of the network. Six additional haplotypes could be placed on the main Bulgarian network if two-step mutational changes in a single locus are also allowed (data not shown). Nine haplotypes, mostly sampled only once, can not be incorporated into the above networks.

The group of Bulgarian Turks demonstrated the highest haplotype diversity (0.985) with 38 different haplotypes among 48 males investigated. Only five haplotypes were sampled more than once. The network analysis resulted in five separate networks: two small ones, consisting of only two haplotypes; two medium and a big network, that connects 12 different haplotypes including the most common (Fig. 26.1b). Two additional haplotypes can be incorporated in the main Turkish network, supposing also two-step changes in one locus (data not shown). Moreover, one of the medium networks can be convincingly connected with the main network, if the existence of an intermediate unsampled haplotype (given in italics on Fig. 26.1b) is supposed. The absence of this haplotype could be due to the small sample size analyzed. Ten Turkish haplotypes remain isolated from the constructed networks.

Twenty-three different haplotypes were observed among 54 Gypsies studied. One haplotype (32221) clearly dominated in the group with a frequency of 26 per cent. Four more common haplotypes (frequency of nine and seven per cent) were also detected. The haplotype diversity in the sample was not considerably reduced (0.911) despite these high frequency haplotypes. All observed haplotypes except for five could be connected in three separate networks (Fig. 26.1c). The biggest one unites the dominant and most of the common YAP- haplotypes and demonstrates a strong founder effect in this haplogroup. The centre of the network lies at the horizontal level 10. Two haplotypes can be added to the main Gypsy network in case two-step mutational events are also included (data not shown). The second network (surrounded by a circle on Fig. 26.1c) connects all sampled YAP+ haplotypes in the group (15 per cent). The third network contains several less common YAP- haplotypes. Three haplotypes can not be incorporated in these networks. The high frequency of one of them (54222; seven per cent) may be a result of genetic isolation (Fig. 26.1c).

To understand better the relationships between the three largest ethnic groups in Bulgaria, a common network of 'adjacent' haplotypes was constructed on the basis of all observed haplotypes. Among 172 males studied, 72 different haplotypes were found. Fifty-eight of them (81 per cent) could be placed on one main and three small separate networks (Fig. 26.2). All observed YAP+ haplotypes form the first network (shown by a circle on Fig. 26.2). The two other small networks consist of mainly 'private' (sampled only once, in a particular population group) YAP- Bulgarian and Turkish haplotypes. The main network unites 39 YAP- haplotypes and appears to have two centres containing the most common haplotypes. In the lower centre of the network (horizontal level 10–11) mainly Gypsy haplotypes dominate, while in the upper centre (level 14–15) — mostly Bulgarian. Turkish haplotypes are dispersed all over the main network. The two sections of the network are not convincingly connected as the intermediate levels 12 and 13 demonstrate a scarcity of haplotypes and especially of common ones. Five haplotypes could be placed on the main YAP- network and two additional haplotypes could be linked separately if two-step mutational changes in a single locus are also allowed (data not shown). Seven haplotypes remain isolated from the constructed common networks.

Discussion

The performed network analysis of 'adjacent' haplotypes showed evidence for homogeneity of Bulgarians (one large network and a small YAP+ branch) and heterogeneity of Gypsies (three networks and an isolated founder haplotype) and Bulgarian Turks (at least four networks). While the heterogeneity of Gypsies could be explained by the complex internal structure of this group (with genetically isolated subgroups: Kalaydjieva *et al.* 1996), the most likely explanation for heterogeneity of the Turkish group is a great extent of admixture, especially with Bulgarians. This is clearly demonstrated by the common network analysis of the three population groups where two of the YAP- networks are shared with Bulgarians. Moreover, the Turkish haplotypes are dispersed all over the main YAP- network, while the Bulgarian and Gypsy haplotypes tend to lie at its upper and lower parts respectively.

The strong Gypsy founder haplotype (32221) is absent in the Bulgarian group, but appears also as a founder haplotype in the group of Bulgarian Turks. Its high frequency supposes existence of many adjacent neighbours; however, it is the less-connected haplotype of the common Turkish YAP- haplotypes. This observation is difficult to explain by admixture between population groups especially when taking into account the genetic isolation of Gypsies. A possible explanation could be the phenomena of 'preferential self-consciousness' (public declaration of other 'nongypsy' ethnicity) spread among Gypsies (Marushiakova & Popov 1993). Some Gypsies prefer to call themselves 'Turks' (rarely 'Bulgarians' and 'Romanians') for mainly social and religious reasons. This declaration of a false ethnicity can not always be identified during sampling. Therefore, it can not be excluded that Gypsy males are present in the group of Bulgarian Turks and to a considerably less extent Bulgarians.

On the basis of the biallelic polymorphism DYS287 (YAP) at least two male lineages (with and without Alu insertion) were convincingly defined in

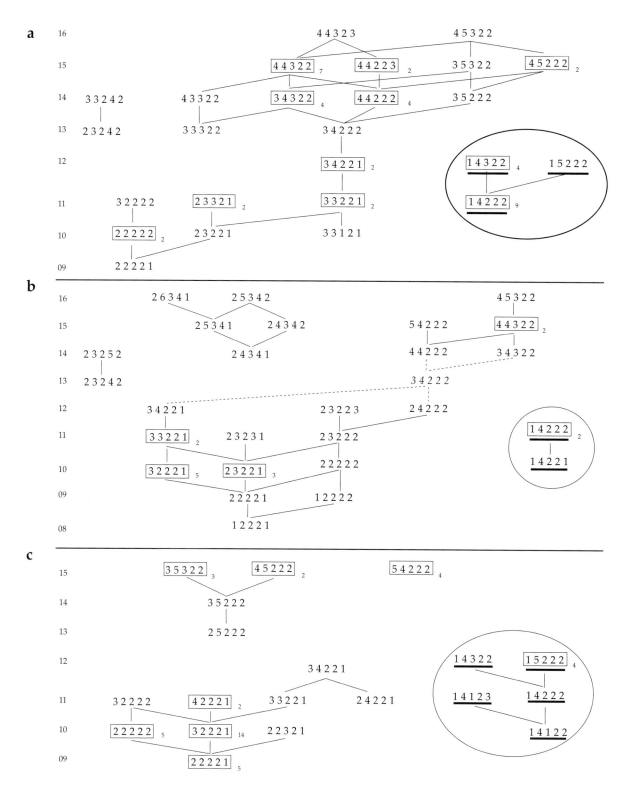

Figure 26.1. *Networks of 'adjacent' (one repeat difference) haplotypes of three Bulgarian population groups: a) Bulgarians; b) Bulgarian Turks; c) Gypsies. Haplotypes sampled more than once in the population are represented by boxes. Subscripts indicate the number of males sharing the same haplotype. Underlining indicates the Alu insertion (YAP+) state. Dotted lines show the links between an unsampled, but hypothetically existing in the population group haplotype (given in italics) and its adjacent neighbours.*

Y-chromosome Compound Haplotypes in Bulgarian Population Groups

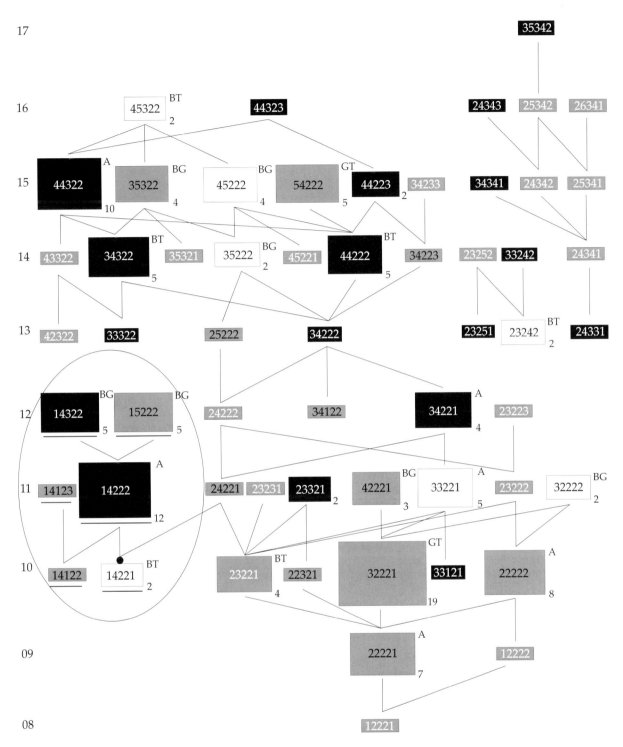

Figure 26.2. *Common networks of 'adjacent' haplotypes of Bulgarian population. Population origin of haplotypes is indicated by font and background colour: white font on black background = Bulgarian mainly; white font on grey background = Turkish mainly; black font on grey background = Gypsy mainly; black font on white background = equally represented haplotypes as denoted by superscript populations. Superscripts indicate where haplotypes are detected in multiple population groups:* [A] *= all three population samples;* [BG] *= Bulgarians and Gypsies;* [BT] *= Bulgarians and Bulgarian Turks;* [GT] *= Gypsies and Bulgarian Turks. Background area around haplotypes is proportional to their frequency. Subscripts indicate the number of individuals sharing the same haplotype. A black dot denotes that both YAP$^+$ and YAP$^-$ states are observed with the microsatellite haplotype.*

the studied Bulgarian population groups.

One network in each population reflects the YAP⁺ insertion. The frequency of YAP$^+$ haplotypes is moderate in Bulgarians and Gypsies (20 and 15 per cent) and low in the group of Bulgarian Turks (six per cent). The Gypsy YAP$^+$ lineage demonstrates the highest diversity (0.786), followed by Turkish and Bulgarian YAP$^+$ lineages with diversities of 0.667 and 0.538, respectively. The high observed frequency of Bulgarian YAP$^+$ haplotypes and their reduced variability could be a result of a relatively recent immigration of the founder haplotype. Although the Bulgarian YAP$^+$ haplotypes are also present in the Gypsy group, their frequencies differ considerably and the founder haplotypes for these lineages seem also to be different.

The small YAP$^+$ lineage detected in the group of Bulgarian Turks is most probably a result of admixture with Bulgarians. It consists of only two haplotypes: the more common one is the Bulgarian YAP$^+$ founder haplotype, while the other Turkish YAP$^+$ haplotype (14221) was found on a Bulgarian YAP$^-$ chromosome and obviously reflects recurrent microsatellite evolution.

The predominant male lineages in the three studied populations are YAP$^-$ (without Alu insertion). The diversity within these haplogroups is considerably higher in comparison with YAP$^+$ lineages (0.984, 0.974 and 0.882 for Bulgarian Turks, Bulgarians and Gypsies, respectively). The performed network analysis reveals heterogeneity within Turkish and Gypsy YAP$^-$ lineages that certainly can not be due only to the small sample sizes and unsampled intermediate haplotypes. The inclusion of other biallelic polymorphisms in our study will permit more precise defining of these YAP$^-$ lineages.

In conclusion, network analysis of six-locus Y-chromosome compound haplotypes was performed among the three largest Bulgarian population groups. At least two male lineages in each population were defined and the variability within them was studied. High heterogeneity in the groups of Bulgarian Turks and Gypsies as well as a great extent of admixture between Bulgarian Turks and Bulgarians were revealed. Strong founder effect was observed in the main male Gypsy lineage and an isolated founder haplotype was also detected. Relatively recent immigration of the Bulgarian founder YAP$^+$ haplotype was supposed.

Acknowledgements

We thank Dr P. de Knijff (Leiden University) for kindly providing of allelic ladders for the investigated microsatellite loci. This work was supported by a TEMPUS grant (Project JEP N°081-21-94/2; to BZ) and a grant (nr G.0241.98; to RD) from the Fund for Scientific Research — Flanders.

References

Cooper, G., W. Amos, D. Hoffman & D.C. Rubinsztein, 1996. Network analysis of human Y microsatellite haplotypes. *Human Molecular Genetics* 5, 1759–66.

de Knijff, P., M. Kayser, A. Caglia, D. Corach, N. Fretwell, C. Gehrig, G. Graziosi, F. Heidorn, S. Herrmann, B. Herzog, M. Hidding, K. Honda, M. Jobling, M. Krawczak, K. Leim, S. Meuser, E. Meyer, W. Oesterreich, A. Pandya, W. Parson, G. Penacino, A. Pérez-Lezaun, A. Piccinini, M. Prinz, C. Schmitt, P.M. Schneider, R. Szibor, J. Teifel-Greding, G. Weichhold & L. Roewer, 1997. Chromosome Y microsatellites: population genetic and evolutionary aspects. *International Journal of Legal Medicine* 110, 134–40.

Decorte, R. & J-J. Cassiman, 1996. Evaluation of the ALF sequencer for high-speed sizing of short tandem repeat alleles. *Electrophoresis* 17, 1542–9.

Hammer, M.F. & S. Horai, 1995. Y-chromosomal DNA variation and the peopling of Japan. *American Journal of Human Genetics* 56, 951–62.

Jobling, M.A. & C. Tyler-Smith, 1995. Fathers and sons: the Y-chromosome and human evolution. *Trends in Genetics* 11, 449–56.

Jobling, M.A., V. Samara, A. Pandya, N. Fretwell, B. Bernasconi, R.J. Mitchell, T. Gerelsaikhan, B. Dashnyam, A. Sajantila, P.J. Salo, Y. Nakahori, C.M. Disteche, K. Thangaraj, L. Singh, M.H. Crawford & C. Tyler-Smith, 1996. Recurrent duplication and deletion polymorphisms on the long arm of the Y-chromosome in normal males. *Human Molecular Genetics* 5, 1767–75.

Jobling, M.A., A. Pandya & C. Tyler-Smith, 1997. The Y chromosome in forensic analysis and paternity testing. *International Journal of Legal Medicine* 110, 118–24.

Kalaydjieva, L., J. Hallmayer, D. Chandler, A. Savov, A. Nikolova, D. Angelicheva, R. King, B. Ishpekova, K. Honeyman, F. Calafell, A. Shmarov, J. Petrova, I. Turnev, A. Hristova, M. Moskov, S. Stancheva, I. Petkova, A. Bittles, V. Georgieva, L. Middleton & P.K. Thomas, 1996. Gene mapping in Gypsies indentifies a novel demyelinating neuropathy on chromosome 8q24. *Nature Genetics* 14, 214–17.

Kayser, M., A. Caglia, D. Corach, N. Fretwell, C. Gehrig, G. Graziosi, F. Heidorn, S. Herrmann, B. Herzog, M. Hidding, K. Honda, M. Jobling, M. Krawczak, K. Leim, S. Meuser, E. Meyer, W. Oesterreich, A. Pandya, W. Parson, G. Penacino, A. Pérez-Lezaun, A. Piccinini, M. Prinz, C. Schmitt, P.M. Schneider, R. Szibor, J. Teifel-Greding, G. Weichhold, P. de Knijff & L. Roewer, 1997. Evaluation of Y-chromosomal

STRs: a multicenter study. *International Journal of Legal Medicine* 110, 125–33.

Marushiakova, E. & V. Popov, 1993. *Gypsies in Bulgaria*. Sofia: Club'90 Publishing House

Nei, M., 1987. *Molecular Evolutionary Genetics*. New York (NY): Columbia University Press.

Roewer, L., M. Kayser, P. Dieltjes, M. Nagy, E. Bakker, M. Krawczak & P. de Knijff, 1996. Analysis of molecular variance (AMOVA) of Y-chromosome-specific microsatellites in two closely related human populations. *Human Molecular Genetics* 5, 1029–33.

Santos, F.R., N.O. Bianchi & S.D.J. Pena, 1996. Worldwide distribution of human Y-chromosome haplotypes. *Genome Research* 6, 601–11.

Underhill, P.A., L. Jin, A.A. Lin, S.Q. Mehdi, T. Jenkins, D. Vollrath, R.W. Davis, L.L. Cavalli-Sforza & P.J. Oefner, 1997. Detection of numerous Y-chromosome biallelic polymorphisms by denaturing high-performance liquid chromatography. *Genome Research* 7, 996–1005.

Chapter 27

MtDNA Sequence Diversity in Three Neighbouring Ethnic Groups of Three Language Families from the European Part of Russia

Vladimir Orekhov, Pavel Ivanov, Lev Zhivotovsky, Andrey Poltoraus, Viktor Spitsyn, Eugeny Ginter, Elsa Khusnutdinova & Nikolay Yankovsky

The hypervariable segment I (HVS I) of mtDNA has been sequenced for 214 individuals from three ethnic groups from the European part of Russia. They are Russians (from three regions: Rjazaní, Kostroma, Kursk; a total of 103 individuals), Mari (from the Morkinsky region of the Mary El republic; 62 individuals), and Tatars (from Elabuga and Almetevsk city; 49 individuals). These groups are almost neighbouring but belong to three different language families: Indo-European, Finno-Ugric and Altaic respectively. Within population genetic diversity (as measured by pairwise difference, D_i) in the case of Mari (3.98) is one of the lovest among Europeans; it is average in Russians (4.27), and highest in Europe for the Tatars (5.36). The distributions of pairwise differences are nearly unimodal for each of the studied groups which may reflect an exponential growth of the populations, with Tatars starting somewhat earlier than the Russians and Mari. Genetic distances (D) between Russians, Tatars, Mari and some other groups have been calculated and a phylogenetic tree was built. Haplogroups typical of Europeans (Richards et al. 1996) have been found in 97 per cent of Mari, 90 per cent of Russians, and 76 per cent of Tatars. Each of the groups contains a component of Mongoloid origin, which reaches up to 14 per cent in the case of the Tatars. At least one quarter (Tatar) to up to one half (Mari) of individuals have individual haplotypes on HVS I which are also presented in one more of the studied groups. The general conclusion is that Russians, Mari and Tatars are relatively related on maternal lines. The relationship is predominantly of common European origin and in small part is of Mongoloid origin.

The territory of the upper and middle Volga river basin is now a crossroad for ethnic groups speaking three different distantly related languages (Indo-European, Finno-Ugric and Altaic language families).

According to archaeological data the first evidence of modern humans inhabiting the territory is dated at least 22,000 years BC, i.e. before the last glacial period in the region. In the Mesolithic to Neolithic periods the territory was repopulated first by different groups of Caucasians, belonging to Baltic tribes and Finno-Ugric tribes (Alekseeva 1999). Tribes of Mongoloid origin started to appear in Eastern Europe in the Neolithic 6000–4000 years BC (Alekseeva 1997). Slavonic groups started to move into the territory in the sixth century AD. The archaeological and anthropological background, along

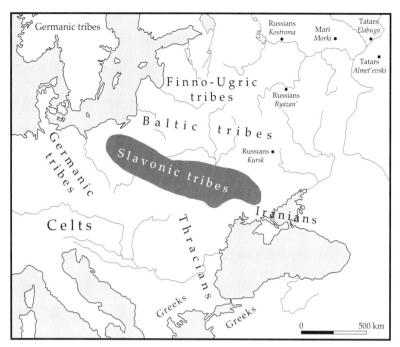

Figure 27.1. *Historical background of Eastern Europe (6th century) and sampling localities. (After Alekseeva 1999.)*

cently described the mtDNA sequence diversity for Russians (Slavonic language of Indo-European language family) (Orekhov et al. 1999). The goal of this study is to establish the data for two other ethnic groups which are Mari (Finno-Ugric language family) and Tatars (one of the Turk languages of Altaic family). Hence, the three ethnic groups are neighbouring in space but speak languages from three of the most distant language families in the area. In the case of these three groups one might expect to find the sharpest genetic differences which might help to make this the most meaningful interpretation of the mtDNA sequence diversity data in terms of ethnogenesis on the studied territory.

with the sampling locations, is summarized on Figure 27.1. The phylogenetic history and relations between the ethnic groups of this region is still not quite clear. The genetic profile of present populations in the area is a combination of profiles of the three main groups.

One of the modern approaches to studying ethnogenesis is based on molecular genetic methods (DNA typing). For this, each ethnic group is described by a specific profile of individual genotypes, which are experimentally established and interpreted. MtDNA is the most useful substrate for DNA typing. It is a highly variable, effectively haploid, highly abundant small circular DNA molecule inherited through the maternal line only. The sequence of the two HVS regions along with RFLP analysis of the remaining part of the DNA molecule is the standard basis for phylogenetic speculations. The analysis of mtDNA polymorphism has already helped to clarify primate speciation history including extinct forms (Krings et al. 1997), worldwide human ethnogenesis (Cann et al. 1987), as well as the relations between neighbouring populations (Sajantila et al. 1995; Mountain et al. 1995).

The molecular-genetic characterization of ethnic groups inhabiting the territory of the upper and middle Volga river basin now may help to reconstruct the ethnogenesis on the territory and relations between groups inhabiting the area. We have re-

Material and methods

Methods of DNA purification and PCR conditions have been described in our previous paper (Orekhov et al. 1999). The sequencing procedure on the Applied Biosystems DNA sequencer model 377 using the 'BigDye Terminator kit' has been performed according to the manual. The following primers were used for amplification of 444 base pair fragments of HVS I: L15997: 5'-CACCATTAGCACCCAAAGCT-3' and H16401: 5'-TGATTTCACGGAGGATGGTG-3' (Ward et al. 1991), where 'L' and 'H' refer to 'light' and 'heavy' strands of mtDNA respectively and numbers indicate the 3' base according to the Cambridge reference sequence of human mtDNA (Anderson et al. 1981).

Standard methods of calculating mean number of pairwise differences (D_i), haplotype diversity (H), and genetic distance (D) have been used and described in our previous article (Orekhov et al. 1999).

To estimate the genetic relations between examined populations and neighbouring groups we used published data for Finns (Sajantila et al. 1995; Richards et al. 1996), Karelians (Sajantila et al. 1995), North Germans (Richards et al. 1996), and Mongolians (Kolman et al. 1996).

Results

The sequence of 377 nucleotides in HVS I mtDNA (positions 16,024–16,400 of mtDNA: Anderson et al. 1981) was unambiguously established and used to characterize the mtDNA diversity for 62 Mari and 49 Tatar individuals.

For the Mari cohort 37 sites were found to be polymorphic among the 377 analyzed sites. Twenty-two different haplotypes have been identified among the 62 Mari individuals (i.e. 35 per cent of the typed individuals were different on the HVSI sequence alone). Nine of the haplotypes (41 per cent) were found only once in the cohort. The most frequent haplotype (19 per cent) for the studied region of mtDNA corresponded to the Cambridge reference sequence (Anderson *et al.* 1981), which is the most widespread among Europeans (up to 20 per cent individuals (Richards *et al.* 1996). Within-population diversity for Mari was found one of the lowest for the European (Richards *et al.* 1996) both for the mean pairwise difference (D_i = 3.98) and for haplotype diversity (H = 0.929).

For the Tatar cohort sixty-one sites were found to be polymorphic among the 377 analyzed sites. Thirty-nine different haplotypes have been identified among the 49 Tatar individuals (i.e. 80 per cent of the typed individuals were different on HVSI sequence alone). Thirty-two of the haplotypes (82 per cent) were found only once in the cohort. The most frequent haplotype (8 per cent) for the studied region of mtDNA corresponded to the Cambridge reference sequence (Anderson *et al.* 1981), as in the case of Russians (Orekhov *et al.* 1999) and Mari. Within-population diversity for Tatars was found to be much higher than for other Europeans, both for the mean pairwise difference (D_i = 5.36) and for haplotype diversity (H = 0.988). The closest index values for Europeans was shown for Spain (D_i = 4.11, H = 0.984) (Richards *et al.* 1996). Though the indexes are lower than for non-Europeans (Middle East D_i = 6.76, H = 0.993: Di Rienzo & Wilson 1991); Mongols D_i = 6.55, H = 0.995: Kolman *et al.* 1996).

Genetic distances (*D*) (Nei & Li 1979) between Mari, Tatars and Russians have been calculated on the basis of the within-population diversity (D_i) and between population diversity (D_{ij}). A phylogenetic tree based on the genetic distance data has been built with NJ algorithm by means of MEGA program (Fig. 27.2).

The distribution of the pairwise differences for the three studied groups is shown in Figure 27.3. For each of the groups the distribution was found to be unimodal (the recess in the case of the Mari does not seem to be statistically significant), which may be explained by exponential growth of the populations after the groups were formed. The peak of the distribution for Tatars is to the right of the Russians and Mari. It may reflect that Tatar population have been formed earlier than the two other groups.

According to the European haplogroup frequencies (Richards *et al.* 1996) Mari, Russians and Tatars are close to other Europeans and far away from Mongols. The share of European haplogroups was found to be 97 per cent for Mari, 90 per cent for Russians, and 76 per cent for Tatars. The fraction of individuals carrying European haplotypes is smaller in the Mari, Russian, Tatar row mainly at the expense of haplogroup 1, which is still the most widespread in each of the groups as well as in other Europeans.

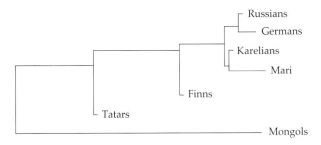

Figure 27.2. *Neighbour-joining tree built according to HVRI sequence data showing the genetic position of Russians, Mari and Tatars among neighbouring populations.*

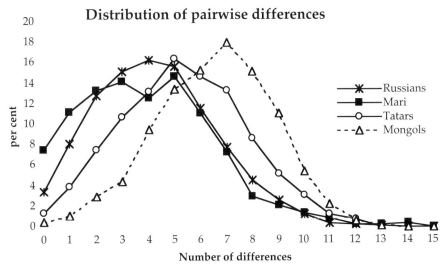

Figure 27.3. *The distribution of pairwise nucleotide differences of Russians, Mari and Tatars. Mongols are shown for comparison as a more distant group.*

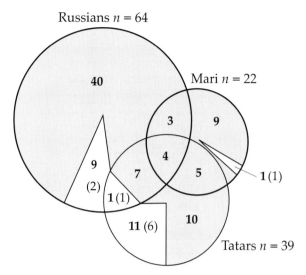

Figure 27.4. *Number of shared mitotypes between Russians, Mari and Tatars.*

The frequencies of non-European haplotypes are increasing in the same row, mainly at the expense of Mongoloid haplotypes.

Comparison of individual haplotypes is another measure of the relations, by maternal lines, between Mari, Russians and Tatars. The maximum number of Mari individuals sharing the same haplotypes as Russians is 52 per cent. A minimum of 25 per cent of Russian individuals have the same haplotypes as Mari. The values for Tatars are within the limits.

The numbers of individual haplotypes shared and unique for the three groups are shown in Figure 27.4. Haplotypes corresponding to European haplogroups (Richards *et al.* 1996) are coloured in grey, non-European are white; the numbers in brackets show clearly Mongoloid haplotypes. The numbers in the overlapping sections of the circles correspond to the minimal number of different mothers common for the groups. There are at least four mothers common for all the three groups. The share of non-European haplotypes increases from Mari to Russians and is large among Tatars (5 per cent, 16 per cent, and 31 per cent, respectively). The non-European component common for each of the groups is of Mongoloid origin (figures in brackets are seen in each of the groups), including one haplotype of Mongoloid origin shared by Russians and Tatars.

Conclusion

The general conclusion of our study is that Russians, Mari and Tatars are fairly related on mother lines. The relation is predominantly of common European origin and in small part of Mongoloid origin.

References

Alekseeva, T.I., 1997. Neolithic population of woodland in Eastern Europe, in *Neolithic of Woodland in Eastern Europe: Anthropology of Sakhtysh Type-sites*, ed. T.I. Alekseeva. Moscow: Nauchny Mir, 18–41.

Alekseeva, T.I., 1999. Ethnogenesis and ethnic history of Eastern Slavs, in *Eastern Slavs: Anthropology and Ethnic History*, ed. T.I. Alekseeva. Moscow: Nauchny Mir, 307–16.

Anderson, S., A.T. Bankier, B.G. Barrel, M.H.L. DeBruijn, A.R. Coulson, J. Drouin, I.C. Eperon, D.P. Nierlich, B.A. Roe, F. Sanger, P.H. Schreier, A.J.H. Smith, R. Staden & I.G. Young, 1981. Sequence and organization of the human mitochondrial genome. *Nature* 290, 457–65.

Cann, R.L., M. Stoneking & A.C. Wilson, 1987. Mitochondrial DNA and human evolution. *Nature* 325, 31–6.

Di Renzo, A. & A.C. Wilson, 1991. Branching pattern in the evolutionary tree for human mitochondrial DNA. *Proceedings of the National Academy of Sciences of the USA* 88(5), 1597–601.

Kolman, C.J., N. Sambuughin & E. Bermingham, 1996. Mitochondrial DNA analysis of Mongolian populations and implications for the origin of New World founders. *Genetics* 142, 1321–34.

Krings, M, A. Stone, R.W. Schmitz, H. Krainitzki, M. Stoneking & S. Pääbo, 1997. Neandertal DNA sequences and the origin of modern humans. *Cell* 90(1), 19–30.

Mountain, J.L., J.M. Hebert, S. Bhattacharyya, P.A. Underhill, C. Ottolenghi, M. Gadgil & L.L. Cavalli-Sforza, 1995. Demographic history of India and mtDNA-sequence diversity. *American Journal of Human Genetics* 56, 979–92.

Nei, M. & W.-H. Li, 1979. Mathematical model for studying genetic variation in terms of restriction endonucleases. *Proceedings of the National Academy of Sciences of the USA* 76, 5269–73.

Orekhov, V., A. Poltoraus, L.A. Zhivotovsky, V. Spitsyn, P. Ivanov & N. Yankovsky, 1999. Mitochondrial DNA sequence diversity in Russians. *FEBS Letters* 445(1), 197–201.

Piazza, A, S. Rendine, E. Minch, P. Menozzi, J. Mountain & L.L. Cavalli-Sforza, 1995. Genetics and the origin of European languages. *Proceedings of the National Academy of Sciences of the USA* 92(13), 5836–40.

Richards, M., H. Côrte-Real, P. Forster, V. Macaulay, H. Wilkinson-Herbots, A. Demaine, S. Papiha, R. Hedges, H-J. Bandelt & B. Sykes, 1996. Paleolithic and Neolithic lineages in the European mitochondrial gene pool. *American Journal of Human Genetics* 59, 185–203.

Sajantila A., P. Lahermo, T. Anttinen, M. Lukka, P. Sistonen, M.L. Savontaus, P. Aula, L. Beckman, L. Tranebjaerg, T. Geddedahl, L. Isseltarver, A. Dirienza & S. Pääbo, 1995. Genes and languages in Europe: an analysis of mitochondrial lineages. *Genome Research* 5(1), 42–52.

Ward, R.H., B.L. Frazier, K. Dew & S. Pääbo, 1991. Extensive mitochondrial diversity within a single Amerindian tribe. *Proceedings of the National Academy of Sciences of the USA* 88, 8720–24.

Chapter 28

Mitochondrial DNA Variability in Eastern Slavs

Boris A. Malyarchuk & Miroslava V. Derenko

Hypervariable segment 1 (HVS-1) sequences of mitochondrial DNA (mtDNA) were determined in 50 individuals of Eastern Slavonic origin, in Russians and Ukrainians. Forty-four sequence types defined by 45 variable positions were found. In general, Russians and Ukrainians show little deviation from other Caucasian data bases of Western and Eastern European descent. However, the results of phylogenetic analysis demonstrate considerable heterogeneity among Russian populations. The Russians presented here are closely related to southern European populations, whereas another published Russian sample from central-northern Russia is clustered together with Volga Finno-Ugric and German populations. The data received allow us to conclude that the Slavonic migrations in the Early Middle Ages from their putative homeland in central Europe to the east of Europe were accompanied by the same mtDNA types characteristic of the pre-Slavonic populations of Eastern Europe. The minor female contribution into Slavonic migrations may also explain the picture observed.

Analysis of mitochondrial DNA (mtDNA) polymorphisms has become a useful tool for human population and molecular evolution studies. The maternal mode of inheritance of mitochondrial genome and the high rate of base substitutions allow us to use the mtDNA polymorphisms for inferring the pattern of prehistoric female migrations and peopling of different regions of the world.

Archaeological and anthropological studies indicate that the colonization of the Eastern European Plain by Slavs began in the Early Middle Ages from the central part of Europe (Niederle 1896; Alekseeva 1973; Sedov 1979). However, anthropologically the Russians, Ukrainians and Belorussians — the main modern ethnic groups of Eastern Slavs — are heterogeneous (Alekseeva & Alekseev 1989). There are several theories on the origin of Eastern Slavs. These theories are based on the morphological, cultural and genetic variation represented in modern and ancient Eastern Slavonic populations. Currently, hypotheses can be classified into two groups: hybridization and transformation theories. Hybridization theories claim that modern Russians, Ukrainians and Belorussians are the result of an admixture of Slavonic tribes, whose homeland was probably in Central Europe, and pre-Slavonic populations of Eastern Europe such as Finno-Ugric (in the northwest and east of Eastern Europe), Baltic (in the west) and Iranic tribes (in the south) (Alekseeva 1973). This theory predicts that Eastern Slavs have genes deriving from both the Slavonic and pre-Slavonic people.

On the other hand, transformation theory posits that Russians as well as Ukrainians gradually evolved from ancient populations of Eastern Europe, at least from the Late Bronze Age (Alekseev 1989). In this theory, in the Early Middle Ages, Slavonic people might have contributed culturally but not genetically to the formation of Eastern Slavonic tribes, imposing a language of the Slavonic group. In this scenario, a model of linguistic replacement and most likely an élite dominance process (Renfrew 1994) is assumed.

These theories of the origin of the Eastern Slavs origin, however, were primarily based on the morphological evidence. Therefore, it is necessary to study the origin and formation of the modern Russians, Ukrainians and Belorussians on the basis of genetic evidence.

The first studies of the mtDNA non-coding hypervariable segment 1 (HVS-1) sequences as well as restriction enzymes analysis in the coding region have allowed us to conclude that the Eastern Slavs are close to the Western European populations (Malyarchuk et al. 1995; Malyarchuk 1997). In this study we present the diversity of the HVS-1 mtDNA sequences data for 50 Russians and Ukrainians from Magadan city, northeastern Siberia. We have also used the additional HVS-1 sequence information from 17 other Caucasian populations, including Russians from the European part of Russia (Orekhov et al. 1999). The results of phylogenetic analysis show a striking heterogeneity of Russian populations. Moreover, a search of genetic traces for putative Slavonic migrations from the west to the east of Europe did not allow us to identify any specific combinations of mitochondrial haplotypes distinguishing Russians from the neighbouring European populations.

Materials and methods

DNA samples from 37 Russians and 13 Ukrainians were obtained from maternally-unrelated individuals. A 573 base pair fragment, encompassing the entire mtDNA first hypervariable segment, was amplified using PCR primers L15926 and H16498 (Di Rienzo & Wilson 1991). Nucleotide sequences between positions 16,000 and 16,400 were determined by use of the Sanger dideoxy chain-termination method and Sequenase enzyme with amplification primers as sequencing primers. The HVS-1 sequences obtained were compared with the Cambridge reference sequence — CRS (Anderson et al. 1981). The phylogenetic relationships of the HVS-1 sequences were analyzed by the median-network method (Bandelt et al. 1995).

For phylogenetic comparison, HVS-1 sequence data from 18 Caucasian populations from Europe and Asia Minor were used including the sequence data presented here. The relationships between them were studied through intermatch-mismatch diversity values using the Sendbs program (provided by Takezaki N., National Institute of Genetics, Tokyo). Several DNA distances between populations were calculated and the neighbour-joining (NJ) tree was built. The distance matrix was also represented by means of principal component (PC) analysis.

Results and conclusion

The high level of the mtDNA HVS-1 sequence heterogeneity was observed in the Eastern Slavonic sample studied: 44 sequence types defined by 45 variable positions were found among 50 individuals. Russians and Ukrainians shared only two HVS-1 sequence types (Fig. 28.1). The probability of two randomly selected individuals from a sample having identical HVS-1 types is 2.7 per cent.

The most frequently occurring sequence type (16069T–16126C) comprised 4 individuals, whereas 40 types were observed in only one individual. In the Russian sample, no individual was found to match the CRS (Anderson et al. 1981) — one of the most frequent HVS-1 sequence in Europe.

The results of sequence type identification (Fig. 28.1) have shown that 92 per cent of them belong to the eight European mitochondrial haplogroups (H, U, K, J, T, I, W, X) with the haplogroup T as one of the most prevalent among Eastern Slavs (its frequency is 22 per cent).

Some peculiarities of the mitochondrial gene pool of the Russians were observed. At the low frequencies (less than 2 per cent) we have found in our data base and in the published data set from the European part of Russia (Orekhov et al. 1999) the HVS-1 sequence types belonging to the Asian-specific haplogroups C and D, Saami-specific haplogroup U5b1 and mtDNA group with HVS-1 motif 16129A-16185T-16223T-16224C-16260T-16298C designated recently as haplogroup Z (Schurr et al. 1999). In Europe this haplogroup has limited distribution among Saami with a frequency of 5.2 per cent (Sajantila et al. 1995). We should note that haplogroup Z was found also with a high frequency (26.2 per cent) in Evens, Tungusic-speaking population of Eastern Siberia (Derenko & Shields 1997). The presence of Asian-specific components in the mitochondrial gene pool of Russians may be explained by their complicated ethnic history including long-lasting interactions with Asians. These genetic data confirm the anthropological view of the presence of Mongoloid traits among the physical anthropological features characteristic of Russians (Alekseeva 1973). In general, Russians and Ukrainians show little deviation from other Caucasian mtDNA data bases of Western and Eastern European descent and we have not found an extensive Mongoloid admixture in the maternal gene pool of modern Russians and Ukrainians. However, in the study of mtDNA sequences only the female lineages are taken into account whereas Mongoloid morphological traits in Russians might have been derived from male migrants.

By means of phylogenetic analysis it has been shown recently (Pult et al. 1994; Sajantila et al. 1995; Comas et al. 1996) that European populations dem-

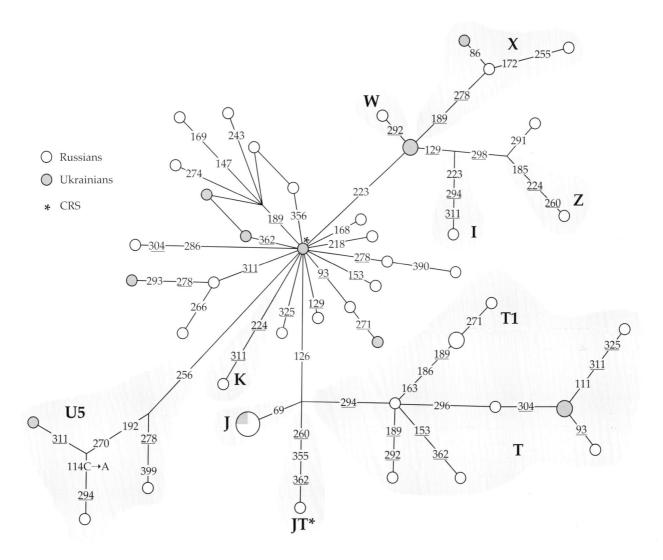

Figure 28.1. *Median network of 44 Eastern Slavonic mtDNA HVS-1 sequence types. MtDNA types are represented by circles, with areas proportional to number of individuals. Links are labelled by the nucleotide positions (nps) in HVS-1 (minus 16,000) to designate transitions; transversions is further specified. Variations of Cs in the unstable region between nps 16,182–16,193 were not included. The node marked * matches the CRS. Ethnic origin is indicated by the scheme (see key). Clusters of HVS-1 sequences comprising named haplogroups are labelled by the shading scheme. Underlining indicates nucleotide positions which have mutated more than once, and reticulations indicate ambiguity in the topology.*

onstrate limited genetic differentiation. Low differentiation suggests that most mitochondrial sequences in Eastern and Western Europe have a local and common ancestry in the Upper Palaeolithic and Neolithic. Nevertheless, the results of phylogenetic (Fig. 28.2) and PC (data not shown) analyses demonstrate considerable heterogeneity among Russian populations. We have not analyzed the Ukrainian mtDNA diversity in the phylogenetic context because of their small sample size. Figure 28.2 shows a NJ tree for 18 populations, which was constructed on the basis of HVS-1 sequences. This population tree demonstrates a major division between populations from Asia Minor and southern Europe (Turks, Sardinians, Tuscans) and the remaining populations, which can be divided into two subclusters: one that includes Bulgarians and Adygei, and another that includes the majority of European populations analyzed. The PC analysis of the HVS-1 sequences displayed patterns of population clustering that were very similar to the phylogenetic tree. It is interesting that both Russian populations do not cluster together

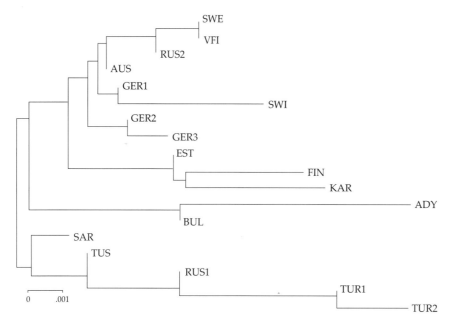

Figure 28.2. *Neighbour-joining tree for 18 European populations, based on mtDNA HVS-1 sequences. Populations coded as: RUS1 = Russians (present data); RUS2 = Russians (Orekhov et al. 1999); TUR1 = Turks (Calafell et al. 1996); TUR2 = Turks (Comas et al. 1996); ADY = Adygei (Macaulay et al. 1999); BUL = Bulgarians (Calafell et al. 1996); SAR = Sardinians (Di Rienzo & Wilson 1991); TUS = Tuscans (Francalacci et al. 1996); GER1 = South Germans (Lutz et al. 1998); GER2 = Bavarians, GER3 = North Germans (Richards et al. 1996); SWI = Swiss (Pult et al. 1994); AUS = Austrians (Parson et al. 1998); FIN = Finns, VFI = Volga Finnic, EST = Estonians, KAR = Karelians, SWE = Swedes (Sajantila et al. 1995).*

In order to create the most appropriate model for the origins of Russians, we examined the HVS-1 sequence variation in modern Russian populations — RUS1 and RUS2, in comparison with published sequence data for 35 Caucasian populations (2163 individuals, in total) from western Eurasia, northern Africa and central Asia. In addition, HVS-1 sequence data on 407 Africans and 538 Asians, including native Siberians, were incorporated into the analysis. Together with the published sequence data for 103 Russians from the central-northern European part of Russia (Orekhov *et al.* 1999), a total of 140 sequences were analyzed in Russians. There were 92 distinct types of HVS-1 sequence in total. Of these, 43 were unique to Russians. However, we have not found any specific combinations of mtDNA types distinguishing Russians from the neighbouring Eastern European populations such as Estonians, Karelians, Volga Finno-Ugric, and Adygei.

Among 49 sequence types shared between Russians and other populations, 21 were found in common among many western Eurasian populations. The maximum numbers of identical sequence types between Russians and other populations in pairwise comparisons were observed between Russians and southern Germans (18 shared types), and northern Germans (15 types), and Austrians (14 types), and Estonians (14 types), and Karelians (14 types). To identify the populations closest to Russians, we have excluded from analysis the most frequent HVS-1 sequence types which were held in common by more than eight populations. As a result, we identified three populations — two Central European (southern Germans and Austrians) and one Eastern European (Volga Finno-Ugric) — which shared maximum numbers of rare HVS-1 types (more than four types) with Russians. Therefore, this kind of analysis has enabled us to detect two weak signals in the mitochondrial gene pool of Russians suggesting their dual (Central and Eastern European) origin and supporting hybridization model on the origin of modern Russians.

and that the Russians presented here (RUS1) are closely related to 'southern' populations such as Turks, Sardinians and Tuscans, whereas the Russians (RUS2) from the central-northern European part of Russia (Orekhov *et al.* 1999) are clustered together with Volga Finno-Ugric and Germanic populations. The Slavonic-speaking Bulgarians occupy the third position on the mitochondrial phylogenetic tree together with Adygei from the Northern Caucasus. It seems that Bulgarians and Adygei are grouped together on the basis of an anthropological criterion because they both are characterized by the Pontic anthropological type (Alekseeva & Alekseev 1989). In general, the results of phylogenetic analysis indicate that Slavonic populations sharing the same language group display a large amount of interpopulation genetic variation. However, in other groups of populations — for instance, in three populations of Germans and in two populations of Turks (Fig. 28.2) — language affiliation and geographic proximity appear to be the major factor contributing to genetic similarities.

The data received allow us to conclude that the Slavonic migrations in the Early Middle Ages from their putative homeland in Central Europe to the East of Europe were mostly accompanied by the same mtDNA types characteristic for the pre-Slavonic populations of Eastern Europe. The minor female contribution into Slavonic migrations may also explain the picture observed. These data, together with the results of phylogenetic analysis, highlight the difficulty involved in the identification of clear coevolutionary patterning between linguistic and genetic relationships in particular human populations — in Slavonic populations, for instance. The evidence from the present study suggests that the assumption of a common Central European origin of Slavs should be tested with studies that include mtDNA and Y-chromosome loci and Slavonic populations from the southern and western Europe. An appropriate linguistic model explaining the distribution of Slavonic languages and culture against the background of genetic heterogeneity of modern Slavonic populations is also required.

Acknowledgements

This work was supported by the Russian Fund for Basic Research (grant 99-06-80430) and the State Program 'Frontiers in Genetics' (grant 99-4-30).

References

Alekseev, V.P., 1989. *Historical Anthropology and Ethnogenesis*. Moscow: Nauka. [In Russian.]

Alekseeva, T.I., 1973. *Ethnogenesis of Eastern Slavs*. Moscow: Moscow State University. [In Russian.]

Alekseeva, T.I. & V.P. Alekseev, 1989. Anthropological view of the origin of Slavs. *Priroda* 1(881), 60–69. [In Russian.]

Anderson, S., A.T. Bankier, B.G. Barrell, M.H.L. de Bruijn, A.R. Coulson, J. Drouin, I.C. Eperon, D.P. Nierlich, B.A. Rose, F. Sanger, P.H. Schreier, A.J.H. Smith, R. Staden & I.G. Young, 1981. Sequence and organization of the human mitochondrial genome. *Nature* 290, 457–65.

Bandelt, H-J., P. Forster, B.C. Sykes & M.B. Richards, 1995. Mitochondrial portraits of human populations using median networks. *Genetics* 141, 743–53.

Calafell, F., P. Underhill, A. Tolun, D. Angelicheva & L. Kalaydjieva, 1996. From Asia to Europe: mitochondrial DNA sequence variability in Bulgarians and Turks. *Annals of Human Genetics* 65, 35–49.

Comas, D., F. Calafell, E. Mateu, A. Pérez-Lezaun & J. Bertranpetit, 1996. Geographic variation in human mitochondrial DNA control region sequence: the population history of Turkey and its relationship to the European populations. *Molecular Biology and Evolution* 13, 1067–77.

Derenko, M.V. & G.F. Shields, 1997. Mitochondrial DNA sequence diversity in three North Asian aboriginal population groups. *Molecular Biology* (Moscow) 31(5), 665–9.

Di Rienzo, A. & A.C. Wilson, 1991. Branching pattern in the evolutionary tree for human mitochondrial DNA. *Proceedings of the National Academy of Sciences of the USA* 88, 1597–601.

Francalacci, P., J. Bertranpetit, F. Calafell & P.A. Underhill, 1996. Sequence diversity of the control region of mitochondrial DNA in Tuscany and its implications for the peopling of Europe. *Amercian Journal of Physical Anthropology* 100, 443–60.

Lutz, S., H-J. Weisser, J. Heizmann & S. Pollak, 1998. Location and frequency of polymorphic positions in the mtDNA control region of individuals from Germany. *International Journal of Legal Medicine* 111, 67–77.

Macaulay, V., M. Richards, E. Hickey, E. Vega, F. Cruciani, V. Guida, R. Scozzari, B. Bonné-Tamir, B. Sykes & A. Torroni, 1999. The emerging tree of West Eurasian mtDNAs: a synthesis of control-region sequences and RFLPs. *American Journal of Human Genetics* 64, 232–49.

Malyarchuk, B.A., 1997. Distribution of mitochondrial DNA markers in Caucasoid populations of Eurasia. *Russian Journal of Genetics* 33(7), 836–40.

Malyarchuk, B.A., M.V. Derenko & L.L. Solovenchuk, 1995. Types of mitochondrial DNA control region in Eastern Slavs. *Russian Journal of Genetics* 31(6), 723–7.

Niederle, L., 1896. *O puvodu slovanu v Praze*. Prague. [In Czech.]

Orekhov, V., A. Poltoraus, L.A. Zhivotovsky, V. Spitsyn, P. Ivanov & N. Yankovsky, 1999. Mitochondrial DNA sequence diversity in Russians. *FEBS Letters* 445, 197–201.

Parson, W., T.J. Parsons, R. Scheithauer & M.M. Holland, 1998. Population data for 101 Austrian Caucasian mitochondrial DNA d-loop sequences: application of mtDNA sequence analysis to a forensic case. *International Journal of Legal Medicine* 111, 124–32.

Pult, I., A. Sajantila, J. Simanainen, O. Georgiev, W. Schaffner & S. Pääbo, 1994. Mitochondrial DNA sequences from Switzerland reveal striking homogeneity of European populations. *Biological Chemistry Hoppe Seyler* 375, 837–40.

Renfrew, C., 1994. World linguistic diversity. *Scientific American* 270(1), 104–10.

Richards, M., H. Côrte-Real, P. Forster, V. Macaulay, H. Wilkinson-Herbots, A. Demaine, S. Papiha, R. Hedges, H-J. Bandelt & B. Sykes, 1996. Paleolithic and Neolithic lineages in the European mitochondrial gene pool. *American Journal of Human Genetics* 59, 185–203.

Richards, M.B., V.A. Macaulay, H-J. Bandelt & B.C. Sykes, 1998. Phylogeography of mitochondrial DNA in western Europe. *Annals of Human Genetics* 62, 241–60.

Sajantila, A., P. Lahermo, T. Anttinen, M. Lukka, P. Sistonen, M.L. Savontaus, P. Aula, L. Beckman, L.

Tranebjaerg, T. Gedde-Dahl, L. Issel-Tarver, A. Di Rienzo & S. Pääbo, 1995. Genes and languages in Europe: an analysis of mitochondrial lineages. *Genome Research* 5, 42–52.

Schurr, T.G., R.I. Sukernik, Y.B. Starikovskaya & D.C. Wallace, 1999. Mitochondrial DNA variation in Koryaks and Itelímen: population replacement in the Okhotsk Sea–Bering Sea region during the Neolithic. *American Journal of Physical Anthropology* 108, 1–39.

Sedov, V.V., 1979. *Origin and Early History of Slavs*. Moscow: Nauka. [In Russian.]

Chapter 29

Mitochondrial DNA Diversity among Nenets

Juliette Saillard, Irina Evseeva, Lisbeth Tranebjærg & Søren Nørby

The Nenets comprise about 30,000 individuals living in subpopulations spread over a vast arctic near-coastal area in central and northwest Russia from the Tamyr peninsula in the east to the Kola peninsula in the west. By sequencing the mtDNA highly variable segment I (HVS I) from 58 individuals we found 24 different sequence types which fall into six major haplogroups (H, U, T, J, C, D). These groups include variants found in Siberian, Central Asian and European (in particular, Finnish-Estonian) populations, which shows that the mitochondrial lineages of present-day Nenets of the Kola peninsula have widely different origins. This testifies to a high degree of admixture during previous migrations of their Asian predecessors to the Kola peninsula.

The study of mtDNA sequence variation, by restriction analysis of the entire molecule as well as by direct sequencing of the coding region, has allowed the definition of population-specific lineages. Four major haplogroups are observed in Siberian and Native American populations (Torroni *et al.* 1993) while Europeans exhibit more variation (Macaulay *et al.* 1999), though 40 per cent of the sequences belong to the haplogroup H, represented by the CRS (Cambridge reference sequence: Anderson *et al.* 1981).

In the present paper, we present the variation of the sequences of the mtDNA hypervariable segment I (HVS I) among 58 individuals from the Nenets Autonomous Area, NAA (Fig. 29.1). The results show a striking mixture of European and Asian sequences indicating diversified origins of the Nenets female lineages.

History

The Nenets live in the polar regions of northeastern Europe and northwestern Siberia from the Kanin Peninsula on the White Sea to the Yenisey delta, occupying the central area of the land of the Samoyed-speaking people. The native land of the Nenets, covering about one million square kilometres, comprises tundra and forest tundra, a country of permafrost, numerous rivers and vast marshy areas.

The Nenets constitute the largest ethnic group among the speakers of the Samoyedic languages. According to the 1990 statistics they numbered up to 35,000 persons, 80 per cent of whom spoke Nenets as their primary language. While the population shows a tendency towards increase, the data shows a decline in native language speakers. The percentage of Nenets within the populations of their native regions is also decreasing.

Anthropologically, the Nenets show clear Mongoloid characteristics. They are commonly of short stature, the average male height being 158 cm. The Nenets of the Arkhangelsk region exhibit somewhat stronger European traits.

The language of the Nenets belongs to the Samoyedic branch of the Uralic languages. Owing to a rather low density of population spread over a vast territory the language is rich in dialects. An overwhelming majority (about 95 per cent) speak the Tundra dialect. The dialectal differences are actually quite minor, thus Nenets from different regions would have no difficulty in understanding each other's speech.

Knowledge and use of Russian grew constantly after the 1930s, particularly during the period of intense russification in the 1970s. Today, all the west-

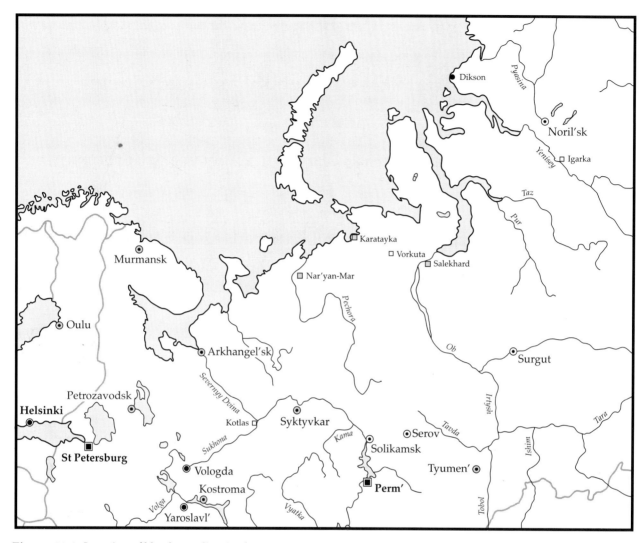

Figure 29.1. *Location of blood sampling in the Nenets Autonomous Area.*

ern Nenets are bilingual, only east of the Ural mountains is an equal knowledge of Russian not yet common.

The present Nenets are mainly reindeer herders, hunters and fishermen. The reindeer — either wild or domestic — represents one of the basic subsistence elements in their culture. The reindeer provides food, clothing, household tools and serves as a means of transport. It is the main unit of social status and wealth.

The Nenets have adopted a nomadic form of reindeer pastoralism. It means that a household travels along with their reindeer stock during their annual migration, thus herding it all year round. Their subsequent expertise in reindeer breeding has been of value to several other peoples.

In the 1950s the Russians began to merge small collective farms. This meant deportations for the Nenets and forced transition from a nomadic to a settled way of life.

Since the 1950s chemical and oil industries have exerted their influence on the life of the Nenets. The main project that changed the lives of the Nenets, was the exploitation of the huge natural gas field in the Yamalo-Nenets national region, which since the early 1960s has caused severe damage to the tundra environment. The life expectancy of Nenets is 45 to 50 years.

Materials and methods

Fifty-eight blood samples have been collected in two different locations of the NAA: Naryan'Mar, the capital, and the settlement Karatayka (Fig. 29.1).

Sequence analysis of the hypervariable mtDNA HVS I has been performed twice for each individual and in both directions.

Amplification of the control region
A 1261 base pair segment, encompassing the entire mtDNA control region, was amplified using the following PCR primers:
5' TTGAGGAGGTAAGCTACATA (H599) and
5' ATACACCAGTCTTGTAAACC (L15907).
mtDNA was amplified directly from whole blood, using a rapid alkaline DNA extraction protocol (Rudbeck & Dissing 1998), with an incubation of 5 minutes at 75°C.

Sequencing
PCR products were purified with spin columns and sequenced on an automated sequencer using standard protocols and the following primers:
5' CTCCACCATTAGCACCCAAAGC (L15975) and
5' TGATTTCACGGAGGATGGTG (H16420).
The sequences obtained were compared with the CRS.

Restriction analysis
In order to make unambiguous haplogroup assignments, it was necessary in two cases to confirm the haplogroup by restriction analysis.

Results

The sequencing of the Nenets HVS I provided a diverse picture of the genetic variation of this population. Twenty-four different haplotypes were observed, due to variation at 32 positions in the segment 16050–16370. Most of the variants are the result of transitions with only two transversions.

The sequences can be clustered into the following six haplogroups, H, U, T, J, C, D, though the J group only harbours one individual.

Figure 29.2 shows the sequences organized in a phylogenetic tree. Most of these sequences have been observed in other populations: Kazakhs, Kirghiz for the C and D group; all over Europe, including Russia, for the other groups (Comas *et al.* 1998; Richards *et al.* 1998; Macaulay *et al.* 1999; Orekhov *et al.* 1999).

Table 29.1 shows the frequency of the different haplogroups, the Asian haplogroups constitute half of the genetic pool.

Conclusion

The current analysis of the mitochondrial HVS I sequences of 58 Nenets reveals the presence of a wide range of female lineages in this population, reflecting a mixture of European and Central Asian line-

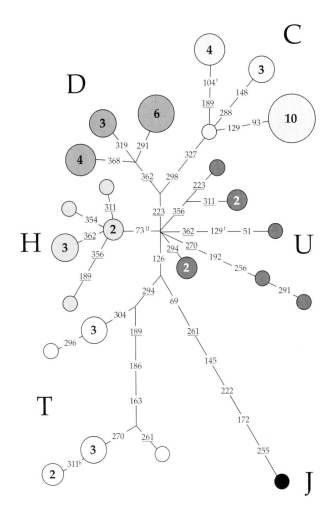

Figure 29.2. *Reduced median network of 58 Nenets HVS I sequences.*

Table 29.1. *Frequency of the different haplogroups.*

Group	Frequency (%)
C	32
D	22
T	17
H	13
U	13
J	2

ages. H, U, T, J are the groups typically found in Europe, while C and D are observed in Asia and the Americas.

The samples are currently analyzed for autosomal as well as Y-chromosomal polymorphisms. The latter results will shed light on the origins of male lineages in the Nenets population. The increasing knowledge of linguistic, anatomical and archaeological data, as well as genetic studies, will provide a more complete view of the history of the peopling of northwest Russia.

References

Anderson, S., A.T. Bankier, B.G. Barrell, M.H. de Bruijn, A.R. Coulson, J. Drouin, I.C. Eperon, D.P. Nierlich, B.A. Roe, F. Sanger, P.H. Schreier, A.J. Smith, R. Staden, & I.G. Young, 1981. Sequence and organization of the human mitochondrial genome. *Nature* 290, 457–65.

Comas, D., F. Calafell, E. Mateu, A. Pérez-Lezaun, E. Bosch, R. Martinez, J. Clarimon, F. Facchini, G. Fiori, D. Luiselli, D. Pettener & J. Bertranpetit, 1998. Trading genes along the silk road: mtDNA sequences and the origin of central Asian populations. *American Journal of Human Genetics* 63, 1824–38.

Macaulay, V., M. Richards, E. Hickey, E. Vega, F. Cruciani, V. Guida, R. Scozzari, B. Bonné-Tamir, B. Sykes & A. Torroni, 1999. The emerging tree of West Eurasian mtDNAs: a synthesis of control-region sequences and RFLPs. *American Journal of Human Genetics* 64, 232–49.

Orekhov, V., A. Poltoraus, L.A. Zhivotovsky, V. Spitsyn, P. Ivanov & N. Yankovsky, 1999. Mitochondrial DNA sequence diversity in Russians. *FEBS Letters* 445, 197–201.

Richards, M., V. Macaulay, H-J. Bandelt & B. Sykes, 1998. Phylogeography of mitochondrial DNA in Europe. *American Journal of Human Genetics* 62, 241–60.

Rudbeck, L. & J. Dissing, 1998. Rapid, simple alkaline extraction of human genomic DNA from whole blood, buccal epithelial cells, semen and forensic stains for PCR. *Biotechniques* 25, 588–9.

Torroni, A., T.G. Schurr, M.F. Cabell, M.D. Brown, J.V. Neel, M. Larsen, D.G. Smith, C.M. Vullo & D.C. Wallace, 1993. Asian affinities and continental radiation of the four founding Native American mtDNAs. *American Journal of Human Genetics* 53, 563–90.

Chapter 30

Genetics and Population History of Central Asia

Francesc Calafell, David Comas, Anna Pérez-Lezaun & Jaume Bertranpetit

We have analyzed the maternally-inherited control region mtDNA sequence and the paternally-inherited Y-chromosome microsatellite haplotypes in four populations from two altitudinal habitats in Central Asia, namely the highland Kirghiz (from Sary Tash, over 3000 m above sea level) and Kazakhs, and the lowland Kirghiz (from the Talas valley) and the Uighur. We tackled three different issues: the origin of Central Asian populations in relation to West Eurasians and to East Asians; the genetic effects of the demographic process of settlement of high-altitude habitats, and the inference of sex-specific characteristics in the population history of Central Asia. When compared to West Eurasians and to East Asians, the pattern of within-population genetic differentiation as well as the genetic distances among populations pointed to an origin for the Central Asian populations as an admixture between already-differentiated West Eurasians and East Asians. By means of a phylogenetic discriminant analysis, we estimated that roughly 6 per cent, 33 per cent and 61 per cent of Central Asian mtDNA samples could have respectively local, West Eurasian and East Asian origins. MtDNA sequences showed high levels of internal diversity in all four Central Asian samples, and were extremely homogeneous across populations. To the contrary, Y-chromosome haplotypes were less diverse in high-altitude populations and much more heterogeneous across populations. The reduction in Y-chromosome diversity at high altitudes may be a consequence of the recent settlement of those habitats, but this by itself cannot explain the different patterns observed for patrilineal and matrilineal markers. Polygyny and a higher male pre-reproductive mortality may contribute to the difference in diversity patterns, which can be mainly explained by a higher female migration rate, probably linked to patrilocal marriage.

Human populations and the genome

Central Asia is a vast territory which may have played a crucial role in the history of mankind. However, it is often regarded as a borderland between East and West, without a singular history, which is not the case (Bowles 1977; Sellier & Sellier 1993). Central Asia, as defined by Soviet scholars, stretches from the Caspian Sea to the modern Chinese boundaries, and it comprises the former Soviet republics of Uzbekistan, Tajikistan, Turkmenistan, Kirghizstan, and part of Kazakhstan; in Western literature, Mongolia, Tibet and Sinkiang (Pinyin Xinjiang, western China) sometimes are included. In this paper, we shall use the first definition plus the westernmost half of Sinkiang. It is a territory of vast contrasts, with most of the land at high altitudes or covered by cold deserts. However, the river basins have been occupied since early times, and the steppes have offered pasture to itinerant shepherds and their flocks.

The role of Central Asia in early human evolution and history is not well established. According to an old, long-dismissed hypothesis, the nearby Altai region would have been the cradle of humankind. It is known that the region was populated during the Lower Palaeolithic, and there is ample evidence of settlement during the Middle Palaeolithic, including Teshik-Tash, the easternmost site from which Neanderthal remains have been recovered. It is not clear, however, whether the region harboured a 'maturation' phase in the evolution of anatomically modern humans, whether it was a thoroughfare in the colonization of Europe and eastern Asia, or whether it was the place where Asian and European groups met after having differentiated (Bowles 1977).

M. Otte (1996) suggested that the population movement that brought the Aurignacian culture (and, presumably, anatomically modern humans and their genes) to Europe originated in Central Asia, given that, in his interpretation, the Aurignacian culture showed traces of adaptation to a steppe environment.

The advent of the Neolithic seems to have had a similar development to that in western Asia, but with an irregular advance: while some tribes adhered to the new form of economy, in vast areas of Central Asia the transition was delayed. Tribes of hunters, fishers and food-gatherers were contemporary with sedentary communities in oases (Sarianidi 1992). This patchy pattern of Neolithization is likely to have weakened the demographic impact of an external wave of advance and thus no significant external population inputs are likely to have been introduced. The domestication of the horse in the steppes (Anthony & Brown 1991) and, subsequently, the development of wheeled vehicles (Anthony & Vinogradov 1995) had a major impact on world history, as mobility increased dramatically and warfare was profoundly changed. Central Asia enters history around the seventh century BC with the Scythians, a people described as having European morphological traits both by ancient Chinese texts and by Herodotus. In the second century BC, the Chinese established a trade route from eastern Asia to the Mediterranean, which would later be known as the Silk Road, and that lasted until it was replaced by sea routes during the European colonial expansion. During the third and fourth centuries AD, Turkic hordes of Siberian origin created a great empire from Mongolia to the Black Sea and replaced the ancient languages of Central Asia; the empire was revived by Genghis Khan in the thirteenth century. Later, the Chinese and Russian empires established their rule over the vast territories of Central Asia.

The population history of Central Asia offers several issues on which population genetics can shed light. One of these is the very origin of Central Asian populations and the role that Central Asia may have played in the first colonization of Eurasia by anatomically modern humans; others refer to the external influences tied to successive cultural changes, be they of known or unknown origin. In all cases a population reconstruction of the past of the extant population is undertaken. This kind of inference is possible because the mechanisms that generate genetic diversity between and within populations have long been known and modelled, and many of those factors, such as genetic drift and migration, are tightly linked to population history. Beyond a theoretical framework, the relevant genetic data can nowadays be produced with relative ease and speed in the laboratory. In particular, we can explore genetic polymorphism in the uniparentally inherited portions of the human genome, that is, the female-transmitted mitochondrial DNA (mtDNA), and the male-transmitted Y chromosome. In the case of mtDNA, genetic analysis is based in the DNA sequence of a fraction of the mtDNA genome, called hypervariable region I (HVRI), in the control region or D-loop. For the Y chromosome, the genetic results consist of the number of times a given three- or four-nucleotide motif is repeated at a particular location in the genome. Those repeated structures are called microsatellites or short tandem repeats (STRs).

Variation in mtDNA and the Y chromosome shares a number of properties that make those genomic regions particularly useful for human population genetics:

i) the whole mtDNA molecule (16,569 basepairs) and the non-recombining region of the Y chromosome (~30 million basepairs) are inherited *en masse* from a single parent. Thus, we can characterize maternal and paternal lineages, and haplotypes (i.e. the combinations of states at each polymorphism) in individuals can be directly inferred. In the rest of the genome, which is found in a double copy and is biparentally inherited, haplotype reconstruction entails typing each individual and both their parents or inferring it using mathematical models. Since recombination is absent and mutation is the only mechanism that generates new variation in those regions, it is feasible to reconstruct the phylogeny of the observed variation. Recently, the possibility that mtDNA molecules exchange information (that is, that they recombine) has been pointed out (Awadalla *et al.* 1999). However, the methods used are controversial (Macaulay *et al.* 1999), and mutation in the mtDNA hypervariable region may be two orders of magnitude faster than recombination. Thus, it is not expected that recombination played a significant role in generating variability at the hypervariable region of mtDNA in the last ~50,000 years.

ii) Mutation rate estimates are available both for mtDNA sequence variation and for STRs in the Y chromosome. Thus, the phylogenies we estimate can be anchored to a timeframe and, since the rate of accumulation of genetic variation (i.e. the mutation rate) is known, the time that any given population has been accumulating variation since a founding event can be estimated.

iii) A few, recognizable mutation events happened after the major continental population groups split and, therefore, most lineages can be attributed to a broad continental origin. Given the existing variation in the mtDNA it is possible to reconstruct the phylogeny of the molecule, that is, the mutational process that gave rise to the observed variation; this process has taken place alongside with human dispersion and migration.

iv) Highly polymorphic markers can be found both in the mtDNA and in the Y chromosome. These markers are particularly sensitive to the demographic processes that lead to gains or losses of genetic variation. For instance, Mateu *et al.* (1997) showed that mtDNA sequences were more diverse in the São Tomé islanders (a former slave tradepost in the Bight of Biafra peopled by the descendants of slaves from all over Africa) than in the nearby Bioko island, with a much older, spontaneous settlement from the mainland.

Figure 30.1. *Location of the populations sampled.*

The present study discusses, in terms of population history, the results of having sequenced the mtDNA hypervariable region I (HVRI) and typed seven microsatellites in the non-recombining segment of the Y chromosome in samples from three Central Asian populations: the Kirghiz, the Kazakh and the Uighur (Fig. 30.1). Two samples were available for the Kirghiz: one from Sary-Tash, a high-altitude settlement at 3200 m above sea level, and another from Talas Valley, at 900 m above sea level. The Uighurs were sampled in the village of Penjim (600 m above sea level), and the Kazakh in the high Kegen valley, at 2100 m. The issue we tackled first (Comas *et al.* 1998) was the population history of the area in relation to the different possible roles of the region in the early dispersal of anatomically modern humans (see above). Then, we changed lenses and focused on the much more recent history of the high-altitude settlements (Pérez-Lezaun *et al.* 1999), trying to understand to what extent the colonization of high-altitude habitats had had an impact on the genetic diversity of those populations. Last but not least the female and male transmitted genetic material were compared in their amount of variation in order to infer sex-specific characteristics in the population history of Central Asia.

Origins of Central Asian populations: source or sink?

Mitochondrial DNA appeared to be a more suitable tool than Y-chromosome STR haplotypes as a means of studying the issue of the ancient origins of Central Asian populations, since more data were available for other populations for comparison (especially East Asian populations) and because the phylogenetic reconstruction can better be achieved in sequence analysis (as in mtDNA) than for repetitive elements.

We obtained a 360-bp HVRI mtDNA sequence for a total of 205 individuals: 55 Kazakhs, 47 highland Kirghiz, 48 lowland Kirghiz and 55 Uighurs. A total of 146 different sequences were found, defined by 108 variable nucleotides.

Genetic diversity within a population can be measured by means of a simple parameter, which ranges from 0 (when all individuals in a sample are genetically equal for the genetic locus analyzed; in biparentally inherited loci, the two copies in each individual must be identical as well) to 1 (when all gene copies in the sample are different from each other). In its specific formulation for DNA sequences, gene diversity is called sequence diversity. Populations that are older, larger, more open (or admixed), and that have not experienced bottlenecks will tend to have higher sequence diversities; on the contrary, younger, smaller, more closed populations, or those which have experienced a recent bottleneck will tend to have smaller sequence diversities. Sequence diversities in the Central Asian populations ranged from 0.984 to 0.995 (Table 30.1). Sequence diversities in 36 European and West Asian populations (Simoni et al. Chapter 13) range from 0.799 to 0.993 with a median of 0.965; in East Asians, sequence diversity ranges from 0.947 in the Ainu to 0.993 in the Chinese. Thus, sequence diversities in Central Asians are among the highest in Eurasia.

As mutations in DNA accumulate with time, the number of nucleotide positions at which they bear different nucleotides gives a rough measure of the relatedness of two sequences. This can be extended to a population sample and the average number of pairwise differences can give an estimate of the relative age at which a population expanded. Thus, the highest averages are found in African populations (Harpending et al. 1993), which has been used as an argument for the replacement hypothesis on the origin of modern humans. The mean pairwise differences in Central Asia range from 5.91 in the Uighurs to 6.64 in the Kazakhs. These values are slightly lower that those in East Asians (6.68–7.51), while they are higher than those in Europeans (3.15–5.03). Thus Central Asians have intermediate average pairwise differences, between Europeans and East Asians.

A direct comparison of the sequences showed that Central Asians share six sequences with the Han Chinese (Horai et al. 1996), five with the Koreans, and four with the Ainu (Horai et al. 1996). Fifteen sequences were shared with Mongolians (Kolman et al. 1996) and nine with the Turks (Calafell et al. 1996; Comas et al. 1996; Richards et al. 1996). However, six of the nine sequences shared with the Turks are also widely distributed across Europe. Central Asians share a total of 17 sequences with Europeans in a data base containing 1420 individuals (Comas et al. 1998). Of these, two sequences are found in Eastern Asians, which, with two other sequences, make up the small set of sequences shared between Europeans and East Asians. Therefore, Central Asians share sequences both with Europeans and East Asians, which present nearly non-overlapping mtDNA sequence pools.

The genetic distance between two populations can also be estimated giving, in a single parameter, an estimate of the degree of genetic similarity. For mtDNA sequences, a measure of distance is obtained by comparing all pairs of sequences made up of one sequence from each population, counting the number of different nucleotides among them and averaging over all possible pairs. This average interpopulation difference is then standardized by subtracting the average within-population pairwise difference. We computed genetic distances among Central Asians, Europeans and East Asians, and found that Central Asian populations presented short genetic distances between each other and intermediate distances both to East Asians and Europeans; these two extreme groups presented the longest genetic distances.

In summary, Central Asians have high pairwise differences, and sequence diversities and genetic distances that are intermediate between Europeans and

Table 30.1. *Diversity parameters in Central Asian populations for mtDNA HVRI sequences and 7-locus Y-chromosome STR haplotypes. N = sample size; k = number of different sequences or haplotypes in the sample; D = sequence/haplotype diversity; pw = average pairwise differences.*

Population	mtDNA HVRI sequences				Y-chromosome STR haplotypes			
	N	k	D	pw	N	k	D	pw
Kazakh	55	45	.990	6.64	49	18	.738	5.52
Highland Kirghiz	47	34	.984	6.12	43	11	.545	6.22
Lowland Kirghiz	48	43	.995	6.48	41	25	.955	8.32
Uighur	55	46	.993	5.91	39	36	.995	7.81

East Asians, with whom they share a number of mtDNA sequences. This set of observations could be compatible with at least two hypotheses:

i) Both Europe and East Asia were first colonized from Central Asia. This would explain the pattern of sequence sharing and genetic distances, but it is incompatible with the average pairwise differences which indicate that Central Asian populations are younger than Eastern Asians. Sequences shared with Europeans are found at moderate frequencies in all four Central Asian samples, and it is difficult to imagine how those could have been brought from Central Asia into Europe without also introducing the East Asian sequences into Europe. Thus, there is no genetic evidence for a Central Asian origin of European modern humans that could have spread with the Aurignacian (Otte 1996).

ii) The Central Asian mtDNA pool is the result of admixture from the previously differentiated Europeans and Central Asians. Again, this would explain sequence sharing, intermediate genetic distances, intermediate pairwise differences and higher sequence diversity.

In order to further assess the admixture hypothesis, we tried to classify each sequence in the Central Asian samples as having originated either in Europe or in East Asia by means of a phylogenetic discriminant analysis. To that effect, we constructed a genetic tree with 104 mtDNA sequences from Europe (Britain and Tuscany) and 113 different sequences from East Asia (Ainu, Han Chinese and Korean samples). We could identify 16 groups of sequences, which encompassed 82.5 per cent of the sequences. Only one group contained significant numbers of both European and East Asian sequences; the remaining 15 were either exclusively European or East Asian. We identified the nucleotide changes that were shared by all sequences in each group and that constituted the putative ancestral sequence of that group; next, we tried to identify those nucleotide changes in

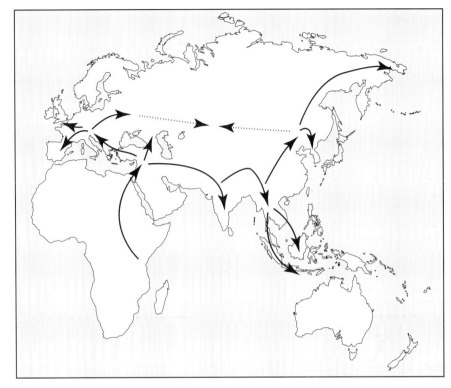

Figure 30.2. *Proposed routes for the dispersion of anatomically modern humans out of Africa (solid arrows) and for the first settlement of Central Asia (broken arrows).*

the Central Asian sequences: if a match was found, the sequence was assigned to that group, and classified accordingly as European or East Asian (Table 30.2). Almost all sequences (93.7 per cent) could be assigned to their broad geographic origins. On average, one third of all Central Asian sequences could be classified as European (from 25 per cent in the lowland Kirghiz to 40 per cent in the Kazakh), while the remaining two-thirds could be considered East Asian.

The genetic analysis has pointed clearly to an admixed origin for the Central Asian populations (Fig. 30.2). The genesis of these populations took place when East and West Eurasian populations had already differentiated in their respective homelands: each have their own mtDNA lineages with slight overlapping. Nonetheless, the present analysis cannot provide a clear answer either concerning the timing or the tempo of the admixture process. It could have happened as a trickle over many generations or as a faster process. It is tantalizing to assign to the Silk Road a main role in channelling migration from the East and West to the region, but it is likely that Central Asian populations had a more ancient origin, as it is well recognized that the roots of extant populations extend far beyond to the past.

Table 30.2. *Fraction of mtDNA sequences in four Central Asian populations that could be attributed to an Eastern Asian or to a European/West Asian origin.*

	Eastern Asian	European	Unassigned
Kazakh	56%	40%	4%
Highland Kirghiz	40%	32%	8%
Lowland Kirghiz	73%	25%	2%
Uighur	55%	35%	10%
Average	61%	33%	6%

Up and away: detecting the genetic effects of the colonization of high-altitude habitats

As explained above, four population samples from Central Asia were analyzed: two from high altitudes (Kazakhs and highland Kirghiz) and two from lower altitudes (Uighurs and lowland Kirghiz). Some of the highland habitats, particularly the Sary Tash area where the highland Kirghiz live, may have been colonized quite recently, maybe just a few centuries ago according to oral tradition. If the highlands were first settled by a small number of individuals, we may still observe the genetic effects of the population bottleneck on present-day highlanders. Then, it is expected that internal genetic diversity is reduced in the highland populations. We compared genetic diversity in the four population samples by means of mtDNA sequences (see previous section) and Y-chromosome microsatellite haplotypes. We analyzed seven microsatellites (an their combinations in each chromosome or individual, known as haplotypes) in the male-transmitted Y chromosome in individuals whose mtDNA sequence was known (Pérez-Lezaun *et al.* 1999). Next, we compare genetic diversity within and between Central Asian populations by means of mtDNA sequences and Y-chromosome STR haplotypes.

MtDNA sequence diversity (Table 30.1) was very similar across Central Asian populations. The most extreme values were found in the highland Kirghiz ($D = 0.984$) and in the lowland Kirghiz ($D = 0.995$). The difference between those two values was not statistically significant ($p = 0.067$, permutation test). The apportionment of genetic variance between and within populations was investigated by means of AMOVA (Analysis of MOlecular VAriance: Excoffier *et al.* 1992), an application of the Analysis of Variance that takes into account the genetic differences among the individuals and the populations being compared. When applied to Central Asian populations, AMOVA revealed that 99.54 per cent of the genetic variance was contained within the populations, while the remaining 0.46 per cent (not statistically different from 0, $p = .094$) could be attributed to differences between populations. AMOVA can be extended to groups of populations in a hierarchical fashion; thus, when we grouped the four populations in two groups according to altitude, the fraction of genetic variance that could be attributed to altitude was estimated at –0.52 per cent (not statistically different from zero), which means that the grouping of those populations according to altitude does not reflect any difference in the mtDNA pool. A similar pattern was found for language, which could account for a meaningless –0.47 per cent of the mtDNA genetic variance. In summary, both lowland and highland Central Asian populations contain high levels of mtDNA diversity, and their mtDNA sequences are highly homogeneous across populations. This observation also implies that we could not detect the effects of natural selection on mtDNA at high altitudes. MtDNA codes for enzymes of the metabolic pathway in which oxygen is used to degrade sugars and obtain energy. If different forms of those enzymes existed and if the low oxygen pressures at high altitude revealed a different performance of those enzymes, the mtDNA molecule coding for the fittest enzymes would have increased in frequency from generation to generation, thus reducing mtDNA sequence diversity. Such a reduction of mtDNA sequence diversity, as stated above, could not be observed.

We analyzed seven Y-chromosome STRs in the same Central Asian samples for which we had mtDNA sequences and obtained complete haplotypes for 172 individuals (Table 30.1). We observed 90 different haplotypes, although in very different numbers across populations. The Uighurs presented 36 different haplotypes in 39 individuals, the 41 lowland Kirghiz exhibited 25 different haplotypes, while 49 Kazakhs had 18 different haplotypes and 43 highland Kirghiz had 11 different haplotypes (Fig. 30.3). The latter presented a given haplotype at a high frequency (67.4 per cent), which was also present in 19.5 per cent of the lowland Kirghiz. The most prevalent haplotype in the Kazakh had four differences in repeat size from that in the highland Kirghiz and was found at a frequency of 51.0 per cent. Haplotype diversities, which are computed in the same way as sequence diversities, are given in Table 30.1. Highland Kirghiz ($D = 0.545$) and Kazakhs ($D = 0.738$) had significantly lower diversities than lowland Kirghiz ($D = 0.955$) and Uighurs ($D = 0.995$). AMOVA showed that 20.6 per cent of the Y-chromosome genetic variation could be attributed to differences among populations (remember that the same value was 0.5 per cent for mtDNA sequences). If we grouped populations by altitude, the fraction of the genetic variance attributable to altitude was –6.0 per cent (not different from zero). However, when we grouped the populations according to language in Kazakhs, Kirghiz and Uighur, language could explain 16.2 per cent of the Y-chromosome genetic variance. In summary, the Y-chromosome STR haplotypes showed a reduction of within-population genetic diversity and clear differences among populations that could be accounted for by language but not by altitude.

The pattern of genetic diversity we observed in male (Y-chromosome) lineages clearly bears the imprint of the reduction of genetic diversity caused by the settlement of high altitude habitats. On the contrary, this reduction is not observed in the female (mtDNA) lineages. A quick inspection of Figure 30.3 clearly shows that highland Kirghiz could have been founded by lowland Kirghiz through a random sample on which the founder effect had a deep action. The same could have happened in Kazakhs in relation to a source population with a genetic composition similar to the Uighur.

Male and female migration patterns

Genetic diversity does not depend directly on the number of individuals in the population but on those that contribute their genes to the next generation (i.e. the effective population size or Ne). Several demographic and social factors could act to reduce the proportion between the male and female effective population sizes and thus explain the discrepancies found between male and female genetic lineages: female lineages are homogeneous across populations, whereas male lineages are not. One of these social factors is polygyny; the populations we studied were traditionally Moslems. However, an ethnographical survey undertaken in 1885 (White 1988) showed that roughly the top 10 per cent of the wealthiest men were polygynous. The average number of wives per polygynous man was not reported, but, if the worldwide average (2.6) is applied, the resulting value is insufficient to explain the relative male to female effective population sizes. Higher prereproductive male mortality (which would also reduce the male effective population size) was not observed in recent decades. The high between-population genetic differentiation observed for Y chromosomes and the homogeneity of mtDNA sequences across populations is compatible with a higher female than male

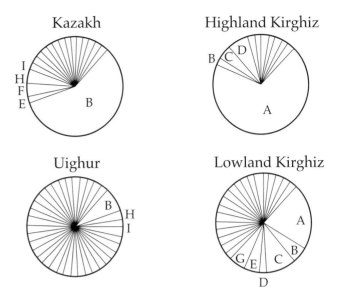

Figure 30.3. *Y-chromosome haplotype frequency distribution in four Central Asian populations. Sectors are proportional to the frequency of each different haplotype found in each population. Only haplotypes shared by more than one population are labelled (Pérez-Lezaun et al. 1999).*

migration between those populations. This has been observed for other populations, such as the Bedouins (Salem *et al.* 1996) and it may be a global pattern (Seielstad *et al.* 1998), linked to the widely diffused social precept of marriage patrilocality (that is, when it is customary that the bride moves to the groom's village). A higher female migration rate could have replenished the genetic diversity lost in the foundation event of highland populations.

In summary, through the analysis of uniparental lineages, we have investigated both the ancient roots of the Central Asian populations and its recent history. We have shown that those populations may have had an admixed origin and that they are unlikely to have been the source for anatomically modern humans in Europe. The traces left by the colonization of high altitude habitats and by a higher female migration rate on the genomes of those populations have also been recognized. This is just an example of how the genetic analysis of extant populations can be used to reconstruct past events.

Acknowledgements

We are grateful to Davide Pettener (University of Bologna, Italy) for having provided us both the Central Asian samples and the questions in their history that genetic analysis can investigate. Funding for this research was provided by Dirección General de Investigación Científico-Técnica (Spain) grants PB95-0267-C02-01 and PB98-1046.

References

Anthony, D.W. & D.R. Brown, 1991. The origins of horseback riding. *Antiquity* 65, 22–38.

Anthony, D.W. & I.P. Vinogradov, 1995. Birth of the chariot. *Archaeology* 28, 36–41.

Awadalla, P., A. Eyre-Walker & J. Maynard Smith, 1999. Linkage disequilibrium and recombination in hominid DNA. *Science* 286, 2524–5.

Bowles, G.T., 1977. *The Peoples of Asia*. London: Weidenfeld & Nicholson.

Calafell, F., P. Underhill, A. Tolun, D. Angelicheva & L. Kalaydjieva, 1996. From Asia to Europe: mitochondrial DNA sequence variability in Bulgarians and Turks. *Annals of Human Genetics* 60, 35–49.

Comas, D., F. Calafell, E. Mateu, A. Pérez-Lezaun & J. Bertranpetit, 1996. Geographic variation in human mtDNA control region sequence: the population history of Turkey and its relationship to European populations. *Molecular Biology and Evolution* 13, 1067–77.

Comas, D., F. Calafell, E. Mateu, A. Pérez-Lezaun, E. Bosch, R. Martínez-Arias, J. Clarimón, F. Fachini, G. Fiori, D. Luiselli, D. Pettener & J. Bertranpetit, 1998. Trading genes along the silk road: mtDNA sequences and the origin of Central Asian populations. *American Journal of Human Genetics* 63, 1824–38.

Excoffier, L., P.E. Smouse & J.M. Quattro, 1992. Analysis of molecular variance inferred from metric distances among DNA haplotypes: application to human mitochondrial DNA restriction data. *Genetics* 131, 479–91.

Harpending, H.C., S.T. Sherry, A.R. Rogers & M. Stoneking, 1993. The genetic structure of ancient human populations. *Current Anthropology* 34, 483–96.

Horai, S., K. Murayama, K. Hayasaka, S. Matsubayashi, Y. Hattori, G. Fucharoen, S. Harihara, K.S. Park, K. Omoto & I-H. Pan, 1996. MtDNA polymorphism in East Asian populations, with special reference to the peopling of Japan. *American Journal of Human Genetics* 59, 579–90.

Kolman, C.J., N. Sambuughin & E. Bermingham, 1996. Mitochondrial DNA analysis of Mongolian populations and implications for the origin of New World founders. *Genetics* 142, 1321–34.

Macaulay, V., M. Richards, E. Hickey, E. Vega, F. Cruciani, V. Guida, R. Scozzari, B. Bonné-Tamir, B. Sykes & A. Torroni, 1999. The emerging tree of West Eurasian mtDNAs: a synthesis of control-region sequences and RFLPs. *American Journal of Human Genetics* 64, 232–49.

Mateu, E., D. Comas, F. Calafell, A. Pérez-Lezaun & J. Bertranpetit, 1997. A tale of two islands: population history and and mitochondrial DNA sequences of Bioko and São Tomé, Gulf of Guinea. *Annals of Human Genetics* 61, 507–18.

Otte, M., 1996. Le bouleversement de l'humanité en Eurasie vers 40.000 ans, in *The Last Neandertals, the First Anatomically Modern Humans: Cultural Change and Human Evolution. The Crisis at 40,000 BP*, eds. E. Carbonell & M. Vaquero. Tarragona: URV, 95–106.

Pérez-Lezaun, A., F. Calafell, D. Comas, E. Mateu, E. Bosch, R. Martínez-Arias, J. Clarimón, G. Fiori, D. Luiselli, F. Facchini, D. Pettener & J. Bertranpetit, 1999. Sex-specific migration patterns in Central Asian populations revealed by the analysis of Y-chromosome STRs and mtDNA. *American Journal of Human Genetics* 65, 208–19.

Richards, M., H. Côrte-Real, P. Forster, V. Macaulay, H. Wilkinson-Herbots, A. Demaine, S. Papiha, R. Hedges, H-J. Bandelt & B. Sykes, 1996. Paleolithic and Neolithic lineages in the European mitochondrial gene pool. *American Journal of Human Genetics* 59, 185–203.

Salem, A-H., F.M. Badr, M.F. Gaballah & S. Pääbo, 1996. The genetics of traditional living: Y-chromosomal and mitochondrial lineages in the Sinai Peninsula. *American Journal of Human Genetics* 59, 741–3.

Sarianidi, V., 1992. Food-producing and other Neolithic communities in Khorasan and Transoxania: Eastern Iran, Soviet Central Asia and Afghanistan, in *History and Civilizations of Central Asia*, vol. I: *The Dawn of Civilization: Earliest Times to 700 BC*, eds. A.H. Dani & V.M. Masson. Paris: Unesco, 109–26.

Seielstad, M.T., E. Minch & L.L. Cavalli-Sforza, 1998. Genetic evidence for a higher female migration rate in humans. *Nature Genetics* 20, 278–80.

Sellier, J. & A. Sellier, 1993. *Atlas des peuples d'Orient*. Paris: Editions de la Découverte.

White, D.R., 1988. Rethinking polygyny: co-wives, codes and cultural systems. *Current Anthropology* 29, 529–58.

Chapter 31

An Indian Ancestry: a Key for Understanding Human Diversity in Europe and Beyond

Toomas Kivisild, Surinder S. Papiha, Siiri Rootsi, Jüri Parik,
Katrin Kaldma, Maere Reidla, Sirle Laos, Mait Metspalu,
Gerli Pielberg, Maarja Adojaan, Ene Metspalu, Sarabjit S. Mastana,
Yiming Wang, Mukaddes Gölge, Halil Demirtas,
Eckart Schnakenberg, Gian Franco de Stefano, Tarekegn Geberhiwot,
Mireille Claustres & Richard Villems

A recent African origin of modern humans, although still disputed, is supported now by a majority of genetic studies. To address the question when and where very early diversification(s) of modern humans outside of Africa occurred, we concentrated on the investigation of maternal and paternal lineages of the extant populations of India, southern China, Caucasus, Anatolia and Europe. Through the analyses of about 1000 mtDNA genomes and 400 Y chromosomes from various locations in India we reached the following conclusions, relevant to the peopling of Europe in particular and of the Old World in general. First, we found that the node of the phylogenetic tree of mtDNA, ancestral to more than 90 per cent of the present-day typically European maternal lineages, is present in India at a relatively high frequency. Inferred coalescence time of this ancestral node is slightly above 50,000 BP. Second, we found that haplogroup U is the second most abundant mtDNA variety in India as it is in Europe. Summing up, we believe that there are now enough reasons not only to question a 'recent Indo-Aryan invasion' into India some 4000 BP, but alternatively to consider India as a part of the common gene pool ancestral to the diversity of human maternal lineages in Europe. Our results on Y-chromosomal diversity of various Indian populations support an early split between Indian and east of Indian paternal lineages, while on a surface, Indian (Sanskrit as well as Dravidic speakers) and European Y-chromosomal lineages are much closer than the corresponding mtDNA variants.

The title of this chapter was not chosen in order to insist that India was the place from where Europe was colonized by modern humans some 40,000–50,000 years ago. It does, however, imply that the understanding of population genetics of contemporary Indians is useful and needed in the attempt to reconstruct the process of the out-of-Africa spread of modern humans.

In our analysis below we largely restrict ourselves to the two sex-linked genetic systems: mtDNA and the Y chromosome. There is also extensive earlier literature on the genetics of Indian populations, based on the analysis of the distribution and frequencies of classical genetic markers (for reviews see Cavalli-Sforza *et al.* 1994; Papiha 1996). Yet the mere size of the Indian population, exceeding 1 billion (and much more if we include Pakistan and Bangladesh, i.e. the whole sub-continent) and its ethnic diversity make it clear that to comprehend the richness of the gene pool of Indians long-lasting systematic efforts are needed. Here we address a number of problems relevant to the placement of Indian genetic lineages within the context of the rest of Eurasia, taking the recent out-of-Africa colonization of the world by modern humans as a starting point.

The root

Most of the archaeological, anthropological and genetic evidence on modern humans (reviewed recently by Disotell 1999; Foley 1998; Harpending et al. 1998; Jorde et al. 1998) supports their recent origin in and spread out from Africa (Stringer & Andrews 1988). While this is generally accepted (but see also Wolpoff 1999), various alternative pathways and modes of this dispersal have been suggested (see, for example, Chu et al. 1998; Hammer et al. 1997; 1998; Jin et al. 1999; Kivisild et al. 1999a; Lahr & Foley 1994; 1998; Templeton 1997).

A route or routes

The peopling of Asia can be interpreted in terms of one or several pathways and also in terms of one, two or multiple migrations out of Africa. The third variable is time. The actual situation becomes even more complicated if there were migrations back to Africa, in between outward migration waves. First the alternative extremes: i) the northern route, over Sinai, leading to eastern Asia through the steppes of Central Asia and southern Siberia and ii) the southern route over southern Arabia, followed by the migration along the coastline of southern Asia (Fig. 31.1). While the northern route model could explain the peopling of the whole Eurasia by a single migration from Africa, the southern route model is interpreted as implying at least two separate Late Pleistocene dispersal events, one leading to the northwest and the other to the east of Eurasia (Cavalli-Sforza et al. 1994; Lahr & Foley 1994). On the first hypothesis, the molecular trees of present-day Eurasian populations are not expected to split at the depth of the time of their coalescence back to Africans. This means that the geographic sequestration can well be invisible in deeper branches and would start to appear in the terminal branches. This is because the eastern and western Eurasian populations should share the basic branches that were already present in the initial population that left Africa. Archaeologically, this model is expected to be expressed in largely uniform transitions in Palaeolithic tool technology.

In contrast, the multiple dispersal models would predict the existence of more than one set of basic genetic lineage groups present in Eurasians. In archaeological terms this model does not foresee uniformity of technology associated with different waves of dispersal of modern humans (Foley & Lahr 1997).

Indians as a key

The geographic position of the Indian populations makes them a good example to study the question of the early dispersal of our ancestors. Unfortunately, it is not clear yet when the subcontinent became inhabited by modern humans. Archaeological and palaeoanthropological records are scanty and limited in details. The time period around 30,000–50,000 BP, when the first signs of modern humans can be traced in Eurasia (Smith et al. 1999) has revealed both Middle (up to around 20,000 BP) and Upper Palaeolithic (starting from c. 30,000 BP) tool assemblages in India (Joshi 1996). So far the earliest fragmentary skeletal evidence (at c. 34,000 C14 BP) of anatomically modern humans comes from Sri Lanka (Deraniyagala 1998). Note that this island was at that time connected with the continent.

In genetic distance trees based on classical genetic markers Indians cluster more closely with western Eurasian populations than with either other Asians or Africans (Cavalli-Sforza et al. 1994). This clustering supports the traditional classification of Caucasoids, according to which most Indian populations are included within this terminologically somewhat unsatisfactory, but widely used grouping, covering linguistically Indo-European-, Finno-Ugric-, Caucasian- and Hamito-Semitic- (Afro-Asiatic) speaking populations plus a few outliers, notably the Basques.

Two major waves of migrations into India have been proposed to account for this greater similarity of Indians with western Eurasians than with Mongoloid people to the east of India. The more widely known scenario is an invasion of nomadic Indo-Aryan tribes around 4000 BP either from the west or from the Central Asian steppes in the north. The other, more recently proposed hypothesis is based on the fact that some 8000–9000 years ago several varieties of wheat and other cereals reached India, presumably from the Fertile Crescent. This hypothesis is also supported by linguistically based suggestions of a recent common root for Elamite and Dravidic languages (Diamond 1997; Renfrew 1989).

We stress that these two hypotheses, which are not mutually exclusive, leave completely open the history of the 'indigenous', pre-Neolithic inhabitants of India, and the question of their contribution to the gene pool of the contemporary Indians. Were they largely replaced by much more recent immigrants or alternatively, was the result of the recent migrations insignificant genetically while perhaps still profound culturally? There are currently about 500 tribal populations scattered over the Indian peninsula and generally thought of as the survivals of the pre-Neolithic Indians (Cavalli-Sforza et al. 1994; Papiha 1996). These tribal populations make up only a minor fraction of the total population of the present-day Indians. Study of their genetic identity would allow one to ask an important question: Do the tribal populations possess genetic lineages absent or rare among the other Indian populations or are they largely genetically identical to the latter, particularly if the differences attributable to drift can be clearly distinguished?

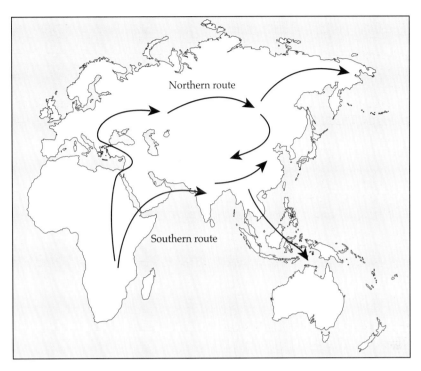

Figure 31.1. *Two alternative routes for out-of-Africa migration.*

Our work: questions, populations and methods

The hypotheses set out above are the basis of our research in the form of two questions. The first is in the placement of the genetic lineages of Indians in the global context. We are trying to understand, in both qualitative and quantitative terms, the extent of the overlap of the genetic lineages found in India with those found elsewhere. Second, provided that such lineage clusters can be reliably reconstructed, can we establish when they diverge from those found elsewhere? Clearly the two questions are connected and the answers we look for depend on a detailed general knowledge about the genetic structure of the other Eurasians as well, and, at least partially, of the Africans.

So far we have analyzed 19 different Indian populations, covering the subcontinent geographically from Punjab and Kashmir to Sri Lanka and western Bengal. For a wider context, we have also investigated southern Chinese, Anatolian, Trans-Caucasian and eastern European populations. Our project has also included different Ethiopian populations.

Two principal experimental approaches have been used: i) the first hypervariable segment of the control region of mtDNA was sequenced and 15 diagnostic RFLP markers were typed from the coding region for haplogroup affiliation; ii) a selection of bi-allelic Y-chromosomal markers were studied together with STR loci. Phylogenetic analysis was performed following the method of reduced median networks (Bandelt *et al.* 1995) using maximum parsimony as the guiding principle. The general schemes of mtDNA and Y-chromosome haplogroups are presented in Figure 31.2.

Mitochondrial DNA

Previous studies have revealed that mtDNA lineage clusters (haplogroups) are specific to large geographic areas (Chen *et al.* 1995; Torroni *et al.* 1996; Wallace 1995). For example, nine mtDNA haplogroups (H, I, J, K, T, U, V, W and X) comprise about 95 per cent of the western Eurasian mtDNA pool, including Mediterranean Africa, whereas haplogroups M, B, F and A are specific for Mongoloid populations. The sub-Saharan African mtDNAs belong largely to a mtDNA supercluster L, further divided as L1, L2 and L3 (Watson *et al.* 1997). L3 is at the root of nearly all mtDNA diversity found outside sub-Saharan Africa.

Haplogroup M as a cluster of the proto-Asian maternal lineages
The most frequent mtDNA cluster found among Indian populations is haplogroup M (Bamshad *et al.*

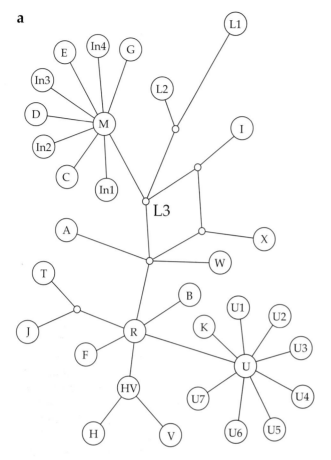
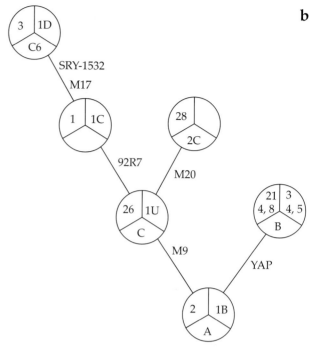

Figure 31.2. *An unrooted tree relating mtDNA (a), according to Torroni et al. 1996 and Chen et al. 1995, and Y-chromosomal (b) haplogroups. (The latter is adapted from (clockwise, starting from noon): Karafet et al. 1999; Previdere et al. 1999; Underhill et al. 1997.)*

1997; Kivisild *et al.* 1999b; Passarino *et al.* 1996). This Asian-specific lineage cluster is widespread among Mongoloid populations (Kolman *et al.* 1996; Torroni *et al.* 1993b; 1994) as well as among Native Americans (e.g. Bonatto & Salzano 1997; Torroni *et al.* 1992; 1993a). Its frequency drops in southeastern Asians, where haplogroups B and F become dominant varieties of mtDNA (Ballinger *et al.* 1992; Wallace 1995). We have observed that the frequency of haplogroup M among Indians is even higher than among eastern Asians (Table 31.1), ranging from 40 per cent in Punjabis to almost a fixation level among Chenchus, a tribal hunter-gatherer community in Andhra Pradesh (97 per cent; $n = 96$).

There is an important finding, having possibly profound implications for the interpretation of the peopling of southern and eastern Asia, and allowing a better understanding of the genetic background of the populations living in Central Asia. This is that the Indian haplogroup M sub-structure (sub-clusters In1–In4 in Fig. 31.2) differs profoundly from that observed in Mongoloid populations (Kivisild *et al.* 1999b). Two conclusions are immediately apparent:

i) the expansion of the super-haplogroup M in Indians and Mongoloids, including the genesis of the region-specific sub-clusters, was clearly separated in space; ii) since then, there has been only very limited gene flow between India and eastern Asia. When did this separation happen? The haplogroup M coalescence time for eastern Asians has been estimated as around 55,000–73,000 BP (Chen *et al.* 1995). Coalescence calculations of Indian haplogroup M diversity (Kivisild *et al.* 1999b; Passarino *et al.* 1996) agree well with this estimate and suggest that the two macro-populations started to expand simultaneously, perhaps due to improved climate between the two major glaciations at about 73,000–65,000 and 24,000–16,000 years ago. The lack of any signs of significant re-migrations of eastern Asians to India is further supported by the scarcity of mtDNA lineages belonging to haplogroups A, B and F in India.

The fact that the Indian and eastern Asian/Mongoloid varieties of haplogroup M are so distinct also allows one to interpret the haplogroup M lineages found in Central Asian populations phylogeographically: virtually all of those described so far belong to the Mongoloid-specific branches of this haplogroup. This again tells us that no large-scale migration from Central Asia to India has occurred, at least any involving the presently Turkish-speaking populations of this area, among whom the frequency of haplogroup M is otherwise close to that in India and in eastern Asians.

Haplogroup U and the proto-western-Eurasian maternal lineages

The absence of haplogroup M in Europeans, compared to its equally high frequency among Indians, eastern Asians and in some Central Asian populations is inconsistent with the 'general Caucasoidness' of Indians. Any relationship between Indians and 'Caucasoids' must therefore be based on qualitative and quantitative data on genetic markers common to Europeans and Indians. Analyzing Indian maternal lineages further, we found that an extensive overlap was provided by haplogroup U. The distribution of haplogroup U (Table 31.1) is a mirror image of that for haplogroup M: the former has not been described so far among eastern Asians but is frequent in European populations as well as among Indians (Kivisild *et al.* 1999a). This reverse analogy goes further: Indian U lineages differ substantially from those observed in Europe and their coalescence to a common ancestor, like that for the haplogroup M lineages, dates back to about 50,000 years (Kivisild *et al.* 1999b). We infer from the fact that Indians and other populations do not generally share mtDNA lineages at the tips of the branches of the global phylogenetic tree with either eastern or western Eurasians that the Indian maternal gene pool has come largely through an autochtonous history since the Late Pleistocene.

The extent of a recent admixture in the Indian mtDNA pool

Our results suggest that the sum of any recent (the last 15,000 years) western mtDNA gene flow to India comprises, in average, less than 10 per cent of the contemporary Indian mtDNA lineages (Kivisild *et al.* 1999a). This fraction clusters closely together with the tips of the western Eurasian mtDNA tree, is higher among western Indian populations (Punjabis) and drops lower in most of the tribal populations studied by us. However, even the high castes share more than 80 per cent of their maternal lineages with the lower castes and tribals. We conclude that the recent enrichment of the Indian mtDNA pool with the western-Eurasian lineages is clearly detectable but had a relatively minor impact.

Table 31.1. *MtDNA haplogroup U and M frequencies worldwide.*

	n	U (%)	M (%)	
W-Europe	172	23.9	0.6	(our data)
E-Europe	329	23.7	0.6	(our data)
The Caucasus	532	23.6	1.5	(our data)
Anatolia	379	24.5	3.9	(our data)
C-Asia	195	8.7	41	(deduced from Comas *et al.* 1998)
India	1061	13	60.1	(our data)
S-China	69	0	37.7	(our data)
NW-Africa	268	13.7	0.6	(Rando *et al.* 1998)
E-Africa	199	2.5	8	(our data)

The Indian Y-chromosomal lineages

East–west affinities

Unlike with the Indian maternal lineages, Indian Y chromosomes belong mainly to the same broad haplogroups present among modern western Eurasians (our unpublished data). In contrast, haplogroup 26 that is known to be highly frequent in eastern and southeastern Asians (Karafet *et al.* 1999), is very rare in Indians. At the same time, the Indian Y-chromosomal pool is rich in derivatives of haplogroup 26, such as 1 and 3 (1C and 1D in Karafet *et al.* 1999, respectively), which are, in turn, rare in the Chinese population.

However, an apparent lack of YAP+ chromosomes in Indians (Karafet *et al.* 1999; Bhattacharyya *et al.* 1999; Santos *et al.* 1999; Thangaraj *et al.* 1999; Underhill *et al.* 1997; our unpublished results) complicates the interpretation of largely western connection of the Indian Y chromosomes. It is well documented that in the European populations haplogroup 21 chromosomes harbouring the Alu insert are widespread, albeit generally at rather low frequency. The occurrence of the insert is more frequent in localities closer to India: the populations of the Middle East and the Turks carry this marker at frequencies about of 10 per cent or higher (Hammer 1994; Skorecki *et al.* 1997; Thomas *et al.* 1998). Yet Indians lack the YAP+ varieties, with the only exceptions being groups documented as being recent immigrants, the Siddis (Thangaraj *et al.* 1999) and Parsees (our unpublished observation). Several demographic scenarios may account for this discrepancy, including the possibility that the common pool of Y chromosomes shared by the western Eurasian and Indian populations lacked YAP+ chromosomes at the time of the split of the two proto-macropopulations.

In interpreting the Y-chromosomal data, we

must first recognize that certain haplogroups, frequent both among Europeans and Indians, are in fact complex aggregates of lineages deriving from central internal nodes of the Y-chromosomal phylogenetic tree. In the extreme case, haplogroup 2 is just an assemblage of lineages that so far cannot be assigned to any terminal node of the tree. Secondly, male and female genetic lineages might have been differently affected by gene flow, genetic drift or bottlenecks in their complex histories.

Returning to the beginning: almost certainly several routes

The main empirical conclusions from our results on Indians using M-haplogroup data are as follows:
1. MtDNA haplogroup M is, as a first approximation, equifrequent in populations living in three wide geographic areas: in southern, central and eastern Asia (+ Native Americans).
2. At the same time, the central and eastern Asian haplogroup M lineages belong to one, and the southern Asian lineages to another subset of haplogroup M.
3. The node — M* — is present both in southern and eastern Asia and the observed diversity implies a late Pleistocene origin.
4. Haplogroup M is virtually absent in western Eurasia and the lineages found so far in Africa belong all to a narrow sub-cluster, phylogenetically far apart from the ancestral nodal sequence.

Thus, the empirical results obtained so far clearly separate southern and eastern Asian populations from the western Eurasians through the sharing of haplogroup M. Its distribution combined with the age of its origin and diversification is best explained by a separate Late Pleistocene migration wave, most probably via the 'southern route'.

We note that the spread of haplogroup M to Eurasia by the southern route was during the final preparation of the manuscript suggested by an additional evidence relating African and Indian mtDNAs (Quintana-Murci et al. 1999).

The second set of conclusions is based on the phylogeography of the mtDNA haplogroup U:
1. Haplogroup U is absent in eastern Asian populations.
2. Haplogroup U is the second most frequent variety of mtDNA in India and in western Eurasia.
3. Indian and western Eurasian haplogroup U varieties differ profoundly; the split has occurred about as early as the split between the Indian and eastern Asian haplogroup M varieties. The data show that both M and U exhibited an expansion phase some 50,000 years ago, which should have happened after the corresponding splits.
4. Haplogroup U frequency is low in Africa with one dominant African-specific variety — U6.

These observations make it unlikely that the immediate ancestors of the carriers of haplogroups U and M left Africa as a common wave of migrants. We suggest that the ancestors of haplogroup U carriers used the 'northern route' because of U is present in both Indian and western Eurasian populations. Yet both these migrations seem to have occurred at comparable time depths.

The third set of conclusions build a bridge over Eurasia:
1. Topologically speaking, all major Caucasoid-specific mtDNA haplogroups (H, T, J and U) as well as the two major eastern Asian-specific haplogroups B and F, derive from a common internal node R*.
2. It seems that the only regions in Eurasia (and, indeed, globally) where one finds a large variety of the 'non-canonical' derivatives of this node are India and southern China. However, no data exist about the populations between these regions.
3. Nowhere is there any extensive overlap of western and eastern Eurasian mtDNA lineages except in Central Asian populations and Indians.

Our tentative general conclusion from the third list of conclusions is that the carriers of the R* node reached India very early and migrated eastwards, possibly together with the carriers of M*. As for M*, the expansion phase came later, leading in southeastern Asia to the formation of two major lineage clusters, F and B. In India, a large variety of derivatives of R* arose, while in western Eurasia its major derivatives that survived consist of haplogroups H, T, J and U. Among the latter, U seems to be the most ancient, explaining why one of its deepest branches is widely present in and specific for Indians.

For Y-chromosomal markers, we are limited to three empirical conclusions:
1. There is almost no direct overlap between the Indian and southeastern Asian (southern Chinese in our experiments) Y chromosomes.
2. The Y-chromosomal haplogroups present in Indians are also frequent in western Eurasians.
3. Not all Y-chromosomal haplogroups typical for western Eurasians are frequent in Indians.

Thus, we have observed a significant difference between the spread of Y-chromosomal and mtDNA markers in southern and southeastern Asians but we are not yet in a position to suggest whether these differences are due to differential migrational patterns or, rather, due to profound differences in the demographic histories of the sex-specific genes, where the survival of the male lineages have undergone severe bottlenecks in some lineages and rapid bursts in others.

Returning now to the expansion of modern humans out of Africa, we suggest a scheme as shown in Figure 31.3. We regard it as minimalist, because many aspects known already are omitted, such like the sharing of mtDNA haplogroup U7 between Indian and Anatolian populations, lack of YAP+ Y chromosomes in India etc. However, data are lacking from such critically important regions as Afghanistan, Iran, Iraq, Myanmar, Thai etc. The ever-improving selection of genetic markers, combined with improvements in the understanding of the driving forces of demographic behaviour of humans and their prehistory and history in general, allow us to expect further progress in our reconstructions of the past. However, despite its shortcomings, we hope that this article shows that study of the Indian mtDNA and Y-chromosomal lineages sheds new light on the general problem of the peopling of the Old World by modern humans.

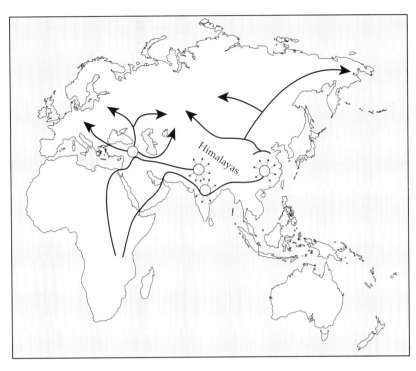

Figure 31.3. *Scheme combining the routes and expansions of modern human early dispersals.*

Acknowledgements

We thank Professor P. Broda and Dr P. Forster for many helpful suggestions in the preparation of this manuscript.

References

Ballinger, S.W., T.G. Schurr, A. Torroni, Y.Y. Gan, J.A. Hodge, K. Hassan, K.H. Chen & D.C. Wallace, 1992. Southeast Asian mitochondrial DNA analysis reveals genetic continuity of ancient mongoloid migrations. *Genetics* 130, 139–52.

Bamshad, M.B., B. Rao, J.M. Naidu, B.V.R. Prasad, S. Watkins & L.B. Jorde, 1997. Response to Spurdle *et al*. *Human Biology* 69, 432–5.

Bandelt, H.J., P. Forster, B.C. Sykes & M.B. Richards, 1995. Mitochondrial portraits of human populations using median networks. *Genetics* 141, 743–53.

Bhattacharyya, N.P., P. Basu, M. Das, S. Pramanik, R. Banerjee, B. Roy *et al.*, 1999. Negligible male gene flow across ethnic boundaries in India, revealed by analysis of Y-chromosomal DNA polymorphisms. *Genome Research* 9, 711–19.

Bonatto, S.L. & F.M. Salzano, 1997. A single and early migration for the peopling of the Americas supported by mitochondrial DNA sequence data. *Proceedings of the National Academy of Sciences of the USA* 94, 1866–71.

Cavalli-Sforza, L.L., P Menozzi & A. Piazza, 1994. *The History and Geography of Human Genes.* Princeton (NJ): Princeton University Press.

Chen, Y.S., A. Torroni, L. Excoffier, A.S. Santachiara-Benerecetti & D.C. Wallace, 1995. Analysis of mtDNA variation in African populations reveals the most ancient of all human continent-specific haplogroups. *American Journal of Human Genetics* 57, 133–49.

Chu, J.Y., W. Huang, S.Q. Kuang, J.M. Wang, J.J. Xu, Z.T. Chu *et al.*, 1998. Genetic relationship of populations in China. *Proceedings of the National Academy of Sciences of the USA* 95, 11,763–8.

Comas, D., F. Calafell, E. Mateu, A. Pérez-Lezaun, E. Bosch, R. Martínez-Arias, J. Clarimón, F. Fachini, G. Fiori, D. Luiselli, D. Pettener & J. Bertranpetit, 1998. Trading genes along the silk road: mtDNA sequences and the origin of Central Asian populations. *American Journal of Human Genetics* 63, 1824–38.

Deraniyagala, S.U., 1998. Pre- and protohistoric settlement in Sri Lanka, in *XIII U.I.S.P.P. Congress, Forli*, vol. V, eds. G. Bermond Montanari, R. Francovitch, F. Mori, P. Pensabene, S. Salvatori, M. Tosi & C. Peretto. Forlì: A.B.A.C.O. s.r.l.,

Diamond, J., 1997. *Guns, Germs and Steel: the Fates of Human Societies*. London: Jonathan Cape.
Disotell, T.R., 1999. Human evolution: sex-specific contributions to genome variation. *Current Biology* 9, R29–31.
Foley, R., 1998. The context of human genetic evolution. *Genome Research* 8, 339–47.
Foley, R.A. & M.M. Lahr, 1997. Mode 3 technologies and the evolution of modern humans. *Cambridge Archeological Journal* 7(1), 3–36.
Hammer, M.F., 1994. A recent insertion of an alu element on the Y chromosome is a useful marker for human population studies. *Molecular and Biological Evolution* 11, 749–61.
Hammer, M.F., A.B. Spurdle, T. Karafet, M.R. Bonner, E.T. Wood, A. Novelletto et al., 1997. The geographic distribution of human Y chromosome variation. *Genetics* 145, 787–805.
Hammer, M.F., T. Karafet, A. Rasanayagam, E.T. Wood, T.K. Altheide, T. Jenkins et al., 1998. Out of Africa and back again: nested cladistic analysis of human Y chromosome variation. *Molecular and Biological Evolution* 15, 427–41.
Harpending, H.C., M.A. Batzer, M. Gurven, L.B. Jorde, A.R. Rogers & S.T. Sherry, 1998. Genetic traces of ancient demography. *Proceedings of the National Academy of Sciences of the USA* 95, 1961–7.
Jin, L., P.A. Underhill, V. Doctor, R.W. Davis, P. Shen, L.L. Cavalli-Sforza et al., 1999. Distribution of haplotypes from a chromosome 21 region distinguishes multiple prehistoric human migrations. *Proceedings of the National Academy of Sciences of the USA* 96, 3796–800.
Jorde, L.B., M.B. Bamshad & A.R. Rogers, 1998. Using mitochondrial and nuclear DNA markers to reconstruct human evolution. *Bioessays* 20, 126–36.
Joshi, R.V., 1996. South Asia in the period of *Homo sapiens neanderthalensis* and contemporaries (Middle Palaeolithic), in *History of Humanity*, vol. I, eds. S.J. De Laet. UK: Clays Ltd. St Ives plc., 162–4.
Karafet, T.M., S.L. Zegura, O. Posukh, L. Osipova, A. Bergen, J. Long, D. Goldman, W. Klitz, S. Harihara, P. de Knijff, V. Wiebe, R.C. Griffiths, A.R. Templeton & M.F. Hammer, 1999. Ancestral Asian source(s) of new world Y-chromosome founder haplotypes. *American Journal of Human Genetics* 64, 817–31.
Kivisild, T., M.J. Bamshad, K. Kaldma, M. Metspalu, E. Metspalu, M. Reidla et al., 1999a. Deep common ancestry of Indian and western Eurasian mtDNA lineages. *Current Biology* 9, 1331–4.
Kivisild, T., K. Kaldma, M. Metspalu, J. Parik, S.S. Papiha & R. Villems, 1999b. The place of the Indian mitochondrial DNA variants in the global network of maternal lineages and the peopling of the Old World, in *Genomic Diversity*, eds. R. Deka & S.S. Papiha. New York (NY): Kluwer/Academic/Plenum Publishers, 135–52.
Kolman, C., N. Sambuughin & E. Bermingham, 1996. Mitochondrial DNA analysis of Mongolian populations and implications for the origin of New World founders. *Genetics* 142, 1321–34.
Lahr, M. & R. Foley, 1994. Multiple dispersals and modern human origins. *Evolutionary Anthropology* 3, 48–60.
Lahr, M. & R. Foley, 1998. Towards a theory of modern human origins: geography, demography, and diversity in recent human evolution. *American Journal of Physical Anthropology* supplement, 137–76.
Papiha, S.S., 1996. Genetic variation in India. *Human Biology* 68, 607–28.
Passarino, G., O. Semino, L.F. Bernini & A.S. Santachiara-Benerecetti, 1996. Pre-Caucasoid and Caucasoid genetic features of the Indian population, revealed by mtDNA polymorphisms. *American Journal of Human Genetics* 59, 927–34.
Previdere, C., L. Stuppia, V. Gatta, P. Fattorini, G. Palka & C. Tyler-Smith, 1999. Y-chromosomal DNA haplotype differences in control and infertile Italian subpopulations. *European Journal of Human Genetics* 7, 733–6.
Quintana-Murci, L., O. Semino, H-J. Bandelt, G. Passarino, K. McElreavey & A.S. Santachiara-Benerecetti, 1999. Genetic evidence of an early exit of *Homo sapiens sapiens* from Africa through eastern Africa. *Nature Genetics* 23, 437–41.
Rando, J.C., F. Pinto, A.M. Gonzalez, M. Hernandez, J.M. Larruga, V.M. Cabrera & H-J. Bandelt, 1998. Mitochondrial DNA analysis of northwest African populations reveals genetic exchanges with European, near-eastern and sub-Saharan populations. *Annals of Human Genetics* 62, 531–50.
Renfrew, C., 1989. The origins of Indo-European languages. *Scientific American* 261, 82–90.
Santos, F.R., A. Pandya, C. Tyler-Smith, S.D. Pena, M. Schanfield, W.R. Leonard, L. Osipova, M.H. Crawford & R.J. Mitchell, 1999. The central Siberian origin for native American Y chromosomes. *American Journal of Human Genetics* 64, 619–28.
Skorecki, K., S. Selig, S. Blazer, R. Bradman, N. Bradman, P.J. Waburton et al., 1997. Y chromosomes of Jewish priests. *Nature* 385, 32.
Smith, D.G., R.S. Malhi, J. Eshleman, J.G. Lorenz & F.A. Kaestle, 1999. Distribution of mtDNA haplogroup X among Native North Americans. *American Journal of Physical Anthropology* 110, 271–84.
Stringer, C.B. & P. Andrews, 1988. Genetic and fossil evidence for the origin of modern humans. *Science* 239, 1263–8.
Templeton, A.R., 1997. Out of Africa? What do genes tell us? *Current Opinions in Genetic Development* 7, 841–7.
Thangaraj, K., G.V. Ramana & L. Singh, 1999. Y-chromosome and mitochondrial DNA polymorphisms in Indian populations. *Electrophoresis* 20, 1743–7.
Thomas, M.G., K. Skorecki, H. Ben-Ami, T. Parfitt, N. Bradman & D.B. Goldstein, 1998. Origins of Old Testament priests. *Nature* 394, 138–40.
Torroni, A., T.G. Schurr, C.C. Yang, E.J. Szathmary, R.C. Williams, M.S. Schanfield, G.A. Troup, W.C. Knowler, D.N. Lawrence, K.M. Weiss & D.C.

Wallace, 1992. Native American mitochondrial DNA analysis indicates that the Amerind and the Nadene populations were founded by two independent migrations. *Genetics* 130, 153–62.

Torroni, A., T.G. Schurr, M.F. Cabell, M.D. Brown, J.V. Neel, M. Larsen, C.M. Vullo & D.C. Wallace, 1993a. Asian affinities and continental radiation of the four founding Native American mtDNAs. *American Journal of Human Genetics* 53, 563–90.

Torroni, A., R.I. Sukernik, T.G. Schurr, Y.B. Starikorskaya, M.F. Cabell, M.H. Crawford et al., 1993b. MtDNA variation of aboriginal Siberians reveals distinct genetic affinities with Native Americans. *American Journal of Human Genetics* 53, 591–608.

Torroni, A., J.A. Miller, L.G. Moore, S. Zamudio, J. Zhuang, T. Droma et al., 1994. Mitochondrial DNA analysis in Tibet: implications for the origin of the Tibetan population and its adaptation to high altitude. *American Journal of Physical Anthropology* 93, 189–99.

Torroni, A., K. Huoponen, P. Francalacci, M. Petrozzi, L. Morelli, R. Scozzari, D. Obidu, M-L. Savontaus & D.C. Wallace, 1996. Classification of European mtDNAs from an analysis of three European populations. *Genetics* 144, 1835–50.

Underhill, P.A., L. Jin, A.A. Lin, S.Q. Mehdi, T. Jenkins, D. Vollrath et al., 1997. Detection of numerous Y chromosome biallelic polymorphisms by denaturing high-performance liquid chromatography. *Genome Research* 7, 996–1005.

Wallace, D., 1995. Mitochondrial DNA variation in human evolution, degenerative disease, and aging. *American Journal of Human Genetics* 57, 201–23.

Watson, E., P. Forster, M. Richards & H-J. Bandelt, 1997. Mitochondrial footprints of human expansions in Africa. *American Journal of Human Genetics* 61, 691–704.

Wolpoff, M.H., 1999. The systematics of *Homo*. *Science* 284, 1774–5.

Part VI

Methodologies in the Application of Molecular Genetics to Archaeology

Chapter 32

Maximum Likelihood Estimation of Genetic Diversity

Gunter Weiss

In population genetics the evolution of sequence data is frequently described by a stochastic process, known as the coalescent. While this process is mathematically well understood, likelihood-based inference of model parameters from sequence data remains a challenge due to the huge state space of the coalescent. A new sampling strategy is introduced to search the state space. Coalescent genealogies are generated repeatedly taking the ancestral information contained in the sequences into account. The likelihood of the data given each genealogy is computed and an average is built, which yields an estimate of the likelihood of the model parameters given the data. The method can be used for maximum likelihood estimation of parameters and for model comparison.

In recent years much effort has been put into collecting DNA samples from human groups classified by ethnology, linguistics, geography or medicine. The development of modern techniques in molecular biology made it possible to generate large amounts of DNA sequence data in an effective and fast way. The typical outcome of such a population study is a set of distinct DNA sequences each having a certain frequency in the sample. In the anthropological and population genetics context the interest concerns the information about human history and prehistory contained in these data. The question arises as to how to extract this information.

If the sequences in the sample were not affected by recombination and no recurrent mutations took place during the evolution of the sequences since their most recent common ancestor, the data can be displayed by a unique treelike structure (Griffiths 1987), equivalent to a maximum parsimonious tree (Fitch 1971). However, many data sets contain sites that are incompatible with these assumptions. In this case a unique tree does not exist, but the data could be represented by a median network for example (Bandelt *et al.* 1996). For data from loci, where recurrent mutations are common (like the hypervariable regions of mtDNA), a graphical representation is no longer feasible. Yet, the reduced median network approach (Bandelt *et al.* 1996) yields a *graphical interpretation* of the data. Based on this interpretation attempts have been made to determine and date demographic events of human populations (Forster *et al.* 1996; Richards *et al.* 1998).

A different approach for inferring aspects of evolutionary history is the use of *coalescence theory* (Kingman 1982a,b; Tavaré 1984; Hudson 1991). Here, the evolution of sequences in a sample is described by a stochastic process, called the *coalescent process*. This process describes, in terms of probability distributions, the ancestral relationship of sequences going backwards in time. The dynamics of this process is driven by the dynamics of the mutational and demographical processes assumed for the evolutionary history of the sampled sequences. This in turn enables one to infer population dynamics from given sequence data by means of a pure statistical estimation procedure.

In the following I will shortly describe concepts, assumptions and some features of the coalescent process and how it can be used for statistical inference. Despite the simple structure of the coalescent computational problems arise if one tries to use the complete data information in a maximum likelihood framework. One possibility for overcoming these difficulties will be outlined.

In the following description of the coalescent proc-

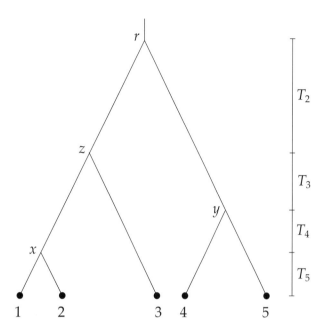

Figure 32.1. *Coalescent genealogy of a sample of five sequences.*

ess we regard, for the sake of simplicity only, the case where selection and recombination do not affect the evolution of sequences (but see Griffiths & Marjoram 1996; Neuhauser & Krone 1997; Wiuf & Hein 1999). Then, DNA sequences sampled from a current population are related by a common genealogy back to a most recent common ancestor some point in time. Along this genealogy evolution has taken place and shaped the DNA sequences via mutations into the set we observe today. The coalescent process is a mathematical description of this genealogy of say n randomly sampled sequences from a large population (see Fig. 32.1)

Tracing the ancestry back from present to past coalescent events occur, i.e. two ancestral lineages, chosen at random, share their common ancestor and are merged in the diagram. This process ends when the last pair of lineages has coalesced. The length of these times $T_n, T_{n-1}, \ldots, T_2$ are stochastic variables and their joint distribution depends on the model of population history assumed. In the simple case of a haploid Wright-Fisher population of constant size N the times T_i during which there are i distinct ancestors in the genealogy follow independent exponential distributions with parameters $i(i-1)/2, i = n, \ldots, 2$, where time is measured in units of N generations. If more complex scenarios of population history are assumed, the distribution of coalescence times can still be computed by scaling of exponential distributions according to the appropriate population size functions (Kingman 1982c; Griffiths & Tavaré 1994b).

Under the assumption of neutrally evolving sequences, the number of mutations is usually modelled by a Poisson process with rate $\theta/2$, where $\theta = 2N\mu$ is a scaled mutation rate and μ the mutation rate per sequence and generation. This parameter θ is a measure of genetic diversity and appears frequently in the context of theoretical population genetics. Therefore, several methods have been proposed to estimate this quantity from genetic data. Many estimates are based on summary statistics such as number of variable positions, mean pairwise differences, pairwise difference distributions or counts in mutational classes (Watterson 1975; Tajima 1983; Slatkin & Hudson 1991; Rogers & Harpending 1992; Fu 1998; Weiss & von Haeseler 1998). However, reducing the limited information contained in the data is not recommendable, and likelihood based methods were developed which use the complete data as information for inference (Griffiths & Tavaré 1994a,b; Kuhner *et al.* 1995; 1998). Additional to the advantageous properties of maximum likelihood estimation, likelihood based methods allow for a consequent testing of the model assumptions which can lead to further refinement steps in the data analysis. The concept of maximum likelihood methods is simple: assume we have a completely parameterized model for the evolution of the sequences in the data set. In the following we restrict ourselves to the simple case of a Wright-Fisher population of constant size where θ is the only relevant parameter. Knowing this scaled mutation rate enables one to compute the probability of observing a certain data configuration D, $P(D \mid \theta)$. In turn, regarding this probability as a function of the unknown parameter given the observed data and maximizing this function in θ yields an estimate of this parameter. The actual data D has its highest probability of being observed if this maximum likelihood estimate were the true value of the parameter. While this statistical principle is easy to formulate, a direct computation of the likelihood $P(D \mid \theta)$ is not possible in general, because the genealogical tree relating the sampled sequences is not observable. But coalescent theory provides a probability measure over the possible genealogies depending on the parameter θ. Therefore, $P(D \mid \theta)$ is computed by conditioning on the unknown genealogies g and intergrating over the space of all genealogies G:

$$P(D \mid \theta) = \int_G P(D \mid g, \theta) P(g \mid \theta) \, dg.$$

The probability $P(g|\theta)$ is determined by the joint distribution of the coalescent times given θ. The computation of $P(D|g, \theta)$ depends on an appropriate probabilistic model for the evolution of the locus under study: for example, a locus with a small mutation rate such that parallel mutations are very unlikely may be modelled by the infinitely many sites model (Watterson 1975), whereas a rapidly evolving locus with mutational hot spots like the hypervariable regions of human mtDNA deserves a rather complex model (Weiss & von Haeseler 1998; Meyer et al. 1999).

Complete integration is impossible due to the huge (infinite) size of the state space G. To overcome this problem Markov Chain Monte Carlo procedures (Kuhner et al. 1995; 1998; Wilson & Balding 1998) and importance sampling methods (Griffiths & Tavaré 1994a,b) have been suggested. Here, I will describe the key ideas of a heuristic search algorithm which was designed to sample efficiently those genealogies which contribute most to the overall likelihood. This sample can be used to approximate $P(D|\theta)$ for a specified value θ. A naïve attempt would be to simulate genealogies using the coalescent prior $P(g|\theta)$ without regarding the data. This would be very inefficient, because $P(D|g, \theta)$ will be very small for almost all of these genealogies. Therefore, data information should be taken into account, when genealogies are sampled. To this end coalescent trees are generated and evaluated as follows:

1. Set $j = n$
2. Sample T_j from an exponential distribution with parameter $j(j-1)/\theta$.
3. Compute for each of the $\binom{j}{2}$ possible pairs of sequences/subtrees the probability of coalescence at time $S_j = \sum_{i=n}^{j} T_i$. This is achieved by using a method similar to the 'pruning' algoritm (Felsenstein 1983).
4. Choose one of these pairs to coalesce at time S_j at random, where the probability for a pair being chosen is proportional to the weight computed in step 3.
5. To condense the data information of the newly generated subtree: create a probability distribution for the unknown sequence at the root of the subtree according to the sequence information of the two merged subtrees.
6. Set $j = j - 1$.
7. If $j > 2$ go to step 2.
8. If $j = 2$ one genealogy g is complete and we can compute $P(D|g, \theta)$.

This algorithm is repeated many times and the result is averaged to yield an approximation of $P(D|\theta)$.

In order to generate a likelihood curve for various θ' from a single value θ, we have to take $P(g|\theta)$ as an importance function into account, and compute

$$P(D|, \theta') \approx \frac{1}{\#g} \sum_g P(D|g) \frac{P(g|\theta')}{P(g|\theta)}.$$

The value of θ corresponding to the maximum of this curve serves as a maximum likelihood estimate of the genetic diversity of the sample. Additionally, we can construct regions of confidence from the likelihood curve assuming that the estimator is approximately normally distributed. Since the algorithm provides an approximation for the distribution of coalescent genealogies, we can answer further questions about the evolution of the sampled sequences. In terms of empirical distributions we get information about the time to the most recent ancestor of the sample, the sequence type at the root of the genealogy, the placement of the root in a tree or network and the time where mutations might have occured. The assumption of a population of constant size through time used for the description of coalescent methods is surely simplistic, when real data should be analyzed. But coalescent models can easily be extented to very general deterministic models of population history. Let $v(t)$, $t > 0$ describe the size of a population as a function of time t going backward in time and define $\Lambda(x) = \int_0^x v^{-1}(t)dt$. Then, the coalescent times T_j^* under a model defined by $v(t)$ can be derived by the following rescaling of the corresponding times T_j, which can be solved recursively (Griffiths & Tavaré 1994b):

$$\sum_{i=n}^{j} T_i^* = \Lambda^{-1}\left(\sum_{i=n}^{j} T_i\right).$$

As another extension one can allow for some structure of the population, such that additional to coalescence and mutation events migrations of individuals between subpopulations are possible. For many of these models maximum likelihood based inference techniques have been published (Griffiths & Tavaré 1994b; Kuhner et al. 1998; Weiss & von Haeseler 1998; Beerli & Felsenstein 1999). However, one has to keep in mind that the information about population history contained in DNA data is limited, especially in a single locus approach. This fact might conflict with the desire of an investigator to obtain a precise picture of the complexities during evolution of a population. Likelihood-based techniques can be helpful in deciding whether an extension of a given model is reasonable given the available data information. Assume we performed a maximum likelihood analysis of the data D under a given model M_0. We extent this model to M_1 such that M_0 remains a special case of M_1. Then, we compute the maximal value of the likelihood of the data under model M_1. The ratio of the maximal likelihoods under the two models M_0 and M_1 is a measure of progress in explaining the data we made by including more complexity in the model. If this improvement is significant, M_0 should be rejected and M_1 should serve as a model for further investigation. If, however, this ratio is close to one, then it is doubtful that an extension to the model M_1 leads to deeper insight of the processes that generated the data. One might then think of extenting M_0 in a different way and performing the same procedure. To make things clear: if a model M_0 is not rejected by the described procedure this does *not* mean that the model is 'correct' or 'true'. It does simply mean that the data is compatible with the model M_0 and there is not enough information contained in the data to justify rejection of the model.

In this paper, I argued for the usefulness of likelihood-based inference methods. Their advantages become especially explicit in cases where complex mutational models are studied. The proposed algorithm can be applied to data from loci with complex mutation structure (e.g. mtDNA data). However, single locus data might not be sufficient to study prehistoric demography. Data from multiple loci should improve our ability to reconstruct aspects of population history and the extension of likelihood based methods to multiple loci data is straightforward. Yet, another promising concept could be the probabilistic incorporation of interdisciplinary information by way of the theory of Bayesian statistics (Bernardo & Smith 1994).

Acknowledgements

Financial support by the European Commission and the Max Planck Gesellschaft is gratefully acknowledged.

References

Bandelt, H-J., P. Forster, B.C. Sykes & M.B. Richards, 1996. Mitochondrial portraits of human populations using median networks. *Genetics* 141, 743–53.
Beerli, P. & J. Felsenstein, 1999. Maximum-likelihood estimation of migration rates and effective population numbers in two populations using a coalescent approach. *Genetics* 152, 763–73.
Bernardo, J.M. & A.F.M. Smith, 1994. *Bayesian Theory*. Chichester: John Wiley & Sons.
Felsenstein, J., 1983. Statistical inference of phylogenies. *Journal of the Royal Statistical Society* A, 146, 246–72.
Fitch, W.M., 1971. Toward defining the course of evolution: minimum change for a specified tree topology. *Systematic Zoology* 20, 406–16.
Forster, P., R. Harding, A. Torroni & H-J. Bandelt, 1996. Origin and evolution of native American mtDNA variation: a reappraisal. *American Journal of Human Genetics* 59, 935–45.
Fu, Y-X., 1998. Probability of a segregating pattern in a sample of DNA sequences. *Theoretical Population Biology* 54, 1–10.
Griffiths, R.C., 1987. *An Algorithm for Constructing Genealogical Trees*. (Statistics Research Report 163.) Australia: Department of Mathematics, Monash University.
Griffiths, R.C. & P. Marjoram, 1996. Ancestral inference from samples of DNA sequences with recombination. *Journal of Computational Biology* 3, 479–502.
Griffiths, R.C. & S. Tavaré, 1994a. Simulating probability distributions in the coalescent. *Theoretical Population Biology* 46, 131–59.
Griffiths, R.C. & S. Tavaré, 1994b. Sampling theory for neutral alleles in a varying environment. *Philosophical Transactions of the Royal Society of London* B 344, 403–10.
Hudson, R.R., 1991. Gene genealogies and the coalescent process, in *Oxford Surveys in Evolutionary Biology*, vol. 7, eds. D. Futuyma & J. Antonovics. Oxford: Oxford University Press, 1–44.
Kingman, J.F.C., 1982a. The coalescent. *Stochastic Processes and their Applications* 13, 235–48.
Kingman, J.F.C., 1982b. On the genealogy of large populations. *Journal of Applied Probability* 19A, 27–43.
Kingman, J.F.C., 1982c. Exchangeability and the evolution of large populations, in *Exchangeability in Probability and Statistics*, eds. G. Koch & F. Spizzichino. Amsterdam: North-Holland Publishing Company, 7–112.
Kuhner, M.K., J. Yamato & J. Felsenstein, 1995. Estimating effective population size and neutral mutation rate fromsequence data using Metropolis-Hastings sampling. *Genetics* 140, 142–30.
Kuhner, M.K., J. Yamato & J. Felsenstein, 1998. Maximum likelihood estimation of population growth rates based on the coalescent. *Genetics* 149, 429–34.
Meyer, S., G. Weiss & A. von Haeseler, 1999. Pattern of nucleotide substitution and rate heterogeneity in the hypervariable regions I and II of human mtDNA. *Genetics* 152, 1103–10.
Neuhauser, C. & S.M. Krone, 1997. The genealogy of samples in models with selection. *Genetics* 145, 519–34.
Richards, M.B., V.A. Macaulay, H-J. Bandelt & B.C. Sykes, 1998. Phylogeography of mitochondrial DNA in western europe. *Annals of Human Genetics* 62, 241–60.
Rogers, A.R. & H. Harpending, 1992. Population growth

makes waves in the distribution of pairwise genetic differences. *Molecular Biology and Evolution* 9, 552–69.

Slatkin, M. & R.R. Hudson, 1991. Pairwise comparison of mitochondrial DNA sequences in stable and exponentially growing populations. *Genetics* 129, 555–62.

Tajima, F., 1983. Evolutionary relationship of DNA sequences in finite populations. *Genetics* 105, 437–60.

Tavaré, S., 1984. Line-of-descent and genealogical processes, and their applications in population genetics models. *Theoretical Population Biology* 26, 119–64.

Watterson, G.A., 1975. On the number of segregating sites in genetical models without recombination. *Theoretical Population Biology* 7, 256–76.

Weiss, G. & A. von Haeseler, 1998. Inference of population history using a likelihood approach. *Genetics* 149, 1539–46.

Wilson, I.J. & D.J. Balding, 1998. Genealogical inference from microsatellite data. *Genetics* 150, 499–510.

Wiuf, C. & J. Hein, 1999. The ancestry of a sample of sequences subject to recombination. *Genetics* 151, 1217–28.

Chapter 33

Sampling Saturation and the European MtDNA Pool: Implications for Detecting Genetic Relationships among Populations

Agnar Helgason, Sigrún Sigurðardóttir, Jeffrey R. Gulcher, Kári Stefánsson & Ryk Ward

In recent years, mitochondrial DNA (mtDNA) markers have been extensively used to throw light on the demographic history of human populations in Europe (Sajantila et al. 1995; Richards et al. 1996; 1998; Simoni et al. 2000). However, while geographic patterns of genetic variation in Europe have been effectively demonstrated for nuclear markers, and seem to be emerging for Y-chromosome markers (Casalotti et al. 1999), large surveys of mtDNA variation in Europe have thus far failed to reveal any such patterns. In this chapter we examine two methods for assessing the current levels of sampling saturation of mtDNA lineages in Europe, and discuss the implications for detecting geographic patterns of mtDNA variation. Our findings indicate that the European mtDNA pool contains many more lineages than have yet been sampled and is in fact severely undersampled. We suggest that this may conceal geographically informative distributions of individual lineages or sub-clusters of lineages, and hinder our ability to reliably identify genetic relationships between European populations using the mtDNA locus.

The apparent lack of geographic patterns in European mtDNA

In a world-wide study of geographic patterns of genetic variation, Seielstad et al. (1998) reported relatively high F_{ST} values for Y-chromosome loci (0.64), and much lower F_{ST} values for mtDNA (0.19) and autosomal (0.14) loci. F_{ST} values represent the proportion of genetic variation that is found between populations as opposed that found within populations, and thus serves as useful index of genetic differentiation between populations. Seielstad et al. (1998) also presented a regression analysis of genetic and geographic distances among European populations, with smaller F_{ST} values for all loci, but similar relative differences between the Y-chromosome, mtDNA and autosomal markers. According to these results, then, there is greater genetic differentiation between European populations at the mtDNA locus than for autosomal loci. However, while there is abundant evidence for geographic patterns of genetic variation at autosomal loci in Europe, such patterns have not so far been observed for mtDNA. The geographic patterns of autosomal variation that have been observed in Europe are typically in the form of clines (Cavalli-Sforza et al. 1994) or sharp genetic boundaries (Barbujani & Sokal 1990). In many cases, researchers have also been able to demonstrate a linguistic association with genetic differences. In contrast, studies of mtDNA lineages in Europe reveal a continent with little or no observable geographic patterns of variation. In the first large-scale European mtDNA analysis, Sajantila et al. (1995) attempted to uncover evidence for the geographic structuring of mtDNA variation and correlation with language distribution. This study reported little ob-

Table 33.1. *Summary statistics for mtDNA lineages from ten European populations.*

Population	HVS1				HVS2			
	Sample size	No. of lineages	θ_k	Private (%)	Sample size	No. of lineages	θ_k	Private (%)
Germans	418	219	237.1	58.0	264	82	40.4	54.9
Norwegians	216	123	117.7	52.8	–	–	–	–
Spanish	192	112	111.6	50.9	–	–	–	–
British	167	98	98.5	50.0	118	48	20.2	39.6
Austrians	117	73	82.0	45.2	99	40	24.5	27.5
Russians	132	74	68.7	44.6	–	–	–	–
Icelanders	447	125	57.2	55.2	346	50	15.8	28.0
Finns	176	74	47.6	37.8	–	–	–	–
Basques	106	53	41.5	32.1	–	–	–	–
Saami	115	25	9.6	40.0	58	6	1.5	16.7

servable genetic differentiation, with the only exception being the Saami, who were clearly differentiated with an unusual distribution of mainly typical European mtDNA lineages. A major phylogeographic analysis of European mtDNA variation by Richards *et al.* (1998) investigated the geographic distribution of mtDNA haplogroups and sub-clusters of lineages within haplogroups. Splitting Europe into sub-regions and isolates (Icelanders and Basques), Richards *et al.* (1998) found subtle geographic patterns in the distribution of the rare haplogroups U3, U4, U6 and I — the first three being predominantly found in southern Europe, the last mainly in northwest Europe. Geographic differences were also found within haplogroups U5 and J. The former contained sub-clusters associated with north, south and northeast Europe, respectively. The latter contained sub-clusters associated with the British Isles, north-central and Alpine Europe, and south Europe, respectively. Yet, these differences were far from unequivocal, and did not provide strong evidence for general geographic patterns. A recent analysis by Simoni *et al.* (2000) applied autocorrelation statistics to mtDNA haplogroup frequencies and lineages defined by a subset of 22 deep phylogenetic substitutions (out of a total of 241 variable sites) from the HVS1 region. This study found no evidence of geographic structure in Europe, apart from the divergence of the Saami and an area of clinal variation around the Mediterranean Sea. Unfortunately, the strategy of omitting the vast majority of variable sites from this study prevented the detection of any geographic dimension to the distribution of sub-clusters of lineages within haplogroups. Overall, then, we are left with a rather bleak picture of the capacity of mtDNA markers to reveal geographic patterns of genetic relationships between European populations. Judging from the results of Richards *et al.* (1998) and Simoni *et al.* (2000), the F_{ST} values reported by Seielstad *et al.* (1998) for European populations cannot be the product of geographic variations in the frequencies of haplogroups or larger phylogenetic clades of lineages. Consequently, they must stem from differences in the distribution of individual lineages or smaller lineage clusters among populations. A better understanding of the nature of the differences between the mtDNA pools of European populations can be gained through an examination of the way in which mtDNA lineages are distributed within and between populations. This is the subject matter of the next section.

Some curious features of mtDNA lineage distribution in European populations

Table 33.1 shows summary data for HVS1 and HVS2 sequences from ten European populations with sample sizes larger than 100. A feature of the data that immediately draws attention is the remarkable number of distinct lineages observed per sample size. Another curiosity is the high proportion of 'private' lineages. A lineage is defined as private when it is only found in a single population from an extended data set of 2969 HVS1 (sites 16,090–16,365) and 1095 HVS2 (sites 63–297) mtDNA sequences (see Helgason *et al.* 2000 and references therein for more details).

Taken together, the observations in Table 33.1 indicate that existing population samples from Europe contain a large number of rare lineages, many of which appear to be unique to populations. It is this high proportion of private lineages that must largely account for the F_{ST} values reported by

Seielstad et al. (1998) for European mtDNA.

Why are there so many private lineages in European populations? For mtDNA sequence data, the signal of a close genetic relationship between two populations lies in an unusually high degree of lineage sharing and/or the tendency for the lineages of the populations to differ by an unusually small number of base-pairs. Conversely, private lineages are indicative of genetic differences between the populations. The existence of such private lineages can be attributed to one of three possible causes: 1) they were generated in one population through unique mutation events and have not found their way into the other population; 2) they existed in both populations in the past, but have since been lost from one of them; 3) they currently exist in both populations, but have only thus far been sampled from one of them. The first two cases represent genuine genetic differences between the populations. The third case, however, results in a misleading inflation of genetic differences. This raises the question: which of the three explanations accounts for the high incidence of private mtDNA lineages observed in European populations?

A particularly useful case with which to begin to tackle this question, and the one that led us to this line of enquiry in the first place, is that of the Icelanders. Iceland is thought to have been settled by a mixture of Norse and British individuals around AD 900 (see Helgason et al. 2000 for more details). This timeframe does not allow for many new mutations to have occurred, leading to the expectation that the vast majority of the mtDNA lineages currently found in the Icelanders ought to be identical by state to the lineages carried by female settlers 1100 years ago. To our great surprise, however, just over half of the Icelandic HVS1 lineages are not found in a comparative sample of 2249 European sequences (of which 857 derive from Scandinavia and the British Isles) (see Helgason et al. 2000). Furthermore, when attempting to measure the genetic divergence of the Icelandic mtDNA pool from that of the rest of Europe, we found that Icelandic sequences differ on average by 0.46 substitutions from their nearest putative European founder lineages. Given a mutation rate of 1 substitution every 20,186 years (as employed by Forster et al. 1996), this implies that the observed genetic difference between the Icelandic and European mtDNA pools would have had to accumulate over a period of roughly 9646 years. This figure is almost nine times larger than the historically and archaeologically supported date of 1100 years. A recent study of 705 mtDNA transmission events from 26 large Icelandic pedigrees revealed a slightly higher mutation rate of 0.32/site/million years, corresponding to roughly 1 substitution every 11,364 years (Sigurðardóttir et al. 2000). Even using this higher rate we still obtain a divergence time of about 5227 years.

Can the extraordinarily large number of Icelandic lineages identified as private and the evident overestimation of the timeframe of Icelandic genetic divergence be due to a combination of Icelandic specific mutations and lineage loss in Europe? These factors are certain to account for some of the private lineages. However, to account for them all in this way would seem to require an unreasonably high mutation rate, coupled with an improbable rate of lineage loss in such large populations. Another possible explanation could be the existence of a hitherto unsampled parental population. However, this is unlikely, given that all historical, archaeological and linguistic evidence limits the source populations of the Icelandic settlers to northwest Europe (and specifically to Scandinavia and the British Isles). As a result, we are left with the prospect that under-sampled European populations are the cause of many of the apparently private Icelandic lineages.

Evidence supporting this latter idea was provided in a recent study by Pfeiffer et al. (1999), which measured the rate of detection of new lineages sampled from a single German village. Through repeated increases in sample size, it was observed that the rate of detection of new distinct lineages from this village showed no sign of decline, even after 700 individuals had been sampled (a total of 317 lineages were obtained). This finding is all the more remarkable given the fact that the sampling locale was a village, where the probability of maternal relatedness (and therefore of individuals having identical mtDNA lineages) is presumably much higher than for a city or larger region.

In the next section we will attempt a quantitative assessment of the sampling saturation of mtDNA lineages from European populations and consider the implications for detecting genetic relationships between populations.

What is the current sampling saturation of the European mtDNA pool?

At the heart of the analysis presented here is the parameter θ_k. This parameter defines the relationship between the sample size and the number of different alleles or lineages observed therein assuming an infinite allele model (Ewens 1972), and represents $2N_{fe}\mu$ (where N_{fe} represents the female effective population size and μ the mutation rate). θ_k is estimated using the formula:

$$E(k) = \theta_k \sum_{i=0}^{n-1}(1/\theta_k + i)$$

where k represents the number of distinct lineages observed in a sample size of n individuals.

The values of θ_k for the ten European populations are shown in Table 33.1. As the mtDNA control region mutation rate should be the same in all populations, differences in θ_k values reflect differences in female effective population size (N_{fe}). N_{fe} represents the harmonic mean of the number of females that have successfully transmitted their mtDNA to female offspring during past generations, with the result that their lineages are observed in the present-day population. As it is based on the relationship between sample size and the number of distinct lineages, θ_k is highly sensitive to the effects of lineage sorting (the pattern by which variation in fertility shapes the transmission of genes between generations over time) and thus tends to reflect effective population size during recent periods of demographic history. In contrast, another widely used measure of $2N_{fe}\mu$, θ_π, is based on the accumulation of mutational differences between lineages and is thus more influenced by ancient demographic histories (Rogers & Harpending 1992).

Ewens' (1972) sampling formula, used to calculate θ_k, dictates that the relationship between sample size and the number of observed lineages will be one of diminishing returns. This is because of the way lineages are distributed. Figure 33.1 shows the observed frequency distribution of lineages from a German population sample ($N = 418$), and the expected distribution according to Ewens sampling formula (calculated using Arlequin 1.1: Schneider *et al.* 1997). This distribution is typical of a European population, with a few moderately frequent lineages and a majority of very rare lineages. Given this sort of distribution, it follows that the rate of detection of new lineages is greatest at the onset of sampling, and that this rate decreases rapidly after the most common lineages have been detected. Owing to the abundance of rare lineages, large sample sizes are needed to detect them.

It should be noted that the observed distribution departs slightly from the expected distribution in that there is a slight excess of the most common lineages and a corresponding deficit of the intermediate lineages. This probably results from a violation of one or more of the three major assumptions underlying Ewens' sampling formula. These assumptions are that: i) selection is not influencing the locus; ii) each new mutation creates a lineage that is distinguishable by state from all others (the infinite alleles assumption); and iii) population size has been sufficiently constant to maintain a steady-state distribution of lineages. In the case of the mtDNA locus, any one of these assumptions might be violated. It is our contention, however, that while such violations af-

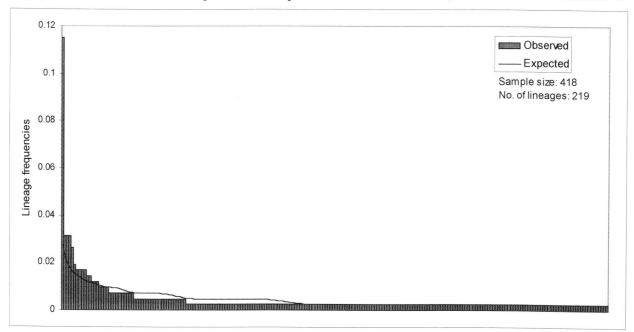

Figure 33.1. *The frequency spectrum of 418 German mtDNA HVS1 sequences.*

fect the absolute values obtained from the use of θ_k, they do not affect the use of this index as a comparative measure among populations. Thus, it seems unlikely that selective forces (if present) have varied much between European populations at the mtDNA locus. As regards the second issue, although recurrent mutations are common in the control region (Meyer *et al.* 1999) and these usually generate a new identifiable lineage because they occur on a novel genetic background created by mutations at other sites. Moreover, as in the case of possible selection effects, even if the infinite alleles assumption is violated, this should affect populations equally and therefore would not diminish the validity of comparing θ_k values among populations. The assumption of constant population size is obviously violated for most European populations. We assume, however, that most European populations have experienced a similar degree of population expansion, and thus once again that the validity of θ_k as a comparative measure among populations remains intact.

Owing to the way in which the index is calculated, it is possible to ask the following question: given the θ_k values obtained from our actual population samples, how many more individuals from each of the populations would we need to sample before we have an adequate picture of the variation contained in its mtDNA gene pool? (cf. Ward *et al.* 1993; Francalacci *et al.* 1996). The precise point at which sampling is stopped is necessarily based on an arbitrary choice. We chose to cease our sampling experiment when, for repeated incremental increases in sample size of ten, we obtained less than one new lineage from the population. The resultant sample size represents a fixed point at which the rate of detection of new lineages is the same for all populations. Figure 33.2 shows the outcome of using Ewens' sampling formula in this way to predict the number of lineages per sample size for seven European populations. Each curve represents the expected number of lineages detected when sample sizes are increased until the cut-off point is reached. As expected, populations with higher θ_k values require larger sample sizes to reach the same level of sampling saturation as populations with lower values of θ_k (i.e. they have longer curves). The black triangles superimposed onto each curve indicate the actual sample sizes for each population. The positions of these triangles relative to the total lengths of the curves provide an *ad hoc* comparative measure of

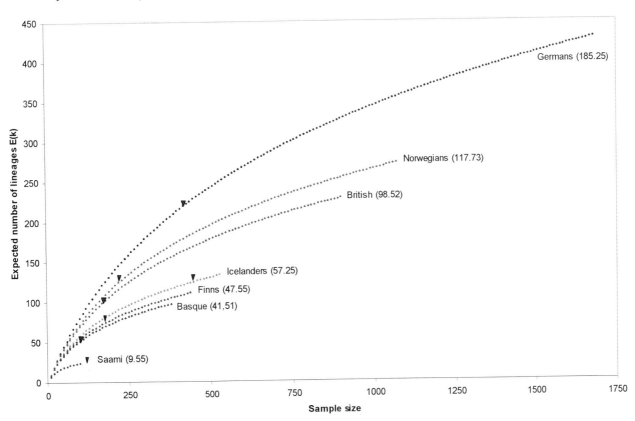

Figure 33.2. *Sampling saturation of mtDNA HVS1 lineages from seven European populations.*

sampling saturation.

Figure 33.2 not only suggests that are there many mtDNA lineages that have yet to be sampled from the larger European populations, but also that sampling saturation varies considerably between populations. Indeed, this analysis suggests that we only have a reasonable idea of the overall mtDNA variation in the Saami and Icelanders. However, the sampling saturation is less complete for the Finns, and we are missing a substantial proportion of lineages from the Basques, British, Norwegians and Germans (the situation is similar for the Austrians, Russians and Spanish, curves not shown). According to our results sample sizes from the latter seven populations will need to be increased by 200–300 per cent in order to reach levels of saturation equivalent to those observed for the Icelanders and Saami.

An alternative method of examining sampling saturation in populations, and a way of independently assessing the validity of our results based on θ_k values, is to compare the frequency of lineages that occur once, twice or more than twice. The rationale underlying this method is as follows. In a moderately sized sample taken from a population containing a large number of lineages (exhibiting an approximate steady-state distribution), one would expect most lineages to be encountered only once or twice. However, if the sample size were to be substantially increased, the frequencies of lineages encountered only once or twice would decrease owing to a rise in the probability of recurrently sampling rare lineages. Extending this line of reasoning, it is plain to see that the frequency of lineages encountered once or twice in a sample should reflect the sampling saturation of the population.

Table 33.2 shows these frequencies for HVS1 lineages from the ten European population samples. Clearly, the frequency spectrum of these population samples is dominated by rare lineages. When the combined proportions of lineages occurring once and twice are compared, it becomes clear that the Saami, Icelanders and Finns have noticeably fewer such lineages than the other populations. This supports the results based on θ_k from Figure 33.2, and suggests that these populations have been relatively thoroughly sampled. Other populations appear to be less thoroughly sampled.

Given the excessive number of rare mtDNA lineages in European populations, it is both practically unachievable and economically unfeasible to attempt to detect them all. Nonetheless, it is important that there be some way of determining when sampling saturation has reached a sufficient level to minimize the stochastic error introduced into estimations of genetic relationships due to pseudo private lineages. At present we do not feel confident to suggest any absolute guideline. A reasonable strategy, however, might be to continue sampling until only about 70–80 per cent of the lineages detected are present once or twice in the sample. Although this will require substantial sample increases for most populations, it should improve the capacity for detecting true lineage sharing among populations — and thereby provide more valid estimations of the genetic relationships between populations.

While the additional lineages sampled will by definition be rare, there is an expectation that the majority will already exist in the present European data set. Treating the combined European data set as a single population, we used θ_k and employed the same sampling criterion as before, and determined the point at which less than one new lineage was detected for incremental increases of ten in sample size. The Icelanders, Saami and Basques were excluded from this analysis, as they are non-typical European isolates with unusually low values of θ_k. The combined data set comprised 1418 sequences with 553 distinct lineages, giving a θ_k value of 332.86. The point at which the curve for the combined data set ended was at a sample size of 3010 with 215 new lineages detected. However, if treated individually, we would expect to sample 876 new lineages from these seven populations. This discrepancy leads to the conclusion that only roughly 24.5 per cent of the new lineages found in each population will not have been previously observed in Europe. Hence, the majority of new lineages detected in each population should reveal new cases of lineage sharing, thus increasing the potential for detecting the true nature of the genetic relationships between populations.

Table 33.2. *The proportion of rare HVS1 lineages in European populations.*

Population	Proportion of lineages occurring:		
	Once	Twice	More than twice
Saami	0.680	0.000	0.320
Icelanders	0.544	0.160	0.296
Finns	0.554	0.203	0.243
Basque	0.698	0.151	0.151
Germans	0.772	0.096	0.132
Russians	0.770	0.122	0.108
British	0.786	0.112	0.102
Norwegians	0.764	0.138	0.098
Spanish	0.786	0.116	0.098
Austrians	0.795	0.137	0.068

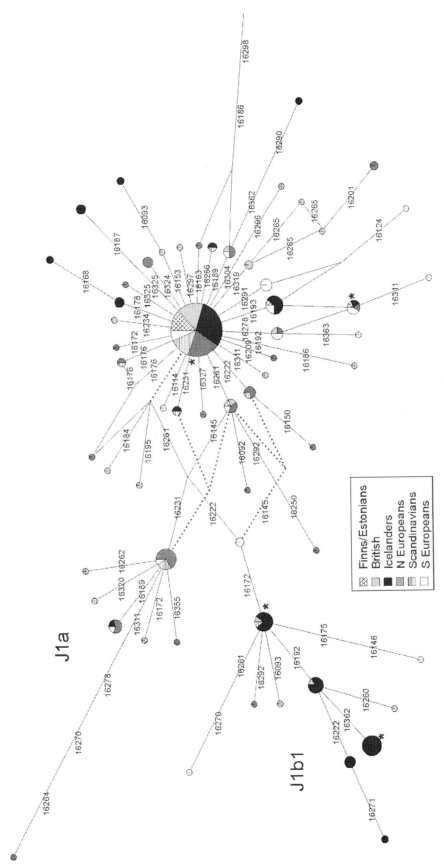

Figure 33.3. *A median-joining phylogenetic network of haplogroup J lineages from different European regions.*

The implications of this are perhaps again best demonstrated by the case of the Icelanders. There are evidently a number of lineages that have yet to be sampled from the British Isles and Scandinavia, and some of these unsampled lineages are certain to be ones currently defined as being unique to the Icelanders. By increasing sample sizes from Scandinavia and the British Isles, and possibly uncovering new cases of lineage sharing between the Icelanders and these parental populations, our capacity to estimate accurately the ancestry of the female settlers of Iceland is sure to be enhanced.

This was certainly the experience when our new data set of 394 Icelandic HVS1 sequences (Helgason et al. 2000) was added to the existing 53 Icelandic sequences previously published by Sajantila et al. (1995) and Richards et al. (1996). To demonstrate this we present a phylogenetic median-joining network of European HVS1 mtDNA lineages belonging to haplogroup J (Fig. 33.3). All lineages belonging to this haplogroup share two transitions at sites 16,069 and 16,126, and account for about ten per cent of the entire European data set. This subset of the data was chosen for the sake of simplicity and manageability. The point is not to provide an overview of the European mtDNA phylogeny, but to show the effect of a sample size increase on patterns of lineage sharing and thereby on changes in the capacity for reliably detecting genetic relationships between populations.

The circles in Figure 33.3 represent lineages and their area is proportional to the overall count of each lineage. In cases where a lineage is found in more than one population, the circle is divided into slices that show the proportional distribution of the lineage count among populations. Lines represent substitutions and their length indicates the number of substitutions between lineages. Substitutions are indicated by their site numbers. In some cases, it has been impossible to determine a unique mutational pathway between lineages owing to the existence of recurrent mutations. Multiple pathways are depicted in the form of parallelogram-like shapes, where the parallel lines represent possible substitutions at the same site. Dotted lines represent unlikely alternative pathways in the network. Populations are grouped into the following geographic regions: Icelanders, Scandinavians (Norwegians, Danes and Swedes), British, Finns and Estonians, North Europeans (Germans, Swiss and Austrians) and South Europeans (French, Italians, Spanish and Basque) (see Helgason et al. 2000, and references therein).

Among the 53 previously published Icelandic sequences, 9 belonged to haplogroup J and represented 4 distinct lineages (marked by asterisks in Fig. 33.3). Three of these lineages are shared with other populations and one appears to be private. With the addition of our new data set, the number of Icelandic haplogroup J lineages increased by 12. Of these, seven are apparently private, two are shared with Scandinavians, one with British, one with southern and northern Europeans and one with Finns and southern Europeans. Furthermore, two of the new private lineages are mutational derivatives of a lineage that is shared only by the Icelanders and the British. It is also interesting to note what appears to be a very clear geographic dimension to the distribution of the lineage clusters labelled J1a and J1b1. Richards et al. (1998) had previously observed an association of J1b1 with the British Isles and J1a with Alpine and North Central Europe. Intriguingly, in our network Scandinavian lineages appear to be moderately frequent in J1a, while this cluster contains only a single Icelandic sequence. In contrast, only one Scandinavian sequence is found to belong to J1b1, in the midst of numerous Icelandic and British lineages. This finding suggests that at least some of the women among the settlers of Iceland probably did originate from the British Isles.

A more comprehensive interpretation of the genetic relationships between the Icelanders and their putative parental populations will of course have to be based on statistical analyses of all the lineages from these populations. Our point here was just to demonstrate how the validity and accuracy of such analyses are likely to be enhanced by data sets that reflect a greater proportion of the total mtDNA diversity contained in European populations. One can only imagine how much more accurate estimates of Icelandic admixture could be if sample sizes from Scandinavia and the British Isles were to be substantially increased. Hopefully, it will not be too long until this becomes a reality.

Conclusions

Migration has been a major feature of human population history in Europe. In spite of this it is possible to detect unmistakable geographic patterns of variation at Y-chromosome and autosomal loci. Why are such patterns so elusive in European mtDNA variation? It certainly does seem to be the case that haplogroup frequencies do not exhibit any obvious geographic patterns (cf. Richards et al. 1998; Simoni et al. 2000). This is not surprising, however, since the mutations that define the haplogroups are ancient and probably predate the formation of most modern

European populations (Sykes 1999). What of individual lineages and sub-clusters of lineages, defined by mutations that could well have occurred somewhere in Europe during the past two or three millennia, and thus have the potential to reveal links between particular populations or geographic regions? Is it the case, as suggested by Seielstad *et al.* (1998), that high rates of female migration have erased old geographic patterns and prevented the formation of new ones? While such factors may have had a homogenizing effect on the European mtDNA pool, we have presented strong evidence in this chapter to suggest that an important part of the answer to the first question could lie in the low sampling saturation of most European populations.

Our findings, based on predictions from θ_k estimates and an empirical investigation of lineage counts, indicate that the mtDNA pools of European populations are dominated by an extremely large number of rare lineages, many of which have not yet been detected. Moreover, the current mtDNA data set provides very little reliable information about relative lineage frequencies within population samples. In other words, we do not yet have a particularly detailed or reliable record of mtDNA variation in most European populations. Consequently, geographic patterns of mtDNA variation exist at the level of individual lineages or lineage clusters, then they may be concealed by this deficiency in the data set. In support of this hypothesis, we have demonstrated how an increase in the sample size of Icelanders brought to light a number of new lineage associations with Scandinavian and British populations (as would be predicted from historical and archaeological evidence) and substantially enhanced a previously recognised geographic dimension to the distribution of two lineage sub-clusters from haplogroup J.

If we are right, then more extensive sampling of large Eurasian populations should provide more precise and reliable information about the genealogical relationships between their mitochondrial gene pools, and might even reveal continent-wide patterns in the distribution of particular lineages or lineage clusters.

References

Barbujani, G. & R.R. Sokal, 1990. Zones of sharp genetic change in Europe are also linguistic boundaries. *Proceedings of the National Academy of Sciences of the USA* 87, 1816–19.

Casalotti, R., L. Simoni, M. Belledi & G. Barbujani, 1999. Y-chromosome polymorphisms and the origins of the European gene pool. *Proceedings of the Royal Society of London Series B-Biological Sciences* 266, 1959–65.

Cavalli-Sforza, L.L., P. Menozzi & A. Piazza, 1994. *The History and Geography of Human Genes*. Princeton (NJ): Princeton University Press.

Ewens, W.J., 1972. The sampling theory of selectively neutral alleles. *Theoretical Population Biology* 3, 87–112.

Forster, P., R. Harding, A. Torroni & H-J. Bandelt, 1996. Origin and evolution of native American mtDNA variation: a reappraisal. *American Journal of Human Genetics* 59, 935–45.

Francalacci, P., J. Bertranpetit, F. Calafell & P.A. Underhill. 1996. Sequence diversity of the control region of mitochondrial-DNA in Tuscany and its implications for the peopling of Europe. *American Journal of Physical Anthropology* 100, 443–60.

Helgason, A., S. Sigurðardóttir, J.R. Gulcher, R. Ward & K. Stefánsson, 2000. MtDNA and the origin of the Icelanders: deciphering signals of recent population history. *American Journal of Human Genetics* 66, 999–1016.

Meyer, S., G. Weiss & A. von Haeseler, 1999. Pattern of nucleotide substitution and rate heterogeneity in the hypervariable regions I and II of human mtDNA. *Genetics* 152, 1103–10.

Pfeiffer, H., B. Brinkmann, J. Hühne, B. Rolf, A.A. Morris, R. Steighner, M.M. Holland & P. Forster, 1999. Expanding the forensic German mitochondrial DNA control region database: genetic diversity as a function of sample size and microgeography. *International Journal of Legal Medicine* 112, 291–8.

Richards, M.B., H. Côrte-Real, P. Forster, V. Macaulay, H. Wilkinson-Herbots, A. Demaine, S. Papiha, R. Hedges, H-J. Bandelt & B. Sykes. 1996. Paleolithic and Neolithic lineages in the European mitochondrial gene pool. *American Journal of Human Genetics* 59, 185–203.

Richards, M.B., V.A. Macaulay, H-J. Bandelt & B.C. Sykes. 1998. Phylogeography of mitochondrial DNA in western Europe. *Annals of Human Genetics* 62, 241–60.

Rogers, A.R. & H. Harpending, 1992. Population-growth makes waves in the distribution of pairwise genetic-differences. *Molecular Biology and Evolution* 9, 552–69.

Sajantila, A., P. Lahermo, T. Anttinen, M. Lukka, P. Sistonen, M.L. Savontaus, P. Aula, L. Beckman, L. Tranebjaerg, T. Gedde-Dahl, L. Isseltarver, A. Di Rienzo & S. Pääbo, 1995. Genes and languages in Europe: an analysis of mitochondrial lineages. *Genome Research* 5, 42–52.

Schneider, S., J.M. Kueffer, D. Roessli & L. Excoffier, 1997. *Arlequin 1.1: Software for Population Genetics Data Analysis*. Genetics and Biometry Laboratory, University of Geneva, Switzerland.

Seielstad, M.T., E. Minch & L.L. Cavalli-Sforza, 1998. Genetic evidence for a higher female migration rate in humans. *Nature Genetics* 20, 278–80.

Sigurðardóttir, S., A. Helgason, J.R. Gulcher, K. Stefánsson

& P. Donnelly, 2000. The mutation rate in the human mitochondrial control region. *American Journal of Human Genetics* 66, 1599–609.

Simoni, L., F. Calafell, D. Pettener, J. Bertranpetit & G. Barbujani, 2000. Geographic patterns of mtDNA diversity in Europe. *American Journal of Human Genetics* 66, 262–78.

Sykes, B., 1999. The molecular genetics of European ancestry. *Philosophical Transactions of the Royal Society of London Series B-Biological Sciences* 354, 131–8.

Ward, R.H., A. Redd, D. Valencia, B. Frazier & S. Pääbo. 1993. Genetic and linguistic differentiation in the America. *Proceedings of the National Academy of Sciences of the USA* 90, 10,663–7.

Chapter 34

Male and Female Differential Patterns of Genetic Variation in Human Populations

Michele Belledi, Lucia Simoni, Rosa Casalotti & Giovanni Destro-Bisol

Y-chromosome and mtDNA markers have been used to compare male and female migrational behaviours. Differential patterns of variation were highlighted, as exemplified by studies conducted on Ethiopians, Basques and Finns. The aim of this work is to compare the level of genetic variation of unilinearly transmitted markers (Y-chromosome and mtDNA), using different families of genetic markers and different statistical methods. Our data base includes HVR-1 sequences (92 populations), 7 Y-linked Short Tandem Repeats loci (55 populations) and 13 Single Nucleotide Polymorphisms (60 populations). Comparisons between HVR-1 sequences and SNPs were possible for 24 populations from 5 continents, whereas comparison between HVR-1 sequences and STRs markers were possible for 21 populations from 5 continents. The analysis of F_{ST}s and migration rates suggest contrasting behaviours for Y-chromosome markers. In fact, Nm values are almost the same in females and males considering HVR-1 and STRs markers, but a different situation is observed when we compare HVR-1 and SNPs. Our results differ substantially from those of Seielstad et al. (1998). In fact we found similar migration rates when we compare mtDNA sequences with Y-linked STRs, and a three–four fold migration rate for females when we compare mtDNA sequences with SNPs. Performing a Mantel test, we found a significant correlation between mtDNA and STRs loci (r = 0.4; p = 0.02) and a low and insignificant correlation between mtDNA and SNPs (r = 0.05; n.s.) variation. These results suggest the importance of comparing polymorphisms with a similar rate of evolution, as Y-chromosome STRs and HVI mtDNA sequences, to obtain more reliable inferences on differential patterns of genetic variation in human populations.

In a recent worldwide study of genetic variation at DNA loci, Seielstad et al. (1998) reported evidence of a substantially higher migration rate in females than males. This chapter is a first important attempt to analyze the distribution of different genetic systems in human populations, but the conclusions by Seielstad et al. (1998) need further confirmation. In fact, as Stoneking (1998) remarked, the data bases of each genetic system differ each from the other in the number of sampled populations. Furthermore, Seielstad et al. (1998) compared slow-evolving genetic systems (Y-chromosome Single Nucleotide Polymorphisms (SNPs), mitochondrial and nuclear restriction fragment length polymorphisms (RFLPs)), but they could not test their conclusions using unilinearly transmitted fast-evolving polymorphisms.

In this study we analyze the genetic variation in a worldwide data set using different markers (sequences of the hypervariable region of mtDNA, SNPs and Short Tandem Repeats (STRs) of the Y chromosome) and statistical methods (F_{ST}, R_{ST} and Φ_{ST}). Then, we assess the correlation between each genetic system and geography. Thereafter, we calculate and compare the migration rate for females and males

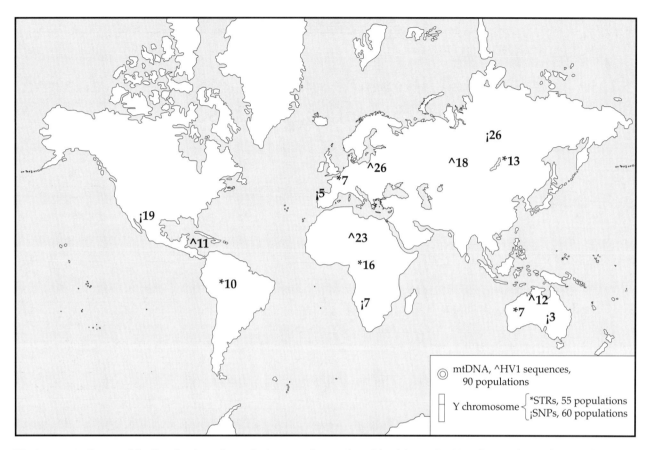

Figure 34.1. *Geographic distribution of population samples analyzed in this study. Numbers refer to the population samples analyzed for each genetic system in each continent (see symbols in the key).*

and discuss the implications of our results. Finally, we suggest how this initial study could be developed.

Material and methods

Our data base (Fig. 34.1) includes sequences from the first hypervariable region of mtDNA (HV1) (90 populations: Comas *et al.* 1998; Handt *et al.* 1997; Destro-Bisol unpublished data), data from seven Y-linked Short Tandem Repeats loci (55 population: Kayser *et al.* 1997; Seielstad *et al.* 1999; Bianchi *et al.* 1998; Pérez-Lezaun *et al.* 1999; Rossi *et al.* 1998; Brinkmann *et al.* 1999; Hurles *et al.* 1998; Horst *et al.* 1999; Caglià *et al.* 1997; Ruiz-Linares *et al.* 1999; Zerjal *et al.* 1997; Destro-Bisol *et al.* unpublished data) and 13 Single Nucleotide Polymorphisms of the Y chromosome (60 populations: Karafet *et al.* 1999).

Genetic variation was analyzed using methods implemented in the Arlequin 1.1 software (Schneider *et al.* 1997). For mtDNA sequences we calculated Φ-statistics (Excoffier *et al.* 1992) using pairwise difference and Kimura 2P (transition/transversion ratio: 1/10; gamma distribution $\gamma = 0.26$) distance methods. For microsatellite data we calculated Φ-statistics based on the sum of squared size differences (RST) (Slatkin 1995) and classical Φ-statistics (Weir & Cockerham 1984). In the case of haplotypic data we used molecular distances based on number of different alleles and on the sum of squared size differences. For Single Nucleotide Polymorphisms (SNPs) we used statistics based on pairwise difference distance methods (Φ_{ST}) and on haplotype frequencies. Mantel tests (Smouse *et al.* 1986) were performed to assess the correlation between genetic and geographic distance matrices.

Results

1. Apportionment of genetic variation

In Table 34.1 we show the apportionment of genetic variation in our data base. Distribution of genetic variation was computed using three distinct hierarchical levels: within populations, among populations within a continent and among continents.

Within-population variation for mtDNA ac-

Table 34.1. *Apportionment of genetic variation in different genetic systems in worldwide populations. Each marker was tested using two genetic distance methods.*

Genetic system (Marker)	N. of Pop.	DISTANCE METHOD	Percentage of genetic variation		
			between continents	within continents	within populations
MtDNA (HV1)	90	PAIRWISE DIFFERENCES	16.4	12.3	71.3
		KIMURA 2 P	17.7	12.8	69.5
Y chromosome (STRs- haplotypic data)	55	N. OF DIFFERENT ALLELES	9	16.5	74.5
		SUM OF SQUARED SIZE DIFFERENCES	20.3	20.6	59.1
Y chromosome (STRs-allele frequ. data)	55	FREQUENCY	5.8	13.2	81
		SUM OF SQUARED SIZE DIFFERENCES	13.2	16.6	70.2
Y chromosome (SNPs)	60	FREQUENCY	22	21.9	56.1
		PAIRWISE DIFFERENCES	39.3	16.9	43.8

Table 34.2. *Genetic variation (F_{ST}) and migration rate (Nv) in worldwide populations estimated using male- and female-mediated markers.* *Φ_{ST}; **R_{ST}

Genetic system	Marker	N. of Pop.	Method	F_{ST}	Nv
Y chromosome	STRs haplotypes	55	N. OF DIFFERENT ALLELES	0.255*	2.92
Y chromosome	STRs haplotypes	55	SUM OF SQUARED SIZE DIFFERENCES	0.408**	1.45
Y chromosome	STRs allele freq.	55	FREQUENCY	0.189	4.28
Y chromosome	STRs allele freq.	55	SUM OF SQUARED SIZE DIFFERENCES	0.298**	2.36
Y chromosome	SNPs	60	FREQUENCY	0.439	1.28
Y chromosome	SNPs	60	PAIRWISE DIFFERENCES	0.562*	0.78
MtDNA	HV1 sequences	90	PAIRWISE DIFFERENCES	0.287*	2.49
MtDNA	HV1 sequences	90	KIMURA 2 P	0.305*	2.28

counts for ~70 per cent of total variation, a value which falls into the range of two previous results based on RFLPs data (65.5–81.4 per cent: Barbujani & Excoffier 1998; Seielstad et al. 1998). Y-chromosome markers display an extended range of variation: from 44 per cent for SNPs, up to 81 per cent for STRs. Fast-evolving markers, such as mitochondrial sequences and microsatellites loci attain a high level of intra-population variation that further increases when molecular information is considered. In contrast, slow-evolving markers seem to be more effective in detecting differences among populations and continents.

2. F_{ST} and Nv estimates
In Table 34.2 we show F_{ST} values and relative Nv (migration rate) estimates obtained with the 'Island model' $F_{ST} = 1/(1 + Nv)$, where Nv is the product of the effective size of a population (N in haploid genetic systems) and of the mutation–migration component (v) (Cavalli-Sforza & Bodmer 1971). If we consider that the mutation rate (μ) is very low and near to zero, we can assume that v is equal to migration (m). Also in this case we found an extended range of migration rate estimates for the Y-chromosome markers (Nvm = 0.78 – 4.28).

3. Choice of population sample composition
We analyzed the fluctuation of F_{ST} values of African, Asian and European mitochondrial gene pools due to the presence of some particular populations in the data base. Eliminating outlier populations (Pygmies, Circumartic and Saami), F_{ST} (at continental level) decreases substantially. For mtDNA, we observed

Table 34.3. *Correlations between Y chromosome/mtDNA/geography. For Y-chromosome STRs data genetic distances were calculated using sum of squared size differences (R_{ST}), for Y-chromosome SNPs using classical F_{ST} based on haplotype frequencies and for mtDNA sequences using a Φ_{ST} based on pairwise difference distance method. Statistically significant values are in bold type.*

Correlation	N. of Pop.	r	p-value
MtDNA (HV1)/Y chromosome (SNPs)	24	0.055	0.319
MtDNA (HV1)/Y chromosome (STRs)	21	0.393	**0.020**
MtDNA (HV1)/geography	24	0.266	**0.021**
Y chromosome (SNPs)/geography	24	0.512	**0.002**
Y chromosome (STRs)/geography	21	0.165	0.095

Table 34.4. *Migration rates in comparable data sets (f = female; m = male).*

rtb Genetic system	N.of Pop.	Nv	Ratio [f]/[m]
MtDNA (HV1) [f]	24	4.13	4.36
Y chromosome (SNPs) [m]	24	0.95	
MtDNA (HV1) [f]	21	3.00	1.29
Y chromosome (STRs) [m]	21	2.33	

that Pygmies account for 23.6 per cent of variation among Africans (23 populations), Circumartic for 39 per cent of variation among Asians (19 populations) and Saami for 41 per cent of variation in Europeans (27 populations). When the same analysis was performed for Y-chromosome markers, we did not find any outlier populations, with Finns (accounting for 56 per cent of variation in European populations) being the only exception. These examples indicate that the evaluation of genetic variability strongly depends on the population samples composition and that populations with a common genetic history are needed when comparing different genetic systems.

4. Correlation between mtDNA, Y chromosome and geography

In order to reduce the confounding effect of variation in population sample composition and size, we selected only those populations analyzed for both genetic systems. Furthermore, we used molecular distances that are robust to the effects of small sample size. A significant correlation between genetic distances obtained from unilinearly transmitted markers would be expected assuming there is no substantial difference between male and female migration rates. Genetic systems characterized by high and similar mutation rates (in the order of 10^{-3}), such as Short Tandem Repeats loci and mtDNA HV1 sequences (Bianchi *et al.* 1998; Forster *et al.* 1996), are well correlated between each other but not with geography (Table 34.3). In contrast, Y-chromosome SNPs, which are slow evolving polymorphisms, are not correlated with mtDNA sequences but are well correlated with geography (Table 34.3).

5. Nvf/Nvm ratio in comparable data sets

Comparing *Nv* obtained from male- and female-transmitted markers, two quite different scenarios become evident (Table 34.4). The first one (HV1 vs SNPs) complies with the conclusions of Seielstad *et al.* (1998), since the female *Nv* largely exceeds that estimated for males (*Nvf*/*Nvm* ratio: 4.13). In contrast, the second scenario (HV1 vs STRs) shows a comparable migration rate for the two genetic systems (*Nvf*/*Nvm* ratio: 1.29).

Conclusions

Quite different *Nvf*/*Nvm* ratios are obtained at worldwide level when HV1-based estimates are compared either with SNPs or STRs (Fig. 34.2). Which is the most reliable scenario? The one based on HV1 mtDNA sequences and Y-chromosome STRs has the advantage of using two genetic systems with a comparable evolutionary rate. However, their low and statistically non-significant correlation with geography and their high level of intra-population variation suggest these results should be considered with caution. Because of their high level of correlation with geography, Y-chromosome SNPs seem to be more suitable tools for migration rate estimates at the worldwide level. However, female-transmitted markers with a similar evolutionary rate should be used for comparison. This was done in the study of Poloni *et al.* (1997), who analyzed variation of mitochondrial and Y-chromosome RFLPs markers. These authors observed that the genetic distances of nineteen populations belonging to four different linguistic families are well correlated (r = 0.529, p <0.001). These results are more in line with our second scenario (r = 0.39, p <0.05) than with the first one (r = 0.05, non-significant.). The validity of our first scenario is probably flawed by differences in the evolutionary rates of the polymorphisms used, while that of Seielstad *et al.* (1998) is probably affected by different population samples composition of the two markers compared.

To develop this initial study, we intend to increase the number of populations and gather data on mtDNA haplogroups, which should be a more useful counterpart for Y-chromosome SNPs. We will also analyze data relative to geographically restricted areas, an application in which fast-evolving polymorphisms have already provided worthwhile results (Pérez-Lezaun et al. 1999).

References

Barbujani, G. & L. Excoffier, 1998. The history and geography of human genetic diversity, in *Evolution in Health and Disease*, ed. S. Stearns. Oxford (UK): Oxford University Press, 41–61.

Bianchi, N.O., C.I. Catanesi, G. Bailliet, V.L. Martinez-Marignac, C.M. Bravi, L.B. Vidal-Rioja, R.J. Herrera & J.S. Lopez-Camelo, 1998. Characterization of ancestral and derived Y-chromosome haplotypes of New World native populations. *American Journal of Human Genetics* 63(6), 1862–71.

Brinkmann, C., P. Forster, M. Schurenkamp, J. Horst, B. Rolf, B. Brinkmann, 1999. Human Y-chromosomal STR haplotypes in a Kurdish population sample. *International Journal of Legal Medicine* 112(3), 181–3.

Caglià A., A. Novelletto, M. Dobosz, P. Malaspina, B.M. Ciminelli & V.L. Pascali, 1997. Y-chromosome STR loci in Sardinia and continental Italy reveal islander-specific haplotypes. *European Journal of Human Genetics* 5(5), 288–92.

Cavalli-Sforza, L.L. & W.F. Bodmer, 1971. *The Genetics of Human Populations*. San Francisco (CA): Freeman.

Comas, D., F. Calafell, E. Mateu, A. Pérez-Lezaun, E. Bosch, R. Martinez-Arias, J. Clarimon, F. Facchini, G. Fiori, D. Luiselli, D. Pettener & J. Bertranpetit, 1998. Trading genes along the silk road: mtDNA sequences and the origin of central Asian populations. *American Journal of Human Genetics* 63(6), 1824–38.

Excoffier, L., P.E. Smouse & L.M. Quattro, 1992. Analysis of molecular variance inferred from metric distances among DNA haplotypes: applications to human mitochondrial DNA restriction data. *Genetics* 131, 479–91.

Forster P., R. Harding, A. Torroni & H-J. Bandelt, 1996. Origin and evolution of Native American mtDNA variation: a reappraisal. *American Journal of Human Genetics* 59, 935–45.

Handt, O., S. Meyer & A. von Haeseler, 1997. Compilation of human mtDNA control region sequences. *Nucleic Acids Research* 26, 126–9.

Horst, B., A. Eigel, T. Sanguansermsri & B. Brinkmann, 1999. Human Y-chromosomal STR types in north Thailand. *International Journal of Legal Medicine* 112(3), 211–12.

Hurles, M.E., C. Irven, J. Nicholson, P.G. Taylor, F.R. Santos, J. Loughlin, M.A. Jobling & B.C. Sykes, 1998. European Y-chromosomal lineages in Polynesians:

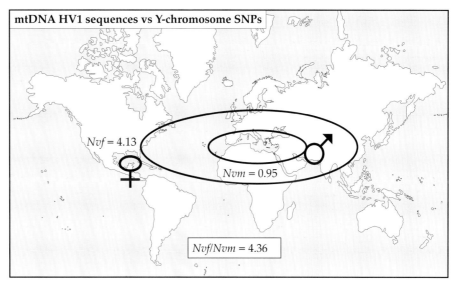
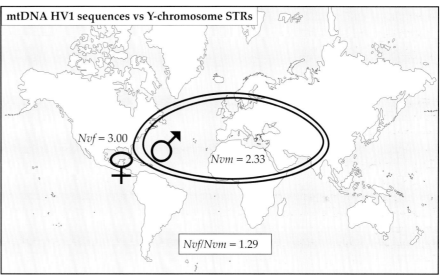

Figure 34.2. *Two different scenarios of female and male migration rates.*

a contrast to the population structure revealed by mtDNA. *American Journal of Human Genetics* 63(6), 1793-806.

Karafet, T.M., S.L Zegura, O. Posukh, L. Osipova, A. Bergen, J. Long, D. Goldman, W. Klitz, S. Harihara, P. de Knijff, V. Wiebe, R.C. Griffiths, A.R Templeton & M.F. Hammer, 1999. Ancestral Asian source(s) of new world Y-chromosome founder haplotypes. *American Journal of Human Genetics* 64(3), 817–31.

Kayser, M., A. Caglià, D. Corach, N. Fretwell, C. Gehrig, G. Graziosi, F. Heidorn, S. Herrmann, B. Herzog, M. Hidding, K. Honda, M. Jobling, M. Krawczak, K. Leim, S. Meuser, E. Meyer, W. Oesterreich, A. Pandya, W. Parson, G. Penacino, A. Pérez-Lezaun, A. Piccinini, M. Prinz, C. Schmitt, P.M. Schneider, R. Szibor, J. Teifel-Greding, G. Weichold, P. de Knijff, L. Roewer, *et al.* 1997. Evaluation of Y-chromosomal STRs: a multicenter study. *International Journal of Legal Medicine* 110, 125–33.

Pérez-Lezaun, A., F. Calafell, D. Comas, E. Mateu, E. Bosch, R. Martinez-Arias, J. Clarimon, G. Fiori, D. Luiselli, F. Facchini, D. Pettener & J. Bertranpetit, 1999. Sex-specific migration patterns in Central Asian populations, revealed by analysis of Y-chromosome short tandem repeats and mtDNA. *American Journal of Human Genetics* 65(1), 208–19.

Poloni, E.S., O. Semino, G. Passarino, A.S. Santachiara-Benerecetti, I. Dupanloup, A. Langaney & L. Excoffier, 1997. Human genetic affinities for Y-chromosome p49a,f/*Taq*I haplotypes show strong correspondence with linguistics. *American Journal of Human Genetics* 61, 10–1035.

Rossi, E., B. Rolf, M. Schurenkamp & B. Brinkmann, 1998. Y-chromosome STR haplotypes in an Italian population sample. *International Journal of Legal Medecine* 112, 78–81.

Ruiz-Linares, A., D. Ortiz-Barrientos, M. Figueroa, N. Mesa, J.G. Munera, G. Bedoya, I.D. Velez, L.F. Garcia, A. Pérez-Lezaun, J. Bertranpetit, M.W. Feldman & D.B. Goldstein, 1999. Microsatellites provide evidence for Y-chromosome diversity among the founders of the New World. *Proceedings of the National Academy of Sciences of the USA* 96(11), 6312–17.

Schneider, S., J.M Kueffer, D. Roessli & L. Excoffier, 1997. *Arlequin ver1.1: a Software for Population Genetic Data Analysis*. Geneva: Genetics and Biometry Laboratory, University of Geneva, Switzerland.

Seielstad, M., E. Minch & L.L. Cavalli-Sforza, 1998. Genetic evidence for a higher female migration rate in humans. *Nature Genetics* 20, 278–80.

Seielstad, M., E. Bekele, M. Ibrahim, A. Toure & M. Traore, 1999. A view of modern human origins from Y chromosome microsatellite variation. *Genome Research* 9(6), 558–67.

Slatkin, M., 1995. A measure of population subdivision based on microsatellite allele frequencies. *Genetics* 139, 457–62.

Smouse, P.E., J.C. Long & R.R. Sokal, 1986. Multiple regression and correlation extensions of the Mantel test of matrix correspondence. *Systematic Zoology* 35, 627–32.

Stoneking, M., 1998. Women on the move. *Nature Genetics* 20, 219–20.

Weir, B.S. & C.C. Cockerham, 1984. Estimating F-statistics for the analysis of population structure. *Evolution* 38, 1358–70.

Zerjal, T., B. Dashnyam, A. Pandya, M. Kayser, L. Roewer, F.R. Santos, W. Schiefenhovel, N. Fretwell, M.A. Jobling, S. Harihara, K. Shimizu, D. Semjidmaa, A. Sajantila, P. Salo, M.H. Crawford, E.K. Ginter, O.V. Evgrafov & C. Tyler-Smith, 1997. Genetic relationships of Asians and Northern Europeans, revealed by Y-chromosomal DNA analysis. *American Journal of Human Genetics* 60, 1174–83.

Chapter 35

Y Chromosomes Shared by Descent or by State

Peter de Knijff

The genetic variation at the Y chromosome was studied among 275 males from four distinct Dutch regions and a pooled random Dutch population sample. For this analysis we used 7 short tandem-repeat (STR)-loci, 3 single-nucleotide-polymorphisms (SNPs) and an ALU ins/del polymorphism (which we, for simplicity will also include as a SNP in this abstract), all from the non-recombining part of the Y chromosome. The 4 SNPs defined 6 distinct Y haplogroups, the 7 STRs defined 145 Y haplotypes, and when SNPs and STRs were combined, 163 distinct Y chromosomes could be identified. Differences and similarities between these regions, based on the relative frequencies of individual alleles and haplogroups/haplotypes, will extensively be discussed. We paid specific attention to those STR-based haplotypes which were shared between SNP-defined haplogroups. For this purpose, we assume that any similarity between such haplotypes is more likely owing to mutation processes (which we define here as identical by state, IBS) rather than because of a direct, albeit ancient, descent (designated identical by descent, IBD). Since Y-chromosome STRs are frequently used for forensic identification purposes it seems not irrelevant to obtain a more detailed knowledge of this phenomenon.

In 1995, the search for a suitable partner for mitochondrial Eve resulted in the discovery of Y-chromosomal Adam who, to Eve's delight, could also be traced back in sub-Saharan Africa some 270,000 ago (Dorit *et al.* 1995; Pääbo 1995). Strangely enough, Adam's discovery was based on the absence of any sequence variation in a part of the ZFY gene among 38 globally dispersed males. How different is the situation nowadays: geneticists can pick their choice from many polymorphic loci on the Y chromosome, ranging from unique-mutation-events (UME; i.e. base-pair substitutions, ALU insertion/deletion polymorphisms, and LINE insertions), short-tandem-repeats (STRs), and a single highly variable minisatellite locus (Jobling & Tyler-Smith 1995; Jobling *et al.* 1997). Most genetically distinct Y lineages can be identified in any given population sample by means of simple PCR-based genotyping methods.

For a forensic scientist, this opens the possibility of identifying male-specific DNA-profiles. With such profiles the identification of matches between mixed crime-scene samples (containing both male and female DNA) and suspects, or, in a missing-person case a match between a Y profile from skeletal remains and possible male relatives, is a matter of a limited number of analyses. For an evolutionary geneticist this wealth of polymorphic Y loci makes it possible to accurately trace and date male-mediated gene flow within and between the continents. For both genetic approaches one of the central questions that remains to be answered is the following: is it possible that seemingly unrelated males share identical chromosome Y profiles? However odd this question may seem, it has been shown that males from completely different populations share identical chromosome Y STR-based haplotypes (de Knijff *et al.* 1997). This similarity could be explained by the relative fast mutation rate of Y-STR loci as a result of which distinct Y lineages become similar by mutation events. In addition, within a given population Y lineages will, in general, be transmitted together with surnames. However, in our modern society, adop-

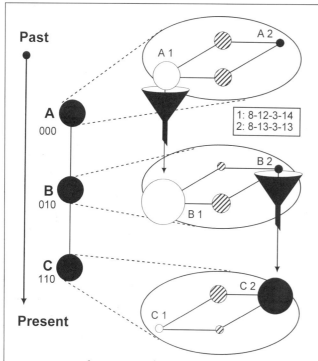

In a male population sample we have genetic information from three different UMEs and four STRs, all located on the Y chromosome. We assume that each UME has a mutation rate which is an order of magnitude slower when compared to each STR-locus. Furthermore, when only UMEs have been scored we define haplogroups, when only STR-loci are scored we define haplotypes, and when both UMEs and STRs are scored we define lineages.

From sequence information obtained from non-human primates we know the ancestral status of each of these UMEs which we designate 0. The derived state of each UME can then be designated 1. Obviously, in our population sample haplogroup A is the most ancestral, oldest haplogroup. Assuming single and non-recurrent mutation events, haplogroup B descents from A, and C descents from B. Hence, haplogroup C is our youngest, most recently derived haplogroup. Haplotype 1 reflects the combination of four STR alleles, each at a single locus, with 8, 12, 3, and 14 repeat-units. Haplotype 2, with the code 8-13-3-13, differs from haplotype 1 in two single-repeat-steps in two different STR loci. Assuming a single-step-mutation model, these haplotypes can be connected into a simple network. Within each haplogroup we can thus identify haplotype networks. In the ideal situation (as above), neighbouring haplogroups share a single STR-haplotype, thereby connecting all Y lineages within our population sample into a single network.

In this example, the mutation leading from a haplogroup A Y chromosome to the first haplogroup B Y chromosome occurred at the STR-haplotype 1 (thus at an A1 lineage). Since among all A chromosomes, type 1 is the most frequent one we assume that A1 is the most ancestral Y lineage in our network. As a consequence of this ultimate bottleneck effect (from all A1 lineages only a single one mutates to the first B1 lineage) B1 is, per definition the ancestral B lineage from which all other B lineages descent. Owing to a rare chance event, the mutation leading to the first C lineage occurred at the most derived B lineage, B2. When two males in our population both have the A1 lineage, we assume they are closely related and share the STR-haplotype 1 identical-by-descent (IBD). In contrast, when two males share haplotype 1, but on a different UME-background (e.g. A1 and C1), they share this haplotype 1 identical-by-state (IBS). Also in the case of e.g. the B2 and C2 lineages, which only differ by a single UME mutation event, we define these two lineages IBS.

Of course, this example illustrates a closed population structure. No deviant Y lineages were introduced into our population by e.g. migration. Also, we suppose to have sampled all distinct Y lineages present in our hypothetical population.

Figure 35.1. *The concept of reasoning and definitions explained.*

tion, the multiple use of specific anonymous sperm-donors, and the habit of using the mother's surname for children are just a few obvious non-genetic reasons why apparently unrelated males (not sharing the same surname) do share identical Y lineages. In this exposé I want to describe the results of a study aimed at finding an acceptable answer for this question.

Materials and methods

A total of 275 Dutch males were initially screened for 10 Y-chromosome-based polymorphisms. These include 6 STR-loci (DYS19, DYS389, DYS390, DYS391, DYS392, DYS393) and 4 UMEs (YAP, 92R7, SRY1532, M9). In DYS389 all four repeat units were scored as DYS389-A, -B, -C, and -D (Rolf *et al.* 1998). Later in our study two additional STR loci (DYS388 and DYS385) were typed to increase the resolution of STR-based haplotyping. Reduced median networks, connecting chromosome Y haplotypes, haplogroups, and lineages, were constructed using Network 2.0 (Bandelt *et al.* 1995).

In Figure 35.1, the theoretical basis for the IBD/IBS approach is explained in detail. In short, distinct Y chromosomes, designated haplogroups, are solely defined on the basis of UMEs. These haplogroups can be connected by means of single-mutation steps.

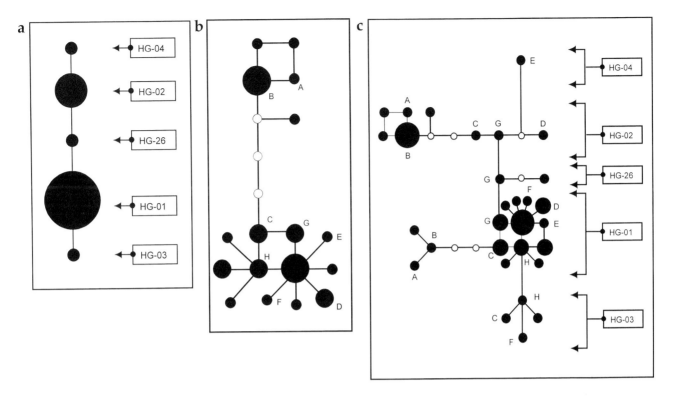

Figure 35.2. *Reduced median networks connecting Y haplogroups (2a), Y haplotypes (2b) and Y lineages (2c) as observed among 273 Dutch males.*

Within each haplogroup, we can identify distinct haplotypes, which are only defined by a number of STR-loci. Also these haplotypes can be connected on the basis of single-step-mutation events. When combined, we identify Y lineages. For the purpose of this paper, Y lineages which are identical for all UME and STR genotypes are defined as identical-by-descent (IBD) and those which are only identical for all STRs are defined as identical-by-state (IBS).

The combined IBS probability, p(IBS), defined as the probability that two Y chromosomes, randomly chosen from the same population, are IBS, was obtained by averaging all haplogroup-specific IBS probabilities, p(IBS)HGX.

Each p(IBS)HGX was estimated as the probability that an STR-defined haplotype, observed in an individual in haplogroup X is shared with at least one other individual in a different haplogroup.

Results and discussion

The genetic variation at the Y chromosome was studied among 275 Dutch males. The 4 UMEs combined, defined 5 haplogroups. The 6 STR loci combined, defined 145 haplotypes. When taken together, these 10 polymorphic Y loci defined 163 distinct Y lineages. Two males with a very rare haplogroup status were excluded from further analyses.

Figure 35.2 illustrates that by combining UMEs and STRs the resolution of chromosome Y-lineage typing increases dramatically. Figure 35.2a illustrates the reduced-median-network connecting the 5 haplogroups. Figure 35.2b illustrates a similar network connecting the 17 most frequent haplotypes (frequencies exceeding 1 per cent; i.e. shared between three males or more). These 17 haplotypes encompassed a total of 128 males (46.5 per cent). When also including the UME-types, 30 different Y lineages can be identified among the same 128 males (Fig. 35.2c). The effect of this increased resolution is most obvious when we consider the 8 haplotypes which are shared between males from different haplogroups. These types/lineages are indicated as A–H. Two seemingly neighbouring haplotypes, C and G (Fig. 35.1a) are observed among males from three distinct haplogroups, 01, 02, and 03 (for C), and 01, 02, and 26 (for G) respectively (Fig. 35.2c). The haplogroup and haplotype frequencies are shown in Table 35.1.

Also indicated in Table 35.1 are the average IBS probability (p(IBS)), and the haplogroup-specific IBS-probabilities (p(IBS)HGX). Not surprisingly, the high-

Table 35.1. *Frequencies and sharing status of chromosome Y haplotypes.*

				Haplogroup		
	All	HG-01	HG-02	HG-03	HG-04	HG-26
Haplotype (see Fig. 35.2):	*Nr. of males with STR-haplotypes shared between haplogroups*					
A	3	1	2	0	0	0
B	22	2	20	0	0	0
C	9	7	1	1	0	0
D	7	6	1	0	0	0
E	5	4	0	0	1	0
F	4	3	0	1	0	0
G	13	9	3	0	0	1
H	13	12	0	1	0	0
Combined	**76**	**44**	**27**	**3**	**1**	**1**
Nr. males with unique STR-haplotypes or shared within haplogroups						
All other haplotypes	197	103	67	12	4	11
Total	273	147	94	15	5	12
p(IBS)	0.008	0.019	0.018	0.003	0.001	0.001

est p(IBS)HGX was observed among the two most frequent haplogroups 01 and 02 (1.9 per cent and 1.8 per cent). We observed a p(IBS) of about 0.8 per cent. It could be argued that we did not use sufficient Y-STR loci to define haplotypes, and/or that UME-genotyping errors were made. Therefore we retyped the UME status in all males with an STR-haplotype shared between haplogroups and conformed that no such errors were made. We also extended our STR haplotype by two additional Y-STR loci, DYS385 (with two variable bands) and DYS388. This indeed reduced p(IBS), from 0.8 per cent to 0.2 per cent (not shown), but still did not resolve all shared haplotypes. As explained in Figure 35.1, we do not expect such a complete resolution. At least the haplotype connecting the haplogroups should (or could) be present in neighbouring haplogroups. It is in this respect interesting to note that the inclusion of DYS385 and DYS388 did not affect the sharing status of haplotype G. This suggests that this haplotype could be the ancestral one connecting the haplogroups 01, 02, and 26.

An average IBS probability of 0.2 per cent may seem irrelevantly low for evolutionary studies, but for forensic purposes it is unacceptably high. In any given Y-STR haplotype data base, most Y lineages will be unique (only observed in a single male), with a frequency in the order of 0.1–1 per cent depending on the size of the data base. However, this means that even a Y-STR haplotype consisting of 8 loci, encompassing 12 variable alleles, has an average p(IBS) approaching the lowest Y-STR haplotype frequency. Evidently, bigger data bases (with 10,000 males or more), other statistical approaches to estimate Y-STR haplotype frequencies from such data bases, and haplotypes with more, and/or more variable loci are necessary to circumvent this problem.

In conclusion, this study clearly illustrates the power of the combined use of Y STRs and UME's to define Y lineages. However, it also illustrates that the IBS sharing status is an hitherto underestimated or even neglected phenomenon which deserve more attention, especially among the forensic scientists using chromosome Y loci.

References

Bandelt, H-J., P.Forster, B.C. Sykes & M.B. Richards, 1995. Mitochondrial portraits of human populations using median networks. *Genetics* 141, 743–53.

de Knijff, P., M. Kayser, A. Cagli, D. Corach, N. Fretwell, C. Gehrig, G. Graziosi, F. Heidorn, S. Herrmann, B. Herzog, M. Hidding, K. Honda, M. Jobling, M. Krawczak, K. Leim, S. Meuser, E. Meyer, W. Oesterreich, A. Pandya, W. Parson, G. Penacino, A. Piccinini, A. Pérez-Lezaun, M. Prinz, C. Schmitt, P.M. Schneider, R. Szibor, J. Teifel-Greding, G. Weichhold & L. Roewer, 1997. Chromosome Y microsatellites: population genetic and evolutionary aspects. *International Journal of Legal Medicine* 110, 134–40.

Dorit, R.L., H. Akashi & W. Gilbert, 1995. Absence of polymorphism at the ZFY locus on the human Y chromosome. *Science* 268, 1183–5.

Jobling, M.A. & C.Tyler-Smith, 1995. Fathers and sons: the Y chromosome and human evolution. *Trends in Genetics* 11, 449–56.

Jobling, M.A., A. Pandya & C. Tyler-Smith, 1997. The Y chromosome in forensic analysis and paternity testing. *International Journal of Legal Medicine* 110, 118–24.

Pääbo, S., 1995. The Y chromosome and the origin of all us (men). *Science* 168, 1141–2.

Rolf, B., E. Meyer, B. Brinkmann & P. de Knijff, 1998. Polymorphism at the tetranucleotide repeat locus DYS389 in 10 populations reveals strong geographic clustering. *European Journal of Human Genetics* 6, 583–8.

Chapter 36

Lactase Haplotype Diversity in the Old World

Edward J. Hollox, Mark Poulter & Dallas M. Swallow

Lactase persistence is a genetic trait present at different frequencies in different populations: it is frequent in northern Europeans and certain African and Arabian nomadic tribes, both groups that have a history of drinking fresh milk. Selection is likely to have had an important role in the establishment of these varying frequencies. We have previously shown that the lactase persistence/non-persistence polymorphism in humans is cis-*acting to the lactase gene (Wang* et al. *1995) and that lactase persistence is associated, in Europeans, with the most common 70 kb lactase haplotype, termed A (Harvey* et al. *1998). We have studied polymorphisms in over 1000 chromosomes to examine 11-site haplotype frequencies in 11 populations that differ in lactase persistence frequency. Our data show that there are four common haplotypes (A, B, C, U) that are not closely related, having a minimum of three sites different from each other. These four haplotypes account for over 80 per cent of all haplotypes in non-African populations, but only account for 48 per cent of all haplotypes in sub-Saharan African populations. The A haplotype is shown to be at a much higher frequency in northern Europeans, where persistence is common, as compared to other populations where persistence is rare. The U haplotype has an unusual distribution, being common in East Asia and sub-Saharan African populations but not observed in Indo-European populations.*

DNA polymorphism data has had an important role in shaping theories on human evolution. Most work has concentrated on non-recombining DNA elements such as mtDNA and Y-chromosomal DNA (Hammer 1995; Vigilant *et al.* 1991), or autosomal regions small enough to exclude significant recombination (Harding *et al.* 1997; Tishkoff *et al.* 1996), but more recently larger sections of autosomal genes, usually associated with diseases, have been analyzed in a variety of populations (Clark *et al.* 1998; Kidd *et al.* 1998; Tishkoff *et al.* 1998).

The population genetics of autosomal genes is complicated by recombination, but studying certain autosomal loci may help reveal the role of selection in human history. For example, the domestication of plants and animals involved a change in diet, and by analyzing genes involved in nutrition we may be able to investigate how that change affected human diversity and infer possible historical events. The lactase persistence/non-persistence polymorphism, in which the intestinal enzyme lactase persists at childhood levels in adults or declines in activity like other mammals, has been of interest to anthropologists and human geneticists for some time. Lactase is required to digest lactose, the main carbohydrate in mammalian milk, and the lactase persistence allele is frequent in northern Europeans and certain African and Arabian nomadic tribes, all groups that have a history of drinking fresh milk.

Since the causative mutation is *cis*-acting to the lactase gene (Wang *et al.* 1995), analysis of variation across the gene may help to reveal the history of the gene and of the human populations themselves. Here we analyze haplotypes across 70 kb of the lactase gene in individuals from several populations that differ in lactase persistence frequency. Studies of

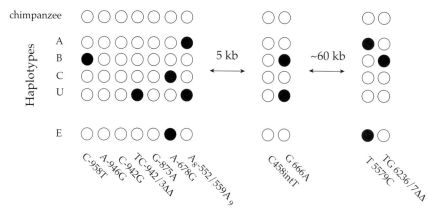

Figure 36.1. *Allelic composition of the four common haplotypes. The relative positions of the polymorphic sites in the gene are indicated by numbers representing nucleotides in the cDNA upstream of the start of transcription, except C458x2T which is intron 1. The polymorphic loci tested are shown as circles, with empty circles representing the original haplotype as defined by sequencing five chimpanzees, and filled circles representing derived alleles. The approximate distances in kilobases between the three groups of loci are shown. One chimpanzee was found to be heterozygous at TG6236/7ΔΔ (data not shown).*

this region in families show that it is small enough to observe linkage disequilibrium across the whole region, yet large enough to observe evidence of historic recombinational events (Harvey *et al.* 1995).

Results

Eleven polymorphic sites, shown in Figure 36.1, were tested as described in Harvey *et al.* 1995. The individuals tested were 52 northern and 53 southern Europeans (Harvey *et al.* 1998), 78 northern Indians, 42 southern Indians (Dravidian speakers), 34 Mordavians (a Finno-Ugric-speaking population in Russia), 40 Russians from Perm, 85 Roma (gypsies) from Slovakia, 51 Chinese living in Singapore, 100 Malay, 41 Japanese from Osaka, 36 Papua New Guineans, 36 Bantu-speaking South Africans from Johannesburg and 15 San from northwest Namibia. Pairwise linkage disequilibria for each pair of loci were calculated using ASSOCIATE (Terwilliger & Ott 1994) and revealed a generally high amount of linkage disequilibrium across the gene. However, population variation in pairwise linkage disequilibria scores was occasionally observed. For example, alleles at the loci A-678G and T5579C are in linkage disequilibrium in all populations except Russians and southern Europeans, where they are in almost total equilibrium.

Allelic phase was determined by allele-specific PCR for T5579C and TG6236/7ΔΔ, by single strand conformation analysis of a single PCR product for sites C458intT and G666A, and by the denaturing gradient gel electrophoresis of a single PCR product for sites C-958T, A-946G, C-942G, TC-942/3ΔΔ and G-875A (Hollox *et al.* 1999). This simplified haplotype determination by allowing a group of phase-known loci to be treated as a single multiallelic locus. Analysis of haplotype frequencies in each population by observation and by maximum-likelihood statistics revealed four common global haplotypes, A, B, C and U (Fig. 36.2). The allelic composition of these haplotypes is shown in Figure 36.1, and the haplotypes are not closely related, each having at least three alleles different from each other. Haplotype A is the most common, and is frequent in northern Europeans, a population with a high frequency of lactase persistence (Fig. 36.2). B is highest in Papua New Guineans, and present in all populations except Bantu. Haplotype C is also present in all populations, but rare in northern Europeans and San.

A further 35 haplotypes are observed in the populations, which are shown as 'others' in Figure 36.2. Twelve haplotypes are unique to the San and Bantu, but a few of these may be artefacts of maximum-likelihood analysis.

By analysis of the allelic composition of the haplotypes, several may be recombinants of the four common haplotypes, such as haplotype E (Fig. 36.1). This was confirmed by analysis of two polymorphic loci 3' to TG6236/7ΔΔ by allele-specific PCR. Haplotype E is quite frequent in southern Europeans and Russians, which reflects the low linkage disequilibrium between A-678G and T5579C in these two populations. Haplotype E presents the fourth combination of the two alleles at these two loci, the other three combinations present in haplotypes A, B and C.

Discussion

This study emphasizes the unusually high frequency of haplotype A in northern Europeans as compared with other populations. We have previously shown that lactase persistence is associated with haplotype

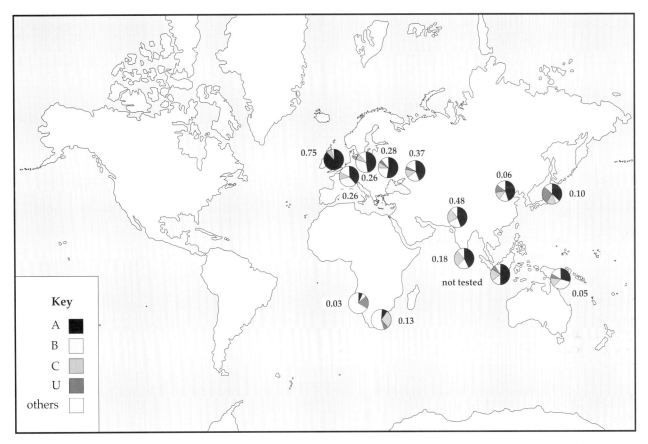

Figure 36.2. *Frequencies of the four common haplotypes in different populations. The numbers by each pie chart represent estimates of the lactase persistence allele frequency in the population from other studies (Flatz 1987). Haplotype frequencies shown are estimated by inspection, except Bantu-speaking South Africans and San where frequencies are estimated by maximum likelihood statistics using the EH program (Terwilliger & Ott 1994). Location and size of each population sample are described in the results.*

A in Europeans (Harvey *et al.* 1998), and we suggest that selection for the linked allele for lactase persistence is responsible for this very high frequency of the A haplotype.

The U haplotype is found at reasonable frequencies in all populations except Indo-European populations, where it is very rare or absent. This unusual distribution parallels the distribution of the YAP haplotype 3 on the Y chromosome, which possibly originated in Asia (Hammer *et al.* 1997). The distribution may also be due to an expansion of Indo-European populations or selection at a locus linked to the U haplotype. The observation of U haplotypes in the Russian population in the Urals may be a result of gene flow from a neighbouring Asiatic population since the Yakut from Siberia have a high frequency of U haplotypes (E.J. Hollox & S. Markova, unpubl. observ.).

The sub-Saharan African samples show much greater haplotype diversity than other populations with many extra haplotypes not seen in the rest of the population samples. This pattern of higher diversity has been observed at other loci and has been interpreted as evidence for an 'out of Africa' model for peopling of the Old World (Ayala & Escalante 1996 for review). Analysis of the unweighted means of haplotype composition shows that the four common haplotypes comprise 88 per cent of total haplotype diversity in non-Africans, but only 34 per cent in sub-Saharan Africans. Both sub-Saharan African populations have a high frequency of other haplotypes, many of which are not observed in non-African populations. This evidence suggests that the four common haplotypes originated in Africa before a single migration into the rest of the Old World, and the distribution arose by genetic drift. Direct selection of alleles is unlikely to account for the loss of haplotype diversity, since there are no alleles shared between all four common haplotypes that are not present in other haplotypes. The migration probably occurred relatively recently since there has not been enough time for recombination to disrupt the common haplotypes.

We hope to extend analysis of the lactase locus by using single nucleotide polymorphisms and microsatellites in European and Middle Eastern populations. This will help to reveal how selection, together with population migration and drift, affected patterns of variation within this genomic segment.

Acknowledgements

We would like to thank Professor T. Jenkins, Dr A. Krause, Dr N. Saha, Dr A. Kozlov, Dr V. Ferak and Dr P. Johnson for samples. This work was funded by the UK Medical Research Council.

References

Ayala, F.J. & A.A. Escalante, 1996. The evolution of human populations: a molecular perspective. *Molecular Phylogenetics and Evolution* 5(1), 188–201.

Clark, A.G., K.M. Weiss, D.A. Nickerson, S.L. Taylor, A. Buchanan, J. Stengard, V. Salomaa, E. Vartianen, M. Perola, E. Boerwinkle & C.F. Sing, 1998. Haplotype structure and population genetic inferences from nucleotide-sequence variation in human lipoprotein lipase. *American Journal of Human Genetics* 63, 595–612

Flatz, G., 1987. Genetics of lactose digestion in humans. *Advances in Human Genetics* 16, 1–77.

Hammer, M., 1995. A recent common ancestry for human Y chromosomes. *Nature* 378, 376–8.

Hammer, M., A.B. Spurdle, T. Karafet, M.R. Bonner, E.T. Wood, A. Novelletto, P. Malaspina, R.J. Mitchell, S. Horai, T. Jenkins & S.L. Zegura, 1997. The geographic distribution of human Y chromosome variation. *Genetics* 145, 787–805.

Harding, R.M., S.M. Fullerton, R.C. Griffiths, J. Bond, M.J. Cox, J.A. Schneider, D.S. Moulin & J.B. Clegg, 1997. Archaic African and Asian lineages in the genetic ancestry of modern humans. *American Journal of Human Genetics* 60, 772–89.

Harvey, C.B., W. Pratt, I. Islam, D.B. Whitehouse & D.M. Swallow, 1995. DNA polymorphisms in the lactase gene: linkage disequilibrium across the 70 kb region. *European Journal of Human Genetics* 3, 27–41.

Harvey, C.B., E.J. Hollox, M. Poulter, Y. Wang, M. Rossi, S. Auricchio, T.H. Iqbal, B.T. Cooper, R. Barton, M. Sarner, R. Korpela & D.M. Swallow, 1998. Lactase haplotype frequencies in Caucasians: association with the lactase persistence/non-persistence polymorphism. *Annals of Human Genetics* 62, 215–23.

Hollox, E.J., M. Poulter, Y. Wang, A. Krause & D.M. Swallow, 1999. Common polymorphism in a highly variable region upstream of the human lactase gene affects DNA-protein interactions. *European Journal of Human Genetics* 7, 791–800.

Kidd, K.K., B. Morar, C.M. Castiglione, H. Zhao, A.J. Pakstis, W.C. Speed, B. Bonné-Tamir, R.B. Lu, D. Goldman, C. Lee, Y.S. Nam, D.K. Grandy, T. Jenkins & J.R. Kidd, 1998. A global survey of haplotype frequencies and linkage disequilibrium at the DRD2 locus. *Human Genetics* 103, 211–27.

Terwilliger, J.D. & J. Ott, 1994. *Handbook of Human Genetic Linkage*. Baltimore (MD): Johns Hopkins University Press.

Tishkoff, S.A., E. Dietzsch, W. Speed, A.J. Pakstis, J.R. Kidd, K. Cheung, B. Bonné-Tamir, A.S. Santachiara-Benerecetti, P. Moral & M. Krings, 1996. Global patterns of linkage disequilibrium at the CD4 locus and modern human origins. *Science* 271, 1380–87.

Tishkoff, S.A., A. Goldman, F. Calafell, W.C. Speed, A.S. Deinard, B. Bonné-Tamir, J.R. Kidd, A.J. Pakstis, T. Jenkins & K.K. Kidd, 1998. A global haplotype analysis of the myotonic dystrophy locus: implications for the evolution of modern humans and for the origin of myotonic dystrophy mutations. *American Journal of Human Genetics* 62, 1389–402.

Vigilant, L., M. Stoneking, H. Harpending, K. Hawkes & A.C. Wilson, 1991. African populations and the evolution of human mitochondrial DNA. *Science* 253, 1503–7.

Wang, Y., C.B. Harvey, W.S. Pratt, V.R. Sams, M. Sarner, M. Rossi, S. Auricchio & D.M. Swallow, 1995. The lactase persistence/non-persistence polymorphism is controlled by a *cis*-acting element. *Human Molecular Genetics* 4, 657–62.

Chapter 37

History of Dairy Cattle-breeding and Distribution of LAC*R and LAC*P Alleles among European Populations

Andrew I. Kozlov & Dmitry V. Lisitsyn

Archaeological and palaeozoological evidence of dairy cattle-breeding in the Neolithic/ Eneolithic, Bronze, Iron and early Middle Ages is compared with the LAC*R (lactase restriction) allele distribution in Europe. Under the influence of the Indo-Europeans, dairy cattle-breeding spread north until the end of the Bronze Age (–3 kya). As a result a focus of high LAC*P allele concentration formed. The northern border of dairy cattle-breeding did not undergo any changes until the Early Iron Age (–1 kya). But the Uralic populations migrated from the northeast and the Altai populations from the southeast and, consequently, this process eroded the LAC*P allele focus. LAC*P allele concentration remained high only in regions close to the North Sea coast. Modern clinal diversity of allele concentrations does not reflect the stages of dairy cattle-breeding development in Europe; rather it shows the history of relationships between Indo-Europeans and representatives of other ethnic groups and cultures.

Physicians, anthropologists and geneticists have actively researched the different reactions to whole milk developed by human organisms. All healthy children under three to five years of age have a highly active digestion of the lactase enzyme, which splits the 'milk sugar' (lactose) the disaccharide contained in milk. Lactose provides about 30 per cent of calories derived from milk.

The LAC*R (lactase restriction) allele provides for the decrease of lactase activity of the small intestine among older children. This phenomenon, which is usually defined as primary hypolactasia, is an evolutionary initial phenotype. It leads to the child's refusal to be breast-fed and increases the chances of the next child being breast-fed properly. Primary hypolactasia is inherited in the autosomal recessive way (Sahi *et al.* 1973).

Primary hypolactasia occurs in different ethnic groups at different frequencies (Flatz 1987). It was shown that the population of the Old World could be divided into three large groups:
1. populations of Central Africa and Southwestern Asia, where lactose intolerance (primary hypolactasia) prevails (LAC*R >0.85);
2. populations of Northern and Central Europe in which stable lactose activity prevails (LAC*R <0.5); and
3. Italian, Spanish and some other populations, with an intermediate frequency of lactose intolerance (0.5 <LAC*R <0.85).

In the European populations no other genetic marker gives as great interpopulation variety as the LAC*R gene (Kozlov *et al.* 1998).

The existence of directed (clinal) variations in the LAC*R gene frequencies in Europe (Kozlov 1995) may be connected with either the history of selection (cultural-genetic hypothesis: McCracken 1971) or with the history of gene migrations (Kretchmer 1972; Kozlov & Lisitsyn 1997).

The selective hypothesis suggests that modern frequencies of LAC*P (lactase persistence) should correlate with the length of the milk cattle-breeding period in this or that ethnic group.

The purpose of our research was to compare the archaeological and archaeozoological evidence of dairy cattle-breeding in the Neolithic/Eneolithic,

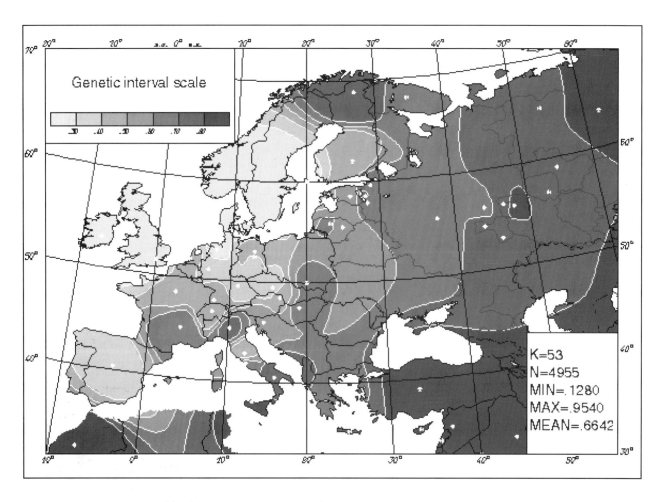

Figure 37.1. *LAC*R gene distribution in European populations.*

Bronze Age, Iron Age and early medieval period with LAC*R and LAC*P alleles distribution in Europe.

Materials and methods

We have summarized archaeological and palaeozoological evidence (Petrenko 1984; Tsalkin 1962, 1964; Kosintsev 1989; Loze 1991) characterizing the spread of dairy (horned) cattle in Eastern Europe and the territory beyond the Urals. As for the data concerning Western Europe, existing summaries have been used (Shnirelman 1989).

The study uses both the data on primary hypolactasia distribution obtained by the authors (Kozlov 1996; Kozlov & Lisitsyn 1996; 1997; Kozlov et al. 1998) and data from publications (reviews: Flatz 1987; Scrimshaw & Murray 1988). Hypolactasia diagnostics methods, as well as the methods used for calculating LAC*R and LAC*P gene frequencies and genetic mapping, were described by Kozlov et al. in 1998. In the map legend (Fig. 37.1) K is the number of original populations indicated on the map by points; N is the number of nodes of a regular grid with interpolated values of the gene frequency; MIN, MAX and MEAN are the minimal, maximal and average gene frequencies for the mapped area. The bottom panel of the map's legend includes an interval scale, corresponding to the selected genetic scale.

Obtained data and analysis of the results

Figure 37.1 demonstrates the modern distribution of LAC*R gene frequencies among various ethnoterritorial groups. Figure 37.2 shows the shift of the northern border of horned dairy cattle spread distribution in Europe in different epochsperiods.

It is assumed that the wild ox or aurochs was domesticated in the Middle East — in Anatolia and the Fertile Crescent (Perkins 1969; Epstein & Mason 1984). In Europe dairy cattle-breeding started not later than 8300 years ago. Under the influence of the Indo-Europeans dairy cattle-breeding spread to the

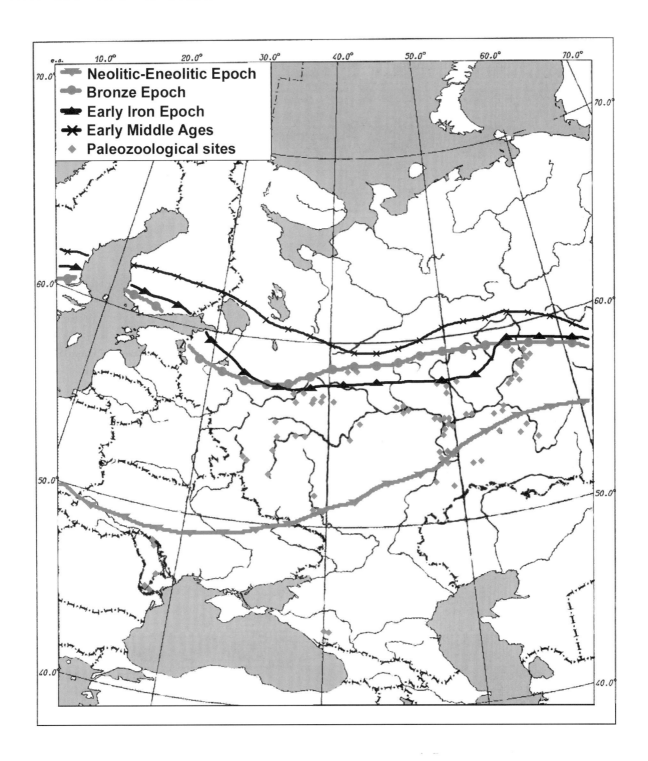

Figure 37.2. *Northern borders of the dairy cattle-breeding in Europe.*

North until the end of the Bronze Age (3000 years ago). According to archaeological and palaeozoological evidence the northern border of dairy cattle-breeding did not undergo any changes prior to the Early Iron Epoch Age (2300 years ago). This situation is depicted on Figure 37.2.

We can suppose that during this relatively stable period a spot of high LAC*P allele concentration formed on the territory of central and northwestern Europe. Selective advantages of the LAC*P allele were created by the ecological situation in the region. The bearers of this allele run smaller risks of suffering from rickets and osteomalacia, because they receive lactose with fresh milk, which promotes assimilation of calcium.

From the Bronze Age up to the Middle Ages there was an eastward migration of Indo-European populations (with comparatively high LAC*P allele concentration). They moved to the East of Europe interacting with other population groups. The lack of signs of constant dairy cattle-breeding in the material culture of ancient Uralic and Altaic peoples makes us suppose that these peoples were characterized by a more ancient genotype with a higher LAC*R allele concentration.

Judging by archaeological evidence, in the Early Iron Age there was a massive shift of Uralic peoples who had lived in the taiga of Eastern Europe and the territory beyond the Urals. They moved southwards and southwestwards until they reached the forest-steppe zone (Markov 1994). Approximately 500 years ago the populations belonging to the Turkic language group started moving to the steppes of Southern Russia. Around 1000 years ago the Uralic and Altaic peoples (Magyars and various Turkic language groups from the South-Russian steppes) dispersed over the territory west of the Carpathians.

The traditional cuisine of nomadic peoples of the Eurasian steppes even now consists primarily of sour milk products. The amount of lactose in these products is from 2 to 10 times less than in whole milk (Schrimshaw & Murray 1988). Modern representatives of the above-mentioned language groups mostly remain LAC*R allele bearers (Fig. 37.1).

As a result of the large-scale migration processes and the interaction of populations with LAC*R and LAC*P genotypes the supposed latitudinal boundaries of the LAC*P allele had to change and consequently came to have modern intricate contours (Fig. 37.1). LAC*P allele concentration remained high only in regions close to the North Sea coast.

Conclusion

In an example of various Saami subpopulations of Finland and Russia we have shown the role of gene inflow in the formation of the distribution of the modern LAC*R and LAC*P alleles (Kozlov & Lisitsyn 1997). Clinal variability in the frequencies of these genes in Europe and beyond can be explained mainly by the influence of the same factor.

The modern diversity of the LAC*R and LAC*P alleles reflects rather more the history of people's migrations than the history of cattle-breeding in Europe. The selective mechanism worked actively at the initial stages of LAC*P gene spreading. Later the distribution of the LAC*R and LAC*P alleles was formed mostly under the influence of migration pressure (gene inflow) on the part of the Uralic and Altaic peoples and their genetic assimilation with Indo-Europeans.

References

Epstein, H. & I.L. Mason, 1984. Cattle, in *Evolution of Domesticated Animals*, ed. I.L. Mason. New York (NY): Longman, 6–27.

Flatz, G., 1987. Genetics of lactose digestion in humans, in *Advances in Human Genetics*, eds. H. Harris & K. Hirschorn. New York (NY): Plenum Press 16, 1–77.

Kosintsev, P.A., 1989. Hunting and cattle-breeding among peoples of forest-steppe zone behind the Urals in the Bronze epoch, in *Formation and Development of Productive Economy in the Urals*, eds. V.D. Viktorova & N.T. Smirnov. Sverdlovsk: Uralic branch of the USSR Academy of Sciences, 84–104. [In Russian.]

Kozlov, A., 1995. The phenocline of primary hypolactasia in Finno-Ugrian populations, in *Papers on Anthropology*, vol. VI, ed. M. Thetioff. Tartu: University of Tartu, 111–15.

Kozlov, A., 1996. *Hypolactasia: Distribution, Diagnostics, Medical Tactics*. Moscow: ArctAn-C Laboratory Edition. [In Russian.]

Kozlov, A. & D. Lisitsyn, 1996. 'The milk habit' (hypolactasia) in Finno-Ugrian Peoples: a crossroad of physical anthropology, ethnology and linguistics. *Finnish-Ugrische Mitteilungen* 18/19, 67–81.

Kozlov, A. & D. Lisitsyn, 1997. Hypolactasia in Saami subpopulations of Russia and Finland. *Anthropologischer Anzeiger* 55 (3/4), 293–9.

Kozlov, A., E.V. Balanovskaya, S.D. Nurbaev & O.P. Balanovsky, 1998. Gene geography of primary hypolactasia in populations of the New World. *Russian Journal of Genetics* 34(4), 445–54.

Kretchmer, N., 1972. Lactose and lactase. *Scientific American* 227(4), 70–78.

Loze, I.A., 1991. Questions of productive economy formation on the territory of Eastern Baltics (cattle-breeding), in *Archaeology and Ethnology of Mari Land*, issue 19, eds. G.A. Archipov & B.S. Soloviyov. Yoshkar-Ola: Institute of History, Language and Culture, 13–22. [In Russian.]

McCracken, R., 1971. Lactase deficiency: an example of dietary evolution. *Current Anthropology* 12, 479–517.

Markov, V.N., 1994. Ananyino problem (certain outcomes and further problems to be solved), in *Questions of Tatarstan Archaeology: Kazan*, issue 1, ed. P.N. Starostin. Kazan: Institute of Language, Literature and History, Academy of Sciences of the Tatarstan Republic, 88–93. [In Russian.]

Perkins, D., Jr, 1969. Fauna of Catal Huyuk: evidence for early cattle domestication in Anatolia. *Science* 164, 177–9.

Petrenko, A.G., 1984. *Ancient and Medieval Cattle-breeding of Middle Volga and the Urals*. Moscow: Nauka. [In Russian.]

Sahi, T., J. Isokoski, J. Jussila *et al.*, 1973. Recessive inheritance of adult-type lactose malabsorption. *Lancet* 2, 823–6.

Schrimshaw, N.S. & E.B. Murray, 1988. The acceptability of milk and milk products in populations with a high prevalence of lactose intolerance. *American Journal of Clinical Nutrition* 48(4), 1079–159.

Shnirelman, V.A., 1989. *The Origin of Productive Economy: the Problem of Primary and Secondary Hotbeds*. Moscow: Nauka. [In Russian.]

Tsalkin, V.I., 1962. *On the History of Cattle-breeding and Hunting in Eastern Europe*. (Soviet Archaeological Materials and Research 107.) Moscow: Nauka. [In Russian.]

Tsalkin, V.I., 1964. Some outcomes of research of animalsí bones remains from archaeological excavations of the Late Bronze Epoch. *KSIA* 101, 24–30. [In Russian.]

Chapter 38

Mitochondrial DNA Diversity and Origins of Domestic Livestock

Daniel G. Bradley

Mitochondrial DNA sequence diversity has proved a rich source of inference about livestock domestic origins. In particular, the reconstruction of phylogenetic patterns in cattle, sheep, pig and water buffalo have revealed a striking duality in domestic sequences. These cluster tightly into two distinct and divergent groups in each case. Additionally these are geographically distributed, with diversity being divided in each case along an east–west axis. Calibrations of time depths corresponding to divergences between these clusters are of an order of magnitude greater than the 10,000-year history of herding. Thus phylogenetic topology, spatial distribution of lineages and quantitative divergence combine to suggest firmly that at least two domestications of divergent strains of the wild progenitor are a feature of the origins of each of the four domesticates.

From facts on the habits, voice, constitution, and structure of the humped Indian cattle, it is almost certain that they are descended from a different aboriginal stock from our European cattle; and some competent judges believe that these latter have had two or three wild progenitors.

Charles Darwin, *The Origin of Species*

Domestication

Darwin recognized the importance of the study of domestic diversity and dedicated the first chapter of *The Origin* to observations about their variation. In recent years, and in several emerging studies their origins are again being re-examined and new layers of information and argument are being added by the molecular analysis of diversity.

The taming and controlled breeding of the major animal domesticates was part of a suite of transformations of human society which formed the Neolithic revolution. This is best understood in the archaeological record of the Near East, in particular the Fertile Crescent, which witnessed the earliest development of an agricultural economy in the world, beginning prior to 10,000 years ago. The biological products of this remarkably innovative region are now ubiquitous and include plants such as barley, the wheats, lentils, plus the four major domestic livestock species, goats, sheep, pigs and cattle (Smith 1995, 255).

Archaeological identification of the sites and times of domestication events is achieved via a number of layers of inference. Obviously, any candidate locale must have been within the range of the wild progenitor at the time of domestication and the case is strengthened where the species has formed a substantial fraction of older hunted remains. Also, patterns of faunal remains under domestication are detectably different from wild captures. Firstly, a suggestion of the presence of domesticates arises where the fraction of the total kill which a species comprises climbs dramatically in later levels. Secondly, size reduction in remains over time (a consequence of both genetic and environmental effects of captivity) is taken as evidence of domestication. Thirdly, patterns within a species kill may be informative, such as sex and age ratios which suggest planned culling rather than the more random distribution expected from a hunted sample. Overall, a convincing candidate region for a domestic centre is one where a staged temporal transition from characteristically wild to domestic faunal patterns may be demonstrated (Meadow 1989).

Phylogeographic inertia

Reconstruction of past population history from genetic data, although fraught with difficulty, presumes that characteristics of modern samples show some continuity with those of similar geography from the distant past. Analysis of mtDNA variation in livestock has advantage in this exercise. Maternal lineages are likely to show some geographical inertia, especially with respect to secondary introgression. This is a result of the high disparity between male and female reproductive variance under managed breeding which means that genetic change has undoubtedly been predominantly male-mediated. For example, African bovine mtDNA display uniformly *Bos taurus* sequences despite a clear and substantial *Bos indicus* introgression to the continent over several millennia (MacHugh *et al.* 1997). Importantly, the rich information inherent in haplotypes invites detailed phylogenetic analysis and sensitive comparisons of genetic diversity between regions.

In the following, I summarize published mtDNA diversity data for four domestic ungulates: cattle, sheep, pig and water buffalo. The striking theme of phylogeographic duality is illustrated and its implications for understanding domestication are discussed.

MtDNA phylogenies

There are four substantial sets of mtDNA data available for production livestock. There are 43 published complete control region sequences published for cattle sampled in disparate locations on three continents (Loftus *et al.* 1994; Mannen *et al.* 1998). The 42 sheep complete control region sequences available all come from a mixed breed sample taken from New Zealand herds (Wood & Phua 1996). Partial pig control region sequences have been analyzed by Watanobe *et al.* (1999) for 11 haplotypes, including both East Asian and European varieties. The water buffalo phylogeny is constructed using 15 cytochrome b sequences (Tanaka *et al.* 1996; Lau *et al.* 1998).

Figure 38.1 summarizes sequence diversity in each domestic ungulate data set using neighbour-joining networks which have been constructed using uncorrected sequence divergences. The most striking feature in each is the double-headed broomstick topology. Sequences invariably cluster into one of two groups, which are separated by a prominent internal branch which in each case is the longest in the phylogeny. Also for each, the separation of the two groupings effected by these bifurcations are supported by a bootstrap value of 100 per cent in 1000 replications.

A second important feature of the diversity is that the two clades in each case have a tendency to be geographically distributed. This distribution in each case is primarily along an East–West division. That division in cattle follows (with some qualification) the division between the European humpless *Bos taurus* and the Indian zebu *Bos indicus* noted by Darwin. The division in pig is also between European samples and those from Asia, in this case mostly animals sampled in Japan. Water buffalo follow a separation between, on one side river buffalo (found in India, Egypt and the Balkans) and on the other swamp buffalo of Southeast Asia. The sheep sample is from New Zealand herds and therefore of necessarily exotic origin. Although there was no clear distribution of the two haplotypes among these source populations, other sources indicate that the ancestral distribution of the two sheep clades are along an east–west Eurasian axis. Hiendleder *et al.* (1998) studied ovine mtDNA variation using RFLP analysis of whole chromosomes and discovered two distinct clades, of similar divergence to those identified here. One of these was found only within their European sheep sample and the second was detected in animals of both European and Central Asian provenance.

The third piece of information conveyed by the phylogenies is the quantitative divergence between the clades. The major internal branch lengths vary between 1.3 per cent sequence divergence (pig) and 4.1 per cent (cattle). Molecular clock calibrations have been applied to each of these data sets and diversities between the two clusters in each have been estimated to correspond to divergence times of the order of hundreds of thousands of (and even towards a million) years (Loftus *et al.* 1994; Hiendleder *et al.* 1998; Tanaka *et al.* 1996; Lau *et al.* 1998; Watanobe *et al.* 1999). It must be noted that there is a considerable potential for error in the calculation of mtDNA substitution rates; a particular complexity is the non-uniformity of rates across different sites, especially in the control region. Nevertheless, these estimates comfortably predate the time depth of domestic history which stretches to only *c.* 10,000 years BP (Smith 1995). Finally, in the four species, the diversity within each sister clade is of similar magnitude, although the numbers sampled in each sometimes differ.

Dual domestication

Two alternative hypotheses about domestic origins may be distinguished by these analyses. One version of the history of domestication holds that the domestication of a wild progenitor population was carried out in only one region for each species; through a set of events delimited in time and space. Where this is invoked for cattle, sheep and pig, that primary domestic centre lies the Fertile Crescent. Water buffalo origins are less well investigated, but a most likely primary centre would be in East Asia, possibly in the Chinese Yangtse valley rice zone (Bellwood 1996). A second version of domestication history holds that the capture and controlled breeding of several of the key livestock species may have taken place in more than one separate location. These multiple domestications may have been independent of each other, or could perhaps consist of one primary and other secondary events.

It is difficult to envisage how a single domestication event could result in any of the phylogenetic patterns shown here. A wild progenitor species would be expected to show some correlation between geography and genetic divergence. Any domestication of beasts from a locale which is restricted — for example to the region of the Fertile Crescent 10,000 BP — would be expected to result in a less divergent genetic sample of mtDNA haplotypes than a process which involved the adoption of maternal lineages over the range of a progenitor species (here often a large portion of Eurasia). The data for each species shows neither a single tightly coalescent pattern nor a collection of multiple, randomly and deeply rooted lineages. Rather these patterns of dual clustering are strongly suggestive of a theme of dual domestication.

It is credible that the diversity within each of the broom-head phylogenetic clusters is derived from the space- and time-delimited sampling of a subset of wild variation which a domestication event would entail. This diversity would also comprise the modest divergence which has undoubtedly accrued within the c. 10,000 years of the history of domestication. The quantitative divergence between each pair of clusters together with their east–west geographical separations provides strong support for at least two domestication centres for each of cattle, sheep, pig and water buffalo.

To adhere to any single origin hypothesis requires either an assumption that the molecular clock calibrations are wildly inaccurate, or that two highly divergent lineages were incorporated together into

Cattle

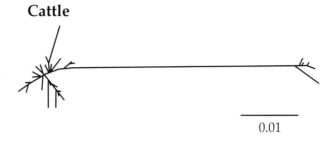

Sheep

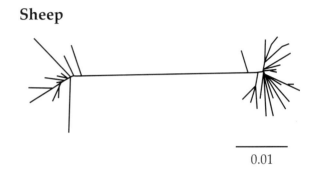

Water buffalo

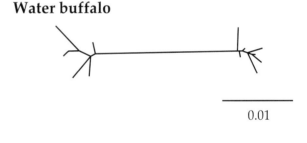

Pig

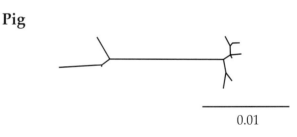

Figure 38.1. *Neighbour-joining networks linking mtDNA sequences from four domestic ungulates. Phylogenies were constructed using sequence divergences calculated without correction and discarding positions with displaying insertions or deletions. Analyses were performed using the PHYLIP programs (Felsenstein 1993).*

the domestic pool, and have persisted at comparable frequencies throughout the history of domestication. The former is a remote possibility and some data

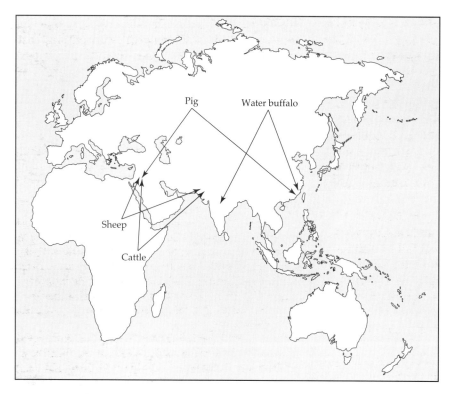

Figure 38.2. *Approximate locations of suggested domestication centres for sheep, cattle, pig and water buffalo.*

which have become, in all but the case of sheep, virtually fixed for one or other type.

Suggested locations of domestication centres for each species are given in Figure 38.2. Different parts of the Near East have been identified as centres for domestication of the pig, sheep and cattle (Smith 1995). Additionally, in accordance with Darwin's assertion and later archaeological works (Grigson 1991), the zebu cattle of India have often been postulated as having a separate domestic origin in that subcontinent. A plausible centre lies in the emerging Indus valley civilizations, where at Mehrgarh, archaeological remains have been documented which are consistent with an ancient transition from wild to domestic usage of local bovines (Meadow 1993). Although, evidence is weaker, this is also a plausible location for a second origin of sheep domestication. Given the current importance and distribution of river buffalo, the Indian subcontinent also serves as one possible location for a domestic origin (Meadow 1996). Early water buffalo remains have also been identified further east in Chinese Neolithic sites in the Yangtse valley rice zone (Bellwood 1996) and these may correspond to a separate origin, that of the morphologically and genetically divergent swamp buffalo. There is also some archaeological indication of early pig remains (of age approaching that of Fertile Crescent finds) in southern China. It seems that the wild boar may have been domesticated in the Near East and at the eastern extreme of its range.

Exceptions which prove the rule

The geographical and morphological separation of the two separate genetic groups in each species is not absolute. Firstly and interestingly, all African cattle sampled belong to the more western *Bos taurus* mtDNA clade, despite the clear morphological and physiological similarities between many African zebu breeds and the Indian *Bos indicus* which form the second clade. The first African cattle were of *Bos taurus* type and it seems that the secondary inward

from direct observation of mtDNA control region mutations in humans has suggested that commonly used calibrations may be underestimates of divergence rate by an order of magnitude (Howell *et al.* 1996); although many are unconvinced by this evidence (Macaulay *et al.* 1997). However, even if all the diversity within a species sample could be post-domestic in origin, this does not explain the tree morphologies incorporating a single predominant internal branch. One must ask, why is there an absence of intermediate branching and why should the bulk of the lineages coalesce within such tight portions of the tree's history? Additionally, when these phylogenies are rooted using outgroup species, the ancestral node falls in each case within the central branch. A fast divergence rate does not in itself allow a single domestication hypothesis.

The second explanation, that of a single domestic incorporation of both divergent clades, also might be possible given that deeply divergent mtDNA lineages may be encountered within geographically close wild species. However, this scenario becomes more laboured when one imposes the necessity of both persisting at appreciable frequencies over a 10,000-year period and also the differential assortment of these into eastern and western populations flow of zebu genetics has not included a substantial contribution of maternal lineages, in contrast to results from Y-chromosome and autosomal assays which confirm a majority *Bos indicus* contribution to many breeds (MacHugh *et al.* 1997).

A more complete correspondence with geography rather than morphology is also a feature in water buffalo. The Sri Lankan buffalo are similar in shape and function to the East Asian swamp type, but in contrast, fall into the the river buffalo mtDNA phylogenetic clade with neighbouring Indian samples.

Pig samples from the British Large White breed yield sequences of both the western and eastern types and, thus far are the only breed to not fall into either one or other mtDNA clade. This concurs with the history of the ubiquitous breed which is known to have a hybrid origin in an improvement programme in the eighteenth and nineteenth centuries which involved both local European and imported Chinese pig varieties.

The more substantial, presumably secondary, mixing of the two sheep mtDNA types between geographical groups may point to aspects of ovine domestic history. Small ruminant livestock in traditional herding societies provide more portable, tradable and smaller units of wealth than the larger domestic ungulates. This may have led to an increased mobility of sheep maternal lineages which has blurred an absolute geographical separation which might once have existed following separate domestications.

Conclusion

In sum, I have argued that the motif of dual domestication is a common one in livestock and may be readily discerned via analysis of mtDNA diversity within species. The strength of the correspondence in phylogenetic patterns between species gives a powerful phylogenetic template against which other species samples may be compared. For example, equine mtDNA diversity is more diffuse and may reflect quite a different, less-localized domestication process (Kim *et al.* 1999). There remains the possibility that further sampling within the livestock species discussed here may yield evidence for additional, separate mtDNA clades and perhaps further domestication events. Also, phylogeographic distribution of diversity within the clusters may allow further inference. For example, colleagues and I have argued previously for an African domestication based on the clustering of European and African haplotypes into two separate (albeit comparatively lowly divergent) *Bos taurus* clades (Bradley *et al.* 1996). One overall theme in these analyses is a pointing eastward away from an exclusive focus on domestic origins in the remarkably innovative Neolithic centres of the Fertile Cresent.

References

Bellwood, P., 1996. The origins and spread of agriculture in the Indo-Pacific regions: gradualism and diffusion or revolution and colonization, in Harris (ed.), 465–98.

Bradley, D.G., D.E. MacHugh, P. Cunningham & R.T. Loftus, 1996. Mitochondrial diversity and the origins of African and European cattle. *Proceedings of the National Academy of Sciences of the USA* 93, 5131–5.

Felsenstein, J., 1993. *PHYLIP (Phylogeny Inference Package) 3.5c edit.* Seattle (WA): University of Washington.

Grigson, C., 1991. An African origin for African cattle? — some archaeological evidence. *African Archaeological Review* 9, 119–44.

Harris, D.R. (ed.), 1996. *The Origins and Spread of Agriculture and Pastoralism in Eurasia.* London: UCL Press.

Hiendleder, S., K. Mainz, Y. Plante & H. Lewalski, 1998. Analysis of mitochondrial DNA indicates that domestic sheep are derived from two different ancestral maternal sources: no evidence for contributions from urial and argali sheep. *Journal of Heredity* 89, 113–20.

Howell, N., I. Kubacka & D. Mackay, 1996. How rapidly does the human mitochondrial genome evolve? *American Journal of Human Genetics* 59, 501–9.

Kim, K.I., Y.H. Yang, S.S. Lee, C. Park, R. Ma, J.L. Bouzat & H.A. Lewin, 1999. Phylogenetic relationships of Cheju horses to other horse breeds as determined by mtDNA D-loop sequence polymorphism. *Animal Genetics* 30, 102–8.

Lau, C., R. Drinkwater, K. Yusoff, S. Tan, D. Hetzel & J. Barker, 1998. Genetic diversity of Asian water buffalo (*Bubalus bubalis*): mitochondrial DNA D-loop and cytochrome b sequence variation. *Animal Genetics* 29, 253–64.

Loftus, R.T., D.E. MacHugh, D.G. Bradley, P.M. Sharp & P. Cunningham, 1994. Evidence for two independent domestications of cattle. *Proceedings of the National Academy of Sciences of the USA* 91, 2757–61.

Macaulay, V., M. Richards, P. Forster, K. Bendall, E. Watson, B. Sykes & H-J. Bandelt, 1997. MtDNA mutation rates: no need to panic. *American Journal of Human Genetics* 61, 983–6.

MacHugh, D.E., M.D. Shriver, R.T. Loftus, P. Cunningham & D.G. Bradley, 1997. Microsatellite DNA variation and the evolution, domestication and phylogeography of taurine and zebu cattle (*Bos taurus* and *Bos indicus*). *Genetics* 146, 1071–86.

Mannen, H., S. Tsuji, R.T. Loftus & D.G. Bradley, 1998. Mitochondrial DNA variation and evolution of Japanese black cattle (*Bos taurus*). *Genetics* 150, 1169–75.

Meadow, R.H., 1989. Osteological evidence for the process of animal domestication, in *The Walking Larder: Patterns of Domestication, Pastoralism and Predation*, ed. J. Clutton-Brock. London: Unwin Hyman, 80–90.

Meadow, R.H., 1993. Animal domestication in the Middle East: a revised view from the Eastern Margin, in *Harappan Civilisation*, ed. G. Possehl. 2nd edition. New Delhi & Oxford: IBH, 295–320.

Meadow, R.H., 1996. The origins and spread of agriculture and pastorlism in northwestern South Asia, in Harris (ed.), 390–412.

Smith, B., 1995. *The Emergence of Agriculture*. New York (NY): Scientific American Library.

Tanaka, K., C. Solis, J. Masangkay, K. Maeda, Y. Kawamoto, & T. Namikawa, 1996. Phylogenetic relationship among all living species of the genus *Bubalus* based on DNA sequences of the cytochrome b gene. *Biochemical Genetics* 34, 443–52.

Watanobe, T., N. Okumura, N. Ishiguro, M. Nakano, A. Matsui, M. Sahara & M. Komatsu, 1999. Genetic relationship and distribution of the Japanese wild boar (*Sus scrofa leucomystax*) and Ryukyu wild boar (*Sus scrofa riukiuanus*) analysed by mitochondrial DNA. *Molecular Ecology* 8, 1509–12.

Wood, N.J. & S.H. Phua, 1996. Variation in the control region sequence of the sheep mitochondrial genome. *Animal Genetics* 27, 25–33.

Chapter 39

Wheat Domestication

Robin Allaby

The transition from a hunter-gatherer to an agrarian society through the rise of crop domestication represents a key point in human prehistory. Current paradigms regarding the biological nature of domestication are themselves rooted in a presupposition that the domestication event was unique for each crop species. At first genetic evidence appeared to support this viewpoint, but progress with cereals such as wheat over the last five years casts doubt on this assumption. Evidence from genetic studies is mounting which supports a pattern of multiple domestications in parallel for many of the domesticate species, including wheat. Consequently, the elucidation of agricultural origins will require the construction of hypotheses and experimental design which do not presuppose an agricultural revolution involving a single domestication event for each species.

The dawn of agriculture represents a transitionary period in history in which humans began to control the floral and faunal environment rather than have their existence dictated by it. This change precipitated the development of human society by allowing the sustenance of high population density giving rise to internal specialization concerned with tasks other than food production. The rise of early society was associated with migration and expansion demography. These population movements not only resulted in the patterns of human allele distribution seen today (Cavalli-Sforza *et al.* 1994) but also in the allele distributions of the associated floral and faunal domesticates. Floral domesticates are of particular interest in this respect. The nature of domestication often renders the plant unable to self-propagate in the absence of human intervention owing to the development of a tough rachis, as in the case of wheat. Consequently, the biogeography of plant domesticates can represent a record of human movement providing an alternative source of genetic data to the humans themselves. In the case of the more primitive domesticates, such a data set may prove to be easier to analyze than humans being less confounded by the ensuing melée of historical movements.

In order to utilize the floral component of human history it is necessary to characterize the domestication process, which is still poorly understood. The transition to agriculture first occurred in the Near East 10,000 years ago, although the precise location of the event is a matter of some archaeological controversy. The northern Levant is favoured by some as the more likely location (Nesbitt 1998; Nesbitt & Samuel 1998), and the southern Levant considered the more likely by others (Jones *et al.* 1998). Each viewpoint presupposes that domestication occurred only once. The underlying assumption that domestication is a single revolutionary event finds its roots in the 'centres of origin' hypothesis first expounded by Vavilov (1926) in which the modern centre of genetic diversity of a crop plant is equated with its geographical origin of domestication. This philosophy, although questioned and reinterpreted over the years (Harris 1996a), has pervaded into current thinking. It is often accepted that each crop plant has a single origin, all cultivated forms deriving from one wild population whose domestication was a discrete event carried out in a single place at a single time (Zohary 1996). In its most extreme form this interpretation is extended to entire crop assemblages, implying that a small group of protofarmers were responsible for putting together an agricultural package meeting all of humanity's needs (Diamond 1997). This revolution-type model has almost certainly had an impact on the interpretation of human movements during prehistory invoking imagery of the first farmers endowed with power over their hunter-gatherer neighbours.

Until recently evidence from genetics gave no reason for questioning this underlying paradigm, and, in fact, appeared to support the monophyletic origins of most crops. The evidence follows three lines of reasoning which are concurrent with a single origin (Zohary 1999). Firstly, domestic crops appear to contain only a fraction of the genetic variation observed in the wild progenitors when their expressed proteins are analyzed. Second, the phylogenetic analyses carried out so far have implied domesticates of einkorn and emmer to be monophyletic (Heun *et al.* 1997; Mori *et al.* 1997). The third, most compelling point is that widespread crossing of domesticates does not result in wild type progeny which would be expected if two domesticates of independent origin were crossed. The reasoning for this is that one would expect two different domestication events to be the result of different mutation events in different genes in each case. A cross would give progeny in which for each domesticate version of a gene the wild type would also be present to counteract, hence a wild phenotype would result.

As our understanding of the genetics of domestication becomes clearer the ensuing picture becomes less simple. The parallel nature of crop domestication has long been apparent in the notable fact that at each of the known domestication centres around the world the major domesticates have all been members of the same family (*Poaceae*). This parallelism has transpired to be quite fundamental. Perhaps the most striking discovery in recent years was made by Paterson *et al.* (1995) who found that in members of *Poaceae* as distinct as maize (*Zea*) and rice (*Oryza*) equivalent genes had acquired mutations to give rise to the domestic phenotype. Consequently the assumption that independent domestications would be achieved by different mutations in different genes is no longer valid, and under these circumstances one might not expect to reproduce the wild phenotype by widespread crossing of domesticates of independent origin. Similarly, it has long been known that very few genes are involved in the production of the domesticate phenotype, however we now know that in the case of some features such as the development of a tough rachis in wheat a solitary gene is involved (Chen *et al.* 1998; Cao *et al.* 1997).

The extreme lack of genetic diversity in the domesticate gene pool predicted from a single domestication event has not been borne out by genetic data. In the case of emmer wheat diversity as measured by the value π, nucleotide diversity, has shown that about half the diversity present in the wild emmer gene pool is present within the domesticate gene pool (Mori *et al.* 1997). A similar situation was observed with barley (*Hordeum*) with similar values of diversity occurring in wild and domesticate forms (Holwerda *et al.* 1986). This observation has now been extended to the identification of alleles within the domesticate gene pool which are extremely divergent and clearly do not originate after domestication occurred 10,000 years ago, but represent ancient diversity (Allaby *et al.* 1999; Talbert *et al.* 1998). The notion that a domestic crop is the progeny of just a single wild plant now seems very unlikely indeed. It may be the case that a population of wild progenitors constitute the ancestry of a domestic crop, and efforts using the coalescent and other models have been used with barley, maize and teosinte to gauge the minimum size of the progenitor population required to give rise to domesticate gene pool observed today (Eyre-Walker *et al.* 1998; Ladizinsky 1998). The consensus of opinion of these studies is that relatively few ancestors are required if no major population bottlenecks occurred subsequent to domestication, Ladizinsky, for instance states a crop spanning 200 hectares is required. These studies are openly speculative and informative, but do require the presupposition of a single geographic origin.

Our understanding of both floral and faunal domesticate origins has developed considerably using three lines of genetic evidence; phenetics, cladistics and biogeography. In this respect studies of domestication have adopted a similar approach to that used in the study of human palaeodemography.

Phenetics and cladistics, although both employ genetics, are based on subtly different types of data and can lead to apparently opposing viewpoints. This has been observed with the study of the Neolithic transition in humans based on essentially phenetic classic markers (Cavalli-Sforza *et al.* 1994) and cladistic studies of mtDNA (Richards *et al.* 1996). The basic difference between the two types of analysis is in the characters used, with cladistics the characters are linked, with phenetics they are not. Phenetics can assimilate seductively large data sets, many genetic loci for instance, allowing a broad overview of genetic relationships. Cladistics is restricted to single gene genealogies or as large a single unrecombining segment of DNA as can be found. Fortunately for the animal kingdom, a rather large and accessible source of cladistic data presents itself in the form of mtDNA and more recently the non-recombining portion of the Y chromosome. Cladistic analysis of mtDNA in animal domesicates has elucidated the multiple domestication of the cow (Loftus et al. 1994; Bradley et al. 1996), pig (Watanabe et al. 1986), dog (Vilà et al. 1997) and in all likelihood the horse too (Lister et al. 1998). Unfortunately, the equivalent loci in plants are not suitable because of extensive recombination. Two studies of wheat evolution have incorporated large phenetic data sets of einkorn and emmer, obtaining genome wide markers using amplified fragment length polymorphisms (AFLP) (Heun et al. 1997) and restriction fragment length polymorphisms (RFLP) (Mori et al. 1997). Both studies show that domesticates most resemble other domesticates on the whole by virtue of the allele combinations present within any single accession. However, this should not be interpreted as a monophyletic origin of the domesticates. Monophyly is a cladistic statement regarding ancestors and daughter taxa. The point can be most clearly illustrated by the RFLP clustering of emmers on the basis of geography, all the wheat samples from Ethiopia for instance form a single 'clade' including accessions of *Triticum dicoccum*, *T. turgidum*, *T. pyramidile*, *T. polonicum*, and *T. abyssinicum* which would all be placed in clades with their own species in a true cladogram. Such a close phenetic relationship is the result of gene flow between these different interfertile species, and as such we should perhaps not be surprised to learn that domesticates form a phenetic group when one considers their sexual isolation from the wild gene pool during the 10,000 years of agriculture and gene flow between domesticate groups. The cladistic analysis of specific gene loci in wheat has repeatedly demonstrated the occurrence of alleles of genes which are undoubtedly ancient, predating the onset of agriculture, either representing multiple domestications or a diverse group of plants being domesticated (Allaby et al. 1999; Talbert et al. 1998). However, cladistics alone is not enough to distinguish between these two scenarios.

In order to examine the geographic history of domesticates one must take in to account their biogeography. To carry this out in a cladistic fashion one must consider the biogeography of single gene genealogies. In wheat the obvious species to choose for study is emmer, the founding crop of all wheats used today. Emmer represents a primitive form of wheat which expanded from the Near East into Europe with the first Neolithic technologies. Usurption of emmer by other wheats such as hexaploid varieties means primitive emmers are likely not have moved greatly from the original trajectory of agricultural spread. Examination of a biallelic system in emmer shows that emmer appears to have expanded from the Near East on two occasions (data submitted) correlating closely with the observation by Richards et al. (1996) of the occurrence of two Neolithic-aged mtDNA haplotypes in western and central Europe respectively.

While the domesticate forms themselves are effectively static geographically, requiring human intervention for propagation, the same is obviously not true for the wild progenitors. The inherently Vavilovian assumption that the location of the wild progenitors today represents the location of the domestication event is at best tenuous. This is especially true in the light of the recent discovery that the Black Sea formed quite suddenly only 7500 years ago (Ryan et al. 1997). This event is likely to have had a large effect on the biome of the Near Eastern region, especially around the northern Levant, probably introducing a more oceanic climate. It is likely to be the case that genetics alone will not be able to pinpoint the location of the first farming communities without the aid of archaeology. To associate the progenital biogeographic centre with the location of domestication also makes an assumption about the nature of the domestication process which may not be valid (Jones et al. 1998), namely is the pressure for cultivation actually within the biogeographical centre where wild stands are at their densist, or closer to the edges of the distribution where the wild resource is rarer? One can, however, identify single or multiple populations of wild forms most closely resembling the domesticates we observe today (Heun et al. 1997). Characterization of wild emmer populations by use of allele frequencies of single gene genealogies and studies of ribosomal DNA are unearthing evidence that the north and south Levant are distinct genetically (data submitted). Each centre appears to be the more likely seat of domestication for the two expansions of emmer respectively.

Evidence is mounting that questions the underlying paradigm of an agricultural revolution taking place in a specific locality, once. More diffuse, less revolutionary, parallel origins should not be discounted that may be indicative of an inevitable consequence of climatic change (Sherratt 1997). This alternative scenario of domesticate origins would have a deep impact on the interpretation of the nature of early human movements. In order to find the truth of the matter a combination of cladistic analyses and biogeography are being applied to the primitive wheats avoiding the presuppositions of a single origin.

Acknowledgements

Thank you to Terry Brown and Martin Jones for helpful discussions.

References

Allaby, R.G., M. Banerjee & T.A. Brown, 1999. Evolution of the High-Molecular-Weight glutenin loci of the A, B, D and G genomes of wheat. *Genome* 42, 296–307.

Bradley, D.G., D.E. MacHugh, P. Cunningham & R.T. Loftus, 1996. Mitochondrial diversity and the origins of African and European cattle. *Proceedings of the National Academy of the Sciences of the USA* 93, 5131–5.

Cao, W., G.J. Scoles & P. Hucl, 1997. The genetics of rachis fragility and glume tenacity in semi-wild wheat. *Euphytica* 94, 119–24.

Cavalli-Sforza, L.L., P. Menozzi & A. Piazza, 1994. *The History and Geography of Human Genes*. Princeton (NJ): Princeton University Press.

Chen, Q-F., C. Yen & J-L. Yang, 1998. Chromosome location of the gene four brittle rachis in the Tibetan weedrace of common wheat. *Genetic Resources and Crop Evolution* 45(5), 407–10.

Diamond, J., 1997. Location, location, location: the first farmers. *Science* 278, 1243–4.

Eyre-Walker, A., R.L. Gaut, H. Hilton, D.L. Feldman & B.S. Gaut, 1998. Investigation of the bottleneck leading to the domestication of maize. *Proceedings of the National Academy of the Sciences of the USA* 95, 4441–6.

Harris, D.R., 1996a. Introduction: themes and concepts in the study of early agriculture, in Harris (ed.) 1996b, 1–9.

Harris, D.R. (ed.), 1996b. *The Origins and Spread of Agriculture and Pastoralism in Eurasia*. London: UCL Press.

Heun, M., R. Schäfer-Pregl, D. Klawan, R. Castagna, M. Accerbi, B. Borghi & F. Salamini, 1997. Site of einkorn domestication identified by DNA fingerprinting. *Science* 278, 1312–14.

Holwerda, B.C., S. Jana & W.L. Crosby, 1986. Chloroplast and mitochondrial DNA variation in *Hordeum vulgare* and *Hordeum spontanuem*. *Genetics* 114, 1271–91.

Jones, M.K., R.G. Allaby & T.A. Brown, 1998. Wheat domestication. *Science* 279, 302–3.

Ladizinsky, G., 1998. How many tough-rachis mutants gave rise to domesticated barley? *Genetic Resources and Crop Evolution* 45, 411–14.

Lister, A., M. Kadwell, L.M. Kaagan, W.C. Jordan, M.B. Richards & H. Stanley, 1998. Ancient and modern DNA in a study of horse domestication. *Ancient Biomolecules* 2, 267–80.

Loftus, R.T., D.E. MacHugh, D.G. Bradley, P.M. Sharp & P. Cunningham, 1994. Evidence for two independent domestications of cattle. *Proceedings of the National Academy of the Sciences of the USA* 91, 2757–61.

Mori, N., T. Moriguchi & C. Nakamura, 1997. RFLP analysis of nuclear DNA for study of phylogeny and domestication of tetraploid wheat. *Genes and Genetic Systems* 72(3), 153–61.

Nesbitt, M., 1998. Where was einkorn wheat domesticated? *Trends in Plant Science* 3(3), 82–3.

Nesbitt, M. & D. Samuel, 1998. Wheat domestication: Archaeobotanical evidence. *Science* 279, 1433.

Paterson, A.H., Y-R. Lin, Z. Li, K.F. Schertz, J.F. Doebley, S.R.M. Pinson, S-C. Liu, J.W. Stansel & J.E. Irvine, 1995. Convergent domestication of cereal crops by independent mutations at corresponding genetic loci. *Science* 269, 1714–18.

Richards, M., H. Côrte-Real, P. Forster, V. Macaulay, H. Wilkinson-Herbots, A. Demaine, S. Papiha, R. Hedges, H-J. Bandelt & B. Sykes, 1996. Palaeolithic and Neolithic lineages in the European mitochondrial gene pool. *American Journal of Human Genetics* 59, 185–203.

Ryan, W.B.F., W.C. Pitman III, C.O. Major, K. Shimkus, V. Moskalenko, G.A. Jones, P. Dimitrov, N. Gortir, M. Sakinc & H. Yüce, 1997. An abrupt drowning of the Black Sea shelf. *Marine Geology* 138(1–2), 119–26.

Sherratt, A., 1997. Climatic cycles and behavioural revolutions: the emergence of modern humans and the beginning of farming. *Antiquity* 71, 271–87.

Talbert, L.E., L.Y. Smith & M.K. Blake, 1998. More than origin of hexaploid wheat is indicated by sequence comparison of low copy DNA. *Genome* 41, 402–7.

Vavilov, N.I., 1926. *Studies on the Origin of Cultivated Plants*. Leningrad: Institut Botanique Appliqué et d'Amelioration des Plantes.

Vilà, C., P. Savolainen, J.E. Maldonado, I.R. Amorim, J.E. Rice, R.L. Honeycutt, K.A. Crandall, J. Lundeberg & R.K. Wayne, 1997. Multiple and ancient origins of the domestic dog. *Science* 276, 1687–9.

Watanabe, T., Y. Hayashi, J. Kimura, Y. Yasuda, N. Saitou, T. Tomita & N. Ogasawara, 1986. Pig mitochondrial DNA: polymorphiusm, restriction map orientation, and sequence data. *Biochemical Genetics* 24(5–6), 385–96.

Zohary, D., 1996. The mode of domestication of the founder crops of southwest Asian agriculture, in Harris (ed.) 1996b, 142–58.

Zohary, D., 1999. Monophyletic vs. polyphyletic origin of the crops on which agriculture was founded in the Near East. *Genetic Resources and Crop Evolution* 46, 133–42.

Chapter 40

Inferring the Impact of Linguistic Boundaries on Population Differentiation and the Location of Genetic Barriers: a New Approach

Isabelle Dupanloup de Ceuninck, Stefan Schneider & Laurent Excoffier

Recent advances in the fields of diachronic linguistics and population genetics offer the possibility to study the correspondence between these two disciplines. In the last few years, several studies have revealed a strong correlation between the linguistic and the genetic differentiation of human populations at a worldwide or at a continental scale. These results may reveal the fact that language differences impede gene flow and that linguistic boundaries act as barriers to the exchange of genes between populations. We propose here to present a new methodology we have developed to quantitative study the impact of linguistic frontiers on gene flow, taking into account a pattern of isolation-by-distance between populations from the same linguistic group. This new method, partly based on the analysis of molecular variance, can also be extended to locate and evaluate the genetic impact of the most important genetic boundaries in a surface of gene frequencies, without need for interpolation. We will show the results of the application of our methodology to the genetic structure of populations in different linguistics families of Eurasia and Africa, using mostly conventional genetic markers, for which enough population samples are available.

Several techniques have been developed to detect in a specific region the presence of genetic boundaries which can be defined as spatial areas where the rate of change of gene frequencies is particularly high. When these boundaries are detected, they are often compared to physical or cultural barriers. The goal is usually to interpret the spatial structure of gene frequencies in the light of known ecological or cultural processes. Monmonier algorithm is a clustering method for identifying zones of sharp gene frequency variation (Monmonier 1973). Wombling is an alternative method for locating abrupt genetic change but requires interpolated surface of gene frequencies (Barbujani *et al.* 1989; Barbujani & Sokal 1990).

In a complementary way, several authors have tried to ascertain the biological impact of boundaries defined on the basis of non-genetic factors. The frontiers are here defined *a priori* and the goal is to test for increased genetic changes at these boundaries (Sokal *et al.* 1988; 1989).

We present here a new method to estimate specifically the impact of linguistic boundaries on gene flow and to establish if language-family boundaries correspond to genetic barriers (Dupanloup de Ceuninck *et al.* forthcoming). The action of linguistic boundaries on population differentiation is here evaluated in two complementary ways. We measure the boundary effect using an approach that is specifically based on an isolation by distance model. This approach is compared to a more conventional analysis of genetic structure under a model that does not assume a geographical patterning of the populations. Since the effect of the language boundary may be heterogeneous in its different sections, we evaluate by our two approaches the action of particular genetic processes on different portions of the boundary and thus evaluate the homogeneity of these processes along this boundary.

The second methodology, we present here, permits us to locate the most important genetic barriers in the distribution area of a set of populations (Dupanloup de Ceuninck et al. in prep). With a simulated annealing strategy, we assess the position of the barriers which maximize the genetic differentiation between the groups of populations defined by these barriers.

Material and methods

Inferring the impact of linguistic boundaries on population differentiation

The method we developed to evaluate the genetic impact of linguistic boundaries does not require us to know the exact geographical location of the boundary under study: it is automatically defined on a geographical map from the mere linguistic affiliation of the sampled populations. A Voronoi diagram is superimposed on the distribution map of the sampling points. It corresponds to a set of polygons which delimits influence areas around each sampling point (Voronoi 1908). The Voronoi diagram is used to find the location of the boundary: we search in this diagram the segments that separate sampling points of the two groups analyzed.

We then fragment the whole linguistic boundary into segments of arbitrary size to evaluate the homogeneity of the evolutionary processes acting on different sections of the linguistic boundary under evaluation. The length of each segment is such as having an approximately equal number of population samples on each side of it.

We first test for a genetic difference between the groups defined by the boundary, taken as a whole or in segments, using an AMOVA approach (Excoffier et al. 1992). The F_{CT} statistic, representing in this context the proportion of total variance due to differences between the linguistic groups is evaluated and tested using a permutation procedure (Excoffier et al. 1992; Schneider et al. 1997). For molecular markers, molecular information is taken into account in the computation of the F_{CT} statistic by using a matrix of pairwise differences between haplotypes (Excoffier et al. 1992).

Our second approach is explicitly based on a model of isolation by distance. Our goal is to estimate the additional genetic distance between the populations located on each side of the boundary that is specifically due to the boundary, taking the geographical distances between the populations into account. An estimator δa of this additional genetic distance is computed from a regression model of genetic distances on corresponding geographic distances and tested by a permutation procedure.

Under the hypothesis of the absence of any spatial genetic structure due to the boundary, we would expect similar genetic distances between populations located at a given distance, irrespective of their location on one or the other side of the boundary. Therefore, the idea is to perform separate regressions of genetic distances on geographical distances at within-group level and at the inter-group level, and to measure the difference in the Y-axis intercepts of the two regressions (the δa statistic, see Fig. 40.1). This new statistic δa thus provides a direct estimation of the additional genetic distance due to the boundary effect.

The statistical significance of the δa statistic is evaluated by randomly allocating the populations to any group and re-computing the δa statistic after each permutation round in order to obtain its null distribution. The p-value of the δa statistic is obtained as the fraction of the cases where a random δa value is equal to or larger than the observation.

Inferring the location of genetic boundaries

We developed a method to locate the most important genetic barriers in a geographical region without need for interpolation. This represents a first advantage over Wombling method which requires the construction of continuous allele frequencies and can thus lead to a less than faithful representation of the data in the case of an unequal distribution of the sample points. Unlike Monmonier and Wombling techniques, our method also presents the advantage of evaluating the genetic variability inside and between the groups of populations located on each side of the genetic barriers.

Contrary to frequent studies where groups of populations are defined *a priori* and the genetic structure is tested afterward, we search for spatial groups of populations such that the proportion of total genetic variance due to differences between these groups of populations is maximal.

We first define arbitrary spatial groups of populations and thus the location of boundaries between these groups. We then modify locally these boundaries using a simulated annealing strategy.

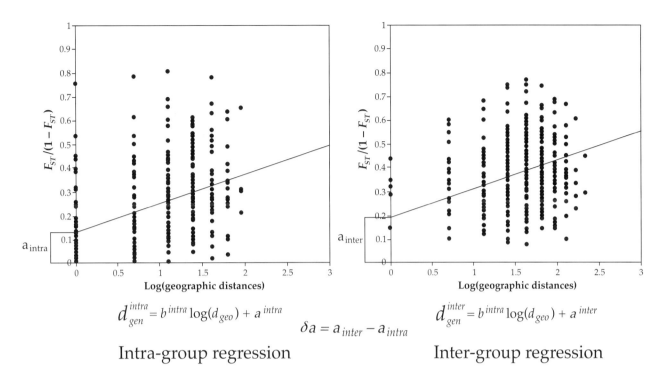

Figure 40.1. *Regression of $F_{ST}/(1 - F_{ST})$ on the logarithm of geographic distances between populations located on each side of a boundary or a segment of boundary (left: intra-group distances, right: inter-group distances).*

In details, we use an iteration procedure to maximize the F_{CT} index between the groups defined by the barriers. At each step of the process, we take a pair of neighbouring populations located on each side of a barrier and with a given probability we move one of the two populations from one group to the other. The barrier is then modified locally. We then compute a new F_{CT} value between the groups located on each side of the new barriers and accept the transformation of the barrier if the new F_{CT} index is larger than the previous one. We can also accept the transformation of the barrier if the new F_{CT} value is smaller than the previous one with a probability which depends on the ratio of the successive F_{CT} values and the number of steps in the simulated annealing process.

The most important genetic barriers are those which maximize the F_{CT} index between the groups of populations defined by the barriers.

Results

Inferring the impact of the Afro-Asiatic and Indo-European linguistic boundary on population differentiation

We evaluate the impact of the linguistic boundary separating the Afro-Asiatic and Indo-European linguistic groups. First of all, we used the RH system data for which the number of populations samples is important enough to enable us to fragment the linguistic boundary into several segments. We also studied the genetic impact of the linguistic boundary between Afro-Asiatic and Indo-European samples tested for the p49a,f/*Taq*I marker on the Y chromosome. For this system, the small number of samples did not allow us to evaluate the impact of different segments of the linguistic boundary. Nevertheless, it is interesting to compare the results of the evaluation of the whole linguistic boundary obtained independently by a classical and a molecular marker.

We have selected 58 Afro-Asiatic and 100 Indo-European samples for the present study (Fig. 40.2) from a data base containing about 600 populations samples tested for the RH system and covering all continental regions (Sanchez-Mazas 1990). We have arbitrarily divided the linguistic boundary into five segments to test the homogeneity of the genetic processes at work along this boundary. The width of each segment is drawn proportional to the F_{CT} index computed for this segment.

The whole linguistic boundary reduces the gene flow between the populations. The Afro-Asiatic and Indo-European cluster are clearly differentiated as indicated by the F_{CT} statistic associated to the whole linguistic boundary (Table 40.1). The δa index give the measure of the additional genetic distance due to

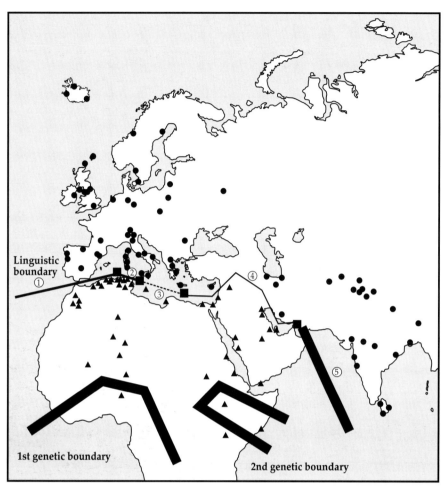

Figure 40.2. *Location and evaluation of the linguistic boundary between populations of Afro-Asiatic and Indo-European linguistic families tested for the RH system. The width of each boundary segment is drawn proportional to the F_{CT} computed for this segment. We also show the location of two genetic boundaries in the repartition area of the Afro-Asiatic and Indo-European populations tested for the RH system (see text).*

Table 40.1. *Results of the evaluation of the genetic impact of the linguistic boundary between populations of the Afro-Asiatic (A) and Indo-European (B) linguistic families.*

	AFRO-ASIATIC CORRELATION coefficient between genetic distances and logarithm of geographic distances	INDO-EUROPEAN CORRELATION coefficient between genetic distances and logarithm of geographic distances	F_{CT}	δa
Rhesus				
Whole boundary	0.292***	0.337***	0.050***	0.075***
Segment 1	0.458***	0.311**	0.064***	0.066***
Segment 2	0.785*	0.148 N.S.	0.044***	0.043***
Segment 3	0.461*	0.158*	0.008	0.128***
Segment 4	0.161*	0.261***	0.016*	0.016*
Segment 5	–0.024	–0.086	0.281***	N/A
Y chromosome				
	–0.050	0.265**	0.097***	0.122***

* $0.01 < P < 0.05$; ** $0.001 < P < 0.01$; *** $P < 0.001$; N/A: Not available

the frontier effect.

The first three segments of the linguistic boundary under evaluation are superimposed on the Mediterranean Sea. From the values of the F_{CT} and δa statistics associated with these segments (Table 40.1), it seems that this ecological barrier does not seem to act as the main isolating factor in the distribution area of the Afro-Asiatic and Indo-European popu-

Figure 40.3. *Location and evaluation of the linguistic boundary between populations of Afro-Asiatic and Indo-European linguistic families tested for the Y-chromosome P49a,f/TaqI haplotypes polymorphism. We show also the most important genetic boundary in the repartition area of the Afro-Asiatic and Indo-European populations tested for this system.*

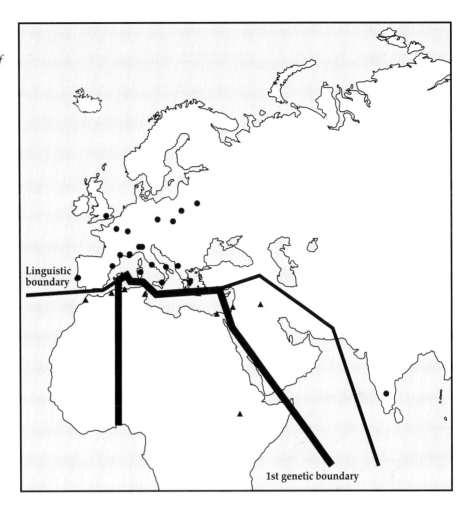

lations. In the Middle East (fourth segment), the contacts between the populations of the two linguistic groups are sufficiently low to cause their weak but significant differentiation. A considerable degree of genetic structure is observed in the eastern region under study (fifth segment). The Afro-Asiatic populations of East Africa and the Indo-European populations of southwestern Asia and India are clearly differentiated.

We have evaluated the genetic impact of the linguistic boundary between 7 Afro-Asiatic and 24 Indo-European samples tested for the p49a,f/*TaqI* polymorphism (Fig. 40.3) (Poloni *et al.* 1997). As for the RH system, the data on this molecular marker show that the gene flow seems significantly reduced across the linguistic boundary according to the F_{CT} and the δa index (Table 40.1). The values of these statistics are greater than the ones computed for the RH system. But since the samples tested for the two markers are not the same, the values of the indexes computed for these two systems are not fully comparable.

Inferring the location of the most important genetic boundary in the distribution area of the Afro-Asiatic and Indo-European groups

Figure 40.2 shows the location of the most important genetic barrier in the distribution area of the Afro-Asiatic and Indo-European populations tested for the RH system. This barrier is associated with a large F_{CT} value ($F_{CT} = 0.282$, $p = 0.004$), four times larger than the F_{CT} value associated with the linguistic boundary between the samples (Table 40.1). The genetic barrier isolates the Hausa sample of Nigeria from the other populations. The genetic structure of the Hausa sample is closer to that of Niger-Congo or Nilo-Saharan samples and as such they are separated from the other Afro-Asiatic samples by the genetic barrier located by our method. If we remove the Hausa sample from our data set, the most important genetic barrier in this region lies between the Issas and Somali samples and the rest of the Afro-Asiatic and Indo-European samples. This barrier is associated with a large F_{CT} value ($F_{CT} = 0.261$, $p < 0.001$). We show again the genetic peculiarities of

Sub-Saharan Afro-Asiatic populations in the total set of populations representing the two language families investigated.

The most important genetic barrier in the distribution area of the Afro-Asiatic and Indo-European populations tested for the Y-chromosome marker under study is shown in Figure 40.3.

This barrier separates four Afro-Asiatic samples from Africa from the rest of the samples. The F_{CT} value associated with this barrier is slightly larger ($F_{CT} = 0.121$, $p < 0.001$) than the one associated with the linguistic boundary (Table 40.1). The two boundaries are almost superimposed. The difference is that the Afro-Asiatic populations of the western part of North Africa and the Middle East join here the Indo-European samples.

Discussion

In this chapter, we present a new method which can provide two independent estimates of the action of linguistic boundaries (F_{CT} and δa) on population differentiation. We controlled by Monte-Carlo simulations the behaviour of the F_{CT} and the δa indexes. These statistics perform well in measuring the impact of simulated boundaries on gene flow when populations of the same group exchange migrants at a higher rate than do populations of different groups. When gene flow rates are equal within and between simulated groups of populations, the F_{CT} index still indicates the presence of a significant barrier. This spurious behaviour is due to the fact that F_{CT} makes no adjustment for spatial pattern; the spatial autocorrelation resulting from isolation by distance biases this F-statistic in the direction of significance. The behaviour of the δa statistic is much more satisfactory in these cases in not detecting a clear enhancer of population differentiation. This new index unfortunately shows an important limitation: when a pattern of isolation by distance in the groups separated by the linguistic boundary analyzed is not observable, the δa statistic cannot be computed and we must base our conclusions on the value taken by the F_{CT} statistic.

Despite this limitation, our method presents the interesting advantage of studying the homogeneity of the genetic processes at work along the evaluated boundary; as we have shown in the case of the Indo-European and Afro-Asiatic populations tested for the RH system, the segmentation of the linguistic boundary and the independent evaluation of each segment of this boundary can provide useful insights into the history of populations.

Our second method permits us to infer the location of the most important genetic barriers in the distribution area of a set of samples. The most important genetic barriers in a specific region are those which maximize the F_{CT} index between the groups of populations defined by the barriers. Simulation results showed that our method is efficient in retrieving the position of barriers between simulated groups of genetically homogeneous populations. Unlike previous methods, this new technique allows us to measure the amount of differentiation among and between the groups located on each side of the genetic barriers recovered.

In the near future, we plan to apply our first method to evaluate the genetic impact of other cultural or ecological boundaries. And since our two methods can be used at any geographical scale, we plan to apply them to other organisms.

Acknowledgements

Thanks to André Langaney for help during the first phases of these work. We are also grateful to Alicia Sanchez-Mazas and Estella Poloni for providing access to their data bases. This work has been supported by Swiss FNRS grants 31-39847.93 and 31-054059.98.

A computer program to test the impact of boundaries on gene flow and to locate the most important genetic barriers in a geographical region as it is presented here is available from Isabelle Dupanloup de Ceuninck upon request.

References

Barbujani, G. & R.R. Sokal, 1990. Zones of sharp genetic change in Europe are also linguistic boundaries. *Proceedings of the National Academy of Science of the USA* 87, 1816–19.

Barbujani, G., N.L. Oden & R.R. Sokal, 1989. Detecting regions of abrupt change in maps of biological variables. *Systematic Zoology* 38, 376–89.

Dupanloup de Ceuninck, I., S. Schneider & L. Excoffier, in prep. Detection of genetic boundaries: a new regionalization method based on the maximization of a genetic distance between groups of populations.

Dupanloup de Ceuninck, I., S. Schneider, A. Langaney & L. Excoffier, forthcoming. Inferring the impact of linguistic boundaries on population differentiation: application to the Afro-Asiatic — Indo-European case. *European Journal of Human Genetics*.

Excoffier, L., P. Smouse & J.M. Quattro, 1992. Analysis of molecular variance inferred from metric distances among DNA haplotypes: application to human mitochondrial DNA restriction data. *Genetics* 131, 479–91.

Monmonier, M.S., 1973. Maximum-difference barriers: an alternative numerical regionalization method. *Geographical Analysis* 3, 245–61.

Poloni, E.S., O. Semino, G. Passarino, A.S. Santachiara-Benerecetti, I. Dupanloup, A. Langaney & L. Excoffier, 1997. Human genetic affinities for Y-chromosome P49a,f/*Taq*1 haplotypes show strong correspondance with linguistics. *American Journal of Human Genetics* 61, 1015–35.

Sanchez-Mazas, A., 1990. *Polymorphisme des systèmes immunologiques Rhésus, GM et HLA et histoire du peuplement humain*. Geneva: Genetics and Biometry Laboratory, University of Geneva.

Schneider, S., J.M. Kueffer, D. Roessli & L. Excoffier, 1997. *Arlequin ver. 1.1: a software for population genetic data analysis*. Geneva: Genetics and Biometry Laboratory, University of Geneva.

Sokal, R.R., N.L. Oden & B.A. Thomson, 1988. Genetic changes accross language boundaries in Europe. *American Journal of Physical Anthropology* 76, 337–61.

Sokal, R.R., N.L. Oden, P. Legendre, M.J. Fortin, J. Kim & A. Vaudor, 1989. Genetic differences among language families in Europe. *American Journal of Physical Anthropology* 79, 489–502.

Voronoi, M.G., 1908. Nouvelles applications des paramètres continus à la théorie des formes quadratiques, deuxième mémoire, recherche sur les paralléloedres primitifs. *Journal Reine Angewandte Mathematik* 134.

Chapter 41

Patterns of Genetic and Linguistic Variation in Italy: a Case Study

Franz Manni & Italo Barrai

The amount of genetic data (sequences, gene frequencies and isonymy) existing about the Province of Ferrara (Italy) makes this area one of the biologically most known worldwide. For this reason we studied the population structure of this territory trying to draw inferences on the underlying demographic processes by comparing archaeological, historical, linguistic, geological and palaeoclimatological data. This multi-level approach has allowed us to date some characteristics of the population structure, beginning from prehistoric times to the Roman Empire and Middle Ages, and to recognize the overlapping of biological, cultural and geographic boundaries. To detect linguistic boundaries inside this area we have turned pronunciation differences into phonetic notation and computed pairwise distances by means of bio-informatic methods for multiple sequence analysis, in order to obtain a distance matrix of the overall pronunciation variability. This kind of approach has enabled us to test the association between linguistic, geographical and genetic distance matrices using the same statistical tests. Results indicate that demographic phenomena can be traced back also in a small area and that, on a microregional scale, recent events may have contributed to determine important aspects of the overall genetic variation.

Aims

Rates of gene frequency change across language boundaries are often significantly higher than those at random locations and many linguistic boundaries are known to occur at physical barriers. Several authors focused on this topic and in many cases correlations between genetics and linguistics were found (Sokal *et al.* 1988; Barbujani & Sokal 1990; Barbujani *et al.* 1996).

We are interested in precisely dating historically, when possible, genetic and linguistic structures. To do so their geographical location must be taken into account because a relationship between a genetic or linguistic structure and changes in geographical features of the territory conformation can be found.

The amount of different data to be processed for this purpose is considerable, so we have focused our attention on one area whose genetic structure has been investigated in detail in the past 50 years, namely the Province of Ferrara[1] (Bianco *et al.* 1952; Zanardi *et al.* 1977; Barrai *et al.* 1984; Beretta *et al.* 1989; Barbujani *et al.* 1996). This area is located in the northeastern part of Italy and includes most of the Delta of the Po, the main Italian river. Its territory is located on an alluvial plain characterized by subsidence since remote times. For this reason marshland took up large areas of land until a century ago, when mechanical reclamation became possible. The availability of detailed studies (Bondesan 1986; 1989; Fabbri 1987) concerning this gradual reversion to wetland and the absence of any mountain or hill in the province, allowed us to recognize that the main constraint to peopling is the hydrography of the area. The Province of Ferrara is an optimum place to start a comparative investigation, considering that linear geographic distances can be calculated without any correction because of the flatness of the plain.

In addition to biological data we considered isonymic information, physical boundaries and linguistic barriers in order to test any possible association between different data. Language, isonymy and genetics enable us to make inferences about the past at different time scales. Indeed isonymy is computed on the distribution of surnames and no conclusions could reasonably be traced for periods previous to the Late Middle Ages, when surnames originated and spread in Italy (Cavalli-Sforza 1982).

From a linguistic point of view the problem is very complex: since we considered only the spelling of some terms and we have not examined philology, caution must be adopted in avoiding drawing scenarios before the beginning of this millennium. For this reason the absence of correlation between different sets of data is as informative as its presence and it prompts us to establish when the origin of a given structure took place.

Methods

Linguistic data

The Linguistic Atlas of Italy (*Atlante Linguistico Italiano*) is the result of an extended survey on the exact pronunciation of some terms in many places through the country. We considered 16 localities, 10 in the Ferrara Province and 6 in the neighbouring areas, in order to detect possible linguistic boundaries and find differences within the area. To make possible a statistical comparison between linguistic and genetic data we developed a quantitative method to represent linguistic differences in the form of a distance matrix.

For this reason we converted the phonetic notations reported in the Atlas to the closest alphabetical spelling. The calculation of pronunciation differences was performed by adapting the alignment method for DNA sequences. Actually, we can consider the alphabet as a code made of 21 bases,[2] and words as sequences. In this way it is possible to find the best alignment between two pronunciation data and identify the equivalent of transitions, transversions, insertions and deletions. We assigned penalties following the scheme presented in Table 41.1, quite similar to the Levenshtein Distance, also called *unit cost model*. The predominant quality of this model is its simplicity and, even if more sophisticated models could be used, it seems suitable when a glottological and phylological approach is not employed.

We analyzed in this way 26 dialect terms and expressions. We obtained 26 distance matrices and from these an overall matrix describing the linguistic variability of the area considered. From that we traced boundaries in a Delaunay Network (see below). We must stress that this underestimates linguistic variability because there is loss of information when the phonetic notation is turned into standard alphabet.

Isonimic data

In 1996 we bought from Topware, a commercial firm, the CD-Rom I-Info© with the files of all telephone users registered by the Italian telephone society Telecom© for that year. The surnames of the individual users were downloaded from the CD-Rom to ASCII files, then to data base files, using the standard software provided by the firm. We considered all users from 24 of the 26 Municipalities of the Province of Ferrara. In this way 117,542 users were selected, with no sex distinction and they constituted a 33 per cent sample of the population of 350,238 residents in the whole province.

Based on their surname distribution, random isonymy between Municipalities i and j (I_{ij}) was estimated as $I_{ij} = \Sigma_k p_{ki} p_{kj}$, where p_{ki} and p_{kj} are the relative frequencies of surname k in Municipalities i and j respectively; the sum is over all surnames. Note that I_{ij} is twice Lasker's coefficient of relationship, R_i (Lasker 1977). For details on computation of Lasker distance see Rodriguez-Larralde *et al.* 1993. A Delaunay Network was constructed from the matrix of Lasker's distances.

Table 41.1. *Scheme of the unit cost model through which linguistic differences were calculated.*

Case	Penalty	Example
Insertion = Deletion	1	Term: Baffi (Moustache) Loc. A: *Baafi*; Loc. B: *Bafi* A) → `baafi` → `.....` B) → `ba.fi` → `..-..` *Penalty: 1* (1 ins./del.)
Transition = Transversion	1	Term: Carne (Flesh) Loc. A: *Caran*; Loc. B: *Caren* A) → `caran` → `...a.` B) → `caren` → `...e.` *Penalty: 1* (1 transit./transv.)

Genetic data

We used the results of a screening on 1364 adults in the Ferrara Province based on seven presumably neutral isoenzyme systems (ACP$_1$, ESD, GLO I, GPT, PGD, PGM$_1$, PGP) typed over all the 26 Municipalities of the Province (Beretta *et al.* 1989). From these data we have calculated a matrix using Nei's standardized distance and from it an unrooted UPGMA dendrogram using the Neighbor© software version 3.2 (Felsenstein 1989) provided with the Phylip© package. Bootstrapping was performed 1000 times with software developed for the purpose of this chapter.

Boundaries

The localities studied were joined by a Delaunay graph (Brassel & Reif 1979), a method used to connect adjacent points in a network. For a description of this method see Barbujani *et al.* 1996.

Results

We detected that isonymy reaches maximum values in populations close to the Po Delta and to the coastline (Fig. 41.1; Table 41.2). This area is also linguistically peculiar and the pattern of phonetic changes enables us to trace boundaries 3 and 4 as shown in Figure 41.1. Thus it is possible to infer that in the populations in the northeastern part of the Ferrara Province cultural and reproductive isolation go together. This finding can be explained by considering the fact that since the early Middle Ages subsidence gave rise to a large expanse of lagoon-building, a real physical barrier to human movements (Bondesan 1986; 1989). In the past century their extension reached a maximum distance of 25 kilometres between the coast and the

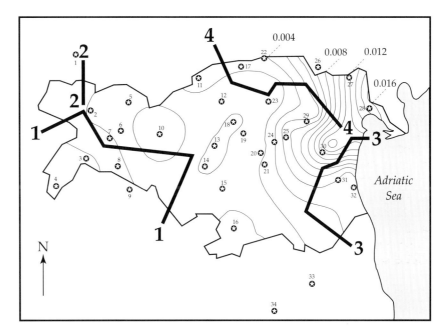

Figure 41.1. *Plan of the Province of Ferrara. Small numbers indicate the location of all the localities considered in our study. Thin solid lines represent an interpolation of isonymic data as shown in Table 41.2. Thick solid lines are linguistic boundaries obtained on a Delaunay Network in the order indicated by numbers (1; 2; 3; 4). Interpolation performed by Surfer© v. 6.1. Numbers refer to Figure 41.3 and Table 41.2.*

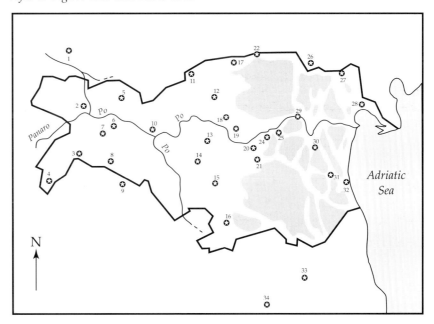

Figure 41.2. *Plan of the Province of Ferrara as in Figure 41.1. The shaded areas correspond to marshlands which existed from the third century AD to the nineteenth century AD. Nowadays only the Comacchio (31) lagoons exist (A). Solid lines indicate main courses of the Po river until the dramatic changes of the eleventh century (Rotta di Ficarolo) when the hydrographic situation became the present one. Nowadays the main flow of the Po can be superimposed to the northern limit of the Province. Numbers refer to Figure 41.3 and Table 41.2.*

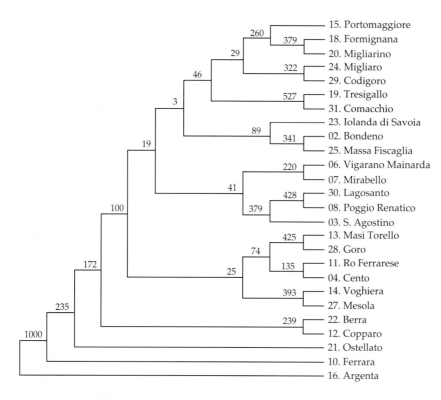

Figure 41.3. *Bootstrap UPGMA tree based on seven enzymatic systems (ACP_1, ESD, GLO I, GPT, PGD, PGM_1, PGP). Nei's std distance (data from Beretta et al. 1989). The numbers reported at the end of branches are a key to the numbering of localities reported in Figures 41.1 and 41.2.*

inner dry lands (Fig. 41.2), bringing a dramatic change in the scheme of settlements as supported by archaeological findings referring to the Late Middle Ages (Visser-Travagli 1987).

Wetlands have played a leading role in phases of colonization because they soon became the habitat of Anopheles; the seriousness of the malaria fever is witnessed by the high rate of heterozygotes for Cooley's Anemia which is still present today in the Province, this rate is at its maximum level (15 per cent) near the coast (Silvestroni & Bianco 1975; Barrai et al. 1984).

Moreover, linguistic limits 1 and 2 (Fig. 41.1) are closely related to the former course of the Po and to the present course of the Panaro river (Fig. 41.2) so that it is possible to argue that the geographic barriers represented by these rivers had effects on linguistic structures.

Argenta (16) occupies a marginal position from the geographic and genetic point of view. From standardized Nei's distance we obtained a tree (Fig. 41.3) tested with bootstrap analysis which shows that significance is achieved only for the first cluster that divides Argenta from all the other localities (100 per cent bootstrap score). Surname distribution is so peculiar that the first Delaunay boundary built on Lasker's distance (not shown) encompasses Argenta. Even if in the linguistic dendrogram (not shown) this town is set apart, no boundaries can be traced because phonetics change sharply through a north–south gradient. Although surrounding wetlands on the West and East of Argenta (Valli di Comacchio, still existing as shown in Fig. 41.2) may have played a role, we cannot infer cultural isolation. It is also interesting to note that in the past the administrative authority moved to Ferrara from Ravenna, to whose diocese Argenta still belongs. This small town maintained economic relationships with both cities for a long time. All the above reasons explain the cultural continuity with the surrounding communities. Actually, even in the past, some geographic continuity along a north–south axis never disappeared.

Comacchio (31) exhibits many linguistic differences since the dialect spoken there is quite different from the rest of the Province. Some scholars would interpret these differences by relating them with the Spina civilization that flourished in the third century BC. A few kilometres from Comacchio, Spina was an important centre with its own fleet: many archaeological findings reveal important commercial exchanges with Athens and other Greek cities as witnessed by the discovery of a bulk of Greek vessels. There is no evidence of a biological relationship, however, between these two populations and it has been demonstrated that the β-thalassemia of the Province of Ferrara is due to a mutation different from that still present in Greece (Facchini 1986). Nevertheless Comacchio is well-defined linguistically, as boundary 3 in Figure 41.1 indicates. With no doubt wetlands isolated Comacchio during the past (Fig. 41.2) and this, as discussed above, brought about a cultural identity or enforced an existing previous one. It must be emphasized that Comacchio was the seat of a diocese until recent times; this fact is relevant because dioceses often reflect ancient social and economic situations.

Conclusions

Usually genetic and linguistic differences can be drawn only on a macroregional scale because the amount of data needed to perform these analysis has to be large and usually the grid of sampling points is too wide to carry out such studies on a microregional scale.

The results presented here show that it is possible to detect linguistic and genetic differences even in an area as small as an Italian Province and that the observed differences attain statistical significance.

By comparing different sets of structural data (historical, archaeological, geological, etc.), belonging to a small territory, it has been possible to draw a consistent picture of the colonization of a small area. Furthermore, the multilevel approach enabled us to date these structures.

On this microregional scale recent phenomena such as geological and ecological changes played a leading role in determining the observed variability because they altered deeply the genetic and linguistic structures of the Province of Ferrara. At least on a microregional scale, the last of the observed cultural and biological differences originated as recently as the last century. These findings suggest that a multilevel approach should be applied when the data set is sufficiently large. Otherwise, recent demographic phenomena could be underestimated and a possibly strong bias could affect inferences about the settlement of people in the area under consideration.

Acknowledgements

We wish to thank M. Chicca, M. Mazzei-Traina and G. Sensi for reading the manuscript and their useful suggestions. We are also in debt to M. Stenico who gave generously his help in statistical analysis and to G. Barbujani for providing useful comments.

Notes

1. 2630 km; 350,238 inhabitants; 26 Municipalities (1997 census).
2. In the Italian alphabet.

References

AAVV, 1995. *Atlante linguistico italiano (Istituto dell')*, vol. I. (Il corpo umano: anatomia, qualità e difetti fisici, protesi popolari.) Rome: Istituto poligrafico e Zecca dello Stato.

AAVV, 1996. *Atlante linguistico italiano (Istituto dell')*, vol. II. (Il corpo umano: funzioni principali, malesseri e affezioni patologiche comuni, malattie principali.) Rome: Istituto poligrafico e Zecca dello Stato.

Barbujani, G. & R.R. Sokal, 1990. Zones of sharp genetic change in Europe are also linguistic boundaries. *Proceedings of the National Academy of Sciences of the USA* 90, 4670–73.

Barbujani, G., M. Stenico, L. Excoffier & L. Nigro, 1996. Mitochondrial DNA sequence variation across linguistic and geographic boundaries in Italy. *Human*

Table 41.2. *Key to the numbering of localities reported in Figures 41.1 and 41.2. The presence of a symbol in columns G, S, L indicates that the place has been used in Genetic, Surname or Language analysis. In column I is reported isonymy computed on surname distribution on a sample whose size is in column Tel users. In the Locality column, between parenthesis, are indicated the Provinces when not Ferrara. Italic characters indicate Municipalities.*

No.	Locality	G	S	L	I	Tel users
01	Ceneselli (Ro)		•		–	–
02	*Bondeno*	•	•		0.00291	5213
03	*S. Agostino*	•	•		0.00429	1861
04	*Cento*	•	•	•	0.00522	9061
05	Casaglia			•	–	–
06	*Vigarano Mainarda*	•	•		0.00225	1887
07	*Mirabello*	•	•		0.00368	1105
08	*Poggio Renatico*	•			–	–
09	Malalbergo			•	–	–
10	*Ferrara*	•	•	•	0.00082	50,543
11	*Ro Ferrarese*	•	•		0.00391	1246
12	*Copparo*	•	•	•	0.00212	5885
13	*Masi Torello*	•	•		0.00322	764
14	*Voghiera*	•	•		0.00323	1226
15	*Portomaggiore*	•	•		0.00225	4213
16	*Argenta*	•	•	•	0.00184	6454
17	Cologna			•	–	–
18	*Formignana*	•	•		0.00348	935
19	*Tresigallo*	•	•		0.00204	1612
20	*Migliarino*	•	•		0.00405	1207
21	*Ostellato*	•	•	•	0.00351	2028
22	*Berra*	•			–	–
23	*Iolanda di Savoia*	•	•		0.00183	1133
24	*Migliaro*	•	•		0.00282	722
25	*Massa Fiscaglia*	•	•		0.00591	1324
26	Ariano (Ro)			•	–	–
27	*Mesola*	•	•	•	0.01161	2183
28	*Goro*	•	•		0.01652	1090
29	*Codigoro*	•	•	•	0.00407	4266
30	*Lagosanto*	•	•		0.01280	1414
31	*Comacchio*	•	•	•	0.00440	10,170
32	Porto Garibaldi			•	–	–
33	Alfonsine (Ra)			•	–	–
34	Lugo (Ra)			•	–	–

Biology 68, 201–5.

Barrai, I., A. Rosito, G. Cappellozza, G. Cristofori, C. Vullo, C. Scapoli & G. Barbujani, 1984. Beta-thalassemia in the Po delta: selection, geography, and population structure. *American Journal of Human Genetics* 36, 1121–34.

Beretta, M., P. Mazzetti, E. Mamolini, R. Gavina, R. Barale, C. Vullo, A. Ravani, A. Franzè, T. Sapigni, E. Soracco, D. Davi, S. Rendine, A. Piazza & I. Barrai, 1989. Genetic structure of the human population in the Po Delta. *American Journal of Human Genetics* 45, 49–62.

Bianco, I., G. Montalenti, E. Silvestroni & M. Siniscalco, 1952. Further data on genetics of microcythaemia or thalassaemia minor and Cooley's disease or thalassaemia major. *Annual of Eugenics* 16, 148–9.

Bondesan, M., 1986. Lineamenti di geomorfologia del basso ferrarese, in *La civiltà comacchiese e pomposiana dalle origini preistoriche al tardo Medioevo: Convegno nazionale di studi storici (Proceedings) Comacchio, 17–19 May 1984*. Bologna: Nuova Alfa Editoriale, 17–28.

Bondesan, M., 1989. Evoluzione geomorfologica e idrografica della pianura ferrarese, in *Terre ed acqua*. Ferrara: Gabriele Corbo Editore, 13–20.

Brassel, K.E. & D. Reif, 1979. A procedure to generate Thiessen polygons. *Geographical Analysis* 11, 289–303.

Cavalli-Sforza, L.L., 1982. Isolation by distance, in *Human Population Genetics: the Pittsburg Symposium*, ed. A. Chakravarti. New York (NY): Van Nostrand Reinhold Company, 229–48.

Fabbri, P., 1987. L'evoluzione del delta padano dall'Alto al Basso Medioevo, in *Storia di Ferrara*, vol. 5: *Il Basso Medioevo XII–XIV*. Ferrara: Gabriele Corbo Editore, 15–42.

Facchini, F., 1986. Aspetti di antropologia protostorica e storica del basso ferrarese, in *La civiltà comacchiese e pomposiana dalle origini preistoriche al tardo Medioevo: Convegno nazionale di studi storici (Proceedings) Comacchio, 17–19 May 1984*. Bologna: Nuova Alfa Editoriale, 29–40.

Felsenstein, J., 1989. PHYLIP: phylogeny inference package vs. 3.2. *Cladistics* 5, 164–6.

Lasker, G.W., 1977. A coefficient of relationship by isonymy: a method for estimating the genetic relationship between populations. *Human Biology* 49, 489–93.

Rodriguez-Larralde, A., I. Barrai & J.C. Alfonzo, 1993. Isonimy structure of four Venezuelan states. *Annual of Human Biology* 20, 131–45.

Silvestroni, E. & I. Bianco, 1975. Screening for microcytemia in Italy: analysis of data collected in the past 30 years. *American Journal of Human Genetics* 20, 198–212.

Sokal, R.R., N.L. Oden & B.A. Thomson, 1988. Genetic change across language boundaries in Europe. *American Journal of Physical Anthropology* 70, 489–502.

Visser-Travagli, A.M., 1987. Profilo archeologico del territorio ferrarese nell'alto medioevo: l'ambiente, l'insediamento e i monumenti, in *Storia di Ferrara*, vol. 4: *L'alto Medioevo VII–XII)*. Ferrara: Gabriele Corbo Editore, 47–106.

Zanardi, P., G. Dell'Acqua, C. Menini & I. Barrai, 1977. Population genetics in the Province of Ferrara, vol I: Genetic distances and geographic distances. *American Journal of Human Genetics* 29, 169–77.

Prospect

Concluding Remarks

Luca Cavalli-Sforza

I left Italy for California almost 30 years ago, but in recent years I have been working part time in Europe, though spending on average more time on the other side of the Atlantic. It is not always easy or possible for me to attend all European meetings of interest, but I was very glad to be invited to take part in this one. Research in human population genetics was always reasonably active in Europe, and I have the impression that the quality and quantity of research going on are definitely increasing. This is a pleasant discovery. In fact the amount of scientific information circulated during this meeting is so high that I find it impossible, and perhaps even sterile, to try to summarize it in the short space available. I would like, instead, to use a little of your patience to suggest possible avenues and plans of future research, in the hope that some of you will be able to invest time and energy into it.

I have been impressed by the example of cooperation given by scientists of a special discipline not at all remote from ours: HLA research. HLA was a simple gene at the beginning, rapidly became a supergene and finally a family of gene families that today explains an important part of immunology. The group of HLA scientists was very good also at undertaking collaborative ventures. 'Workshops' were organized in four-year cycles on specific themes of common interest, agreed upon by participants who shared the tasks of providing the data for the next Workshop. A major inspirer of these programmes was and is Jean Dausset, the man who was responsible for finding the HLA gene. At the beginning there was only one, and for this discovery he received the Nobel prize. Many other enthusiastic and brilliant scientists have constantly contributed to the choice and implementation of these cooperative programmes. There is much to be learnt from the organization and enthusiasm of HLA researchers.

The absolute need of cooperation in our field became clear to me more than ten years ago. A survey of published protein polymorphims carried out with Paolo Menozzi and Alberto Piazza revealed how difficult it is to analyze satisfactorily an enormous body of data on gene frequencies collected in 70 years of research of human population genetics. Because of the total lack of coordination of efforts among research workers, the many protein genes studied in about 2000 populations form an almost empty data matrix, extremely difficult to study. The transition from protein to DNA testing began in earnest after 1981 and, though slow in the first years, became a major breakthrough in the nineties. In 1984 I started, in collaboration with the Kidds, a collection of cell lines that could provide indefinite amounts of DNA to research workers. With the very modest means that were available we could only produce a relatively small number of cell lines from some interesting world populations (15–60 individuals for each of a few dozens of them). In late 1991 I was helped by several colleagues to start planning a collaborative effort, called the Human Genome Diversity Project (HGDP), for widening the collection and distributing DNA samples to research workers. This programme proceeded well until, little more than a year later, a Canadian private organization, RAFI, began attacking us with totally false accusations. According to this politically inspired slandering campaign we were interested in exploiting indigenous people for personal financial benefit, with support from pharmaceutical companies, and developing patents for vaccines made from cell lines. Incidentally, their campaign caused NIH to withdraw a patent of a viral vaccine, which RAFI attributed to our group. Thanks to their previous fax and e-mail connections with a network of indigenous populations, their vicious efforts were also successful in spreading panic among some indigenous groups, especially in America and Australia, and getting a lot of media attention for their organization. The HGDP survived their attacks but there were serious delays in our programmes.

Coordinating research on DNA rather than on protein genes is in principle much easier, but the level of coordination and sharing among laboratories is still minimal, and limited to personal contacts of small groups. But we can announce today a step forward which I hope will be of interest to many of you. As most of you will know, Jean Dausset generated in Paris a laboratory called CEPH (Centre pour l'Etude du Polymorphisme Humain), which started in the eighties a magnificent effort, collecting cell lines from about 60 families with parents, grandparents and many children. This collection was ideal material for studies of crossing-over, from which DNA could be prepared and distributed to laboratories involved in testing linkage of disease genes with genetic markers, and willing to communicate their results to CEPH. Analyzing these data, CEPH generated the current excellent set of human chromosome linkage maps. This programme is close to completion, and it was clear that the same instrumentation and organization could become useful today for HGDP purposes. Dausset has always been interested also in the population genetics aspect of HLA, and it was easy to establish a collaborative project between CEPH and the HGDP. We have almost finished collecting 1000 cell lines already existing in eight laboratories all over the world, including, of course, the Kidds' and ours. These lines are donated to CEPH, which grows the cell lines to produce DNA and will soon proceed to distribute it to interested, non-profit research laboratories. The world collection thus generated has contributions from all five continents, but no new cell lines were made for the purpose, and vast regions are still imperfectly represented. Hopefully in the future it will be possible to increase the collection and fill the major gaps. Almost fifty laboratories have already shown interest in receiving DNA from the cell lines and in returning results to CEPH. A DNA data base will thus be generated and available to participating laboratories. Research workers who are interested in obtaining DNA samples from CEPH should write now to Howard Cann.

This is an important beginning, but there is a clear need for extending greatly the central collection. It is very likely that national banks of cell lines will be established by those individual countries which aspire to contribute to the research effort. For instance, Qasim Mehdi of Islamabad has been assembling a collection of over 1000 cell lines from 10 important Pakistani populations. Almost 20 per cent of these cell lines were already donated to CEPH for the HGDP-CEPH collection. Research workers who want numbers of individuals of Pakistani populations greater than those available at CEPH, should request directly the cooperation of Mehdi. This seems a very good scheme and similar ones are followed also by other countries. Israel has a permanent collection of Israeli cell lines and makes them available to research workers. China is developing a permanent collection. It could probably be enough if a fraction of 10 to 20 per cent of cell lines of developing national banks were available to the central repository.

Some problems need further consideration. Which populations should be collected? What size of samples of individuals per population? Which sample of markers should be tested? Beginning with the second problem, originally it was thought 25–50 cell lines per population (unrelated individuals) would be enough, but some investigations require more individuals. For these purposes one might collect a larger number of blood samples of 15–20 ml (perhaps 5 to 10 more than the cell lines), to be kept as such at low temperature, or stored as DNA. CEPH may be able to offer free storage for these samples. An embarrassing thought is that, today, males are much more informative than females, because they allow us to study Y chromosomes, mtDNA, as well as X chromosomes in the haploid condition. One cannot totally neglect sampling females, but today, my suggestion would be to sample more males (perhaps 70–80 per cent).

Which populations should one collect? There are two different modes of sampling populations, and I think both are useful. When there are strong reasons in favour of studying specific populations, which represent isolates of special historical or anthropological interest, they should be sampled for their own sake. But they can hardly be taken to represent the country where they reside. On the other hand, in order to have a sample representative of a country, one should stratify it according to regions distinct on the basis of geography. It may be useful to exemplify the composition of some of the samples in the present HGDP-CEPH collection. The Pakistani population (almost 100 million) is most satisfactorily represented in the national bank and both sampling criteria indicated above are approximately satisfied, in that at least one usually large, well known, traditional, linguistically distinct population has been sampled from every major region. There is an approximately equal number of cell lines for all major populations. The Chinese population is made of 1.1 billion Han, and 55 ethnic non-Han minorities. The latter total about 100 million. The Han are rather different in north and south China, and tend to show considerable similarity to the local ethnic minorities. Obviously a certain amount of genetic exchange has taken place between the majority and the indigenous groups, but there are still important differences. It is not an easy job, of course, to represent one-fifth of the human species with a very small number of individuals. We are compromising by having less than 100 Han, from the north and from the south, and small samples (of 10 individuals each) of ethnic minorities of which there already exist cell lines. Most Israeli samples are of the special populations type. Practically all important groups are represented with different numbers. For France there is a small, but geographically stratified sample, with one or two individuals of local parentage from each of 15 traditional regions. Other European countries represented are Italy, where we are still collecting four special populations (two in the north, one in the centre and two in the south) and Great Britain (Orcadians). I would hope that the European representation in the sample will increase soon, and that it will be possible to fill the other present major gaps and imbalances of the world collection.

Which markers to use? Like others, at one time I fell into the trap of thinking one should choose them and recommend that one uses a standard set, but I realized it is much too early to make suggestions. It seems important that HGDP-CEPH collaborators are free to use the markers they are interested in. There is still need for much experimentation on markers. But it is essential that HGDP-CEPH collaborators test as many of the first collection of 1000 DNAs as possible, ideally all, or at least a good sample as geographically varied as possible. Moreover, it is clear that for every geographic region there will have to be optimal markers, but the data base to be built will contribute important results also on this point.

I would like to come back to what I said at the beginning: HLA researchers have shown the way to setting up global research cooperations on well chosen themes and in carrying them through at a remarkably fast pace. Concerning Europe, I can think of a few programmes that might be particularly timely from our point of view and locally meaningful. Let me single out two examples of potential cooperative themes of potential interest to this group, also remembering that our convener is a well known British archaeologist. Great Britain has an important share of the relevant monuments.

I am thinking of the extraordinary geographic distribution of megalithic monuments, which is mostly, though not exclusively, European. The people responsible for their dissemination are totally mysterious. They were architects and builders, engineers, astronomers, certainly excellent farmers, missionaries and preachers. They were most probably also good navigators given the geographic distribution of their monuments, mostly along coasts and on many islands, especially but not only in the Mediterranean. They may well represent an early example of an 'élite dominance' (an expression introduced in linguistics by Renfrew to refer to one of the mechanisms of language replacement). This, if strictly true, would make it especially difficult to find genetic traces of their movements, because people involved in the migrations might be basically few. The chance of finding genetic traces of specific migrations like this one depends essentially on the ratio of the number of immigrants to that of earlier (and later) settlers. But the size of the megalithic monuments suggests that founders of new colonies could not have been too few, even if they were good at persuading large numbers of local people, or finding slaves, to help with their building activity.

Had I not witnessed, in the last few years, the growth of analytic power of the Y chromosome, and how fast one can accumulate new Y markers for solving special problems, I would not suggest this research. I already tried more than thirty years ago to give a genetic contribution to the problem of megalithic monuments, on a small scale, and limited to south Italy. With the benefit of hindsight, it is not too surprising if, with the poor markers then available, we failed. I believe today it is feasible. The archaeological information is fully available. It is clear that the research should be made by collecting samples of local people in rural areas where these monuments are particularly rich, and in control populations from adjacent areas where monuments are typically absent. It is only by examining several such regions and comparing the people now living in the areas where megaliths are common, with non-megalithic neighbours, that one can hope to contribute to the problem. To ensure finding 'autochthonous' peoples, especially for the Y chromosome, it is useful and easy to select samples on the basis of local surnames. Surnames are certainly five or ten times younger than megalithic monuments, but they at least help to exclude foreign immigrants of the last five or ten centuries, who are likely to generate the greatest noise. The finding of a few possibly rare Y markers in several megalithic areas and not in the adjacent ones could give some ideas of where these people came from. A little luck is necessary for finding the right markers, but observations reported at this meeting by Peter Underhill, and others which came up in continuation of this work show that this hope is not unreasonable. Naturally all other markers might be useful, obviously including HLA, and it would be good to involve also the HLA organization. DNA from bones, of course, could contribute important information.

There is another major mystery in European archaeogenetics: who were the Celts? There is an Austrian/Swiss connection (Hallstatt and La Tène) but it may be too late to be really useful for understanding their origins. This is of course another problem. But it is possible that the study of megalithic monuments could provide information useful also for understanding the Celtic problem, if nothing else because the respective geographic areas are largely overlapping, and there may be a reason for it.

Europe is blessed by the most developed archaeological and linguistic knowledge in the world. The European Union has started contributing, though modestly, to foster genetic investigations which could greatly help the study of its history and prehistory. It is to be hoped that the EU will intensify the help it has started providing to these initiatives. But EU work would be greatly simplified if at least the major European countries developed HGDP bases and banks for exchanging DNA and information with other countries, following the example of Pakistan, China and Israel.

I conclude that this was a very good meeting. In particular, I found the encouragement given to young researchers especially novel and valuable. It was done in an efficient way which I hope future organizers will remember and imitate. I think we should all be grateful to Colin Renfrew for putting so much thought and care into organizing this symposium. And we should look forward to starting soon some cooperative activity around important themes.